Transgenic Crop Protection
Concepts and Strategies

Transgenic Crop Protection
Concepts and Strategies

Edited by

Opender Koul
Insect Biopesticide Research Centre, Jalandhar, India

and

G. S. Dhaliwal
Punjab Agricultural University, Ludhiana, India

Science Publishers, Inc.
Enfield (NH), USA   Plymouth, UK

SCIENCE PUBLISHERS, Inc.
Post Office Box 699
Enfield, New Hampshire 03784
United States of America

Internet site: *http://www.scipub.net*

sales@scipub.net (marketing department)
editor@scipub.net (editorial department)
info@scipub.net (for all other enquiries)

Library of Congress Cataloging-in-Publication Data

Transgenic crop protection: concepts and strategies/edited by Opender Koul and G.S. Dhaliwal.
    p. cm.
  Includes bibliographical references and index.
  ISBN 1-57808-302-8
    1. Transgenic plants. 2. Plants—Disease and pest resistance—Genetic aspects. I. Koul, Opender. II. Dhaliwal, G.S.

  SB123.57.T723 2004
  631.5'233—dc22

                                                        2004042861

ISBN 1-57808-302-8

© 2004, Copyright Reserved

All rights reserved. No part of this publication may be reproduced, stored in a retrieval system, or transmitted in any form or by any means, electronic, mechanical, photocopying or otherwise, without the prior permission of the publisher. The request to produce certain material should include a statement of the purpose and extent of the reproduction.

Published by Science Publishers Inc., Enfield, NH, USA

Printed in India

PREFACE

The world population is poised to become a staggering nine billion in another 50 years, a factor which will aggravate food insecurity in the developing countries. Enhancement of production on the available land needs to be achieved in a sustainable manner without causing disruption to the environment. Thus, crop varieties with higher yield and yield stability are the need of the hour. However, insects, diseases and weeds continue to threaten the sustainability and account for more than 40% loss in agricultural production.

The massive and often indiscriminate application of pesticides has resulted in severe pollution of the environment and contamination of food commodities. This destruction has spurred the scientists and the agrochemical industry alike to devise safer alternatives of pest control. Adoption of integrated pest management (IPM) strategies encompassing the development of resistant germplasm, safer pesticides and improved agricultural practices has not resulted in an appreciable increase in yield due to pest infestation in crops over the last two decades. Breeding with wild genotypes possessing an inherent resistance towards pests has been hampered by the intrinsic limitations of breeding and relatively slow progress in this area. Biological-based pest management using predators and parasitoids is at best, able to supplement the other crop protection measures, only in part in a few specific cases.

Recent advances in cellular and molecular biology have opened new avenues for the production of genetically-engineered (transgenic) plants with new genetic properties. Development of transgenic crops by transferring the genes for pest resistance from microbes or other plants or synthesis of new insecticidal genes is a promising approach for effective pest control. Deployment of transgenic crops provides a wider window for a comprehensive use of IPM technology since the site of action of transgenes in target species is different from that of the synthetic pesticides. The area under transgenic crops has increased globally from 2.8 million ha in 1996 to estimated 67.7 million ha in 2003 in 16 countries. However, strategies for development of transgenic crops employing different types of genes, their potential benefits and practical concerns regarding widespread commercial cultivation in different agricultural backgrounds merit due consideration. The present book focuses on such critical issues in the development of transgenic crops, their problems and prospects, both in developed and developing countries.

The first chapter sets the stage for the book and provides an overview of the progress in the development of transgenic crops and their role in plant protection. *Bacillus thuringiensis* (Bt) proteins have been applied on crops extensively and in 1990s, the first commercial transgenic crops appeared which capitalized on the safety of Bt insecticidal proteins. Thus, most research and development so far has focussed on the transfer of genes expressing Bt Toxins. In a similar vein, the second chapter comprehensively describes the transgenic crops expressing Bt proteins, their current status and the challenges facing their deployment. Plant lectins are another group of carbohydrate-binding proteins toxic to pests and, accordingly, transgenic crops expressing plant lectins have been developed. The next chapter describes the insect-resistant transgenic crops expressing such lectins. The present volume also deals with genetically-modified herbicide tolerant crops and transgenic crops that resist diseases; particularly the rice crop has been taken as a specific example in chapter 5. As we know that resistance development is an unavoidable problem with insects for any control strategy, therefore, the development of resistance in pests to transgenic plants is no exception. Accordingly, chapter 6 describes the mechanisms and management strategies for the development of such a resistance against transgenic plants and the specific example of Bt corn in the USA has been dealt with in chapter 8.

Delivery of gene products to plants basically involves three broad strategies: the gene product administered as a foliar spray; cloning the gene into the plant genome; and involving microorganisms for the delivery. Endophytic microorganisms represent a niche which they occupy in plants that can be exploited for the expression of Bt genes in the plants, with features more acceptable than Bt transgenic plants. Chapter 9 describes the role of such transgenic microbes and endophytes in crop protection. It was also imperative to look into the impact of such transgenic crops on non-target animal species as also on human and environmental health. These aspects have been comprehensively dealt with in chapters 10 and 11. In recent years, there has been a certain uncertainty and conflict between proponents and critics of the application of gene technology in agriculture. Chapter 11 discusses the regulation of genetically-modified crops with a scientific perspective so as to show how these controversies could be resolved. This will help in convincing the farmers to use such a technology without any fear. At this stage, we are not sure whether transgenic crops for small farmers is a dream or a nightmare—therefore, it makes the content of chapter 12.

As gene revolution is considered to possess tremendous potential in augmenting the quantity and the quality of food produced, the issue of large-scale commercial cultivation of transgenic crops is fraught with seemingly complex ecological, economic and social ramifications, especially in the agricultural scenario of the developing world. Therefore, the last chapter of this book is an attempt to discuss certain existing perspectives on the significant issues relating to transgenic crop cultivation in the developing world.

We are thankful to all contributors for the meticulous job they have done in preparing their respective chapters within the stipulated period. We are also grateful to the reviewers of various manuscripts for taking their time to give useful suggestions for the improvement of the chapters. It is hoped that the book will prove to be of immense use to all those who are concerned with the pros and cons of transgenic crops. It will also stimulate further research in this vital area and prove indispensable to teachers and students. The book will provide a tool to pest management experts to guide them towards sustainable management of crop pests.

April 30, 2004

OPENDER KOUL
G. S. DHALIWAL

CONTENTS

Preface	v
The Contributors	xi

1. **Transgenic Crop Protection: An Introduction**   1
   Opender Koul and G. S. Dhaliwal

2. **Transgenic Crops Expressing Bt Proteins: Current Status, Challenges and Outlook**   15
   John H. Benedict and Dennis R. Ring

3. **Insect-Resistant Transgenic Crops Expressing Plant Lectins**   85
   Jesusa Crisostomo Legaspi, B. C. Legaspi, Jr. and M. Sétamou

4. **Genetically-Modified Herbicide-Tolerant Crops-A European Perspective with a United Kingdom Emphasis**   117
   K. Berry, P. J. W. Lutman, L. A. P. Lotz and C. Kempenaar

5. **Transgenic Rice for Disease Resistance**   157
   Swapan K. Datta

6. **Development of Resistance in Pests to Transgenic Plants: Mechanisms and Management Strategies**   177
   Charles F. Chilcutt and Marshall W. Johnson

7. **Gene Pyramiding: A Transgenic Approach to Enhancing Resistance Durability in Plants**   219
   K. R. Rajyashri and Madan Mohan

8. **Current Resistance Management Strategies for Bt Corn in the United States**   261
   Sharlene M. Matten, Richard L. Hellmich and Alan Reynolds

9. **Role of Transgenic Microbes and Endophytes in Crop Protection**   289
Sarvjeet Kaur, Rhitu Rai and Aqbal Singh

10. **Impact of Transgenic Bt Crops on Non-Target Animal Species**   307
Graham Head and Galen Dively

11. **Regulation of Genetically-Modified Crops: A Scientific Perspective**   325
Kees Hulsman

12. **Transgenic Crops for Small Farmers: A Dream or a Nightmare**   351
Yolanda Masieu Trigo

13. **Ecological, Economic and Social Perspectives on Transgenic Crop Protection: Path for the Developing World**   373
Sarvjeet Kaur

Index   *407*

THE CONTRIBUTORS

John H. Benedict
Texas A & M University and
Texas Agricultural Experimental Station
Corpus Christi, TX 78406-9704, USA
E.Mail: john.benedict@sbcglobal.net

K. Berry
Rothamsted Research
Harpenden, UK

Charles F. Chilcutt
Department of Entomology
Texas A&M University System, Route 2
Box 589 Corpus Christi, TX 78406-9704, USA
E.Mail: c-chilcutt@tamu.edu

Swapan K. Datta
Plant Breeding, Genetics, and Biochemistry Division
International Rice Research Institute
DAPO Box 7777, Metro Manila, Philippines
Email: SDATTA@CGIAR.ORG

G. S. Dhaliwal
Department of Entomology
Punjab Agricultural University
Ludhiana-141 004
E.Mail: gsd251@redifmail.com

Galen Dively
Department of Entomology
University of Maryland
College Park, MD 20742, USA
E.Mail: gs7@umail.umd.edu

Graham Head
Monsanto LLC
700 Chesterfield Pkwy North
St. Louis, MO 63198, USA
E.Mail: graham.p.head@monsanto.com

Richard L. Hellmich
USDA–ARS, Corn Insects and Crop Genetics Research Unit
Department of Entomology, Genetics Laboratory, c/o Insectary
Iowa State University, Ames, IA 50011, USA
E.Mail: rlhellmi@iastate.edu

Kees Hulsman
Australian School of Environmental Studies
Faculty of Environmental Sciences
Griffith University, Nathan. Q. 4111, Australia
E.Mail: k.hulsman@mailbox.gu.edu.au

Marshall W. Johnson
Department of Plant and Environmental Protection Sciences
University of Hawaii at Manoa, 3050 Maile Way
310 Gilmore Hall, Honolulu, HI 96822, USA
E.Mail: mjohnson@unkac.edu

Sarvjeet Kaur
National Research Centre on Plant Biotechnology
Indian Agricultural Research Institute, New Delhi 110 012, India.
Email: sk@primeindia.com

C. Kempenaar
Plant Research International
Wageningen University, The Netherlands

Opender Koul
Insect Biopesticide Research Centre
30 Parkash Nagar, Jalandhar-144 003, India
E.Mail: koul@jla.vsnl.net.in

B. C. Legaspi, Jr.
USDA-ARS, Center for Biological Control
Florida A&M University
Tallahassee FL 32307, USA
E.Mail: bclegaspi@netscape.net

Jesusa C. Legaspi
USDA-ARS, Center for Biological Control
Florida A&M University
Tallahassee FL 32307, USA
E.Mail: jlegaspi@nettally.com

L. A. P. Lotz
Plant Research International
Wageningen University, The Netherlands
E. Mail: L.A.P.Lotz@plant.wag-ur.nl

P. J. W. Lutman
Weed Ecology group
Plant and Invertebrate Ecology Division
Rothamsted Research
Harpenden, UK
E.Mail: peter.lutman@bbsrc.ac.uk

Sharlene M. Matten
United States Environmental Protection Agency
Office of Pesticide Programs
Biopesticides and Pollution Prevention Division (7511C)
1200 Pennsylvania Ave., NW, Washington, D.C., 20460, USA
E.Mail: matten.sharlene@epa.gov

Madan Mohan
Plant Resistance Group
International Centre for Genetic Engineering and Biotechnology
Aruna Asaf Ali Marg, New Delhi-110 067, India
E.Mail: mohanm@de112.vsnl.net.in

Rhitu Rai
National Research Centre on Plant Biotechnology
Indian Agricultural Research Institute, New Delhi-110 012, India

K. R. Rajyashri
Plant Resistance Group
International Centre for Genetic Engineering and Biotechnology
Aruna Asaf Ali Marg, New Delhi-110 067, India

Alan Reynolds
United States Environmental Protection Agency
Office of Pesticide Programs
Biopesticides and Pollution Prevention Division (7511C)
1200 Pennsylvania Ave., NW, Washington, D.C., 20460, USA

Dennis R. Ring
Louisiana State University Agricultural Center
Baton Rouge, LA 79894-5100, USA
E. Mail: Dring@agctr.lsu.edu

M. Sétamou
Texas Agricultural Experiment Station
2415 East Hwy 83, Weslaco, TX 78596, USA
E.Mail: msetamou@icipe.org

Aqbal Singh
National Research Centre on Plant Biotechnology
Indian Agricultural Research Institute, New Delhi-110 012, India
Email aqbals@hotmail.com

Yolanda Massieu Trigo
Fuentes 18A, Col. Toriello Guerra, C.P. 14050
México, D.F., México
E.Mail: yolanda_massieu@hotmail.com

1

TRANSGENIC CROP PROTECTION: AN INTRODUCTION

OPENDER KOUL* AND G. S. DHALIWAL**
*Insect Biopesticide Research Centre, Parkash Nagar
Jalandhar-144 003, India
**Department of Entomology, Punjab Agricultural University
Ludhiana-141 004, India

INTRODUCTION

A transgenic crop is a crop that has been improved upon by the incorporation of a gene from another species. Recent advances in molecular techniques have led to the development of methods for the genetic transformation of a wide range of plants. A major advantage that genetic manipulation offers over conventional plant breeding is the ability to improve specific characteristics—including resistance to pests—without risking the food or feed safety of the crops. Since the technique involves the transfer of genes between unrelated taxa, it greatly increases the pool from which desirable crop improvement traits may be selected.

The demand of the day is to improve both productivity and sustainability of agriculture. To achieve this end, it is pertinent to assess the contribution of transgenic technology to sustainable agriculture. Transgenic technology should be used to increase the production of main food staples, improve the efficiency of production, reduce the environmental impact and provide access to food for small-scale farmers, according to the Transgenic Plants and World Agriculture Report 2000, National Academy Press, Washington DC, prepared under the auspices of the Royal Society of London, the US National Academy of Sciences, the Brazilian Academy of Sciences, the Chinese Academy of Sciences, the

Indian Science Academy and the Third World Academy of Sciences (www.nap.edu/html/transgenic). For this purpose, various techniques of genetic transformation have now vastly expanded and most of the transgenics today are first generation transgenics that were developed with genes from very distant organisms, mostly prokaryotes (Pental, 2003). An interesting example of the use of such transgenes is towards the improvement of insect resistance using insecticidal genes from the bacterium *Bacillus thuringiensis*. Similarly, herbicide-resistant crops and tospovirus-resistant crops are recent improvements in the use of transgenics during the last 5 to 6 years. With the development of high throughput technologies in gene sequencing, there is also a possibility of mining genes of high agronomic value and introducing them into crop varieties via genetic transformation. Genomes of two higher plants, arabidopsis and rice, have already been sequenced and information from these genomes would allow characterization of resistance-conferring genes to related plant species and genera (Pental, 2003). Such mining of genes also could pave the way for mining of alleles for resistance to diseases and pests in crops which are well adapted to dry land agriculture but suffer from a large number of biotic stresses.

DISTRIBUTION

First field trials of transgenic crops were conducted in the USA and France, but the People's Republic of China was the first country to commercialize transgenic crops in the early 1990s with the introduction of virus-resistant tobacco. In 1994, the USA followed when the US company 'Calgene' obtained approval to commercialize the genetically-modified Flavr Savr® delayed ripening tomato. From then onwards, the development and use of transgenic crops has gained momentum. Today, more than 20 genetically-engineered crop hybrids and varieties are being commercialized and the number is likely to increase steadily in the coming years. The introduction of transgenic crops, commonly called genetically-engineered (GE) crops has taken an early lead in major field crops like corn, soybean, cotton, canola and potatoes. The global area under GE crops has increased from 2.8 million hectares in 1996 to 58.2 million hectares in 2002 (James, 2003). This represents a 21-fold increase in six years (Table 1.1). The number of countries planting these crops went up to 14; eight of them are industrialized countries and six of them developing nations. According to the 'Global Review of Commercialized Transgenic Crops: 1999', the trend in use of GE crops indicates that industrial countries account for 82% and developing countries for 18%.

Table 1.1 Global area of genetically-improved crops planted in the period 1996-2002

Year	Million hectares	Million acres	Increase (%) from 1996
1996	2.8	7.1	
1997	12.8	32.5	78.1
1998	27.8	69.5	89.9
1999	39.9	98.6	92.9
2000	44.2	109.2	93.7
2001	52.6	130.0	94.7
2002	58.2	143.8	95.2

Source: James (2003)

Despite the on-going debate on the safety of transgenic crops, particularly in Europe, millions of farmers in both industrial and developing countries continue to increase their planting of transgenic crops. It is estimated that 3.5 million farmers grew transgenic crops for their health and economic advantages in 2001 and as the population keeps on growing globally—mostly in Asia, Africa and Latin America—transgenic crops represent promising technologies that can make a vital contribution to global food, feed and fiber security (James, 2003).

In fact, 2001 was the first year in which transgenic crops exceeded the 50 million hectare mark, which is equivalent to more than 5% of the total land area of China (956 million hectares) or the United States of America (81 million hectares) and more than twice the land area of the United Kingdom (24.4 million hectares). There was a 19% increase in the area of transgenic crops between 2000 and 2001, a figure almost twice the corresponding increase in the previous year (11%). Four countries grew 99% of the global transgenic crops in 2000—two developing countries (Argentina and China) and two industrialized nations (USA and Canada). In 2001, the USA grew genetically-improved crops on 35.7 million hectares, followed by Argentina (11.8 million hectares), Canada (3.2 million hectares) and China (1.5 million hectares). In percentage terms, the USA grew 68% of the global transgenic crops, followed by Argentina with 22%, Canada with 6% and China with 3%. However, another 10 countries (South Africa, Australia, Mexico, Bulgaria, Uruguay, Romania, Spain, Indonesia, Germany, and France) account for less than 1% each. Thus, major changes in the global share of transgenic crops have been correlated with an increase in the area of transgenic soybean (63%) and maize (19%) in 2001 that contribute to major share of transgenic crops. In 2001, the global hectarage of the herbicide-tolerant soybean was estimated to have increased by 20% to 7.5 million hectares. However, transgenic corn in 2001 was estimated to have decreased by about 500,000 hectares. The global area of transgenic cotton in 2001 was estimated to have increased by about 28% year-over-year, the most significant increase being in China, which tripled its insect resistant *Bacillus thuringiensis* (Bt) cotton area from 0.5 million hectares to 1.5 million hectares in 2001. In the USA, this increase was from 72 to 77%, in Australia by 33%, and the size of area planted remained unchanged year-over-year in Mexico, Argentina and South Africa (James, 2003). Considerable progress has been made in developing insect-resistant transgenic crops in India. Approval and commercial cultivation of Bt cotton in 2002 and its perceived benefits would certainly initiate more interest and activity in both public and private sector in India. On the whole, transgenic crop area as per cent global area of principal crops in 2001 is estimated as 46% for soybean, 20% for cotton, 11% for canola and 7% for maize. Although Bt-based transgenics for insect pest management have made considerable impact (see Chapter 2), there are other transgenics being studied in detail presently.

Transgenic Herbicide Tolerant Crops

Herbicide tolerance may be an inherent character of a plant, but can also be introduced by selection, mutation or genetic modification, including introduction of transgenes. Herbicide tolerance, introduced by genetic modification in crops, currently involves two herbicides: phosphinothricin (or glufosinate) and glyphosate. Both chemicals are broad-spectrum herbicides that make no distinction between weeds and crops. By the introduction of transgenes, crops can be given the ability to tolerate the presence of such toxic chemicals. Transgenic phosphinothricin tolerance originates from the microbe *Streptomyces viridochromogenes* and several other *Streptomyces* species. Phosphinothricin is an amino acid-like compound, produced naturally by specific *Streptomyces*, which inhibit the key enzyme in nitrogen metabolism glutamine synthetase (GS). The *Streptomyces* protect themselves from the toxic effects by producing an enzyme that deactivates the phosphinothricin. This enzyme is called phosphinothricin-N-acetyltransferase (PAT). The gene coding for PAT is called the *bar* gene, which was successfully introduced, in a large number of crops, making them tolerant to phosphinothricin, and thus transgenic herbicide-tolerant crops.

Similarly, glyphosate is a single tertiary amine, which inhibits the enzyme 5-enolpyruvylshikimate-3-phosphate synthase (EPSPS). This enzyme operates in the shikimate pathway that yields aromatic amino acids and secondary plant products. To obtain transgenic glyphosate tolerance, two strategies are being adopted:

(i) Introduction of EPSPS encoding genes with a reduced affinity to glyphosate after isolation from a number of microbial and plant sources; and

(ii) Introduction of a glyphosate-degrading enzyme known as glyphosate oxidoreductase (GOX). The GOX gene was isolated from an *Achromobacter* bacterial strain (Nap, 1999).

Both types of genes have been successfully introduced into canola, chicory, cotton, maize and soybean and commercialized by Monsanto, AgrEvo, Bejo Zaden and Novartis (now Syngenta) in various countries. This transgenic approach makes it possible to selectively use these hitherto broad-spectrum herbicides to control weeds in agricultural crops: the transgenic crop survives, whereas the competing weeds get eliminated. Recently, highly regenerable basal segment calli have been employed as the target tissue for genetic transformation of Indian varieties of bread wheat and emmer wheat. The *bar* gene conferring herbicide tolerance was introduced in one-month-old wheat calli employing both particle bombardment and *Agrobacterium*-mediaed transformation strategies (Chugh and Khurana, 2003). These studies have paved the way for the introduction of agriculturally-desirable traits in commercially-important wheat crops.

Herbicide-tolerant soybean was the dominant transgenic crop grown commercially in the USA, Argentina, Canada, Mexico, Romania, Uruguay and South Africa, occupying 33.3 million hectares in 2001. On the whole, 77% of the 52.6 million hectares of all transgenic crops were herbicide-tolerant soybean, corn and cotton in 2001. The estimates show that herbicide-tolerant crops increased by 24% between 2000 and 2001. There are 5 other crops that occupy the remaining 23% of global transgenic crop area, which include herbicide-tolerant canola (5%), herbicide-tolerant cotton (5%), Bt/herbicide-tolerant cotton (5%), herbicide-tolerant maize (4%) and Bt/herbicide-tolerant maize (3%).

Transgenic Tospoviruses Resistant Crops

Tospoviruses comprise a relatively new group of viruses, which attack many economically important crops worldwide. Research efforts have been initiated globally in order to improve crop resistance to these viruses. The annual worldwide loss due to tospoviruses is estimated at US $ 1 billion. Peanut bud necrosis virus (PBNV) damages peanuts especially in Asia, with annual losses estimated at more than US $ 89 million. The increased incidence of tospoviruses in several parts of the world since the 1980s has been attributed to the introduction and/or proliferation of its insect vectors; several species of thrips. Unlike some other plant virus-vector relationships, the fidelity between tospoviruses and the thrips species that transmit them is limited. For example, tomato spotted wilt virus (TSWV) can be transmitted by at least seven different species of thrips, whereas peanut bud necrosis tospovirus (PBNV) and watermelon silver mottle tospovirus (WSMV) are transmitted only by *Thrips palmi*.

Once the thrips larvae acquire a virus, the virus replicates and survives through the developmental stages of the insect. Only virus ingestion by first instar thrips larvae can lead to the adults that transmit the virus. A virus ingested by the adult is not transmitted due to the presence of a midgut barrier to virus acquisition in adult insects. Infective adults retain the ability to transmit the virus for life. However, the virus is not transmitted from an adult to its offspring. Tospoviruses as well as their thrips vectors have a wide host range. For example, TSWV is known to infect about 650 plant species in 45 families (Pappu, 1997).

Transgenic resistance to TSWV has been achieved by introducing specific TSWV DNA sequences and, in some cases, impatiens necrotic spot virus (INSV) sequences. In order to broaden the resistance to several tospoviruses, tobacco has been transformed with NP gene sequences from three tospoviruses: TSWV, groundnut ring spot tospovirus (GRSV) and tomato chlorotic spot tospovirus (TCSV). Several transgenic lines expressing all of these three genes exhibited resistance to challenge infection by the respective tospoviruses (Pappu, 1997).

While most efforts to introduce genetically-engineered resistance to tospoviruses have focused on tobacco, other crops of commercial value that have been transformed (Table 1.2) include tomato, lettuce, groundnut and ornamental crops. Projects are underway to introduce resistance to PBNV in groundnut in India, to watermelon silver mottle virus (WSMV) in watermelon in Taiwan, to TSWV in groundnut and to Burley and Flue-cured tobacco in the USA.

Table 1.2 Transgenic crops resistant to tospovirus infections

Crop	Introduced sequence	Greenhouse/field test
Tobacco	NP gene of TSWV	Both
	NSm of TSWV	Greenhouse
	NP gene of TSWV, TCSV, GRSV	Greenhouse
Tomato	NP gene	Both
Groundnut	NP gene of TSWV	Both
Chrysanthemum	NP gene of TSWV	Greenhouse
Lettuce	NP gene of TSWV	Both

Various research collaborations exist to achieve an increased resistance to tospoviruses in crops. For instance, collaboration between the University of Georgia and ICRISAT resulted in cloning of portions of viral genome and the subsequent construction of the NP gene for plant transformation. Similarly, other collaborations among researchers in Africa, Europe, Asia and the Americas have resulted in the production of high quality virus-specific antibodies and reliable diagnostic methodologies to detect and differentiate various tospoviruses.

ADOPTION

Adoption rates of transgenic crops are expected to increase, but will be subject to the resolution of issues in relation to labeling freedom of choice and other considerations. Public acceptance constraints, however, mainly apply to Europe. In North America, companies have developed different strategies to market their new transgenics, i.e. sell them as traditional hybrids and varieties. Others charge a separate fee for the transgenic technology on the basis of contracts. A review of the products being tested in field trials confirms that the R & D pipeline is full of new transgenic products that are likely to be available in the near future. The number of countries growing transgenic crops has increased from one in 1992, to 6 in 1996, to 9 in 1998 and to 12 in 1999 (ISAAA, 2000) and 14 in 2001. The adoption rate is increasing with a corresponding increase in the available technology. For example, Monsanto's Bt cotton technology was initially developed for large-scale cotton farming systems in the USA and Australia and not for the small-scaled diversified systems in China. But today, Monsanto's variety 33B is currently used in Hebei, which has an economic advantage over the local non-Bt cotton varieties, for which insecticide application is required. According to Monsanto, farmers in China benefited from yield increases and lower insecticide use, resulting in an overall benefit of US $ 560 per hectare. China turned to biotechnology for the improvement of crops in the mid-80s and more than 100 laboratories across that country have become involved in the effort. Accordingly, in 1997, China began producing biotech crops commercially, with the planting of roughly 1 million acres of improved transgenic crops. China is accelerating the commercialization of Bt maize/corn, soybean and rice. In 2001, this nation produced an estimated 270 tones of Bt corn seed, 30 tones of Bt soybean seed and 200 tones of Bt rice seed for planting.

Similarly, Cuba's economy still relies heavily on sugarcane for export revenues, while simultaneously, it urgently needs an increase in the local food production per unit of land. For both purposes, Cuba is combining low input organic agriculture with crops improved with transgenes. Transgenic crops under development in Cuba include sugarcane, potato, papaya, tomato, maize, citrus, pineapple, etc., and emphasis is on both laboratory and field tests for insect, virus, fungal and herbicide resistance.

The Japanese Health Ministry has so far approved 29 transgenic biotech varieties including seven crops, corn, soybean, rapeseed, potatoes, cotton, sugar beet and tomatoes. The Agricultural Ministry plans to provide $ 10.3 million to fund private sector R & D in biotech crops in 2000. India follows closely with the Indian Council of Agricultural Research announcing a major project to develop genetically-improved pest-resistance cotton, rice and pigeon pea.

In the next five years, at least 7 Asian countries are expected to adopt transgenic pest and herbicide-resistant crops: India, Indonesia, Philippines, Thailand, Vietnam, Singapore and Malaysia. So far, they have conducted field trials of those transgenic crops, which they most desire to commercialize in the next few years. On this basis, adoption rates are expected to increase with time. The major reasons for the high adoption rates of transgenic crops are that they offer more convenient and flexible management, higher productivity and economic returns and a safer environment through decreased use of conventional pesticides. This will also increase the income of the farmers in developing countries. For instance, it has been estimated that genetically-improved Bt cotton could earn Thai farmers upto 275 baht per rai more than growing conventional cotton. Those who grow 100,000 rai of Bt cotton will be able to save 30 million baht in cotton imports.

If we look at the overall sales of agriculture biotechnology products in the USA, the graph projects US $ 100 million in 1995, an increase to US $ 304 million in 1996 (Hruska, 1996) and a prediction to continue growing at 20% per annum. The global market for agricultural biotechnology was less than US $ 500 million in 1996 and is projected to soar to $ 25 billion by 2010 from an estimated $ 3 billion in 2000 (ISAAA, 2000). These data clearly indicate that the adoption rate of transgenic crops is increasing steadily and includes insect-resistant cotton, maize, rice, tomato and soybean; virus and bacterial disease-resistant papaya, sweet pepper, potato and tobacco; and herbicide-tolerant canola, corn, cotton and soybean. Whereas single traits currently predominate, it is noteworthy that double traits have already been introduced in canola and are expected in several of the major crops in the near future (James, 1998).

TRANSGENIC CROPS AND IPM

Overall, the application of biotechnology to IPM has so far been quite conservative, having focused largely on the improvement of existing crop protection products or technologies by the use and manipulation of viral and bacterial genomes to improve crop resistance to insect pests, diseases, and herbicides. Transgenic plants or plant-associated bacteria seek to improve on what can already be achieved less effectively by topical application of a particular microorganism, particularly Bt The technical objectives of many of these manipulations are directed at improving the performance of an engineered product relative to its wild type competitor by:

- broadening the spectrum of pests targeted by the product;
- increasing the level of control of the target pests;
- increasing the speed of action of the product; and
- improving the delivery of the product to the pest.

Engineering genes for Bt toxins into plants is an ingenious method to delivering these toxins to pests which might naturally avoid them, such as insects which feed inside the plants. From an IPM perspective, the technology has more similarities to plant-resistance breeding. Therefore, biotechnological innovations which improve the persistence or efficiency of the biological processes—by improving survival or transmission rates of pathogens, or facilitating broadly-based quantitative crop resistance to pests—will be valuable to the sustainable IPM of the future. In fact, transgenic crops will help in improving diagnostic systems which allow researchers to recognize desirable genes useful in reducing potential pest damage. Also, with the benefit of the IPM experience, biotechnology stands to contribute greatly to sustainable pest management (Waage, 1996). However, detailed accounts of Bt transgenic crops and IPM-based concepts are given in Chapter 2 of this volume.

RESISTANCE MANAGEMENT FOR TRANSGENIC CROPS

Insects have the potential to develop resistance to transgenic crops because the plants maintain the constant killing level throughout. Therefore, unlike topical sprays which are inactivated over a short period of time, the selection pressure of transgenic plants on susceptible pest population will be much higher.

Resistance management is a way of sustaining the effectiveness of a pest control tool or methodology. It tries to delay or prevent adaptation in pest species by managing the factors that may contribute to resistance development. Its main goal is the preservation and management of genetic resources, i.e. the genes that are responsible for the susceptibility of a pest to the pest control tool or tactic (susceptible genes). This will require cohesive effort and commitment and participation by farmers, pesticides or seed supplies, and regulators to help prevent insect resistance through maintenance and proactive management.

Resistance management programmes rely on four key strategies:
- Mortality source diversification.
- Use of refugia and reduction of selection pressure.
- Prediction and monitoring of resistance.
- Policy implementation.

Mortality source diversification will help pests in not adapting as quickly if they are faced with more than one mortality mechanism, i.e. two or more highly fatal toxins with different modes of action could be combined in the same plant. Similarly, refuges may be an effective way to reduce selection pressure by providing an area for habitation and immigration of susceptible insects. In a transgenic development scheme, this can be achieved by providing a refuge of non-transgenic plants by way of a seed mixture of transgenic and non-transgenic plants; a spatial mixture, or field-to-field mosaic that results in a patchwork of completely transgenic and completely non-transgenic plots; and/or a temporal mixture, or season-to-season sequence that alternates between transgenic and non-transgenic plantings (Liu and Tabashnik, 1997).

There are some other factors such as gene pyramiding, which is an important strategy for resistance management. Many of the candidate genes that have been used in genetic transformation of crops are either too specific or are only mildly effective against the target insect pests. In order to convert transgenics into a more effective weapon in pest control, it is necessary to deploy genes with different modes of action in the same plant. This will increase the multi-gene, multi-mechanistic resistance (Sharma and Ortiz, 2000). It has been suggested that *cry1*A(c) and *cry1*F can be expressed together in transgenic plants for effective control of *Helicoverpa armigera* (Hubner) (Chakrabarti et al., 1998), to increase the durability of resistance. Similarly, activity of Bt in transgenic plants against pests can be enhanced by the addition of genes expressing serine protease inhibitors (MacIntosh et al., 1991), tannic acid (Gibson et al., 1995), and CpTi genes (Zhao et al., 1998).

Destruction of carryover populations that have been exposed to Bt crops in previous generations also poses an important component of resistance management. Therefore, appropriate agricultural practices need to be followed that reduce the carryover of pests from one season to another, including appropriate crop rotations, observing a crop-free season, and other resistance management strategies (see chapters 6,7 and 8).

POTENTIAL ADVANTAGES/DISADVANTAGES

The unprecedented rapid adoption of transgenic crops during the last 7 years reflects the significant multiple benefits realized by large and small farmers. There is a growing body of evidence that clearly shows the improved weed and insect pest control attained with transgenic crops. The introduction of transgenic crops may have a positive impact, most importantly the potential reduction in chemical pesticide use. Cotton, for example, has received traditionally high levels of insecticide applications of which the high economic, environmental and health costs have been well documented. Thus, the availability of transgenic cotton could permit the reintroduction of the crop into the area where such production was stopped or reduced either due to the high cost of insecticides or due to various environmental implications.

A second benefit of transgenic crops is the ease of their implementation, because no new practices need to be learned for the basic use of technology. The whole technology is 'all in the seed'. Therefore, the only challenge is to get the seed into the hands of the farmers.

There is also an advantage of conservation of soil moisture, structure, nutrients and control of soil erosion through no or low-tillage practices as well as improved quality of ground and surface water with less pesticide residues. As per the 1999 estimates, the global economic advantage to farmers was worth $ 700 million and shared equally between developed and developing countries. There is cautious optimism that transgenic crops will continue to grow through 2003-2007 period.

However, certain uncertainties, potential disadvantages and threats have been envisaged. Gene manipulation through biotechnology can cause problems as it is gray area and needs to be addressed systematically with pragmatic approach, especially in developing countries where most of the farmers are either poor or marginal. Possibly, the greatest ecological hazard of transgenic crops is the creation of new weeds and the erosion of genetic diversity due to the exchange of genetic material between transgenic crops and their native wild relatives. The possible gene flow between maize and teosinte is of great concern in Mexico, and has recently been studied by the International Centre for the Improvement of Maize and Wheat. The report recommends that quantitative studies be carried out on the potential gene flow to the maize genus *Zea* before liberating transgenic maize varieties, and that experimentation with transgenic crops take place under the strictest security measures to prevent gene flow (Quist and Chapela, 2001). The report may not have a substantial bearing in the present scenario but once large-scale cultivation of transgenics is undertaken, the possibility of genetic exchange between land races and transgenic material cannot be summarily ignored and it is essential to secure germplasm in the gene banks globally. Toxic effects of Bt corn pollen on the monarch butterfly, *Danaus plexippus* (Linnaeus) (Losey et al., 1999) and *Chrysoperla carnea* (Stephens) (Hilbeck et al., 1998) have been reported and are a matter of concern.

Another big concern is the potential rapid development of resistance in pest populations, as seen in case of Bt toxins. This would not only lead to ineffective Bt plants, but also affect the efficacy of sprayed Bt formulations. That is why resistance management systems are necessary to be included on the regulatory systems.

RATIONAL USE OF TRANSGENICS

Although transgenic crops are yet on the threshold of commercialization globally, there is little to stop farmers from introducing them in various regions. Thus, any debate about barring their entry has probably already been overtaken by events. However, it is necessary to adopt a rational approach and needs to follow certain criteria:

- The existence of new transgenic crops and the possible consequences of their use should be converted into a documented information for regional policy makers, regulators, law makers, private industry, extension services, NGOs and consumers. Special attention should be given to food safety, resistance management, and gene flow. Local systems, their constraints and socio-economic implications should be strictly considered before adopting of any transgenic material.
- Regional policies should be developed towards transgenic crop use. Specific attention is required to the agroecology of the cropping systems in the region. Policies, laws and regulations should be harmonized.
- The risks of gene flow and its effects need to be studied carefully.
- Given the difficulties of enforcement of regulations, the best option might be to regulate seed distribution. This will help in resistance management as well.
- Cost/benefit analysis is necessary for the introduction of transgenic crops.

- It is necessary to undertake a proactive approach to stimulate the use of the right gene for the right reason, and in the right way. Regional research institutions can play an important role in developing, testing and recommending management practices appropriate to the transgenic crop technology, local production systems and pests.

ACKNOWLEDGEMENTS

Authors are thankful to Prof. John H. Bennedict, Texas A&M University, USA for his suggestions on the earlier draft of this chapter.

REFERENCES

Chakrabarti, S. K., Mandaokar, A.D., Ananda Kumar, P. and Sharma, R.P., 1998, Synergistic effect of Cry1Ac and Cry1F δ-endotoxons of *Bacillus thuringiensis* on cotton bollworm, *Helicoverpa armigera*. Curr. Sci., **75**, 663-664.

Chugh, A. and Khurana, P., 2003, Herbicide resistant transgenics of bread wheat (*T. aestivum*) and emmer wheat (*T. dicoccum*) by particle bombardment and *Agrobacterium*-mediated approaches. Curr Sci., **84**, 78-83.

Gibson, D. M., Gallo, L. G., Kransoff, S. B. and Ketchum, R. E. B., 1995, Increased efficacy of *Bacillus thuringiensis* sub.sp. *kurstaki* in combination with tannic acid. J. Econ. Entomol., **88**, 270-277.

Hilbeck, A., Baumgartner, M., Fried, P. M. and Bigler, F., 1998, Effects of transgenic *Bacillus thuringiensis* corn fed prey on mortality and development time of immature *Chrysoperla carnea* (Neuroptera: Chrysopidae). Environ. Entomol., **27**, 480-487.

Hruska, A.J., 1996, Transgenic crops in Central American agriculture. Biotech. Dev. Monitor, **29**, 7-9.

ISAAA, 2000, Area Planted with Transgenic Crops up in 1999. GE-ISAAA report, 29 June 2000.

James, C., 2003, Global review of commercialized transgenic crops. Curr. Sci., **84**, 303-309.

James, C, 1998, Global status and distribution of commercial transgenic crops in 1997. Biotech. Dev. Monitor, **35**, 9-12.

Liu, Y. B. and Tabashnik, B. E., 1997, Experimental evidence that refuges delay insect adaptation to *Bacillus thuringiensis*. Proc. Roy. Soc. London, **264B**, 605-610.

Losey, J. E., Raynor, L. S. and Carter, M. E., 1999, Transgenic pollen harms monarch larvae. Nature, **399**, 214.

MacIntosh, S.C., Stone, T.B., Jokerst, R.S. and Fuchs, R.L, 1991, Binding of *Bacillus thuringiensis* proteins to a laboratory-selected line of *Heliothis virescens*. Proc. Natl. Acad. Sci. USA, **88**, 8930-8933.

Nap, Jam-Peter, 1999, A transgene-centred approach to the biosafety assessment of transgenic herbicide-tolerance crops. Biotech. Dev. Monitor, **38**, 6-11.

Pappu, H.R., 1997, Managing tospoviruses through biotechnology: progress and prospects. Biotech. Dev. Monitor, **32**, 14-17.

Pental, D., 2003, Transgenics for productive and sustainable agriculture: some considerations for the development of a policy framework. Curr. Sci., **84**, 413-424.

Quist, D. and Chapela, I. H., 2001, Transgenic DNA introgressed into traditional maize land races in Oaxaca, Mexico. Nature, **414**, 541-543.

Sharma, H.C. and Ortiz, R., 2000, Transgenics, pest management, and the environment. Curr. Sci., **79**, 421-437.

Waage, J. K., 1996, Integrated pest management and biotechnology: An analysis of their potential for integration. In G.J. Persley (ed.), Biotechnology and Integrated Pest Management, CAB International, Oxon, UK, pp. 36-60.

Zhao, J.Z., Shi, X.P., Fan, X.L., Zhang, C.Y., Zhao, R.M. and Fan, Y.L., 1998, Insecticidal activity of transgenic tobacco co-expressing Bt and CpTI genes on *Helicoverpa armigera* and its role in delaying pest resistance. Rice Biotechnol., **34**, 9-10.

TRANSGENIC CROPS EXPRESSING Bt PROTEINS: CURRENT STATUS, CHALLENGES AND OUTLOOK

JOHN H. BENEDICT* AND DENNIS R. RING**

*Texas A&M University and Texas Agricultural Experiment Station
Corpus Christi, TX 78406-9704, USA
**Louisiana State Univesrsity Agricultural Center
Baton Rouge, LA, 79894-5100, USA

INTRODUCTION

Applications of biotechnology have existed as long as people have grown crops and unknowingly used microorganisms in the process of making bread, cheese, alcoholic drinks, and tanning leather. Today, biotechnology is stimulating many positive changes in agriculture (Penn, 2000; James, 2002a; Chrispeels and Sadava, 2003). Some of the recent tools of biotechnology, especially genetic engineering, have revolutionized the speed of crop improvement and availability of valuable new traits. A current example is the use of Cry proteins found in the bacterium, *Bacillus thuringiensis* Berliner (Bt), to improve crops by suppressing injury from crop pests—known as Bt crops. The first generation of commercial Bt crops has surpassed our expectations for pest control, economic return and safety. The purpose of this chapter is to discuss the challenges and current status of Bt crops, their integration into insect pest management programs, their benefits and risks, and where we can bring light to the subject, their social-cultural, environmental, and economic impacts. We will also discuss some future Bt crop products and the new Bt traits they may possess. We hope this chapter will bring the reader a clearer and more factual perspective on Bt crop products, their value and contributions to society.

BACKGROUND TO Bt CROPS

The bacterium, *B. thuringiensis,* is an insect pathogen that has been used as a topical spray—much like traditional synthetic chemicals—to control insect pests on many crops worldwide since the early 1960s (Hall and Menn, 1999; Benedict and Altman, 2001). Bt insecticides are classified as biopesticides because biological organisms produce the Cry proteins. The Cry proteins are also known as delta-endotoxins. Different strains of this bacterium are unusual in that they produce different Cry proteins with insecticidal activity towards specific insect pests. The DNA sequences are known for more than 100 of these Cry proteins (Crickmore et al., 1998). A number of companies produce sprayable biopesticides using different strains of *B. thuringiensis* that go on to produce various Cry proteins, each targeted to control a select few species of insect pests (Baum et al., 1999). Some companies have genetically-engineered various species of bacteria to produce Cry proteins targeted for different pest species.

The greatest acceptance and use of these sprayable Cry proteins has been for control of mosquitoes and a limited number of caterpillar pests (i.e. Lepidoptera) in forestry, vegetable crops, organic farming, greenhouse crops and home gardens. However, the usefulness of Cry proteins in commercial spray applications has been limited due to their relatively high cost, rapid environmental inactivation, poor crop coverage, and less than desired level of pest control, especially when compared to other less expensive conventional chemical insecticides (Benedict and Altman, 2001). But the compelling reason for using live or dead *B. thuringiensis* sprays is safety—safety to the applicators, beneficial insects that help to naturally control pests, merchants, consumers and the environment. They are among the safest pesticides known. Foliar sprays of *B. thuringiensis* can be applied on crops up to the day the crop is harvested and sold to the consumer. We have all consumed this *Bacillus* bacterium, and its Cry proteins and DNA repeatedly on uncooked vegetables, especially salads, for most of our lives without suffering from any ill effects (US EPA, 2001a).

In the mid 1990s, the first commercial transgenic crops appeared which capitalized on the safety of *B. thuringiensis* insecticidal proteins (Table 2.1) (Carozzi and Koziel, 1997; Krattinger, 1997; Peferoen, 1997). The major differences between Bt crops and topical sprays of *B. thuringiensis* are: (i) Bt crops produce the insecticidal Cry protein inside all plant cells, whereas Bt sprays are present on the surfaces of the plant; and (ii) the *B. thuringiensis* bacterium is not present in Bt crop cells, whereas Bt sprays contain the bacterium, and are present on plant surfaces following application. Due to this new method of producing and delivering Cry proteins, the Environmental Protection Agency now classifies Bt crops as PIPs, for Plant Incorporated Protectants (http://www.epa.gov/pesticides/biopesticides/pips/pip_list.htm). PIPs are also classified as biopesticides. Thus, transgenic Bt crops are really a different method of applying the Cry proteins. The plant produces the Cry protein in most of the plant cells, where it is needed for pest control, rather than man spraying the bacterially-produced Cry protein on the surface of the plant. This change in method of application has resulted in a quantum leap in the effectiveness of Cry proteins to control insect pests—pest suppression in Bt plants now competes with or exceeds that of the best conventional insecticides. However, a very serious limitation of sprayable Bts and current Bt crops is that they provide a high level of pest suppression for only a few of the many pests attacking the crop. Despite their limitations, adoption of Bt crops by farmers around the world has been surprisingly quick (See Chapter 1).

Despite these benefits, considerable controversy has developed globally over the safety and use of Bt crops. It is surprising since the Cry proteins found in Bt crops have been used by organic and conventional farmers globally for more than 40 years, without health or environmental problems. It is to be expected that environmentalists, organic farmers and the public would welcome an improvement in an old product. Instead, commercial production of transgenic Bt crops has provoked a host of changes and turmoil in insect pest management, environmental safety, socio-cultural arenas, commodity markets, production economics and companies in the pharmaceutical, agrochemical and seed industries. In fact, environmental issues, consumer attitudes, and the needs and desires of farmers and special interest groups, like GreenPeace (Thompson, 2002), have had a significant impact on the social acceptance of food products from Bt crops, and the use of Bt crop production technology by farmers. Public response to these issues appears to be more driven by fear than facts (Morris and Bate, 1999).

Today, the greatest challenge for Bt crops is gaining enough public support and acceptance in foods so their use and value can be fully explored and integrated into crop production systems globally.

CURRENT COMMERCIAL Bt CROPS

The first-generation of Bt crops, cotton, corn/maize and potatoes, were approved in 1995 (Shelton et al., 2002) and planted for commercial production in 1996 (Table 2.1). The major players selling Bt crops were Monsanto (based in St. Louis, MO, USA), Syngenta Crop Protection (based in Basel, Switzerland and then known as Novartis) and Aventis (now merged with Bayer CropScience and based in Monheim am Rhein,

Table 2.1 Bt crops and insect control proteins (PIP[a]) registered[b] by country and date, for commercial production as of March 2003

Crop/Company	Event	PIP	Country	Year approved[c]
YieldGard® Corn/ Monsanto	MON810	Cry1Ab	Argentina	1998
			Bulgaria	2000
			Canada	1997
			EU	1998
			Philippines	2002
			South Africa	1997
			Spain	2001
			USA	1996
YieldGard® Corn/ Syngenta	Bt 11	Cry1Ab	Argentina	2001
			Canada	1996
			Japan	1996
			USA	1996
Knockout® Corn/ Syngenta	176	Cry1Ab	Argentina	1998
			Canada	1996
			USA	1995
Herculex® 1 Insect Protected Corn/ Pioneer Hi-Bred-DuPont, Mycogen Seeds-Dow	TC 1507	Cry1Fa2	Canada	2002
			USA	2001
YieldGard® Rootworm Corn/Monsanto	MON863	Cry3Bb1	Canada	2003
			USA	2003
New Leaf Potato®/ Monsanto	Many	Cry3A	Canada	1995
			Romania	1999
			USA	1995
Bollgard® and Ingard® Cotton/Monsanto	MON531, MON757	Cry1Ac	Argentina	1998
			Australia	1996
			China, Hebei[d]	1997
			India	2002
			Indonesia	2001
			Mexico	1997
			South Africa	1997
			USA	1995
Bollgard® II Cotton/ Monsanto	Stacked MON15985 MON531	Stacked[e] Cry2Ab2 Cry1Ac	Australia	2002
			USA	2002
Bt Cotton/CAAS[d]	GK	Fused[f] Cry1Ab Cry1Ac	China	1998
Bt Cotton/CAAS	sGK	Fused[g] Cry1Ab Cry1Ac Stacked CpTi	China	1999

Table 2.1 contd.

Table 2.1 contd.

ᵃ PIP stands for Plant Incorporated Protectant; here, it is the Bt insecticidal protein expressed by the specific Bt transgene present in the crop.

ᵇ Registration means that the crops can be grown and harvested. These registrations do not necessarily include approval for import or food use in the country. See current list at http://www.epa.gov/pesticides/biopesticides/pips/pip_list.htm and http://www.agbios.com/dbase.php?action=Synopsis

ᶜ Approval dates refer to the first Bt variety approved. Additional Bt genes/events may be covered by separate approvals.

ᵈ Hebei was the first province to approve Bollgard cotton in China. This was followed by approvals in Anhui, Shandong and other provinces.

ᵉ The cry1Ac and cry2Ab genes were inserted into the same parent line, called 'stacked genes', to produce commercial crops that express both Cry1Ac and Cry2Ab insect control proteins in the same plant.

ᶠ The cry1Ab and cry1Ac genes were joined together then inserted into the parent line, called 'fused genes', to produce commercial crops that express both Cry1Ab and Cry1Ac insect control proteins in the same plant from one insertion event.

ᵍ A third gene, *cipt*, was stacked in plants previously transformed with the fused cry1Ab/cry1Ac gene. The *cipt* gene expresses a trypsin inhibitor from cowpea. The trypsin inhibitor can interfere with susceptible insect pest digestion, resulting in reduced growth and survival.

Germany), and their numerous seed producers and distributors. However, there has been and continues to be considerable collaboration, trading of technology and patents, and consolidation among companies, especially seed companies. Today, additional companies, universities and government agencies worldwide are developing Bt spray products and PIPs at an increasing rate. New competitors have entered the Bt crop protection market. For example, the large USA-based partnerships of Dupont with Pioneer Hi-Bred Seeds, Dow AgroSciences with Mycogen Seeds and Dow AgroSciences with Phytogen. Jointly, they are marketing a newly-registered Bt corn product, Herculex®1 expressing Cry1F, and expect to have other new products registered in the next two years. These Bt crop products should increase competition in the Bt crop market.

Each of the first-generation Bt crop products possesses a single synthetic Bt gene that produces one of five Bt δ-endotoxins, Cry1Ab, Cry1Ac, Cry3A, Cry9C or Cry1F, in order to control the target pest. Crop expression of the Bt protein is controlled by a promoter gene, and other genes that were introduced into the plant cells with the *cry* gene when the plants were transformed (Perlak et al., 1990, 1991). The Cry proteins are constitutively expressed throughout the plant in all the tissues continuously throughout the season.

A number of factors affect protein expression in various plant tissues. Key among these are: insertion event; tissue age; tissue regeneration process that produced the transformed Bt plant; the specific *cry* gene and the cassette of foreign genes, especially the promoter, that were inserted into the plant; the genetic background into which the *cry* gene was introduced (i.e. specific plant genotype); and the environment in which the Bt crop is grown (Sachs et al., 1996; 1998; Coviella et al., 2000; Benedict and Altman, 2001). All of these factors influence the efficacy of the Bt crop in controlling the target pests.

Following their introduction, the first major change in Bt crop products was the combining—in the same crop plant—of the Bt gene and a transgene so as to provide resistance to a herbicide; either bromoxynil (e.g. Buctril®), glyphosate (e. g. Roundup®) or glufosinate (e.g. Liberty®) (See Chapter 4). Other genes conferring resistance to various herbicides have been introduced into crops but have not been adopted as widely as resistance to glyphosate and glufosinate herbicides. The combining of multiple transgenic traits in the same crop variety—either as separate genes inserted into the plant in separate transformation events or a single transformation event—has been called stacking or pyramiding. For example, select YieldGard® corn hybrids possessing the Bt gene *cry1Ac* have been genetically improved by stacking the Bt gene with the *cp4-epsps* transgene, for resistance to RoundUp®. In gene constructions where the herbicide resistance gene has been joined to the Bt gene and inserted in a single event, some people prefer to call the construct 'fused' genes. Since it results in multiple traits stacked in the same variety, so we will continue to apply the term stacked or pyramided to indicate multiple transgenic traits expressed in the same crop variety, regardless of the transformation specifics.

The objective of stacking multiple *cry* genes in the same crop variety is no different than that of the conventional crop improvement programs operating worldwide; it is to add additional desirable traits (Pfeiffer, 2003). However, some of the methods used to create today's improved transgenic Bt and herbicide-stacked varieties are different than conventional breeding. The most obvious difference in using plant transformation is that only the desired Bt gene and its supporting genes/ DNA sequences (e.g. markers, promoters, and ending sequences) are transferred to the current variety. In contrast, in conventional breeding, when a cross is made between a current variety and another genotype possessing a desired trait, it usually results in thousands of undesired genes being transferred to the current variety along with the desired trait. These undesirable traits are eliminated through an expensive and time-consuming breeding program, but even then some undesirable genes may remain (Simmonds and Smartt, 1999; Pfeiffer, 2003).

The greatest improvement in Bt crops to date has been achieved by the stacking of multiple insect resistant *cry* genes in the same crop variety. There are compelling reasons for improving insect control of Bt crops in this manner:

- To broaden the spectrum of insect pests controlled, called target pests. Single Bt proteins to date have had a very narrow range of those pest species they were effective in controlling.
- Improving the level of control of the target pests, called pesticidal efficacy. Current single Bt proteins provide a high level of efficacy for no more than one or two pests in a crop. There are commonly three or more insect pests that attack a particular corn or cotton field during the season (Metcalf and Metcalf, 1993; Benedict, 2003). These pests are commonly from different taxonomic insect groups, such as beetles (Coleoptera), bugs (Homoptera and Hemiptera) and caterpillar pests (Lepidoptera); most of which are not controlled by first-generation Bt proteins. Conventional insecticides may be required to control the pests beyond the control of Bt proteins.
- To decrease the likelihood that the target pests will develop resistance to the Bt proteins in Bt crops (see Chapter 8). Considerable thought and effort have been expended to develop the Insecticide Management (IRM) plans that are in place today in order to reduce the possibility that target insect pests will adapt to the current first-generation Bt control proteins in crops (Gould, 1998; Roush, 1999; US EPA, 1999; US EPA, 2003c). One of the expectations of these IRM programs is that additional insect control genes will be stacked in Bt crops, since most experts believe that this may be one of the most effective means of delaying resistance development (Gould, 1998; Roush, 1999). By stacking a crop variety with multiple Bt proteins along with other insect control chemistries and mechanisms, the target pest is unlikely to adapt to all these mechanisms at once.

Crops possessing two or more insect control proteins stacked in the same variety can be thought of as the second-generation transgenic Bt insect control products (PIP's). Within the next 4 years, many new crop varieties will enter the commercial market with stacked insect control proteins (see below Future Bt Crops). It is believed that they will have an even more profound effect on crop productivity and the ways of managing insect and other pests in agriculture than the first-generation Bt crops. This is based on the above reasons and the performance of the first registered second-generation product, Bollgard® II expressing stacked Cry1Ac and Cry2Ab (see Cotton below). The first-generation products were limited by the narrow range of pests controlled, their efficacy towards the target pests and the higher price of the technology compared to many competing topical insecticides.

Corn/Maize

Bt corn is grown on about 9.9 million ha. worldwide and is by far the most widely-grown Bt crop in the world (Table 2.2). Bt genes have been placed in field corn, flint, popcorn and sweet corn (James, 2003). The first-generation Bt proteins engineered into corn were Cry1Ab and Cry9C. Monsanto's YieldGard® corn, event MON 810, expressing Cry1Ab, is the most widely-grown Bt crop and transformation event today. It represents >85% of all the Bt corn grown in the world, about 7.7 million hectares (19 million acres) in 2001 (Shelton et al., 2002; James, 2002b). Another dominant Cry1Ab event is Syngenta's Bt 11 sold under the NK-brand as YieldGard® (trademark of Monsanto company). Syngenta's Cry1Ab event 176, known as Knockout®, has a very small portion of the current Bt corn market. Event 176 is not expressed in the corn kernels and thus, these structures are not protected from target insect pests. More importantly, the level of expression tends to be lower in fresh leaves of 176 than events MON810 and BT11. In general, the higher the concentration of the Bt protein in the fresh tissues, the greater the level of target insect pest control.

The Cry9C Bt corn seed product, event CBH-351, was marketed as StarLink® by Aventis CropScience (based in France and formerly called AgrEvo, but now a part of Bayer CropScience), in 1998 through 2000. It is no longer in commercial production (US EPA, 2000). In October of

Table 2.2 Global area of conventional and transgenic Bt maize and cotton crops in 2002 (millions ha)

Crop	Total crop area	Bt crop Area	% Bt crop
Corn/Maize	140	9.9	7.1
Cotton	34.8	4.6	13.2
Total	174.8	14.5	8.3

Source: Modified from James (2002b).

2000, Aventis requested that the EPA cancel their registration for StarLink® because Cry9C protein and Cry9c DNA were detected in food products made from corn. The EPA registration had approved the use of StarLink® in the USA for animal feed products only, not in human food. Public interest groups and scientists became concerned that Cry9C protein might be a food allergen; however, further studies have shown that Cry9C does not pose a risk to public health.

Current Bt corn products are targeted at caterpillar pests known collectively as corn borers (caterpillars are the immature stage of moths and butterflies). Worldwide, the European corn borer (ECB), *Ostrinia nubilalis* (Hubner), is by far the most damaging insect attacking corn

(Fig. 2.1). Other important caterpillar pests in corn are the Asiatic corn borer, *Ostrinia furnacalis* (Guenee); southwestern corn borer, *Diatraea grandiosella* Dyar; corn earworm, *Helicoverpa zea* (Boddie); fall armyworm, *Spodoptera frugiperda* (J. E. Smith); and black cutworm, *Agrotis ipsilon* (Hufnagel) (Metcalf and Metcalf, 1993; Shelton et al., 2002). ECB and other borers damage corn by tunneling into stalks and ears, feeding on the silk and tassels, causing plants to lodge and reducing the flow of sap and nutrients, resulting in overall yield loss (Metcalf and Metcalf, 1993). Insect injury to corn kernels and ears by pests such as corn earworm increases the incidence of some fungal diseases that produce toxins (mycotoxins), such as aflatoxin and fumonisin, that are very toxic to man and many of our domestic animals (Munkvold et al., 1999; James, 2001; CAST, 2003). The presence of these mycotoxins in food and feed reduces the grain quality and value to the grower. Current evidence suggests that growing Bt crops reduces mycotoxins in grain under certain conditions (Munkvold et al., 1999; Dowd, 2001). Mycotoxins are a very serious health risk, especially among subsistence farmers in warm countries worldwide. Fumonisins cause cancer, and liver and brain damage in farm animals and man. Hopefully, second-generation Bt corn hybrids will sustain even less insect injury and have lower mycotoxin concentrations than the present ones.

The level of target insect suppression provided by current Bt crops varies according to the pest species attacking the crop, transformation event, hybrid background, and environmental factors that influence concentration of Bt protein in plant tissues that need protection (Williams

Fig. 2.1 European corn borer larva and tunnel injury inside a corn stalk (*Source:* G.P. Munkvold, Iowa State University)

et al., 1998; Lauer and Wedberg, 1999; Clark et al., 2000). The Cry1Ab protein levels for events BT11 and MON810 in fresh leaf tissue range from 7 to 30 µg/g, and in grain, from 0.2 to 5 µg/g, with the event MON810 at the lower end and event BT11 in the upper end of the range (http://www.agbios.com/).

The level of Cry1Ab protein expression in leaves of event 176 is much lower, ranging from 0.6 to 3.02 µg/g, with no detectable Cry protein expression in grain. The concentration of Cry protein in tissues is the dose of protein that the insect pest receives as it eats. Thus, the higher the dose per bite, the more effective the event is likely to be in controlling the target pests; that is assuming the crop varietal background affects on toxicity of the Cry protein are equal among events. In leaf tissue, Cry1Ab provides good to excellent control of European, Asiatic and southwestern corn borers, is less effective for corn earworms and fall armyworms, and has little or no effect on black cutworms (Wiseman et al., 1999; Shelton et al. 2002). As suggested by the protein concentration data above, the level of target pest suppression is less for kernels in ears than for green leaf tissue. These data and commercial production experiences with these events in Bt corn suggest that increasing the expression of Cry1Ab in leaves and kernels above 30 µg/g fresh tissue could significantly improve control of the target lepidopteran pests.

A promising new research finding is that BT11 and MON810 reduced the infestations of Indian meal moth, *Plodia interpunctella* (Hubner) and Angoumois grain moth, *Sitotroga cerealella* (Olivier), in stored grain (Sedlacek et al., 2001). In tropical and subtropical developing countries stored grain pests attack and destroy on an average 30% of the grain (Mengech et al., 1996; Benedict, 2003). For subsistence farmers, any reduction in losses to stored grain would be desirable. However, the effectiveness of Bt corn in suppressing stored grain pests has yet to be fully proven in the commercial corn storage and subsistence farming.

Another way to improve the level of target pest control with Bt crops, as well as broadening the spectrum of pests controlled, is by inserting a more active and broad-spectrum Cry protein, such as Cry1F compared to Cry1Ab, into the crop. The event TC1507 expresses Cry1F, and was recently approved for commercial production in Canada and the USA. It is marketed under the trade name of Herculex® 1 by the seed partners of Dupont and Dow AgroSciences for control of corn borers, fall armyworm, black cutworm and some suppression of corn earworm (US EPA, 2001b; AGBIOS, 2003). Field performance of Herculex® 1 is being tested by academic researchers and on commercial farms in 2003, following which we should know better how it compares to YieldGard® hybrids expressing Cry1Ab.

Possibly, the most novel new Bt corn event is MON863, known under the Monsanto trade name as YieldGard® Rootworm Corn (Monsanto, 2003; US EPA, 2003a; 2003c). This event was approved February 2003 in the USA for control of a group of closely-related beetle pests, *Diabrotica* spp., collectively called corn rootworms (western, *D. virgifera virgifera* LeConte; northern, *D. barberi* Smith & Lawrence; southern, *D. undecimpunctata howardi* Barber; and Mexican, *D. v. zeae*) (Metcalf and Metcalf, 1993). The larvae of corn rootworms feed on the roots of corn, reducing plant growth and yield (Fig. 2.2). Western and northern corn rootworms are considered to be the most economically-important pests of corn in much of the USA corn belt, costing growers an estimated one billion dollars in crop losses and insecticide control costs. Event MON863 possesses the Bt transgene *cry3Bb1* and expresses the Cry3Bb1 protein constitutively throughout the plant, including silk, grain and pollen, at concentrations of ranging between 10 and 81 µg/g fresh plant tissue, depending upon the tissue examined and the age of tissue (http://www.agbios.com/). Typically, the concentration of all Cry proteins in most of the crops examined decreased as tissues became senescent. Concentrations of Cry proteins in Bt corn are thought to remain relatively uniform from seedling emergence until the crop begins to mature several weeks after anthesis. However, there may be brief drops occurring at anthesis or other times when photosynthate demand by the crop is high, such as times of environmental stress.

Historically, corn rootworms have been controlled by crop rotation and at-planting insecticide applications. About 18% of the 31 million USA corn ha. (76 million acres) are treated with insecticides for these insects (James, 2002a). Due to recent adaptations in the biology of some

Fig. 2.2 Corn roots showing corn rootworm injured non-transgenic roots (Untreated) and uninjured roots of transgenic Cry3Bb1 corn (YieldGard® Rootworm). (*Source:* T. DeGooyer, Monsanto Company)

rootworm species, using crop rotation from corn to soybeans then back to corn is no longer as effective for their control. Thus, corn hybrids carrying MON863 have the potential to reduce conventional at-planting insecticide use and costs, and increase yields and income for producers. Field data suggest that Cry3Bb1 is more effective than conventional insecticides at reducing rootworm injury to roots and increasing yield.

Since rootworms must feed on Bt corn roots in order to ingest a lethal dose of Cry3Bb1 protein, under heavy infestations there is some light scaring visible on Bt corn roots. However, in field studies where infestations were light, this injury was insignificant compared to the injury on untreated or insecticide treated roots. Moreover, yields of Cry3Bb1 corn were 5 to 10% greater than the conventional insecticide-treated non-Cry3Bb1 corn (Monsanto, 2003). The first commercial plantings of MON863, CryBb1 hybrid corn began in 2003.

Cotton

Bt cotton was first grown in 1996, in Australia as Ingard®, and in the USA as Bollgard® (Fitt and Wilson, 2000; Benedict and Altman, 2001; James, 2002a). These first-generation Bt cotton varieties were developed by Monsanto and their seed partners to express the Cry1Ac protein (Table 2.1). By 2002, Bt cotton crops were grown in nine countries, seven developing countries (China, India, Indonesia, Argentina, Mexico, South Africa and Colombia), and two industrial countries (USA and Australia) (James, 2002b). In 2002, Bt cotton was grown on 4.6 million ha. (11.4 million acres), which is 13.2% of all the cotton grown in the world (Table 2.2). Worldwide, several different Cry1Ac events are grown commercially. In the USA and some other countries, Bollgard® event MON531 was used by seed company breeding programs to develop the first commercial Bt cotton varieties. In Australia and other countries, Ingard® event MON757 was used. In 2001, MON531 was the dominant Bt cotton event worldwide. In the USA, Bollgard® MON531 was grown on 34% of the total 6.2 million cotton ha. (15 million acres) in 2001 (Edge et al., 2001; James, 2002a).

China has been growing Bollgard® cotton since 1997, and has also developed the new events, GK, expressing fused Cry1Ab and Cry1Ac proteins, and sGK, expressing the fused Cry1Ac-Cry1Ab and stacked CpTi proteins (James, 2002b; Shelton et al., 2002). GK and sGK were developed by the public sector Chinese Academy of Agriculture Sciences (CAAS) in Beijing, China. The Bt proteins are expressed by a fusion gene, in which the DNA sequences for *cry1Ac* and *cry1Ab* genes are joined, called 'fused,' and inserted in the plant's genome as a single transformation event (James, 2002b). The cotton varieties designated sGK have an additional gene, *cpti*, expressing the cowpea trypsin inhibitor protein, CpTi. This gene is stacked with the *cry1Ab/cry1Ac* fusion gene. The CpTi protein actually attacks other proteins in the insect pests' gut and disrupts digestion, resulting in malnutrition/starvation and death (see also Chapter 3). The rationale for adding the *cpti* gene was to broaden the spectrum and level of insect pests controlled, and to add a third insect toxin to reduce the possibility of selecting for insect pests that are resistant to the Cry1Ac and Cry1Ab toxins in these Bt cotton crops (see Chapter 8) (Shelton et al., 2000, 2002).

On a worldwide basis, these first-generation Bt cotton products target caterpillar pests, many of which are closely-related species (Fitt and Wilson, 2000; Benedict and Altman, 2001; James, 2002a). More than 15 species of caterpillar pests attack cotton worldwide, and each country has 3 or more species that cause serious crop loss (Matthews and Turnstall, 1994). Commonly, these pest species are called bollworms (*Earis* spp., *Helicoverpa* spp., *Heliothis* spp. and *Pectinophora* spp.), leafworms [*Alabama argillacea* (Hubner)], cutworms (*Agrotis* spp.), or armyworms (*Spodoptera* spp.). Unfortunately, many of these are either unaffected or poorly controlled by the Cry1Ac protein in first-generation Bt cottons. In the USA, the target caterpillar pests for Bt cotton are bollworm, *H. zea*, (which is also a very damaging pest of corn, called the corn earworm when attacking corn), pink bollworm, *Pectinophora gossypiella* (Saunders), and tobacco budworm, *Heliothis virescens* (Fabricius). The tobacco budworm and pink bollworm are very well controlled by Bollgard® cotton, whereas the cotton bollworm is controlled satisfactorily except during the bloom stage when feeding on the reproductive parts of flowers. Pollen, anthers, pistils and senescing flower petals tend to have a lower concentration of Cry protein than other plant parts (Fig. 2.3) (Greenplate, 1997, 1999, 2003; Greenplate et al., 2003a,b).

Pink bollworm occurs worldwide and may be the most injurious pest of cotton, destroying upto 70% of the crop in developing countries. This pest is difficult to control because the caterpillars tunnel into flowers, fruits and seeds where they feed and develop to adulthood. Applying conventional topical insecticides to control them once they are in the tunnels is ineffective, whereas the systemic constitutive production of Cry1Ac protein throughout Bt cotton is an ideal method of delivering the insecticidal protein to control pink bollworms and other cryptic feeding caterpillar pests of cotton.

The introduction of Bt cotton in 1996 and 1997 was very timely since the pink bollworm and the tobacco budworm in the USA, and the American bollworm, *H. armigera*, in China and Australia had become resistant to many of the conventional insecticides used to control them

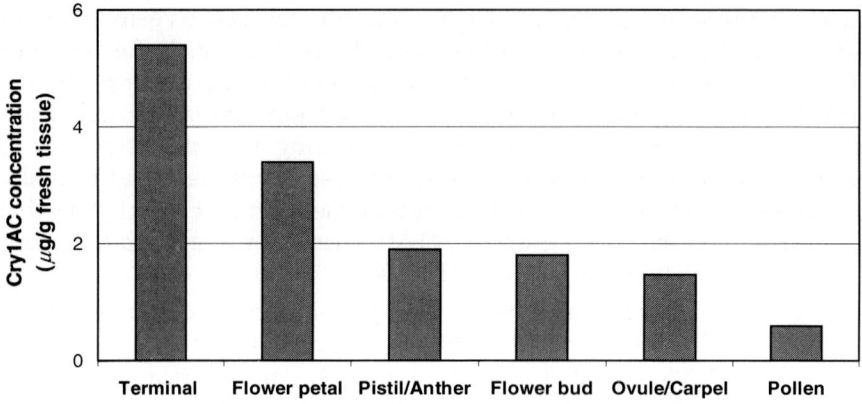

Fig. 2.3 Concentration of Bollgard® Cry1Ac in fresh tissue of NuCOTN33B cotton (After: Greenplate, 1999)

(Fitt and Wilson, 2000; Benedict and Altman, 2001). Bt cottons offered a reasonable solution to the problem.

Bt cotton, along with event MON531 or MON757, expresses Cry1Ac proteins in leaves at about 12.6 µg/g of fresh leaf tissue and 9.9 µg/g of fresh seed tissue (http://agbios.com/). Very little Cry protein can be detected in the pollen, and none in the nectar. Studies of Bt protein expression show that concentration varies as much as 3- to 5-fold during the season in the same plant organ.

Also, MON531 generally has a lower concentration than MON757 early in the season, whereas later in the season, MON531 concentration increases and MON757 decreases, in such a manner that MON531 has the higher concentration (Greenplate, 1997, 2003; Greenplate et al., 2003a, b). The highest concentration of Cry1Ac in the leaves is reached early in the plants' growth, before flower buds develop, so that by the time the plants are blooming, and fruits have begun to mature in mid to late season, the Cry1Ac concentration in leaves starts to decline (Greenplate, 1999; Olsen and Daly, 2000). Researchers have shown that the level of control of American bollworm, bollworm and fall armyworm is positively correlated with the concentration of Cry1Ac protein in the cotton plant—i.e. the higher the concentration, the better the control (MacIntosh et al., 1990; Adamczyk et al., 2001; Greenplate et al., 2003a; 2003b). However, the concentration of natural secondary plant compounds such as tannins and terpenoids also interact with Cry proteins to influence the level of pest control in unexpected ways (Sachs et al., 1996; Olsen and Daly, 2000; Benedict and Altman, 2001). The killing power of terpenoids and Cry proteins appears to be additive, whereas the influence of tannins may be antagonistic and reduce the effectiveness of the Cry protein (Olsen and Daly, 2000; Greenplate, et al., 2003b). A third manner in which stacked multiple toxins in the plant can affect each other and the level of control is called 'synergism' in which the sum of the control provided by the two toxins is greater than simply adding their killing power together. In fact, one chemical may be non-toxic on its own, but in the presence of the other chemical, the percentage of pests killed is much higher than would be expected from either chemical alone or combined (Benedict and Altman, 2001; Greenplate et al., 2003b). In spite of these complexities, we find it amazing how active the Cry1Ac protein in Bt cotton is in controlling susceptible insects. For example, a MON757 plant growing in the field would have on average only 200 µg of Cry1Ac protein in the entire plant (fresh tissue), which amounts to about 30 g/ha. (12.2 g/acre) (based on 148,258 reproducing plants/ha.), yet will provide nearly 100% control of pink bollworm and tobacco budworm.

The second-generation Bt crops with stacked *cry* genes, like those in China's events GK and sGK, and Monsanto's Bollgard® II have been clearly shown to improve the level of control and broaden the spectrum of caterpillar pests controlled (Fitt and Wilson, 2000; James, 2002b; Monsanto, 2003). This is due to multiple Cry proteins—each with a different range of target insects—and to the increased levels of the total Cry toxin, for those insect pests that are susceptible to both proteins. Bollgard® II was created by inserting a synthetic gene from *Bacillus thuringiensis*, Cry2Ab, expressing Cry2Ab protein, into the Bollgard® cotton variety, DP50B, already expressing Cry1Ac through event MON531, creating the new stacked gene event 15985 (Monsanto, 2002). Bollgard® II is much more effective than Bollgard® at controlling all the caterpillar pests of cotton worldwide, especially the American bollworm (Fig. 2.4), beet armyworm, fall armyworm, and bollworm (Perlak et al., 2001; Greenplate et al., 2003b). Laboratory and field studies show that Bollgard® II increased mortality for bollworm from 84.2 to 92.2%, fall armyworm from 16.1 to 100%, beet armyworms from 50.1 to 94.9% and soybean looper from 1.2 to 97.4%. Control of bollworm, *H. zea*, on flowers is also increased on Bollgard® II (Gore et al., 2003). Australian and Chinese producers are particularly interested in the increased control of American bollworm provided by Bollgard® II, to the point where the Australia government is considering raising the national Bt cotton acreage cap from 30% (for Bollgard® and Ingard®) to 70 or 80% for Bollgard® II (Perlak et al., 2001; James, 2002a). This high level of pest control is due, in part, to the very high concentration of Cry2Ab protein in Bollgard® II as compared to Cry1Ac in Bollgard® (Fig. 2.5).

In fact, much of the increased control of caterpillar pests from the two-gene construct may be simply due to a higher dose of total insecticidal protein in each bite—approximately 10-fold higher than

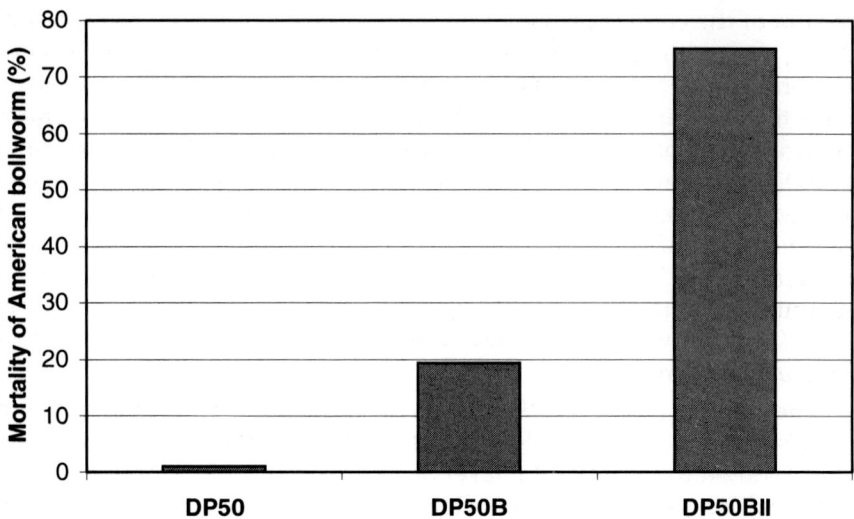

Fig. 2.4 Percent mortality of American bollworm newly-hatched caterpillars after 5 days of feeding on cotton flower buds of three varieties, non-Bt DP50, DP50B (Bollgard®) expressing Cry1Ac, and DP50BII (Bollgard® II) expressing Cry1Ac and Cry2Ab2 (stacked) (Modified from Monsanto Company, 2002)

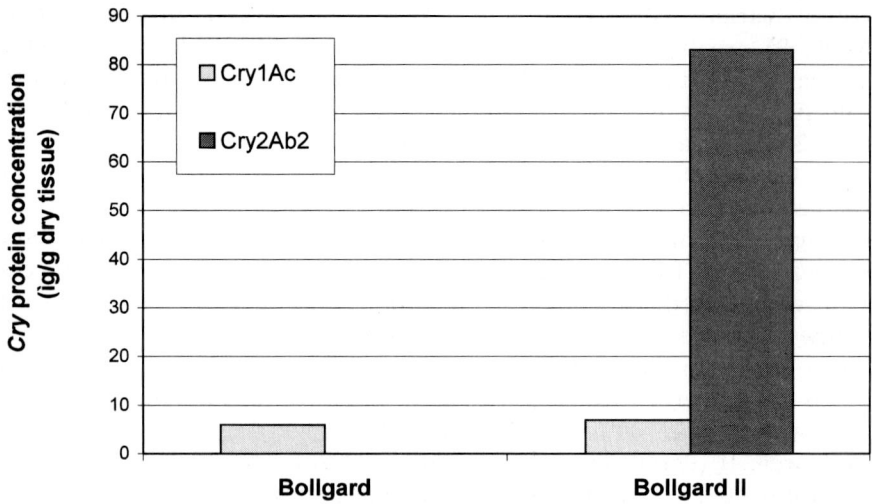

Fig. 2.5 Concentration of Cry1Ac and Cry2Ab2 in dry tissue of Bollgard® and Bollgard® II cottons (Modified from Monsanto Company, 2002)

Bollgard®. This two-Bt gene/protein technology represents a major advance in insect pest control as compared to Bollgard® and should further reduce the use of conventional insecticides, enhance biological control and management of resistance to Cry proteins, and further increase farm profit and yield.

Potato

The first commercial fields of Bt potatoes were grown in the USA in 1996 with a number of different Monsanto potato transformation events expressing Cry3A protein under the name of New Leaf Potato® (Shelton et al., 2002). The Cry3A protein, like the new Cry3Bb1 protein in YieldGard Rootworm® corn, is active against certain beetle species (Coleoptera). The major beetle pest of potatoes in the USA is the Colorado potato beetle, *Leptinotarsa decemlineata* (Say) (Metcalf and Metcalf, 1993). Adults and immatures of this pest feed on the foliage and stems, at times completely defoliating the plants and greatly reducing the size and yields of potato tubers. These first-generation Bt potato varieties are very effective at controlling injury from the Colorado potato beetle and replacing the conventional insecticides used for beetle control. The introduction of Bt potatoes was timely in the sense that the Colorado potato beetle had developed resistance to most conventional insecticides. Commercial Bt potatoes were very effective in controlling Colorado potato beetles and replacing conventional insecticide sprays. However, in response to marketing and political pressures by the public, many food producers, such as Gerber, Novartis, McDonalds and Frito-Lay have chosen not to buy or use Bt potatoes in their products (Shelton et al., 2002). As a result, Monsanto Company stopped marketing Bt potatoes in 2001. Despite the many desirable benefits and minimal risks of Bt crops (discussed below), the question is: what will the future of Bt potatoes be?

FUTURE Bt CROPS

Bt crops and the traits of the future can be predicted with some accuracy by examining the Experimental Use Permits issued by EPA for this year as also the past two years (US EPA, 2003b). These permits are issued to companies and universities for testing new Bt crop traits that are in the late stages of development towards becoming commercial registered products. If we combine these experimental traits with other experimental traits discussed in the published literature (Krattiger, 1997; James, 2002a,b; Shelton et al., 2002), a short list of likely new products emerges (Table 2.3). Most of these products should enter the commercial market in the next two to three years. In addition, many academic and public institutions are developing transgenic Bt crops that may become commercial (Krattiger, 1997; Pardey and Wright, 2003). In fact, the list of crops that have been transformed with a *cry* gene insertion (i.e. Bt crops) now include apple, broccoli, cabbage, canola/rapeseed, corn, cotton, peanut, rice, soybean, tobacco, and tomato (Penn, 2000). However, the only products that are predicted to become commercial, with some degree of confidence, are listed in Table 2.3. Among these, we expect that the most effective Bt proteins/events and stacked second-generation products, in the more agronomicly-attractive hybrids/varieties, will eliminate the less competitive, first-generation, single-gene products. We expect that eventually, most crops will have varieties/hybrids with multiple insect-resistant Bt transgenes stacked with a number of additional insect, disease and nematode pest-resistant genes from diverse sources (see Chapters 3, 4, and 5).

Cotton

Two promising new Bt cotton traits are WideStrike® from Dow AgroSciences-PhytogenSeeds (Dow AgroSciences, 2003; Leonard et al., 2003) and Vip® from Syngenta (Syngenta, 2003). WideStrike® cotton expresses stacked Cry1F with Cry1Ac proteins, using two new events, which dramatically increase the spectrum of pests controlled compared to first-generation Bollgard® expressing Cry1Ac protein only. Field data from 75 trials in the USA show that WideStrike® cotton controls cotton bollworm, tobacco budworm, pink bollworm, armyworms and loopers—the entire spectrum of caterpillar pests of cotton. This should be an excellent product, competitive with Bollgard® II and superior to Bollgard®.

The new Vip® cotton expresses a novel Bt Cry protein, Vip3A, discovered by Syngenta scientists in 1994, with activity against a broad range of caterpillar pests (Estruch et al., 1996; Yu et al., 1997; Mascarenhas et al., 2003; Syngenta, 2003). Preliminary data indicate that Vip® cotton provides control of most of the caterpillar pests of cotton. Additional field testing and commercial production should determine how effectively it controls all caterpillar pests worldwide compared to the Bollgard® II, China's sGk and WideStrike®.

The Vip proteins are unique in that they are naturally produced by strains of *B. thuringiensis* outside the cells during the vegetative stage of bacterial cell growth—hence the name Vip for vegetative insecticidal proteins. As Vip proteins are found outside the bacterium, they are called exotoxins, whereas Cry proteins (e.g., Cry1, Cry2 and Cry3) are found inside the cells during the sporulation stage of some *B. thuringiensis* strains, and are called delta-endotoxins (Hall and Menn, 1999). Vip proteins are not found in a crystal form and thus do not need solubilization in the insect midgut before they can be activated by the

Table 2.3 Bt crops and insect control proteins (PIP[a]) currently under development[b], by crop and developer as of March 2003

Crop/Company or Country	PIP	Anticipated commercialization
WideStrike® cotton/PhytogenSeeds—Dow AgroSciences	Stacked[c] Cry1F Cry1Ac	2004
Vip® cotton/Syngenta	Vip3A	2004
YieldGard® Plus corn/Monsanto	Stacked Cry1Ab CryBb1	2004
Corn/Dow AgroSciences-Mycogen Seeds and Pioneer HiBred	Fused[d] Cry34Ab1 Cry35Ab1	2004
Soybean/Monsanto	Cry1Ac	2005
Rice/China, IRRI[e]	Cry1Ab	2004
Rice/China	Fused Cry1Ab Cry1Ac	2004

[a] PIP stands for Plant Incorporated Protectant; here it is the Bt insecticidal protein expressed by the specific Bt transgene present in the crop.

[b] These products/Cry proteins are listed in Experimental Use Permits issued by the US EPA (http://www.epa.gov/pesticides/biopesticides/regtools/frnotices2003.htm) and/or in the published literature.

[c] The *cry1F* and *cry1Ac* genes were inserted into the same parent line, called 'stacked genes', in order to produce commercial crops that express both Cry1F and Cry1Ac insect control proteins in the same plant.

[d] The *cry34Ab1* and *cry35Ab1* genes were joined together and then inserted into the parent line, called 'fused genes', to produce commercial crops that express both Cry34Ab1 and Cry35Ab1 insect control proteins in the same plant from one single insertion event.

[e] IRRI is the acronym for the International Rice Research Institute in Los Banos, Philippines. It is a part of the Consultative Group on International Agricultural Research, an internationally-sponsored and publicly-funded research organization (Pardey and Wright, 2003).

midgut proteases to become toxic, as required by naturally-produced Cry protein crystals from *B. thuringiensis* (called protoxins). However, most current transgenic commercial Cry proteins are not expressed in the Bt plant as a protoxin crystal, but rather, as a synthetic version of the active protein fragment found in the insect's gut when digestive fluids process the native Bt crystal.

There are no DNA sequences or structural similarities (homologies) between Vip and Cry Bt proteins. Due to this lack of similarity, researchers hope that the Vip and Cry proteins have different target sites in the insect gut (also called binding sites or receptor sites), where they cause pores to form, followed by ionic imbalance, cell rupture, breakdown in the epithelial lining, gut paralysis, bacterial sepsis and, finally, death. Having different binding sites is significant for insect management because the two proteins, i.e. Cry and Vip, could be combined to increase the level of control, broaden the spectrum of pests controlled, and reduce the likelihood of target pests developing resistance to either protein (see Chapters 6, 7, and 8)—a great addition to the tools used in the management of insect resistance. Vip proteins also add valuable new insecticidal proteins in the form of tools for the insect pest management toolbox.

Corn/Maize

Dow AgroSciences also has a new Vip product for corn that expresses a binary insecticidal protein that is actually two toxins, Vip1, now named Cry34Ab1, and Vip2, named Cry35Ab1 (Warren, 1997; Moellenbeck et al., 2001; http://www.biols.susx.ac.uk/Home/Neil_Crickmore/Bt/index.html). This binary Bt toxin is generally referred to as Cry34/35Ab1, because the two proteins occur together in nature and are both required for maximum insecticidal activity. When first discovered to have insecticidal activity against corn rootworm larvae (*Diabrotica* spp.), these Vips were called 149B1 proteins. Transgenic Bt corn hybrids have been developed by Dow AgroSciences and their seed partners, Mycogen Seeds and Pioneer Hi-Bred Int., with the fused gene, *cry34/cry35Ab1*, that expresses the binary Cry34Ab1 and Cry35Ab1 insecticidal proteins. This Bt crop product will compete for the same corn rootworm market, in North America and northern Mexico, as the Monsanto YieldGard® Rootworm corn product that was just approved for commercial production (see above). The mode of action of the Cry proteins in these two corn rootworm products is similar to Cry1- and Vip3-insecticidal proteins; they attack and destroy the integrity of the midgut membrane and cellular epithelial lining (Moellenbeck et al., 2001).

A new second-generation product will be Monsanto's YieldGard® Plus corn, a stacked gene Bt crop, expressing Cry1Ab for corn borer control and Cry3Bb1 for control of rootworms (Monsanto, 2003). As with most of these Bt crop products, growers will also have the choice of stacked Bt traits, food quality traits and herbicide-resistance traits, to meet their desires and best management practices. Corn hybrids appear to be taking the lead in providing the grower with the widest choice of traits to choose from compared to other crops.

Rice

Commercial Bt rice has been slower to develop than Bt corn and cotton despite rice's importance to mankind. Rice is one of the world's major food crops, with 40% of the human population dependent upon it. Rice provides 20% of the per capita energy and 15% of the per capita protein for humans worldwide. This staple crop is grown on more than 145 million hectares, of which 90% lies in Asia (Rice and Choo, 2000). Despite great advances in crop protection in recent years, a number of cryptic caterpillar pests are poorly controlled on rice, reducing annually yields on average 5 to 10%—and occasionally even 60 to 95% (Ye et al., 2000; Wang et al., 2002). Approximately 50% of all insecticide sprays in Asian countries are targeted at these caterpillar pests of rice, especially the yellow stem borer, *Scirpophaga incertulas* (Walker), striped stem borer, *Chilo suppressalis* (Walker), and the leaffolders, *Cnaphalocrocis medinalis* (Guenee) and *Marasmia patnalis* Bradley (Alinia et al., 2000; Shelton et al., 2002; Wang et al., 2002). The average yearly losses caused by yellow stem borer and leaffolders worldwide are estimated at 10 million tons of rice (Herdt, 1991) despite the widespread use of foliar insecticide sprays. Conventional insecticide sprays have a number of undesirable effects, including: increasing production costs and human health risks among rice farmers and their families; disrupting the rice agroecosystem; and contributing to secondary pest outbreaks and the development of insecticide-resistant pests. Many rice farmers are resource poor subsistence farmers in developing countries in Asia, Africa and South America, who could greatly benefit from Bt crops that are safer than many conventional insecticides for the farmer to use and do not disrupt the natural biological control of rice pests.

Stem borer caterpillars are cryptic insects, living inside the rice leaf sheaths and stems, eating the tissues and damaging the vascular system—resulting in stunted, deformed growth and reduced yields. Leaffolders feed on the leaves, defoliating the plants and thus reducing plant photosynthate available for growth and grain production. Bt rice offers the potential to reduce crop losses to these pests, while replacing conventional insecticides and reducing costs. China (Wang et al., 2002), the International Rice Research Institute in the Philippines (Alinia et al., 2000), and other publicly-funded organizations (Park et al., 2001) are all developing Bt rice crops for the developing world. This is a very noble and valuable effort to help feed a growing population in subsistence farming communities worldwide (Cohen, 2003; Sadava, 2003). Various Chinese universities and research institutes, in cooperation with IRRI (International Rice Research Institute in the Philippines) and universities in many countries have transformed rice by inserting the *cry1Ab* gene alone (Ye et al., 2000) and a fused *cry1Ab/cry1Ac* gene (Tu et al., 2000) into conventional rice varieties and hybrids (Alam et al., 1999) (Table 2.3). Initial field and laboratory tests of these Bt lines against yellow stem borer and striped stem borer have shown good season long control (Alinia et al., 2000; Tu et al., 2000; Ye et al., 2000; Wang et al., 2002). Results of commercial field trials should be available soon.

Soybean

At least one company, Monsanto, is testing Bt soybean varieties that express Cry1Ac protein to control caterpillar pests. Soybeans worldwide are attacked by a number of caterpillar pests (Metcalf and Metcalf, 1993). In the US, the likely target pests are soybean looper, stem borer and velvetbean caterpillar. These insects feed on leaves, flowers and fruits, and can dramatically reduce yields when their densities are high.

Bt CROP ADOPTION, USE AND BENEFIT

Farmers worldwide have rapidly adopted pest-resistant Bt crops, with Bt varieties reaching >14 million ha. worldwide in 2002 (Table 2.2; Fig. 2.6; see also Chapter 1) (Krattinger, 1997; James, 2002a,b). Bt corn was the most widely-grown Bt crop, occupying 9.9 million ha. (This figure includes 2.2 million hectares of Bt stacked with herbicide tolerant transgenes) in the USA, Canada, Argentina, South Africa, Spain, and Germany—in descending order from highest Bt hectares to lowest. Globally, Bt cotton occupied 4.6 million ha. (This includes 2.2 million ha. of Bt stacked with herbicide tolerant transgenes.) in the USA, China, Australia, India, Argentina, Mexico, South Africa, Indonesia, and Columbia—in descending order. Both developing and highly industrialized nations have adopted Bt crops with the USA leading with 6 million ha. of Bt corn and 2 million ha. of Bt cotton, in 2001 (Gianessi et al., 2002). The most obvious benefit and reason for farmers rapidly adopting Bt crops is the economic value. In 2001, the global benefit to farmers growing Bt cotton was >US$ 750 million above what non-Bt cotton would have netted (James, 2001, 2002a).

Case Histories of Adoption and Benefit

In 1992, American bollworm outbreaks, insecticide-resistance, insecticide applications and crop losses, all cost China approximately US$ 1.2 billion across her economy. Due to these losses China's cotton industry was reduced by 40% and China began to import cotton. In 1999, 1.5 million small farmers in China adopted pest-resistant Bt cotton varieties on their farms, which averaged 0.4 acres per farm (James, 2000). The primary reasons were effective control of the insecticide-resistant American bollworm (*H. armigera*), insecticide savings, and the dramatically increased income and personal safety (Pray et al., 2002). In 1999, Chinese farmers growing Bt cotton saw yield increases averaging 25% over conventional cotton, with less than one insecticide spray on the Bt cotton

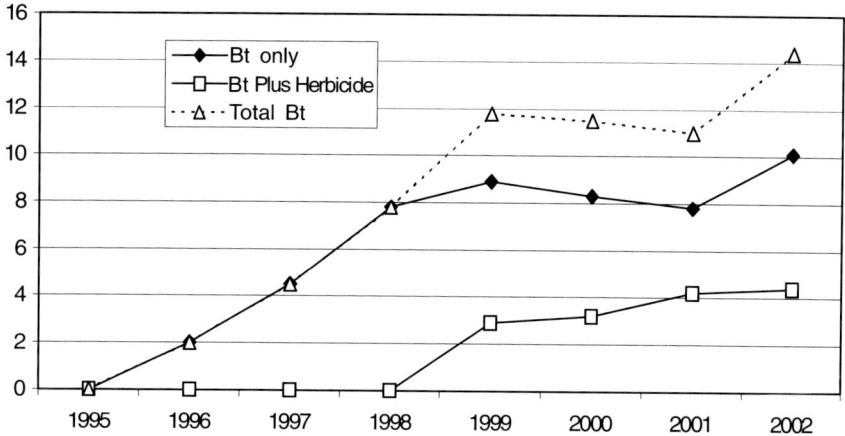

Fig. 2.6 Global area of transgenic crops with Bt only and stacked Bt plus herbicide resistance in millions of ha (Modified from James, 2002a, b)

and 14 sprays on the conventional. In fact, about 70% of Chinese Bt cotton farmers applied no insecticide sprays on their Bt cotton in 1999. They also observed a 25% increase in the natural enemies of cotton pests on Bt cotton. By 2001, there were over 4 million cotton farmers in China farming >4 million hectares of Bt cotton (James, 2002a). For China's subsistence farmers, Bt cotton has proved to be an incredibly beneficial and valuable IPM technology.

In 2002, some 1,500 Indian subsistence farmers grew their first commercial Bt cotton crops and produced yields of 20 to 30% above non-Bt fields. Bt cotton reduced insecticide sprays by half, from 6 applications to 3, during the season (James, 2002b). In the USA, Bt cotton has also been shown to reduce insecticide use by an average of 2.2 sprays, while increasing yields by about 10% and farm profits about $40 to $85 per hectare, as compared to conventional cotton (Benedict and Altman, 2001; James, 2002a; Shelton et al. 2002).

Based on many case studies worldwide (James, 2002a; Qaim and Zilberman, 2003), we find that the motivating reasons farmers are adopting Bt crops can be summarized as follows: (i) improved control of insecticide-resistant and cryptic target caterpillar pests; (ii) reduced conventional insecticide costs; (iii) increased yields; (iv) increased farm profit; and (v) various combinations of the above reasons. Other less economically definable, but equally motivating benefits to the farmers are: (vi) better natural biological control of all cotton pests; (vii) reduced environmental contamination; (viii) safer farm environment in which to live and work; (ix) less risk of crop loss from insect pests; (x) less risk of lawsuits from insecticide-related accidents; (xi) improved relationships with neighbors, especially home/city/school owners; (xii) greater peace of mind in one of the most stressful occupations; and (xiii) ease of implementing pest control measures, i.e. more 'user-friendly', compared to application of conventional insecticides.

Moreover, Bt crops are contributing social and economic benefits to the developing countries because Bt farmers, and not the developers of the Bt technology, are the primary beneficiaries of Bt crops (James, 2002a, b; see Chapters 12 and 13). In China, the economic gain for resource-poor Bt cotton farmers was US$ 500/hectare in 2001 alone. In fact, of the > 5 million Bt cotton and corn farmers globally in 2001, over 70% were small resource-poor farmers in China and South Africa. In South Africa, where 50% of the cotton farmers are women, growing Bt cotton allows them more time for domestic activities, caring for children and self-improvement; and more income for education of children and purchasing much-needed farm, home and health care products and services. Growing Bt crops also reduces insecticide poisonings of farmers and others living in rural communities. Poisonings are common in subsistence farming where many factors contribute to high and frequent exposure, particularly the use of hand sprayers and lack of proper safety equipment. For example, in China 10 to 30% of cotton farmers suffer from pesticide poisoning each year (Pray, 2002). Bt cotton has reduced poisonings by 50% or more for farmers in China.

Unlike industrial countries, most developing countries import more food than they produce, and have a higher percentage of the total population employed in agriculture as either small resource-poor farmers practicing subsistence farming or the rural landless who are dependent on agriculture for survival—70% of the world's 1.3 billon poorest people live in rural subsistence farming communities (James, 2002b). Further, in developing countries, agricultural employment is more than 50% of the total employment. By 2010, the rural resource-poor in developing countries are expected to increase to 3 billion globally. Helping small subsistence farmers to improve their production of food, fiber and feed, through biotechnology and conventional means, is critical for rural communities worldwide. In these subsistence-farming communities, crops are not only the principal source of food for the community, but crop production is also the principal source of livelihood for every one in the community. Increased crop productivity acts as an engine for economic growth and alleviation of the constraints of hopelessness, hunger and poverty. It empowers subsistence farming communities to improve their human condition. Bt crops have a very beneficial influence on the resource-poor communities they have reached—however, much more effort is needed than just growing Bt crops (see Chapters 9 and 11) (James, 2001, 2002a; Thompson, 2002; Cohen, 2003).

MANAGING INSECT PESTS ON Bt CROPS

Despite today's best management efforts, plant-feeding insects reduce the amount of food and fiber produced in agriculture. Under conditions that are ideal for pests to grow and reproduce, pest populations can build up rapidly, because of their very short life cycle from egg to reproducing adult, generally averaging 30 days. Crop losses from pest outbreaks occur regardless of whether the crops are grown in organic, conventional high-input, or subsistence farming systems (Denno and McClure, 1983; Ferro, 1987; Mengech et al., 1996; Benedict, 2003). On a worldwide basis, these insect pest outbreaks reduce yields on average 10%, but may range from 0% in one field to >75% in another, even with our best management practices. Insects also attack stored grain and root crops and reduce their quality and quantity. Stored crop losses in tropical developing countries average 30%, whereas losses are less than 10% in temperate industrialized countries. The intensity of pest infestation and loss varies as per the field, crop, geographic production area, and year. Outbreaks of plant-feeding insects occur in natural plant communities as well as managed agricultural systems (Berryman, 1987), and are thought to be due to:

- Physical environment becoming more favorable for pest reproduction, survival and growth.
- Food plants becoming more abundant and/or nutritious for pests.
- Natural enemies of the pest becoming less common.
- Diseases of the pest becoming less common.
- Competing species becoming less common.
- Environmentally-induced physiological or reproductive or genetic changes in the pest make it better adapted to the environment and/or crop.
- A combination of the above factors, which is most common.

Agricultural activities over the centuries have increased the opportunities for pest outbreaks in crops. Not surprisingly, many of the farming practices that enhance yields can also contribute to increased pest problems by providing optimum conditions for pest reproduction, survival and population growth. The practices most responsible for outbreaks are: (i) growing crops as monocultures that provide large areas of the pest's favorite food; (ii) improving crop suitability for the pest with irrigation, fertilizer, and herbicide; (iii) growing the crop in areas where natural enemies of the pest are limited or nonexistent; (iv) growing the crop in new areas and having native species of insect pests adapt to the crop; (v) accidentally introducing pest species from one geographic area of the world to other geographic areas where the crop is grown, and having them adapt to the new cropping environment; (vi) accidentally removing the natural enemies of the pests through the use of farming practices such as cultivation, irrigation, herbicides, monoculture, plant growth regulators, and broad-spectrum insecticides; and (vii) the frequent use of the same insecticide to control the same pest year after year, resulting in the pest population adapting to the insecticide so that it is no longer toxic, a process known as insecticide resistance. A resistant pest can feed, grow, and reproduce normally in the presence of the insecticide. The widespread use of broad-spectrum insecticides can be one of the most important factors encouraging insect pest outbreaks. When the natural enemies are removed, the crop pest may rapidly return to even more damaging numbers than before it was treated, in a manner known as pest resurgence. Loss of natural enemies may also induce the outbreak of secondary pests. Secondary pests are those pests that were held in check by natural enemies and caused no economic damage until broad-spectrum insecticides were applied. Pest resurgence and/or outbreaks of secondary pests result in the need for even more insecticide sprays throughout the growing season. This insecticide-induced paradigm has been called the pesticide syndrome, and has occurred from time to time in most production systems worldwide (Doutt and Smith, 1971). However, if insecticides are properly managed, most of these insecticide-induced problems can be minimized or avoided. One of the most effective ways to reduce conventional insecticide use on crops and allow natural enemies to become more abundant and effective at controlling crop pests that are insecticide induced, is to develop pest-resistant varieties that target the key pests and those resistant to insecticides—current transgenic Bt crops do this!

Despite the above limitations of insecticides, they have been and will probably continue to be the most cost-effective method to quickly stop pest outbreaks that threaten to destroy crops. However, in response to the insecticide-induced pest problems just discussed, and the environmental and human health concerns that developed in the 1960s as a result of widespread use of chlorinated insecticides, the chemistries of today's insecticides and the way they are used in pest management programs have changed dramatically from the days when DDT was the insecticide of choice for every insect problem (Table 2.4). Today, insecticides are seen as one of many strategies that pest managers have integrated into pest management programs (Casida and Quistad, 1998; Kogan, 1998; Pfadt, 1985). Further, most long-lasting insecticides in the chlorinated hydrocarbon class, such as DDT, have been banned from use in agriculture in the USA and Europe (http://www.epa.gov/oppfead1/

Table 2.4 Classes of insecticides and miticides used in agriculture throughout history

Class	Source	Examples	First used
Inorganics	Natural, earth	Arsenic, sulfur, boric acid	1000BC to 1900s
Botanicals	Natural, plants	Nicotine, pyrethrum, rotenone, azadirachtin	1850s to 1990s
Chlorinated hydrocarbons	Synthetic organic	DDT, chlordane, endosulfan	1940s to 1950s
Organophosphates	Synthetic organic	Parathion, malathion, diazinon	1950s to 1980s
Carbamates	Synthetic organic	Aldicarb, carbaryl, carbofuran	1950s to 1980s
Pyrethroids	Synthetic pyrethrum	Allethrin, permethrin, cyhalothrin	1950s to 1990s
Neonicotinoids	Synthetic nicotine	Imidacloprid	1993
Macrocyclic lactones	Natural & synthetic microbial toxins	Emamectin, spinosad	1990s
Diacylhydrazines	Synthetic hormones	Tebufenozide	1998
Crystal proteins	Natural & synthetic microbial toxins	Cry1Ab, Cry1Ac, Cry3A	1950s
Pyrroles	Semi-synthetic microbial toxins	Chlorfenapyr	1998

international/piclist.htm). Such insecticides were replaced gradually beginning in the 1950s and 1960s with short-lived, weeks to months, environmentally safer, organophosphate and carbamate insecticides—safer because they are short lived, do not build up in food chains, and thus do not move into natural habitats, biomagnify in food chains, and affect wildlife as some chlorinated hydrocarbons did. However, many organophosphates and carbamates are acutely toxic to man and animals, resulting in human poisonings and disruption of biological control of crop pests by killing their natural enemies (see Pest-Resistant Crops and Biological Control below) (Croft, 1990). In the 1970s and 1980s, a number of these insecticides were, in turn, replaced by pyrethroid-based insecticides with less acute toxicity to humans and mammals, and are short lived and safe to the wildlife outside of the agroecosystem. Today, pyrethroids are being replaced with novel short-lived compounds that are even safer to man and the environment.

These short-lived compounds comprise more selective (i.e., killing only the target pest) microbial toxins, new natural and synthetic insecticide chemistries, and transgenic insect pest-resistant Bt crops. This gradual change in insecticide chemistry from one chemical class to another is due to the discovery of chemistries that are safer to humans, wildlife and beneficial insects, and more effective at controlling the target pest.

Synthetic pyrethroids, derived from natural plant insecticides, are still widely used today to control insect and mite pests in agriculture, the home, and on pets and livestock because of their effectiveness and safety to man and wildlife. Their major drawback is that they are broad-spectrum and kill the pest's natural enemies coincidentally with the target pest.

Several new classes of insecticide chemistry, especially the macrocyclic lactones (e.g. Proclaim®, Tracer®), and diacylhydazine insecticides (i.e. Confirm®, Intrepid®, and Mach 2®), along with the crystal proteins (i.e. Bt crops and Bt spray products), appear to be the safest pest control known yet to man, wildlife and natural enemies of pests, while maintaining high levels of efficacy against target pests (Casida and Quistad, 1998; Dhadialla et al., 1998). The diacyclhydrazine insecticides belong to a new class that mimic the insect-molting hormone and cause immature insects to prematurely molt and die. Dow AgroSciences has recently acquired the diacyclhydrazine insecticides, and the rights to manufacture and sell them (James, 2002a). This new 'green' chemistry increases its portfolio of insecticides and Bt crop products. Diacyclhydrazine insecticides are very selective, controlling only a small number of insect species; thus, limiting their usefulness and frequently requiring that a second insecticide or miticide from a different insecticide class be applied to control the two to five pests commonly present in a crop. The diacyclhydrazine insecticides can be considered to be biopesticides along with all sprayable microbial insecticides, entomopathogenic nematodes, baculovirus, and some pheromones when used to control pest populations.

Integrated Pest Management

In the last 50 years, an ecology based philosophy and science for managing insect pests has emerged, known as integrated pest management (IPM)—in part due to the environmental and safety risks, and expense and problems of relying solely on conventional insecticides for pest control (Stern et al., 1959; Metcalf and Luckmann, 1994; Persley, 1996; Kogan, 1998). The key to successful IPM is to understand the factors that regulate population fluctuations of the pest, in a particular region and environment, and to implement specific strategies to suppress them.

There is no general recipe for pest control that works for every pest in every production system. In IPM, multiple strategies of suppressing insect pests are woven together in a prescribed program that relies heavily on natural control factors such as pathogens, predators, parasites, and weather (King et al., 1996) (Table 2.5). However, among these pest control choices, there are differences in effectiveness, economics, safety and risks. IPM integrates natural control with ecology based pest-suppressing strategies, such as cultural practices and pest-resistant crops, to avoid pest outbreaks and keep insect pest injury at economically-acceptable levels—this is the goal of IPM. Conventional insecticides are used only when other sustainable methods (like biological and cultural control) have been exhausted, and then only when justified by expected crop losses (Stern et al., 1959; Pfadt, 1985; Casida and Quistad, 1998). Other key components of IPM include: (i) the biological and economic knowledge base used to make management decisions; (ii) a continuous, in-field, monitoring program of insect pest densities and their injury (factoring in densities of natural enemies and their effects on pests); (iii) a monitoring system for detecting the development of insecticide-resistant pests; and (iv) management practices and theory to avoid or delay insect pests from becoming resistant to insecticides, Bt crops and other valuable IPM strategies.

A central guidepost in IPM decision-making is the economic injury level; this is the point in monitoring pest injury or pest density where the amount of injury or density justifies the cost of an additional tactic to suppress further injury (Benedict et al., 1989; Higley and Pedigo, 1996). Action should be taken at this point or earlier, if possible, to suppress further injury that would result in a significant yield loss at the end of the season when the crop is harvested. The level of pest injury or density at which action is taken to suppress further injury and avoid economic loss at the end of the season is the economic threshold level. These pest management guideposts have been established for many pest species and crops in developed countries and can usually be found in pest management literature produced by the agricultural extension service.

Over the past 50 years, IPM programs have been developed through the combined efforts of universities, extension services, and many public and private organizations for most crops in industrialized countries with highly-mechanized agriculture (Harris, 2000,2001); but have been less developed and understood among subsistence farmers in developing countries—especially in the tropics. This is due, in part, to the extensive training and experience in crop agronomy, entomology, pest management and agroecology that are essential for effective specialists in IPM (Frisbie and Adkisson, 1985; Kogan, 1998). FAO and other public organizations have developed programs that train subsistence farmers in

Table 2.5 Strategies used in integrated pest management

Biological Control
- Introduction and permanent establishment of natural enemies.
- Repeated mass release of natural enemies (predators and parasites).
- Conservation or enhancement of natural enemies.
- Application of mass cultured parasitic nematodes and bacterial, fungal or viral diseases of pests (i.e. biopesticides).

Chemical Control
- Insecticides.
- Behavior-disrupting pheromones that attract or repel.
- Pest hormones that disrupt growth, survival or reproduction.
- Pesticide Resistance Management.

Cultural Control
- Planting, harvesting, and irrigation timing.
- Cultivation and tillage practices.
- Crop rotations and disposal of crop residues.
- Trap crops.
- Pest-resistant varieties.
- Intercropping and strip harvesting.

Mechanical and Physical Control
- Screens, traps, barriers, heating, cooling, light, and energy types to attract, repel or kill pests.

Reproductive and Genetic Control
- Chemicals to alter pest reproduction.
- Mass release of sterile individuals.
- Introduction of deleterious genes into pest population.

Source: Modified from Benedict (2003)

IPM philosophy and practices. However, more IPM training is required (Mengech et al., 1996; FAO, 2003). Also needed are more efforts to craft specific IPM programs for local communities in developing countries. Further, necessity exists for new economic thresholds and injury levels for Bt crops. Some scientists believe that the greatest need is for IPM to become more ecologically based at a global level, an effort which requires a better understanding and integration of biological control and pest-resistant crops within the unique and highly dynamic crop production systems of the world. These scientists believe that today's IPM practices have progressed little over the past 30 years, in part because pest managers continue to rely heavily on the use of conventional insecticides (Ehler and Bottrell, 2000; Epstein and Bassein, 2003). Some scientists are concerned that Bt crops will be used as insurance against pest attack, especially Bt corn to control European corn borers—rather than as needed—based on the use of appropriate economic thresholds (Obrycki et al., 2001). We predict the current change to pest-resistant Bt crops from conventional insecticides to be critical to transitioning IPM to a more biological and ecological-based practice as envisioned by the originators of IPM (Stern et al, 1959; Frisbie et al., 1989; Persley, 1996; Gianessi et al., 2002).

In fact, pest-resistant crops are essential and foundational to establishing IPM practices that reduce grower dependence on broad-spectrum insecticides and increase the use and effectiveness of biological control. The most compelling drivers causing this change are: (i) the mandate by EPA that developers of Bt crops are to increase the number of pest-resistant traits stacked in Bt crops—this includes traits with different modes of action and target sites in the pest, which may be from many sources; (ii) the EPA mandate that the efficacy of target pest suppression be increased over a period time; and (iii) the emphasis by all stakeholders on the reduction or elimination of conventional insecticides with pest-resistant Bt crops and improved opportunities for biological control. We believe these drivers of change offer a turning point in the progress of IPM, i.e. for the primary tools in pest control to become pest-resistance crops and biological control, thus reducing our reliance on conventional broad-spectrum insecticides as the primary IPM strategy to suppress pests and, correspondingly, increase yields.

How to Manage Pests on Bt Crops

The question of how to best manage insect pests that attack Bt crops has been of concern to all involved in Bt crop production. This issue is being addressed through academic, extension service and industry studies and experience with Bt crops in field trials and commercial production (Moore et al., 1999; Benedict, 2003). In most respects, IPM programs for Bt crops are no different from conventional crops (for examples of extension service IPM program guideline see, Ostlie et al., 1997; MSUES, 2001; TAES, 2003; UIES, 2003). In reality, most of the current cotton and corn IPM program components can be applied to IPM programs for Bt crop management. However, because Bt crops have altered the amount and pattern of target pest injury, the areas of greatest concern are how to scout Bt crops for pests, and whether Bt crops should ever be treated with topical insecticides to control the target pests. The most difficult decision in managing pests on any crop is and always has been deciding how much injury can the crop tolerate at any given time during the growing season before economic loss results and additional control is required. When target pest injury on Bt crops reaches levels that if left uncontrolled would result in a significant economic loss at harvest, then the producer is justified in taking additional action to control the pest. For example, he/she could make a mass release of predators or parasites of the pest, or apply a selective insecticide that kills only the pest injuring the crop, or an application of a broad-spectrum insecticide. Just as you or I, the producer will choose the curative strategy that is most effective, risk free (i.e. does not create new problems like a secondary pest outbreak), easy to carry out, and inexpensive.

Field research and seven years of experience with commercial production of Bt crops shows that under some conditions—such as when extremely high densities of target pests attack the crop, or crop plants are stressed, resulting in lowered natural plant-pesticides and/or Cry protein dose, or pests attack the plant or a plant structure when it is more susceptible to injury, Bt crops may sustain sufficient injury to justify taking any action. Typically, producers begin to suffer economic loss that justifies action to stop it when the pest injury is between 5 and 15% of the fruit, or 25 to 50% of the leaves, depending upon a number of biological and economic factors, such as the type of crop and the market the producer is selling to. For example, the housewife buying at the market will tolerate less injury and/or insects present on the produce than the buyer of grain that is going into cattle feed. Over many years, researchers have studied the relationships between in-season pest injury, and losses in yield and quality at the end of the season, and they have found many biological and environmental factors can affect the final yield (Higley and Pedigo, 1996). For example, 15% injury to the flower buds of cotton early in the season may result in a 5% yield loss at the end of the season, or under some circumstances, even a yield increase. Some of the factors that influence the final yield outcome are, crop variety/hybrid; duration of pest feeding; whether the pest becomes more abundant as the season progresses, thus causing more injury; influence of natural enemies and diseases on the abundance and injury of the pest through the season; weather; and management practices such as irrigation, plant growth regulators and fertilization of the crop. Use of economic injury levels based on predictions of yield losses at the end of the season based on in season pest injury and density is inaccurate at times, but it is the best guidepost available to producers and their crop managers in order to make insect management decisions during the growing season.

At this time, Bt crops are managed the same way as non-Bt crops but with some surprising twists. Producers/IPM crop consultants determine months before the growing season whether they need Bt crops. The planting-seed for Bt crops is more expensive than non-Bt seeds, so if Bt crops are not needed, they are not planted. The difficulty is that the decision to plant a Bt variety must be made before the crop is planted and pest injury is present. Various predictive tools are used, such as sampling of adult ECB pest densities in fall or spring before planting using pheromone-baited traps, temperature-driven computer models of pest population growth, and past history of ECB crop injury and yield loss. These tools can aid the producer in deciding to plant a Bt hybrid (Ostlie et al., 1997). However, trying to predict ECB or any pest's presence and level of crop injury months in advance is frequently inaccurate and producers usually make one to two types of errors: (i) decide not to plant a Bt crop in a year when the pest outbreaks and a Bt crop is needed; and (ii) decide to plant a Bt crop in a year when the pest does not cause economic damage. Both errors can result in yield and profit losses.

Producers and their IPM consultants also plan and implement other strategies to control and/or avoid pest problems, such as date of planting, crop rotations, fertility levels, crop maturity and harvest dates, to name a few (Table 2.5). The most widespread use of Bt hybrids/varieties is to plant them in locations where target pests like European corn borers, or insecticide-resistant tobacco budworms or American bollworms, or pink bollworms have repeatedly caused significant injury, yield reductions and profit losses in the previous growing seasons, and are expected to continue being a problem. In fact, at times in the past, these pests, especially the pink bollworm in cotton, have caused such extensive and repeated crop losses in some production areas that growing cotton or corn was abandoned. Today, Bt hybrids/varieties are being grown in some of these areas, allowing the producer to again grow the crop profitably.

Commonly, Bt crops are also planted near locations where conventional insecticide sprays cannot be used to control target pests, e.g. schools, residential neighborhoods, and along waterways. Bt crops are also planted in government/producer eradication/suppression programs designed to entirely eliminate pests like the boll weevil and pink bollworm from cotton production regions. A nation-wide program for eradication of pink bollworm was begun in 2001 in the USA (Allen et al., 2002; El-Lissy, 2002). The rationale for use of Bt crops in these programs is that caterpillar pests are commonly suppressed by natural enemies, known as parasites and predators. However, in eradication/suppression programs where a wide area is sprayed with conventional insecticides, the natural enemies are killed and outbreaks of non-target caterpillar pests can be common and devastating—use of Bt crops to control them has worked well and should be even more effective with the new second-generation Bt corn and cotton. In fact, first-generation Bt cotton has been so effective that it caused an area-wide suppressing of pink bollworm populations in several regions of Arizona, over the past ten years—an unexpected benefit of Bt cotton not seen with conventional insecticide sprays (Carriere et al., 2003).

Once producers have decided where and how many acres of Bt crops they will plant, they must also plan on a nearby non-Bt crop of the same type, e.g. non-Bt corn for Bt corn; and in the correct number of hectares to meet the EPA's mandated refuge requirement (see Chapter 8; Gould, 1998; Roush, 1999; Matten, 2000). These non-Bt plantings are refuges for the target pests of the Bt crop. The refuge acts as an insectary and produces susceptible adult moths of pests. These moths are susceptible to the Cry protein in the Bt crop because they have never been exposed to it. Refuge managers hope that these susceptible moths will mate with any rare Cry-resistant moth that might survive on the Bt crop. The theory being that offspring from the mating of a resistant and susceptible moth will be susceptible to the Cry protein in the Bt crop, thus keeping all target pests susceptible to the Bt crop from year to year. This EPA mandated practice is called insecticide resistance management (IRM) and appears to be working as planned for all the target caterpillar pests of cotton and corn, since no resistance to Cry proteins has developed following 7 years of Bt crop production. However, the effectiveness of current IRM programs is almost impossible to measure. Surprisingly, the pink bollworm in Arizona cotton has tended to become more susceptible to the Cry1Ac toxin in cotton following 7 years of exposure, but the reasons for this are not understood (Sims et al., 2002).

Once the Bt and refuge crops have been planted, IPM specialists regularly scout individual fields, on a weekly or biweekly basis in order to monitor the presence, density and level of injury for all the pests. This is the same practice that occurs on all non-Bt crops managed with current IPM methods. Based on the regular monitoring of pests and their injury, producers and their consultants make pest management decisions. The levels of pest densities or injury to the crop are used to determine if the Bt crop requires additional tactics for pest control, e.g. selective insecticide spray, including the target pests of the Bt crop. In fact, producers and their IPM consultants are advised by extension and researchers to check for all pests, including the target caterpillar pests, even though they may see no injury but may find eggs, and newly-hatched caterpillars and their superficial feeding injury. Newly-hatched caterpillars must feed on the Bt crop so as to acquire a fatal dose of the Cry protein.

A number of genetic and environmental factors have been identified that can influence the survival of target caterpillar pests and their injury to Bt crops throughout the growing season (Benedict and Altman, 2001; Leonard et al., 2003). Crop consultants and producers should be aware of these factors because they can result in more pest injury and crop loss than expected. The most important factors are:

- Variety/hybrid genetic background
- Bt gene construct and insertion event
- Stage of plant growth when attacked
- Agronomic practices and weather

Each of these factors influences injury by affecting plant toxicity to the pest. The genetic background, Bt gene construction, and insertion event, all influence Cry protein expression and thus Cry concentration in the Bt crop (Sachs et al., 1998). Further, all plants—including wild and cultivated crops—possess natural defensive chemicals and morphological structures that are designed to repel and kill pests (Harborne, 1982; Smith, 1989; Kogan, 1998; Sofos, 1998). This means that some varieties/hybrids are naturally more toxic to pests than other varieties/hybrids. Moreover, the Cry proteins work in concert with these natural plant defenses to lower pest survival and injury. However, there are complex interactions of Cry toxins and native plant toxins that may increase or decrease pest mortality compared to what we might expect from Cry proteins alone (Altman et al., 1996; Sachs et al., 1996).

In addition, the production of Cry proteins and natural plant defenses is a function of plant genetics and plant metabolism; and therefore any factor such as nutrition, sunlight, temperature, soil moisture, plant hormone balance, or insect injury, that alters plant physiology, can influence expression of Cry proteins and natural defenses, thereby increasing or decreasing insect pest survival and injury compared to what we might expect. Keep in mind we do not know how all of these factors interact to effect pest injury to current Bt crops. However, some of factors have been observed operating in commercial production. For example, in the USA, during periods of rainy weather, where sunlight and temperatures were severely reduced and soils were saturated, caterpillar injury on Bollgard® cotton in some situations increased to the point where the cotton had to be sprayed with insecticide (Benedict and Altman, 2001). Also, during blooming of Bollgard® cotton, bollworm injury to fruit was higher than expected and had to be treated with insecticides. Consequently, published extension IPM guides for managing bollworm have been modified to recommend that IPM specialists scout flowers for bollworms and treat with insecticides when economic thresholds are reached (Gore et al., 2003; Leonard et al., 2003).

Injury and survival of bollworms in blooms are higher compared to leaves and flower buds because the Cry protein concentration is lower in the bloom, especially in the pollen and pistil (Fig. 2.3). Also, since the concentration of total Cry protein is 10-fold higher in Bollgard® II cotton, we expect that sprays of insecticide for any target caterpillar pest will be unlikely (Fig. 2.5).

How do we Stop Pests From Becoming Resistant to Bt Crops?

Insect pests may have the potential to develop resistance to the Cry proteins in Bt crops, just as they do to conventional insecticides or natural plant insecticides in pest resistant crops, or probably any strategy we can develop (Smith, 1989; Roush and Tabashnik, 1991) (see also Chapters 6, 7, and 8). Although the development of pest resistance to natural or manmade insecticides is common, it is by no means certain to occur for every pest and insecticide. Even when researchers have shown that an insect pest has traits present in their populations to resist an insecticide, this expected resistance in the field may not occur. Usually, these traits, or genes, are very uncommon in natural pest populations. Maybe only one insect in a million or billion possesses the trait that would allow it to survive in the presence of a specific toxin, such as Cry1Ac; this trait is linked to many other traits in the insect's genome. When pest populations are repeatedly exposed to insecticides over many generations, these traits interact to determine whether the frequency of the specific resistance trait will become even more frequent in the population. A multitude of biological and environmental factors place selective pressures on the pest population and ultimately determine what will be the most successful and frequent combination of traits at any particular time. Pest adaptation to environmental factors is a constantly-changing and highly dynamic process among insects with many generations per year and hundreds to thousands of offspring per female. However, based on more than 500 real-world cases, we are in a position to make some generalizations about the development of pesticide-resistant insects (Georghiou, 1986; Roush and Tabashnik, 1991; Denholm et al., 1999; Roush, 1999; Clark and Yamaguchi, 1999).

As discussed, the development of resistant pests is dependent upon a number of biological and ecological factors such as: (i) how effective the insecticide or pest-resistant crop is at killing the pest—usually the more effective it is, the more likely and quickly resistance develops; (ii) how widely the particular insecticide or pest-resistant crop trait is used—the more widely it is used, the more likely and quickly resistance develops; (iii) how mobile, widely distributed, and how many plant species the pest feeds on—the less mobile and widely distributed the pest is, and the fewer plant species it feeds upon, the more likely and quickly resistance develops; (iv) how long and frequently the pest population is exposed to the insecticide or pest-resistant crop trait—the longer the exposure and the more generations exposed, the more likely and quickly resistance develops; and (v) how many insecticide-resistant traits are present in the pest population, how frequent they are, and how they are inherited—the more traits present, the more frequent they are and vis-à-vis, the more easily they are inherited, the more likely and quickly resistance develops. However, even if all these factors favor the development of resistance we still cannot accurately predict 'if' or 'when' this resistance will develop. We can only say that the likelihood of developing pesticide-resistant pests is great. Some insect species have a long history of developing resistance to any insecticide used against them. The most notorious examples worldwide are the Colorado potato beetle, which is resistant to 37 compounds; the diamondback moth, *Plutella xylostella*, which is resistant to 51 compounds and the green peach aphid, *Myzus persicae*, which is resistant to 71 compounds (Georghiou, 1986; Georghiou and Lagunes-Tejada, 1991; Vasquez, 1995). Predicting that resistance will develop to any new insecticide used against one of these pests can be done with some confidence.

Many experts believe that the best way to delay resistance is by reducing the selective pressure on the pest population to adapt to the toxin by maintaining a large proportion of the population as susceptible. The question of how to manage transgenic Bt crops in order to minimize the selection pressure on target pests is a perplexing and controversial question, for which there is not a single best answer. Several approaches have been suggested to reduce the selective pressure asserted by an insecticide that causes the pest population to adapt to it over a period of time (Gould, 1998; Roush, 1999). These approaches are being used in IRM programs to avoid developing pests that are resistant to the Cry proteins in Bt crops. Such approaches are:

Mixtures where two or more toxins (insecticides) are combined in the same field by mixing varieties that produce different toxins, or even combined (pyramided) in the same plant.

Rotations where different toxins are used in rotation—one year with one toxin-producing variety and the next year with another toxin-producing variety.

Mosaics are special patchworks of toxin-producing varieties where varieties with different toxins are planted in adjacent fields.

Refuges, as described above, are varietal plantings without pesticidal plants. These plantings are expected to produce susceptible pests that mate with any rare partially- or fully-resistant individuals produced on a pesticidal-crop. Pest populations are expected to maintain susceptibility over time through the continual 'dilution' of rare resistance genes.

High dosage is the use of pesticidal-crops that produce a dose of the pesticide in their tissues that is several fold higher than needed to kill all susceptible pest insects in a population.

These approaches may be combined in numerous ways. Refuges can be used with mixtures and planted in mosaics. Mixtures and rotations of conventional synthetic spray insecticides have been used with limited success to delay the development of insecticide resistance in some insect pests (Fitt and Wilson, 2000).

Something very unusual happened when the EPA first registered Bt cotton, Bt corn, and Bt potato for commercial production in the mid 1990s. The developer of these products, Monsanto Company, was required to file insecticide-resistance management plans in order to delay the development of target pest resistance to the specific Cry proteins present in their Bt crops (Table 2.6) (see also Chapter 8). Since then, the EPA has made IRM plans a requirement for registration of all Bt crops. These plans have been developed through the cooperative efforts of industry, academic, and governmental scientists and are based on the best theory and real-world data that exists. These plans act as useful proactive tools to avoid/delay the loss of valuable insect control products.

All insecticides are valuable resources, whether natural or synthetic, plant produced or industrial factory produced, and their loss is a serious and expensive event to farmers, the insecticide industry and society. An obvious incongruity of the USA pesticide regulation is the lack of IRM considerations/requirements for conventional insecticides and sprayed Bt biopesticides. Sadly, the diamondback moth has recently developed resistance to sprayed Bt products on vegetable crops in Florida and Hawaii (Tabashnik et al., 1998; Zhao et al., 2001). As far as we know, this is the first insect pest to ever develop resistance to a Cry protein in the field, although Bt products have been commonly used as sprays to control field and horticultural crop pests for more than 40 years (Frankenhuyzen, 1993; Hall and Menn, 1999; Zhao et al., 2001). We wonder whether diamondback moth resistance to sprayed Bt proteins could have been avoided if an IRM program for sprayed Cry proteins had been implemented earlier.

As you can see (Table 2.6), IRM plans for Bollgard® cotton require a gradual phasing in of all the known tactics to avoid/delay the development of pests that are resistant to Bt crops. The stacking of multiple Cry proteins and increasing Cry protein dosage in Bt crops is a

Table 2.6 Current Monsanto deployment strategy to minimize the development of pests with insecticide resistance to Cry1Ac in Bollgard® cotton

---
Short Term (at commercialization 1996-97)
- High dose expression of crystal protein to control insects heterozygous for resistance alleles.
- Refuges of non-Bt cotton to produce susceptible insects (4 unsprayed non-Bt acres per 100 Bt acres, or 25 sprayed non-Bt per 100 Bt acres).
- Agronomic practices that minimize insect exposure to crystal protein in Bt cotton.
- Integrated pest management strategies to increase beneficial insect effectiveness, and reduce conventional insecticide use.
- Monitoring target insect populations for susceptibility to crystal protein.
- Report on Bt cotton performance, especially any 'failures'. Investigate the cause.

Medium Term (2-5 years after commercialization)
- Continue above plus:
- Combine 2 insecticidal genes within the plant with different target sites/modes of action for the target lepidopterans.

Long Term (> 5 years after commercialization)
- Continue all short and medium-term strategies plus:
- Incorporate natural plant resistance to insect traits into Bt cotton.
- Incorporate other novel insecticidal genes into cotton that control the insects.

Source: Modified from Monsanto Co. (2003)

very positive first step to improving the IRM. Plants in nature do not rely on a single genetic trait (monogenic resistance) to resist a pest, but rather, on many resistance traits (polygenic resistance), so we would be smart to take a tip from what works in nature and is supported by population genetics, and develop polygenic resistance in crop varieties with at least two pest-resistant traits pyramided together in each variety. Pyramided traits are expected to be most effective when combined with refuges and a fully-implemented integrated pest management program that uses all available pest suppression methods in a truly holistic crop management approach.

ISSUES OF RISK: FACTS, FACTOIDS AND FEAR

The public controversy and negative responses to the field-testing, commercial production and human consumption of genetically-modified organisms (GMOs), especially Bt crops, have been widespread. This is surprising for a technology that offers so much safety and benefit (Cannon, 2000; Gianessi et al., 2002; Thompson, 2002; Stewart and Wheaton, 2003) (see Chapters 7, 9, 11, 12, and 13 for further discussions of risks). Nothing we do in any aspect of our lives is without risk. However, some special interest groups regard GM crops as unacceptably risky, and work tirelessly to influence governmental decision makers to obstruct GM crop use in production agriculture. Some politicians in Europe have been elected after running on an anti-GM, pro-organic crop platform (Thompson, 2002). This is surprising to many, since GM crops have been grown and consumed safely for seven years; that *Cry* proteins have been sprayed on conventional crops and forests for over 40 years; and because of the many benefits of Bt crops to farmers, the environment and society discussed earlier. Moreover, today, Bt crops are replacing millions of kilograms of conventional insecticides worldwide and have the potential to replace much of the broad-spectrum conventional insecticide sprayed on crops. These first Bt crops have so reduced the use of conventional insecticides worldwide in cotton and corn, replacing 10 to 100% depending on the specific field situation, that a revolution is taking place in the traditional agrichemical industry. Most agrichemical manufactures are now agrichemical/biotechnology/seed firms, commonly called 'life science companies', that emphasize GM crops as a growing component of their crop management product portfolio (James, 2002a).

The replacement of broad-spectrum insecticides that are hazardous to wildlife, the natural enemies of crop pests, and human health, with safe Cry proteins are the very changes in crop production practices that society and many scientists have worked towards since 1962 when Rachel Carson wrote her book, 'Silent Spring' (Carson, 1962). She focused public attention on the available data showing injury to certain predatory bird species through biomagnification of DDT, a chlorinated synthetic insecticide, in their food chains. Carson then predicted in very vivid and certain fashion that the continued use and production of synthetic chemicals, especially synthetic pesticides, would cause wildlife disasters that would decimate bird populations, and cause an epidemic of cancer in mankind (Carson, 1962) (See Gold et al., 2002 and Logomasini, 2002, for a factual assessment of these claims and predictions). Carson was correct in her concern for DDT's damage to certain bird species and we can thank her for alerting us to this problem. Although some of her predications did not come true, public concern in the 1960s for these and other environmental issues gave rise to the environmental movement, and a number of special interest groups (Moore, 2003).

Today, many of the media and special interest groups have inflamed public opinion against Bt crops just as they have against conventional insecticides and conventionally-grown crops (Gold et al., 2002; Thompson, 2002). Why do we see such vocal and vehement resistance to Bt crops is hard to understand, when they can improve insect pest suppression, biological diversity in agroecosystems, sustainability of agriculture, human and environmental safety, production of food and feed per unit of land, farm income, and reduce poverty and malnutrition for millions living in subsistence farming communities worldwide. A number of reasons have been identified, such as: (i) fear of the unknown or of change; (ii) lack of control, i.e. being forced to eat Bt crop products since they were/are unlabeled; (iii) only experts can know the real risks; (iv) Bt crops are synthetic products of the industry, not natural and organic; (v) Bt crops are the products of big multinational companies that cannot be trusted; (vi) they are not culturally and/or philosophically unacceptable; (vii) they are not compatible with vested interests; (viii) the public cannot easily observe the benefits of Bt crops; (ix) they conflict with religious views; and (x) individuals with low tolerance to risk taking are adverse to change (Diamond, 1999; Thompson, 2002; Stewart and Wheaton, 2003).

Acceptance and adoption of any new technology and concepts are almost always met with some fear and apprehension by peoples and societies (Diamond, 1999; Bailey, 2002). History is replete with examples: people resisted Columbus who said that the world was round; they resisted the automobile for the horse and mule; they resisted the train, pasteurized milk, the air plane, conventional plant breeding, hybrid crop plants, electricity, electric lighting in Britain, and modern medicine versus blood letting, to name a few. Raising questions of risk about any new technology are perfectly legitimate, and reasonable thinking suggests that these risks require thorough scientific evaluation before any new technology can be accepted and implemented. The difficulty is in convincing the public of the truth once the facts are known. In this regard, recent studies have clearly shown that the public's fear of pesticides in foods, and their belief that organic foods are safer and healthier than conventional farmed foods are unfounded. Yet many people continue to believe the unsubstantiated or exaggerated claims of danger from residues of synthetic pesticides reported in the popular press, and live in unnecessary fear (Morris and Bate, 1999; Gold et al., 2002; Thompson, 2002; Trewavas, 2003).

The principles and practices for assessing the risks of Bt crops and other technologies have been clearly established in countries belonging to the Organization for Economic Cooperation and Development (OECD), which includes the most developed and some developing countries (CGIAR, 1999; Persley and Siedow, 1999; James, 2000; Thompson 2002). These principles and practices are applied on a case-by-case basis, and have been published in a series of OECD reports over the past decade. Most countries have used these guidelines to develop biosafety testing protocols to assess and manage risk in the development and use of Bt crops (See Chapter 11). Potential risks can be grouped as: (1) risks to human health; (2) risks to the environment; and (3) social and ethical risks to societies and cultures. To date, Bt crops have been found safe to man and the environment, and beneficial to individuals and societies (World Health Organization, 2000; US EPA, 2001a; Carpenter et al., 2002; Gianessi et al., 2002; James, 2002a).

If Cry proteins and Bt crops are so safe and beneficial, then why the nearly unexpected outcry from the European public and activist organizations? (Bailey, 2002; Conko and Prakash, 2002; Thompson, 2002). The reasons for opposing Bt crops could be any one of the 10 reasons outlined above, but for most opponents, it is some combination of these issues (Nill, 2002; Thompson, 2002; Stewart and Wheaton, 2003). However, the media and activist organizations, especially GreenPeace, have dramatically magnified the risks and motivated fierce opposition to Bt crops, in many cases, by preying on the public's fear and mistrust of big business, government bureaucrats, synthetic chemicals and conventional farming systems (US House Agriculture Committee, 2002). The most powerful tool of the Bt opponents is what we will call "factoids" to mislead the public on the true risks, safety and benefits of Bt crops compared to conventional crops and insecticides. Factoids are pieces of unverified or inaccurate information that are presented by an individual or organization to be factual—often as part of a publicity effort—and the pieces of misinformation are then accepted as true because of frequent repetition and source. The following is a quote from the website of GreenPeace-Australia that contains a factoid (www.greenpeace.org.au/truefood): 'It's time for True Foods. Multinational chemical companies are messing with our food. Through genetic engineering (GE), they are creating food crops that could never occur in nature. Now these GE crops have found their way into the food chain and on to our plates. Our food is under threat.'

Our food is not under threat or attack. Multinational companies that produce Bt crops are made up of people just like you and I, and they and their families eat these crops just like everyone else. Today's food in developed nations is the safest, most carefully tested and monitored food in the history of mankind. Furthermore, cancer is declining and human life span is increasing (McClintock et al., 1995; CAST, 2001; Gold et al., 2002; Thompson, 2002; Stewart and Wheaton, 2003).

Factoids Concerning Bt Crops

Some of the untrue claims and factoids circulating in the press and elsewhere about Bt crops are: (i) farmers are losing money growing biotech crops; (ii) USA soy and corn exports have collapsed since biotech crops were introduced; (iii) biotech crops are less hardy than their non-biotech counterparts; (iv) biotech crops have not increased yields as promised; (v) biotech crops have failed to reduce pesticide use; (vi) Bt corn is threatening the monarch butterfly; (vii) only the USA grows significant areas of biotech crops; (viii) biotech crops benefit only the biotech companies; (ix) antibiotic-resistance marker genes in Bt crops will create antibiotic-resistant bacteria in humans; (x) biotech crops are unnecessary, because organic farming can produce similar amounts of food without using any chemicals; and (xi) developing countries have not benefited from USA biotechnology (James, 2002a; Nill, 2002; Thompson, 2002; Stewart and Wheaton, 2003; Trewavas, 2003). None of these claims are true.

A number of possible motives, agendas and behaviors have been suggested for the opponents of Bt crops (James, 2002a; Nill, 2002; Thompson, 2002; AgBioWorld, 2003; Stewart and Wheaton, 2003). Some writers have suggested that certain EU individuals and organizations may be motivated by self interest, since excluding Bt crop imports from the EU could be viewed as a way for the EU to control commodity markets and the price of grain products in Europe. GreenPeace is supported in part by the organic growers; they receive approximately US$100 million per year from the organic farming movement. Could Bt crops threaten the organic farming movement, since the *Cry* protein in Bt crops could qualify them as legitimate organic crops (if they meet other certified organic requirements)? The *Cry* protein in Bt crops is the same active protein that organic growers use to control insects. The natural/organic food business has grown into a big business; just in the USA it grossed more that US$ 6 billion dollars last year (Thompson, 2002). Could the organic food industry see Bt crops as a threat?

The most widespread opposition to genetically modified (GM) crops (e.g. Bt crops) is among the countries of the EU. However, during the past several years, the EU seems to contradict itself by stating that it is not accepting GM foods from Bt crops because of the risks, while at the same time EU imports of USA soybeans and corn gluten have increased 7 to 15% per year for the last three years (EU soybean imports were up 14% in 2002 over 2001, and up 15% in 2001 over 2000; corn gluten imports up 7% in 2002 over 2001) (Nill, 2002; USDA, 2002). These EU imports of soybeans and corn gluten are shipments of mixed GM and nonGM products—the USA is not segregating shipments and the EU is importing and eating GM products.

Based on a recent study by Bonny (2003), a combination of factors heavily influenced by environmental extremism has lead to the current widespread European opposition to GM crops. The primary factors are: (i) public focus on the potential risks of GM crops, due greatly to the extensive public criticism by activist organizations, some politicians and the news media; (ii) the inadequacy of answers provided by government agencies to these criticisms; (iii) European governments' indecision on the safety of GM crops; (iv) public's lack of trust in EU government agencies and politicians (unlike the USA, where public trust is very high in regulatory agencies like the EPA, USDA and FDA) (Thompson, 2002); (v) inability of the public to clearly see and participate in the benefits of GM crops, resulting in the public drawing up an unfavorable risk-benefit balance; and (vi) the public's fear and objection to the current evolution of agriculture and the functioning of society (i.e. limited trust in institutions and firms, and the desire for an imagined utopian lifestyle). These six factors have all crystallized around GM crops.

Some people believe that the opposition to Bt crops by certain environmental organizations runs deeper than their spoken concerns for the monarch butterfly and food safety; rather, it is driven by environmental extremism, the most powerful political philosophy in the world today (Bailey, 2002; Logomasini, 2002; Thompson, 2002). For example, many scientists who strongly supported GreenPeace in the early 1970s have become disillusioned, blaming the organization's abandonment of science and logic for goals of environmental extremism. Recently, a science advisor to GreenPeace, Dr William Plaxton (Professor of Biology at Queen's University, Ontario Canada) resigned, citing fear-mongering and non-scientific attacks on Bt crops as some of the reasons (Thompson, 2002). More significantly, one of the original founders and 15-year veteran of GreenPeace, Dr Patrick Moore, recently quit the organization for these very reasons (see Moore's website http://www.greenspirit.com/, Environmentalism for the twenty-first century). Dr Moore states that pagan beliefs, environmental extremism and junk science were influencing the movement's public policy. Further, he states that GreenPeace environmental extremists tended 'to abandon science and logic and to get the priorities completely mixed up through the use of sensationalism, misinformation and downright lies.' He states that environmental extremists inside and outside GreenPeace are: (i) Anti-human—they believe humans are a cancer on the earth; (ii) Anti-science and technology; (iii) Anti-trade—they are against globalization, free trade and the original 'whole earth' vision; (iv) Anti-business—they believe all corporations are corrupt and profits are unacceptable, that the liberal democratic free market-based model for trade is unacceptable, and they would phase out consumer-based industrial capitalism; and (v) Anti-civilization—they have a view of mankind living off the land in a utopian subsistence-based garden of Eden. We find these views very problematic and disconcerting, because they mean that environmental extremists do not value human life, current civilizations or freedom.

These beliefs—if held by many, especially government regulators and politicans—are a danger to democratic society and to current civilization. For example, would the environmental extremists have the hungry of this world starve to death for their dream of a subsistence-based utopia? Today, many African farmers are 'living the dream' promoted by some environmental extremists, but they aren't liking it. In fact, they are questioning why they should suffer? (Mabonga, 2003).

The reader can see from this brief discussion that opponents of Bt crops have many reasons for their position. In summary, their opposition is based on fears, or personal philosophies, or self-interest or a combination of these. Perhaps much of this could be overcome if the public was better informed on the true risks and benefits of Bt crops.

Bt Crop Effect on Non-target Organisms: Case of the Monarch Butterfly

Prior to the registration of the first Bt crop (event 176 corn) in 1995, the EPA evaluated studies on the effects of Cry proteins on a number of non-target organisms, including aquatic invertebrates, birds, earthworms, fish, honey bees, lacewings, ladybugs, springtails, and parasitic wasps (US EPA, 1995). EPA concluded from these studies that Bt corn had 'no unreasonable adverse effects to humans, non-target organisms or the environment' (US EPA, 1995). However, in 1999, several scientists published a letter stating that when they fed monarch butterfly caterpillars on the leaves of milkweed plants, which the scientists had sprinkled with Bt corn pollen, that the caterpillars grew slower and had higher death rates than caterpillars fed on leaves sprinkled with pollen from non-Bt corn (Losey et al., 1999). The public response was quick and harsh, condemning the registration and commercial production of Bt crops (Sears et al., 2001; USDA ARS, 2002; Shelton et al, 2002). Subsequently, the EPA issued a 'data call in' notice to registrants of Bt corn requesting industry, researchers and all interested parties to submit information and comments for agency use in evaluation and potential deregistration of corn hybrids containing Cry proteins.

In response, more than 40 scientists and many representatives from industry, EPA and other regulatory agencies, public non-profit organizations and academia formed working groups and held public meetings to formulate a response to the Losey et al. (1999) data. Following two years of experiments, meetings and research reporting, and the expenditure of considerable sums of public and private moneys, the scientists and EPA concluded that the impact of Bt corn pollen from current commercial hybrids was negligible (Sears et al., 2001). They point out that determining risk of Bt crops to non-target organisms requires consideration of both the level of expression of the suspected toxicant in the Bt crop, and the likelihood of exposure to the toxicant in the real world. Their specific conclusions were: (i) other researchers attempting to duplicate the Losey et al. experiments were unable to find any difference in growth or mortality between monarch caterpillars fed on milkweed leaves sprinkled with Bt pollen or non-Bt pollen—in other words, the Losey et al. research findings could not be repeated; (ii) no toxic effects on caterpillars could be found for pollens of many corn hybrids, or Bt events (except 176, which was removed from the market in 2003), at densities of Bt pollen grains on milkweeds that would normally be encountered in the field; (iii) temporal overlap is low between time of pollen shed (10 days in July or August), and the time when caterpillars are naturally present on milkweed plants—overlap ranges from 15 to 60% depending upon weather and latitude; (iv) spatial overlap is low because areas of Bt corn production have low densities of milkweed plants compared to non-agricultural habitats; and (v) Bt corn market penetration is low, representing only 19% of the total USA corn crop. We would add that when Bt corn is used to replace conventional insecticides—which we know kill monarch caterpillars (Sears et al., 2001)—that growing Bt corn may, in fact, increase the survival of monarch caterpillars compared to growing non-Bt corn, sprayed with insecticides.

The case of the monarch butterfly clearly demonstrates that evaluations of risk from Bt crops must be made with sound scientific data, which have undergone rigorous review by peer scientists. Moreover, any governmental regulations made for Bt crops must be based on these sound scientific data, rather than formulated as a reaction to media coverage and public outrage.

Two more cases of potential negative effects of Bt corn on non-target insects have become of concern to the public and scientific communities—the black swallowtail butterfly and the lacewing (Williams et al., 1998; Wraight et al., 2000; Shelton et al., 2002). As with the monarch butterfly, sound field studies indicate negligible risk from Bt corn to either insect. The researchers working with lacewing larvae (Hilbeck et al., 1998), a predator of European corn borer, point out that Bt crops 'are still more environmentally friendly than most if not all chemical insecticides.'

Marvin Harris (personal communications) suggests the existence of a potential indirect effect of Bt crops on certain natural enemies. He wonders if some natural enemies of pests targeted by Bt crops may be reduced in densities on Bt crops. Logical thinking suggests that because there are fewer of the preferred caterpillar prey/host for natural enemies to feed on, there will be fewer natural enemies present. We do not know of published data to support this for any natural enemy species found in Bt crops in commercial production.

In conclusion, we can find no serious negative impacts of Bt crops on non-target organisms. This is a valuable improvement over most conventional insecticides that may have a wide range of negative impacts on non-target organisms. Many native animals live in or near Bt crops or pass through them. This list includes birds, butterflies, bees, deer, rabbits, reptiles, amphibians and many beneficial insects that help suppress crop pests by feeding on them. Bt crops offer mankind a great improvement in our ability to protect the presence and balance of these organisms in agroecosystems over the traditional use of broad-spectrum insecticides. Let us not lose sight of this amid the contradictory rhetoric and criticism of Bt crops by some in the media and some special interest groups (see chapter 10 for further discussion on non-target organisms).

Food Safety of Bt Crops

Food safety of Bt crops is regulated by governmental agencies worldwide (Thompson, 2002; Higgins and Chrispeels, 2003; Stewart and Wheaton, 2003). Food, food ingredients, and feed produced by GM plants are subject to strict testing and evaluation procedures in the countries where they are experimentally and commercially grown. In the USA, three agencies coordinate to insure the safety—the United States Department of Agriculture (USDA), the Food and Drug Administration (FDA), and the EPA. They have the legal and regulatory power to ensure food and feed safety. The products of GM crops must have demonstrated safety equal to non-GM crops. Safety testing and evaluation procedures are generally initiated seven to ten years before a new Bt crop can be commercialized. These safety testing procedures for Bt proteins fall into six areas: (i) measuring Bt protein concentrations in various parts of the plant throughout the season; (ii) testing the plants' growth and development characteristics to insure that the transformed plants are normal; (iii) measuring Bt plant chemical and nutritional status to ensure that no significant nutritional changes have occurred; (iv) determining the stability of the Bt genes, and other active inserted foreign genes such as marker genes or promoter genes that actively express foreign proteins in the transformed plant, and their pattern of inheritance; (v) food and feed toxicity and allergenicity of the new Bt protein(s); and (vi) potential routes of human exposure and amount of exposure.

Regulatory agencies and scientists agree that the process by which a new gene is introduced into a crop plant (i.e. gene gun, *Agrobacterium tumefaciens* or conventional plant breeding) will not potentially cause the transgenic plant to be harmful, but it is the new gene product, in this case specific Cry proteins, that must be evaluated on a case-by-case basis (McClintock et al., 1995). Various additional data are collected and tests conducted, such as effects of cooking and food processing on the Cry protein, and the potential intake and dietary impact of the Bt crop food. Toxicity is evaluated based on the relationship of the Bt protein to other known toxins, and from extensive mouse/rat/rabbit/catfish/hog/ chicken/cow feeding trials. These feeding studies include the evaluation of: (i) long-term, e.g. 90 days of feeding on low sublethal doses, known as chronic exposure, that would identify cancer-causing abilities of the Bt protein if present; and (ii) short-term high dose exposures, known as acute exposure, to determine the Bt protein's ability to kill.

Toxicity of Cry proteins

The mode of action of Cry proteins in current Bt crops (Table 2.1), as discussed above, is for the trypsin-resistant Cry protein to selectively bind to the target sites localized on the brush border membranes of midgut epithelial cells in susceptible insect species. Toxicity is directly attributable to the presence of specific binding sites in susceptible insects. Without these sites, insects are not susceptible to Cry protein intoxication. More to the point, there are no binding sites for any Cry proteins from *B. thuringiensis* on the surface of mammalian intestinal cells. Therefore, livestock and humans are not susceptible to these proteins. In other words, to the best of our knowledge and experience, Bt proteins are non-toxic to humans and livestock. Further, the low potential for toxicity of plant-expressed Cry proteins is demonstrated by: (i) the lack of amino acid sequence homology (i.e. similarity) with known protein toxins in the protein databases; (ii) rapid digestion of the Cry proteins in simulated gastric juices; and (iii) lack of toxicity in feeding studies to laboratory animals (http://www.agbios.com/; McClintock et al., 1995; Betz et al., 2000; US EPA, 2001a; Thompson, 2002).

There is no toxicity attributable to our consuming foreign DNA (i.e. consuming Bt genes or any organisms genes), since it is composed of the same base amino acids as DNA of non-Bt crop plants. DNA is the same in all living organisms and would not differ from what is already ingested in our diet (In humans, the daily dietary intakes of RNA and DNA are typically in the range of 0.1 to 1.0 g, respectively.). Further, the Cry gene sequences are similar to DNA sequences that humans have been consuming on Bt sprayed crops for 40 years with negligible health effects. Nor do they differ in safety from the Cry proteins used in the spray products (US EPA, 1988).

At this time, the only human exposure to Cry proteins in Bt crops is in the consumption of whole kernel Bt corn. In the case of Bt cotton, a process that includes heat, solvent and alkali treatments refines the oil; any DNA or Cry protein would be removed and destroyed during processing. Cotton lint is pure cellulose (i.e. lacks any protein or DNA) and is processed with heat and solvents that would remove and destroy any protein or DNA present.

Allergenicity of Cry proteins

Food allergies are adverse physiological responses to an otherwise harmless food or food component (IFIC, 2001; Stewart and Wheaton, 2003). The body's immune system responds abnormally to specific protein(s) in the food. The most common types of food allergies are mediated by allergen-specific immunoglobulin E (IgE) antibodies from the immune system. The proteins acting as allergens have short lengths of amino acids on their surfaces that cause mammals to produce these IgEs. These short lengths of amino acids have been used to identify more than 200 food proteins responsible for food allergies. IgE-mediated reactions are known as 'immediate hypersensitivity reactions' because symptoms occur within minutes to a few hours after ingestion of the challenging food. Another type of true food allergy is cell-mediated, which involves sensitized tissue-bound lymphocytes rather than antibodies. In cell-mediated allergies, the onset of symptoms usually occurs more than 8 hours after ingestion of the offending protein. Symptoms are varied and may occur either in combination or alone, and range from runny nose, swelling or itching lips or mouth and/or throat, nausea, vomiting, cramping, diarrhea, itching skin, hives and asthma to anaphylaxis and death.

Almost all food allergens are proteins, the most common of which are found in peanuts, soybeans, cow's milk, eggs, fish, crustacea, wheat, and tree nuts. These sources account for more than 90% of all food allergens. Food allergens affect approximately 2.5% of the population in developed countries and a higher percentage in developing nations.

Potential food allergenicity is a concern with the commercial production of Bt proteins in corn but of little concern for Bt cotton, since human exposure in food in the latter case is negligible. When new proteins are introduced into a food crop, the following criteria are used by regulatory agencies to evaluate its potential allergenic risk in food (FAO/WHO, 2001): (i) What is the source of the transferred genetic material? Does it contain known allergens? (ii) DNA sequence homologies between known allergens and the newly-introduced crop proteins are assessed—are they similar? (iii) Immunoreactivity of the *Cry* protein with IgE is tested if the newly-introduced protein is derived from a known allergenic source or if it has sequence homology with a known allergen (The IgE is derived from the blood serum of appropriate allergic individuals.). (iv) Effect of pH and/or digestion on the novel protein is measured because most allergens are resistant to gastric acidity and digestive proteases. (v) Heat and processing stability are determined because heat-labile allergens in foods that are eaten cooked are of little concern.

In summary, the allergenicity studies for most current Cry proteins in commercial Bt corn (i.e. Cry1Ab, Cry1Fa2, Cry3Bb1) demonstrate they have no relationship to known allergenic sources, do not possess immunologically-relevant sequences with similarity to known allergens, and do not possess other characteristics of known protein allergens. Further, these proteins undergo rapid breakdown under in vitro digestive conditions that mimic human digestion, and have a long history of no allergenic responses in commercial use as insecticide sprays.

Facts and factoids about cancer, pesticides and synthetic chemicals

Fear of exposure to potentially cancer-causing synthetic chemicals such as pesticides and most manmade chemicals is probably the greatest 'environmental' concern for the general public today. Many environmentalists and special interest groups claim and believe that we have a cancer epidemic (Logomasini, 2002; Gold et al., 2002; Stewart and Wheaton, 2003). However, data from the National Cancer Institute clearly shows that cancer is not increasing in the USA, except for lung cancer from smoking. In fact, in the USA, overall cancer death rates have declined 19% since 1950 (excluding lung cancer). Cancer death rates have also declined in Canada and the UK (Gold et al., 2002; Logmasini, 2002; Trewavas, 2003). Particularly revealing are studies that show farmers have significantly lower cancer rates than the general population, although they have the greatest exposure to pesticides (Trewavas, 2003). From these data, we can surmise that synthetic chemicals, including pesticides, have not caused a cancer epidemic. More to the point, no study has ever shown that synthetic pesticides have caused a single case of cancer (Gold et al., 2002). The facts also show that environmental pollution accounts for only 2% of all cancer cases, whereas tobacco use accounts for 30% and dietary choices account for 35% of annual cancer deaths—thus we can avoid 75% or more of all cancer risks by our own choices (Ames et al., 1995). In reality we consume large quantities of natural pesticides every day in the foods we eat—approximately 1,500 mg per day (Gold et al., 2002). On average, these natural pesticides make up 5-10% of the fruits, vegetables, nuts and grains that we consume as a part of our daily diet (Harborne, 1982; Ames et al., 1990a,b; Sofos, 1998). Plants have evolved these pesticidal chemicals to defend against the many bacteria, virus, nematodes, arthropods and mammals that feed on them (see following section, Pest-Resistant Crops and Biological Control). At least 10,000 natural pesticides and their breakdown products occur in the human diet (Ames et al., 1990a,b). Natural plant pesticides also provide many of the flavors and smells of the spices and foods we eat. In addition they constitute a source for many medical drugs, homeopathic medicines, addictive drugs (like caffeine, cannabis, cocaine, nicotine, and opium), and commercial pesticides (Harborne, 1982; Higgins and Chrispeels, 2003).

A small number of these natural pesticides, 72, have been tested in standard rodent carcinogen tests just as tests conducted on commercial synthetic pesticides, and about 57% were carcinogens (Ames et al., 1990a,b; Coots, 1994; Gold et al., 2002). This is about the same percentage of carcinogens as these tests turn up with synthetic chemicals. In these high-dose rodent carcinogen studies, both synthetic and natural pesticides are carcinogens, mutagens, teratogens, and clastogens. For example, these tests showed that 21 natural pesticides found in roasted coffee were rodent carcinogens. Thus, any concern over a possible cancer risk from minute residues of synthetic pesticides in food is unnecessary, since natural pesticides in food make up 99.99% of the toxin load we are exposed to every day.

More to the point, toxicologists challenge the relevance of these high dose rodent carcinogen tests to fairly assess the risk we face with daily exposure to much lower doses (Gold et al., 2002). They do not believe that a high dose test indicates that these natural and synthetic chemicals will cause cancer at the lower doses that we are exposed to every day.

The human physiology has evolved highly effective general mechanisms to detoxify and avoid the toxic effects from most natural and synthetic toxins, since toxins have always been a part of our environment and diet. These general mechanisms operate to detoxify synthetic chemicals just as they do to natural plant pesticides.

The principle here is that the dose makes the difference between a substance having a toxic effect, no affect, or a beneficial effect. For example, many people each year are hospitalized due to overdose of vitamins (many vitamins are natural plant chemicals), and we could erroneously conclude from these high dose poisonings that vitamins are toxic and should not be consumed, when in reality, vitamins consumed at low doses are beneficial and essential to human health and survival. In fact, many of the natural plant-produced pesticides eaten at low doses prevent cancer (Ames et al., 1990a,b; TIACT, 1998; Higgins and

Chrispeels, 2003). This fact is significant because the high-dose rodent carcinogen test results are misleading us into believing that exposure to low doses of many natural and synthetic chemicals is a health hazard, when actually these chemicals may be a benefit or of no risk—another system of testing the health risk of chemicals is required.

Some of the misconceptions and incorrect factoids about food safety and synthetic chemicals are: (i) synthetic chemicals at normal environmental exposure levels are an important cause of human cancer; (ii) reducing pesticide residues on food is an effective way to prevent diet-related cancer; (iii) synthetic chemicals are the primary source of human exposure to potential cancer hazards; (iv) the toxicology of synthetic chemicals is different from that of natural chemicals; (v) cancer risks to humans of any chemical can be assessed by standard high-dose animal cancer tests; (vi) synthetic chemicals are greater cancer hazards than natural chemicals; (vii) pesticides and other synthetic chemicals disrupt human hormones; and (viii) the current regulation of low, hypothetical risks is effective in advancing public health (Gold et al., 2002; Logomasini, 2002; Trewavas, 2003). None of these claims are true.

In view of what we have discussed in this section on issues of risk, the greatest challenge to commercial production of Bt crops is to obtain public and regulatory acceptance on a global scale, of the true benefits and risks of this technology, especially in the EU. One of the highest hurdles in meeting this challenge is overcoming the pervasive nature of the 'factoids' discussed above and the lack of discernment among those who accept them as fact.

The greatest risk with Bt crops is not exploring and developing the full value of this technology to the benefit of mankind. Our fervent desire is that environmental extremism will not become a pervasive philosophy in any society, and our governmental regulators and politicians will see that it can be a deceptive, cruel, and discriminatory belief system. Moreover, we hope governmental agencies will establish public policy on the bases of truth, not misinformation and misguided outrage.

CONTRIBUTION OF Bt CROPS TO SUSTAINABLE AGRICULTURE

Farming principles and practices are dynamic in nature and have been changing since the beginning of farming (Diamond, 1999; Acquaah, 2002; Shaver, 2003). Today, farming approaches are evolving rapidly in response to new developments in technology, information, and regulations. A central issue in modern agriculture is how to maintain or improve upon the long-term viability of agroecosystems globally (Carroll et al., 1990; Chrispeels and Sadava, 2003). Primary causes for the current loss in agroecosystems viability are: (i) soil erosion; (ii) mining of soil nutrients; (iii) salinization of soils; and (iv) loss of genetic and faunal diversity. Agriculture can also cause environmental and social problems because it directly affects human population distribution and livelihood (Cohen, 2003; Machuka, 2003).

The EPA has identified agriculture as the largest non-point source of surface water pollution in the 1980s. The primary pollutants of concern were herbicides, insecticides and nitrates from fertilizers, which also were detected in groundwater in many agricultural regions (National Research Council, 1989). Due to growing public concern over potential losses in long-term agroecosystem viability, due to the above causes, and public concern about potential environmental pollution, alternative farming practices were developed and adopted. These practices and philosophies are generally classified as Alternative Agriculture (National Research Council, 1989). Other descriptions include organic farming, regenerative agriculture, low-input agriculture and sustainable agriculture.

The philosophy and practices of sustainable agriculture emphasize that a unit of farm land is first an ecosystem, and this ecosystem should be managed to maintain or increase the long-term productivity and health of the nutrients, microorganisms, soils, plants, and animals inhabiting this system (Francis et al., 1990; Hatfield and Karlen, 1994; Shaver, 2003). Sustainable agriculture is agriculture that: (i) enhances environmental quality and the resource base on which agriculture depends; (ii) provides for basic human food and fiber needs; (iii) is economically viable, and (iv) enhances the quality of life for the farmer and society as a whole. It differs from traditional farming in that sustainable agriculture minimizes use of: (i) conventional synthetic insecticides and preplanting herbicides, (ii) conventional inorganic fertilizers; (iii) flood irrigation; and (iv) conventional tillage practices. Sustainable agriculture explores the use of; (i) crop rotations to manage problems with weeds, diseases, insects and other pests; (ii) minimum or no-till practices and contour plowing to reduce soil erosion; (iii) mulching of soil to conserve water; (iv) integrated pest management that relies heavily on the biological control of pests rather than conventional chemical control methods; and (v) the use of genetically-improved crops, including transgenic Bt crops that are resistant to insect pests and diseases (Benedict, 2003; Shaver, 2003; Machuka, 2003).

Expectations of agricultural leaders and decision makers in our society are that genetically-improved crops, especially transgenic insect, herbicide, and disease resistant crops will: (i) reduce our current reliance on conventional insecticides, fungicides and herbicides; (ii) increase the use and effectiveness of biological control of crop pests;

(iii) reduce the need for excessive tillage for weed control; (iv) increase the use of integrated pest management; (v) reduce environmental contamination from toxic agrichemicals; and (vi) increase the profitability and sustainability of world agriculture. We believe the current trends in adoption of Bt crops and the global reductions seen in conventional insecticide sprays on Bt crops show a clear progress toward a more sustainable agricultural system. In the USA, insecticide use for 2001 on Bt cotton and Bt corn was reduced by an estimated 1.87 and 2.6 million pounds of active ingredient, respectively (Gianessi et al., 2002). In China, which produces 40% of the world's cotton, 2001 estimates of insecticide use reduction on Bt cotton was 157 million pounds of the formulated product. These are reductions from 30% to 70% of current insecticide usage on conventional corn and cotton.

Reducing or eliminating broad-spectrum insecticide applications increases the stability, i.e. sustainability of the agroecosystem in a number of ways, including: (i) allowing predators and parasites of pests to respond naturally in a density-dependent manner to the presence and increase in numbers of pest insects (King and Coleman, 1989); (ii) eliminating or minimizing insecticide-induced outbreaks of pests; (iii) eliminating or minimizing outbreaks of pests that are resistant to insecticides; (iv) allowing for the survival and increasing effectiveness of insectivorous birds that respond to pest densities; (v) allowing the development and release of biological agents to control pests; (vi) encouraging the development of Bt crops with multiple stacked traits for improved efficacy and broadening the spectrum of pests controlled because government-mandated IRM programs require it (Table 2.6) (see Chapter 14; Matten, 2000); (vii) increasing the species and genetic diversity of crop plants through IRM plantings of Bt and non-Bt varieties/hybrids, and planting patterns of mosaics with untreated non-Bt crops; and (viii) conserving the biodiversity by increasing crop productivity per unit of land, thereby increasing the possibility of confining crop production to the current 1.5 billion ha. of global cultivable land where sustainable agriculture can be practiced. Some of these benefits have yet to be realized, documented, and reported as fact, but enough evidence exists to strongly suggest that most are true.

PEST-RESISTANT CROPS AND BIOLOGICAL CONTROL

Sustainability of natural and managed ecosystems is achieved, in part, by the combination of pest-resistant plant traits and biological control of pests (Price, 1986, 1997; Smith, 1989; Carroll et al., 1990). In the presence of a pest, pest-resistant varieties sustain less pest injury and produce greater yields than susceptible varieties. Bt crop varieties are pest-resistant. Properly chosen, pest-resistant varieties can also improve the effectiveness of natural enemies of pests, in the process, enhancing biological control (Fig. 2.7). For over 100 years, pest-resistant varieties have been developed by public and private research organizations worldwide in order to reduce pest problems (Smith, 1989; Pfeiffer, 2003). These organizations use many sources for resistant genes to improve crops, including wild populations of crop species, gene banks (facilities that collect and maintain genetic diversity for each crop species and closely-related species) (Clement and Quisenberry, 1999) and, more recently, widely-divergent species through biotechnology.

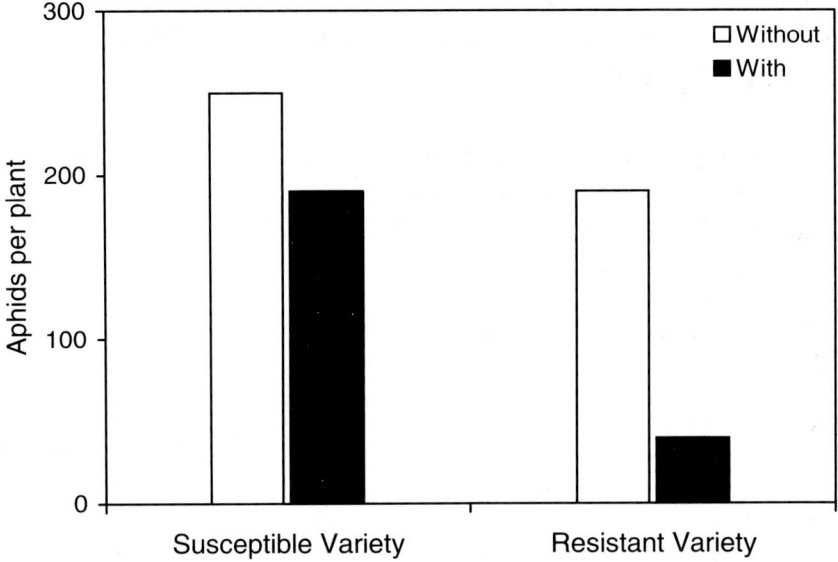

Fig. 2.7 Population size of Greenbug aphid, *Schizaphis graminum* (Rondani), feeding on a susceptible or a greenbug-resistant variety of barley, either without (white bars) or with (black bars) the greenbug parasite, *Lysiphlebus testaceipes* (Cresson) (Modified from Carroll et al., 1990)

Plants, in nature and cultivation, produce a wide array of chemical defenses—more than 50,000 compounds known as secondary plant metabolites or secondary plant chemicals—many of which repel, kill or reduce pest growth and reproductive potential (Harborne, 1982; Price, 1997; Sofos, 1998). Plants also possess morphological and physiological defenses to herbivorous pests (Table 2.7). Plants and herbivores have become specialists. Each plant species synthesizes a limited number of defensive chemicals (may be 50 to 100), and each herbivore species has adapted over a period of time to feed on and detoxify only a narrow range of plant toxins.

Table 2.7 Plant defenses against herbivores

Defenses	Effects on pests
Chemical (secondary metabolites) 　Nitrogen compounds 　Phenolics 　Polyacetylenes 　Proteins 　Terpenes 　Others	Poison and/or repel pests; or reduce pest growth rates and survival by various toxic effects or impairing digestion; may also increase pest susceptibility to natural enemies
Morphological 　Bark and cell wall thickening 　Spines and hairs 　Waxy and silica covered surfaces	Physical barriers that prevent injury, and reduce pest feeding, survival and reproduction; also may contain toxic chemicals
Physiological 　Compensate for yield loss	Plant is able to grow and reproduce in spite of pest injury

Source: Modified from Benedict (2003)

In natural plant communities, herbivore injury to plants is suppressed by a combination of defensive chemical and morphological characters. We have only begun to understand and use these defensive traits in combination with Cry proteins to suppress insect pest injury to crops. Wild plant species have utilized multiple traits for defense against insect herbivores for millions of years. It is a durable method that minimizes the potential for insect herbivores to quickly adapt to the multiple toxins. Stacking pest-resistant genes in the same crop plant is a method that plant breeders can capitalize on in order to increase the durability of Bt crops and the sustainability of agroecosystems. However, just as wild plant species and their herbivores interact over time, resulting in plants slowly producing new defenses and herbivores slowly adapting to them, man is engaged in the same warfare; as pests slowly adapt to the older pest-resistant varieties, plant breeders produce new pest-resistant varieties. In nature, this herbivore adaptation is believed to take many years, perhaps thousands (Price, 1997). Man's quest to reduce pest losses in agriculture and his daily efforts to cope with pest adaptation are also long term in nature, with no end in sight.

Natural biological control of pests is the ideal complement to pest-resistant varieties. Biological control agents put the brakes on pest population growth by responding quickly to increases in pest density, aggregating and eating more pests, and reproducing on those plants where the pests are becoming more abundant. This natural enemy response to pest density is called a density-dependent response and the more pronounced it is, the more effective the natural enemies.

Commonly, when either biological control or a pest-resistant variety is used alone, such a single strategy does not provide a high enough level of control for every pest in a farmer's field, and an insecticide or other strategy may be required. Frequently, the pests that are most poorly controlled by natural biological control are migratory pests or introduced foreign pests, like the European corn borer. Biological control agents can be roughly divided into three groups: (i) Native parasites, predators and diseases that occur naturally in the geographical region where the farmer grows his/her crop, known as natural biological control or natural enemies of pests. (ii) Introduced predators, parasites or diseases collected from foreign lands, commonly the native home of an introduced pest, are released and established in a new geographic crop production region where these biological control agents never existed before. This source of agents is known as classical biological control. After release, the introduced natural enemies are found in association with the introduced pest, and survive by feeding on the introduced pest. Eventually, the populations of the introduced pest and introduced natural enemy reach a new population balance at a lower density and hopefully the introduced pest no longer causes outbreaks and serious economic losses. Natural biological control agents such as predators and parasites respond to pest density of specific pest species in the crop since they are adapted to feed upon them and, thus, may not attack pests introduced from foreign countries. The aim of classical biological control is to permanently lower the population density of the introduced foreign pest and thus eliminate or reduce the need for conventional broad-spectrum insecticides. (iii) Predators, parasites or diseases are mass-produced commercially in factories or insectaries and then applied to the crop, like biopesticides, when pests require additional control measures. This source of biological control agents is commonly called augmentation. However, of the three general approaches to insect biological control, augmentation is the least sustainable because it requires the regular or periodic purchase of commercially-cultured natural enemies. Nonetheless, in some pest situations, this method is a highly efficacious, cost effective, and environmentally sound approach to pest management.

Predators are insects that capture and quickly kill and eat pests, rather like a wolf would capture and eat a rabbit. Parasitic insects (commonly known as parasitoids) behave differently, in that the female parasite deposits one to many eggs or immature offspring in or on the body of a single pest, the immature parasites slowly feed and grow over a period of days to weeks on the living pest, eventually killing it and becoming a free-living adult. Adult parasitic insects are commonly small wasps (Hymenoptera) that feed on nectar, and are some of the most abundant and effective of our natural enemies. Because of their cryptic life histories and small size (most are less than 1/4 inch long), parasites and predators can be easily overlooked by crop consultants untrained in their evaluation.

Biological control and pest-resistant crop varieties are two of the most valuable, ecological, sustainable, inexpensive, and successful long-term pest management strategies available to the farmer. In many cases, the farmer needs to do nothing to make insect-resistant plants and biological control work on his farm except plant an insect-resistant variety/hybrid and minimize the use of conventional broad-spectrum insecticides. Together, these two strategies form the foundation for IPM programs and set the stage by influencing pest and beneficial insect presence, abundance and level of pest injury throughout the season. As discussed earlier, the use of conventional insecticides has frequently impaired the effectiveness of biological agents (Doutt and Smith, 1971; Croft, 1990; Ehler and Bottrell, 2000). Production of stacked gene pest-resistant Bt crops that reduce the use of conventional broad-spectrum insecticides and encourage beneficial insects to flourish and reach their full capacity to suppress insect pests is a quantum leap towards more ecologically-based and sustainable production agriculture. Improving the effectiveness of plant resistance to pests and biological control is essential for IPM to realize its full potential as a long-term ecologically-based insect pest suppression system.

OVERVIEW AND CONCLUSIONS

The commercial production of transgenic Bt corn and cotton crops that produce their own insecticidal Cry proteins is clearly a new paradigm in managing insect pests. This approach towards insect pest management is unique in the sense that the crop plant produces and delivers the insecticide in a manner that is harmless to man, to beneficial fauna that naturally suppress pest insects, and to the environment. Moreover, it is unique for a biopesticide to achieve such high efficacy and long-term suppression, especially against cryptic lepidopteran insects like the pink bollworm that live and feed inside plant tissues. The widespread and rapid adoption of this technology marks a significant departure from our past dependence on broad-spectrum synthetic topical and systemic insecticides. Worldwide, more than 6 million farmers have adopted Bt crops on 8% of all corn and 17% of all cotton land in production in 2002 (James, 2003). These first-generation Bt crops possess a single Cry protein and have a very limited number of insect pests they can effectively control. Second-generation Bt crops, entering the commercial markets in 2003 through 2005, are significantly better at controlling the pests targeted by the first generation products and they are also effective against a wider range of pests, making them even more valuable to farmers.

There are a number of compelling reasons for growing Bt crops. Some of these are: (i) They increase crop yields and thus fulfill the need for global food, feed and fiber security. (ii) They conserve biodiversity by increasing crop productivity per unit of land, helping to keep crop production confined to the current 1.5 billion ha. of global cultivable land where sustainable agriculture can be practiced (James, 2002b). (iii) They encourage more efficient use of external inputs and create a more sustainable environment by reducing conventional insecticide use. (iv) They improve the health and safety of the farming environment for farmers, farm families, and others living in rural communities. (v) They provide economic benefits to farmers and farm communities globally; thus improving their health, education and welfare. This is empowering technology for subsistence farmers, and especially women farmers, in developing countries in an effort to reduce the effects of poverty, hunger and hopelessness, and improve their lives. (vi) Bt crops also provide economic benefits to consumers globally (by lowering prices of agricultural products), seed suppliers of Bt crops (by increasing the value of planting seed) and countries and governments that allow the commercial production of Bt crops (by increasing crop production and income per unit of land, increasing health and income of citizens, and providing more commodities for export trade in international markets).

Bt crops are valuable tools to effectively and economically reduce crop losses to insect pests in farming, but they are only one of the tools in the IPM toolbox. To be most effective, Bt crops must be properly integrated with other tools in a carefully crafted IPM package designed for specific production systems. This is due to the unique set of biological, educational, economic, and cultural constraints faced by individual farmers in their respective farming communities. Moreover, for Bt crops to be most effective in delivering pest control and increased crop yields, government regulators and the public must work together in harmony and goodwill towards this end.

For subsistence farmers in developing countries, Bt crops have much to offer. However, for the full benefit of Bt crops to be realized, these resource-poor farmers need much more. Specifically farmers and agricultural communities in developing nations need: improvements in infrastructure; credit for small farmers; progressive changes that target women; stable societies and economic policies; anti-corruption policies; medical assistance; technology transfers; and greater and better aid, trade, investment, and cooperation from developed countries. Fighting hunger and poverty require much more than the benefits of Bt crops (Thompson 2002; Lacy, 2003; Cohen, 2003).

Despite several potential dangers recently reported by researchers (e.g. Monarch butterfly), and used by some activist organizations and the news media to vilify Bt crops, Cry proteins in Bt crops and topical Bt sprays are still among the safest insecticides to man and the environment, ever devised. Some opponents of Bt crops would have the world apply the precautionary principle that states, 'take no action unless you are certain that it will do no harm' (Thompson, 2002; Lacy, 2003). We can never prove that a technology is absolutely harmless; thus, the use of this principle would result in our never growing Bt crops. But not using Bt crops also has risks! Without Bt crops, there is a greater reliance on conventional insecticides which have known toxic effects on people and wildlife. These toxins disrupt biological control of insect pests, hinder development of more biologically-based IPM programs, and may not provide the increases in yields that are required to feed an expanding human population. Today, the majority of regulators and the public believe that replacing conventional insecticides with Bt crops is a reasonable and relatively risk-free course. For the last seven years, they have been proven right. However, global opposition to Bt crops is growing, especially in Europe where the public views Bt crops as a focus for most of what they oppose in global society from global warming through corporate greed to corporate farming. The greatest challenge to the continued deployment and development of Bt crops is coping with this rising global opposition. The greatest risk with Bt crops may be failing to effectively deploy this very promising, safe, beneficial, and biologically-based pest control tool. The loss of Bt crops from production agriculture would have profoundly damaging economic and environmental consequences globally, resulting from our return to conventional insecticides and the loss of much natural habitat to farming in order to meet the needs of an increasing human population. This is especially true as we anticipate the next generation of Bt crops, which offer radical improvements in pest control over current Bt crops, and a level of season-long pest suppression that is truly remarkable. Since the development and use of Bt crops is so new and unique, we are just beginning to understand how biological it is, and how just valuable it is to improving and realizing the full potential of Integrated Pest Management and crop improvement to meet societies needs (Persley and Siedow, 1999; James, 2002a).

We encourage the effective education of the public on the safety, value and benefits of Bt crops as an ecologically- and biologically-based pest control tool. Only through knowledge and understanding of the facts will the public gain confidence in the safety and benefits of Bt crops and foods.

ACKNOWLEDGMENTS

We wish to thank Debbie Richter and Carolyn Villanueva, and Drs Todd DeGooyer, Les Ehler, Leigh English, John Greenplate, Marvin Harris, Opender Koul, Roger Leonard, Brad Minton, Walt Mullins, Roy Parker, Eric Sachs, Marlin Rice, Tony Shelton, Wally Thingelstad, and Jennifer Thompson for their able assistance in the development and review of this chapter.

REFERENCES

Acquaah, G., 2002 Principles of Crop Production: Theory, Techniques, and Technology. Prentice Hall, Upper Saddle River, NJ, USA.

Adamczyk, J .J., Hardee, D. D. and Adams, L. C., 2001, Correlating differences in larval survival and development of bollworm (Lepidoptera: Noctuidae) and fall armyworm (Lepidoptera: Noctuidae) to differential expression of *Cry*IA delta-endotoxin in various plant parts among commercial cultivation of transgenic *Bacillus thuringiensis* cotton. J. Econ. Entomol., **94**, 284-290.

AGBIOS, 2003, Essential biosafety data and background for event TC1507. AGBIOS Company, Merrickville, Ontario, Canada. http://www.agbios.com/dbase.php?action=ShowProd&data=TC1507&frmat=LONG

AgBioWorld, 2003, AgBioView Newsletter. AgBioWorld Foundation, Tuskegee Institute, AL, USA. http://www.agbioworld.org/biotech_info/pr/jesuits.html and http://www.agbioworld.org

Alam, M. F., Datta, K., Abrigo, E., Oliva, N., Tu, J., Virmani, S. S. and Datta, S. K., 1999, Transgenic insect-resistant maintainer line (IR68899B) for improvement of hybrid rice. Plant Cell Reports, **18**, 572-575.

Alinia, F., Ghareyazie, B., Rubia, L., Bennett, J. and Cohen, M. B., 2000, Effect of plant age, larval age, and fertilizer treatment on resistance of a *cry1Ab*-transformed aromatic rice to lepidopterous stem borers and foliage feeders. J. Econ. Entomol., **93**, 484-493.

Allen, C. T., Herrera, S. E., Smith, L. E. and Patton, L. W., 2002, Status of the pink bollworm suppression/eradication program in Texas. Proc. Beltwide Cotton Conferences. National Cotton Council of America, Memphis, TN, USA.

Altman, D. W., Benedict, J. H. and Sachs, E. S., 1996, Transgenic plants for the development of durable insect resistance. Ann. N.Y. Acad. Sci., **792**, 106-114.

Ames, B. N., Profet, M. and Gold, L. S., 1990a, Dietary pesticides (99.99% all natural). Proc. Natl. Acad. Sci. USA, **87**, 7777-7781.

Ames, B. N., Profet, M. and Gold, L. S., 1990b, Nature's chemicals and synthetic chemicals: comparative toxicology. Proc. Natl. Acad. of Sci. USA, **87**, 7782-7786.

Ames, B. N., Gold, L. S. and Willett, W. C., 1995, The causes and prevention of cancer. Proc. Natl. Acad. Sci. USA, **92**, 5258-5265.

Bailey, R., 2002, Global Warming and Other Eco-Myths, Prima Publisher-Random House, NY, USA.

Baum, J. A., Johnson, T. B. and Carlton, B. C., 1999, *Bacillus thuringiensis* natural and recombinant bioinsecticide products. *In* F.R. Hall and J.J. Menn (eds), Methods in Biotechnology, Vol. 5: Biopesticides Use and Delivery. Humana Press, Totowa, NJ, USA, pp. 189-209.

Benedict, J.H., 2003, Strategies for controlling insect, mite, and nematode pests. In: M. J. Chrispeels and D. E. Sadava (eds), Plants, Genes, and Crop Biotechnology, 2ed., Jones and Bartlett Publishers, Inc., Sudbury, MA, USA, pp. 414-445.

Benedict, J. H. and Altman, D. W., 2001, Commercialization of transgenic cotton expressing insecticidal crystal proteins. In: J. Jenkins N. and Saha S. (eds), Genetic Improvement of Cotton: Emerging Technologies, Science Publishers, Inc., Enfield, NH, USA, pp. 137-2001.

Benedict, J. H., El-Zik, K. M., Oliver, L. R., Roberts, P. A. and Wilson, L. T., 1989, Economic injury levels and thresholds for pests of cotton. In: R. E. Frisbie, K. M. El-Zik and L. T. Wilson (eds), Integrated Pest Management Systems and Cotton Production, John Wiley & Sons, NY, USA, pp. 121-153.

Berryman, A. A., 1987, The theory and classification of outbreaks. In Barbosa P. and Schultz J. C. (eds), Insect Outbreaks, Academic Press, NY, USA, pp.3-30.

Betz, F. S., Hammond, B. G. and Fuchs, R. S., 2000, Safety and advantages of *Bacillus thuringiensis*-protected plants to control insect pests. Reg. Toxicol. Pharmacol, **32**, 156-179.

Bonny, S., 2003, Why are most Europeans opposed to GMOs? Factors explaining rejection in France and Europe. Electronic J. Biotechnol., **6**, 50-71. http://www.ejbiotechnology.info/content/vol6/issue1/full/4/index.html

Cannon, R. J. C., 2000, Bt transgenic crops: Risks and benefits. Integrated Pest Management Reviews, **5**, 151-173.

Carozzi, N. and Koziel, M., 1997, Advances in Insect Control: The Role of Transgenic Plants. Taylor & Francis, Bristol, PA, USA.

Carpenter, J., Felsot, A., Goode, T., Hammig, M., Onstad, D. and Sankula, S., 2002, Comparative Environmental Impacts of Biotechnology-derived and Traditional Soybean, Corn, and Cotton Crops, Council for Agricultural Science and Technology, Ames, IA, USA. http://www.cast-science.org/cast/biotech/pubs/biotechcropsbenefit.pdf

Carriere, Y., Ellers-Kirk, C., Sisterson, M., Antilla, L., Whitlow, M, Dennehy, T. and Tabashnik, B., 2003, Long-term regional suppression of pink bollworm by *Bacillus thuringiensis* cotton. Proc. Natl. Acad. Sci. USA, **100**, 1519-1523.

Carroll, C. R., Vandermeer, J. H. and Rosset, P., 1990, Agroecology, McGraw-Hill, NY, USA.

Carson, R., 1962, Silent Spring, Hamilton, London, UK. (Reprinted in 2002, Houghton Mifflin Co., NY, USA).

Casida, J. E. and Quistad, G. G., 1998, Golden age of insecticide research: past, present, or future? Annu. Rev. Entomol., **43**, 1-16.

CAST, 2001, Evaluation of the U.S. regulatory process for crops developed through biotechnology. Issue Paper 19. Council for Agricultural Science and Technology (CAST), Ames, IA, USA, 14 pp. http://www.cast-science.org/

CAST, 2003, *Mycotoxins:* Risks in plant, animal, and human systems. Report 139. Council for Agricultural Science and Technology (CAST), Ames, IA USA. 199 pp. http://www.cast-science.org/

CGIAR, 1999, Summary Report of a Consultative Group on International Agricultural Research (CGIAR)/NAS International Conference on Biotechnology. Washington, D.C., October 21-22, 1999. CGIAR Secretariat, The World Bank, Washington, DC, USA, 10 pp.

Chrispeels, M. J. and Sadava, D., 2003, Development, productivity, and sustainability of crop production. In: M. J. Chrispeels and D. E. Sadava (eds.), Plants, Genes, and Crop Biotechnology, 2 ed., Jones and Bartlett Publishers Inc., Sudbury, MA, USA, pp. 52-75.

Clark, J. M. and Yamaguchi, I., 1999, Agrichemical Resistance: Extent, Mechanism, and Detection, American Chemical Society Symposium Series, No. 808, Oxford Univ. Press, NY.

Clark, T. L., Foster, J. E., Kamble, S. T. and Heinrichs, E. A., 2000, Comparison of Bt maize and conventional measures for control of the European corn borer. J. Entomol. Sci., **35**, 118-128.

Clement, S. L. and Quisenberry, S. S., 1999, Global Plant Genetic Resources for Insect-Resistant Crops, CRC Press, Boco Raton, FL.

Cohen, M. C., 2003, Food security: why do hunger and malnutrition persist in a world of plenty? In: Chrispeels M. J. and. Sadava D. E (eds.), Plants, Genes, and Crop Biotechnology, 2ed., Jones and Bartlett Publishers, Inc., Sudbury, MA, USA, pp. 76-99.

Conko, G. and Prakash, C. S., 2002, The attack on plant biotechnology. In: Bailey R. (ed.), Global Warming and Other Eco-Myths, Prima Publish-Random House, NY, pp. 179-217.

Coots, J. R., 1994, Risks from natural versus synthetic insecticides. Annu. Rev. Entomol., **39**, 489-515.

Coviella, C. E., Morgan, D. J. W. and Trumble, J. T., 2000, Interactions of elevated CO_2 and nitrogen fertilization: effects on production of *Bacillus thuringiensis* toxins in transgenic plants. Environ. Entomol., **29**, 781-787.

Crickmore, N., Zeigler, D. R., Feitelson, E., Schnepf, E., Van Rie, J., Lereclus, D., Baum, J. and Dean, D. H., 1998, Revision of the nomenclature for the *Bacillus thuringiensis* crystal proteins. Microbiol. Mol. Biol. Rev., **62**, 807-813. (The most current nomenclature can be found at: http://www.biols.susx.ac.uk/Home/Neil_Crickmore/Bt/index.html).

Croft, B. A., 1990, Arthropod Biological Control Agents and Pesticides, John Wiley & Sons, NY.

Denholm, I., Pickett, J. A. and Devonshire, A. L., 1999, Insecticide Resistance: From Mechanisms to Management, CAB International Publishing, Cambridge, MA, USA.

Denno, R. F. and McClure, M. S., 1983, Variable Plants and Herbivores in Natural and Managed Systems, Academic Press, NY.

Dhadialla, T. S., Carlson, G. R., and Le, D. P., 1998, New insecticides with ecdysteroidal and juvenile hormone activity. Annu. Rev. Entomol., **43**, 545-569.

Diamond, J., 1999, Guns, Germs, and Steel: The Fates of Human Societies, W. W. Norton and Company, NY.

Doutt, R. L. and Smith, R. F., 1971, The pesticide syndrome–diagnosis and suggested prophylaxis. In: Huffaker C. B. (ed.), Biological Control, Plenum Press, NY, pp. 3-15.

Dow AgroSciences, 2003, Dow AgroSciences Selects WideStrike® as Brand Name for Cotton Insect Protection Trait. News Release. http://www.dowagro.com/newsroom/news/010803.htm

Dowd, P. F., 2001, Biotic and abiotic factors limiting efficacy of Bt corn in indirectly reducing mycotoxin levels in commercial fields. J. Econ. Entomol., **94**, 1067-1074.

Edge, J. M., Benedict, J. H., Carroll, J. P. and Reding, K., 2001, Bollgard cotton: An assessment of global economic, environmental, and social benefits (CONTEMPORARY ISSUES). J. Cot. Sci., **5**, 121-136. http://www.jcotsci.org/2001/Issue02/html/page121.html

Ehler, L.E. and Bottrell, D. G., 2000, The illusion of integrated pest management. Issues Sci. Technol., **16**, 61-64. http://www.nap.edu/issues/16.3/ehler.htm

El-Lissy, O., Staten, R. T. and Grefenstette, B., 2002, Pink bollworm eradication plan in the U.S, Proceedings Beltwide Cotton Conferences, National Cotton Council, Memphis, TN, USA.

Epstein, L. and Bassein, S., 2003, Patterns of pesticide use in California and the implications for strategies for reduction of pesticides. Annu. Rev. Phytopathol., **41**, 23.1-23.25.

Estruch J. J., Warren, G. W., Mullins, M. A., Nye, G. J., Craig, J. A. and Koziel, M. G., 1996, Vip3A, a novel *Bacillus thuringiensis* vegetative insecticidal protein with a wide spectrum of activities against lepidopteran insects. Proc. Natl. Acad. Sci. USA, **93**, 5389-5394.

FAO, 2003, Global integrated pest management facility at Food and Agriculture Organization of the United Nations. http://www.fao.org/ag/agp/agpp/IPM/gipmf/home.htm

FAO/WHO, 2001, Evaluation of Allergenicity of Genetically-Modified Foods. Report of a joint FAO/WHO Expert Consultation on allergenicity of foods derived from biotechnology. 22-25 January 2001. Food and Agriculture Organization of the United Nations, Rome, Italy.

Ferro, D. N., 1987, Insect pest outbreaks in agroecosystems. In: Barbosa P. and Schultz J. C. (eds), Insect Outbreaks, Academic Press, NY, pp. 195-215.

Fitt, G. P. and Wilson, L. J., 2000, Genetic engineering in IPM: Bt cotton. In: Kennedy G. and Sutton T. (eds), Emerging Technologies for Integrated Pest Management— Concepts, Research and Implementation. Proceedings of a Conference, Raleigh, NC, USA. American Phytopathological Society (APS Press), St. Paul, MN, pp. 108-125.

Francis, C. A., Flora, C. B. and King, L. D., 1990, Sustainable Agriculture in Temperate Zones. John Wiley & Sons, NY.

Frankenhuyzen, K. V., 1993, The challenge of *Bacillus thuringiensis*. In: P.F. Entwistle, J.S. Cory, M.J. Bailey and S. Higgs (eds.), *Bacillus thuringiensis*, An Environmental Biopesticide: Theory and Practice, John Wiley & Sons, NY, pp. 1-35.

Frisbie, R. E. and Adkisson, P. L., 1985, IPM: Definitions and current status in U.S. agriculture. In: Hoy M. A. and Herzog D. C. (eds), Biological Control in Agricultural IPM Systems. Academic Press, NY, pp. 41-51.

Frisbie, R. E., El-Zik, K. M. and Wilson, L. T., 1989, Integrated Pest Management Systems and Cotton Production, John Wiley & Sons, NY.

Georghiou, G. P., 1986, The magnitude of the resistance problem. In: Pesticide Resistance: Stategies and Tactics for Management, National Academy Press, Washington, DC, pp. 14-43.

Georghiou, G. P. and Lagunes-Tejada, A., 1991, The occurrence of resistance to pesticides in arthropods. An index of cases reported through 1989, FAO, Rome, Italy.

Gianessi, L. P., Silvers, C. S., Sankula, S. and Carpenter, J. E., 2002, Plant Biotechnology: Current and Potential Impact for Improving Pest Management in U.S. Agriculture. An Analysis of 40 Case Studies. National Center for Food and Agricultural Policy, Washington, DC. http://www.ncfap.org/

Gold, L. S., Slone, T. H., Manley, N. B. and Ames, B. N., 2002, Misconceptions About the Causes of Cancer. The Fraser Institute, Centre for Studies in Risk, Regulation and Environment. Vancouver, British Columbia, Canada. 141 pp.

Gore, J., Leonard, B. R. and Gable, R. H., 2003, Distribution of bollworm, *Helicoverpa zea* (Boddie) injured reproductive structures on genetically engineered *Bacillus thuringiensis* var. *kurstaki* cotton. J. Econ. Entomol., **96**, 699-705.

Gould, F., 1998, Sustainability of transgenic insecticidal cultivars: Integrating pest genetics and ecology. Annu. Rev. Entomol., **43**, 701-726.

Greenplate, J. T., 1997, Response to reports of early damage in 1996 commercial Bt transgenic cotton (Bollgard®) plantings. Soc. Invert. Pathol. Newsletter, **29**, 15-18.

Greenplate, J. T., 1999, Quantification of *Bacillus thuringiensis* insect control protein Cry1Ac over time in Bollgard cotton fruit and terminals. J. Econ. Entomol., **92**, 1377-1383.

Greenplate, J. T., 2003, Personal communications, Monsanto Company, St. Louis, MO, USA.

Greenplate, J. T., Mullins, J. W., Penn, S. R. and Bickel, A., 2003a, Assessment of Cry1A levels in transgenic cotton varieties destined for commercialization: 2001 gene equivalency trials. J. Cot. Sci. (In press). http://journal.cotton.org/

Greenplate, J. T., Mullins, J. W., Penn, S. R., Dahm, A., Rech, B. J., Osborn, J. A., Rahn, P. R., Ruschke, R. L. and Shappley, Z. W., 2003b, Partial characterization of cotton plants expressing two toxin proteins from *Bacillus thuringiensis*: relative toxin contribution, toxin interaction, and resistance management. J. Appl. Ent., **127**, 340-347.

Hall, F. R. and Menn, J. J., 1999, Biopesticides, Use and Delivery, Vol. **5**, Methods in Biotechnology, Humana Press, Totowa, NJ.

Harborne, J. B., 1982, Introduction to Ecological Biochemistry. (2nd. ed.). Academic Press, NY.

Harris, M. K., 2000, Impact of integrated pest management on academia in the United States. Am. Entomol., **46**, 217-220.

Harris, M.K., 2001, IPM, what has it delivered? Plant Disease, **85**, 112-121.

Hatfield, J. L. and Karlen, D. L., 1994, Sustainable Agriculture Systems. CRC Press Inc., Boca Raton, FL.

Herdt, R. W., 1991, Research priorities for rice biotechnology. In: G. S. Kush and G. H. Toenniessen (eds.), Rice Biotechnology, CAB International, Wallingford, UK, pp. 19-54.

Higgins, T. J. and Chrispeels, M. C., 2003, Plants in human nutrition and animal feed. In: Chrispeels M. J. and Sadava D. E. (eds), Plants, Genes, and Crop Biotechnology, 2nd. ed., Jones and Bartlett Publishers Inc., Sudbury, MA, pp. 152-181.

Higley, L. G. and Pedigo, L. P., 1996, Economic Thresholds for Integrated Pest Management, University of Nebraska Press, Lincoln, NE, USA.

Hilbeck, S., Baumgartner, M., Freid, P. M. and Bigler, F., 1998, Effects of *Bacillus thuringiensis* corn-fed prey on mortality and development time of immature *Chrysoperla carnae*. Environ. Entomol., **27**, 480-487.

IFIC, 2001, Understanding Food Allergy. IFIC Review. International Food Information Council Foundation, Washington, DC. http://ific.org/relatives/17300.PDF

James, C., 2000, Global Review of Commercialized Transgenic Crops: 1999. ISAAA Briefs No. 22. ISAAA, Ithaca, NY. http://www.isaaa.org/

James, C., 2001, Global Review of Commercialized Transgenic Crops: 2000. ISAAA Briefs No. 23. ISAAA, Ithaca, NY. http://www.isaaa.org/

James, C., 2002a, Global Review of Commercialized Transgenic Crops: 2001 Feature: Bt Cotton. ISAAA Briefs No. 26. ISAAA, Ithaca, NY. http://www.isaaa.org/

James, C., 2002b, Preview: Global Status of Commercialized Transgenic Crops: 2002. ISAAA Briefs No. 27. ISAAA, Ithaca, NY. http://www.isaaa.org/

James, C., 2003, Global Review of Commercialized Transgenic Crops: 2002 Feature: Bt Maize. ISAAA Briefs No. 29. ISAAA, Ithaca, NY. htpp://www.isaaa.org/

King, E. G. and Coleman, R. J., 1989, Potential for biological control of *Heliothis* species. Annu. Rev. Entomol., **34**, 53-75.

King, E. G., Phillips, J. R. and Coleman, R. J., 1996, Cotton Insects and Mites: Characterization and Management. The Cotton Foundation Reference Book Ser. No. 3, The Cotton Foundation, Memphis, TN, USA.

Kogan, M., 1998, Integrated pest management: Historical perspectives and contemporary developments. Annu. Rev. Entomol., **43**, 243-270.

Krattiger, A. F., 1997, Insect resistance in crops: A case study of *Bacillus thuringiensis* (Bt) and its transfer to developing countries. ISAAA Briefs No. 2, ISAAA, Ithaca, NY. http://www.isaaa.org/

Lacy, P. G., 2003, Deploying the full arsenal: Fighting hunger with biotechnology. SAIS Review, **23**, 181-202. http://www.agbioworld.org/pdf/peterlacy.pdf

Lauer, J. and Wedberg, J., 1999, Grain yield of initial Bt corn hybrid introductions to farmers in the northern corm belt. J. Prod. Agric., **12**, 373-376.

Leonard, B. R., Gable, R. and Gore, J., 2003, 2002 Update: Bt transgenic cotton. Cotton Incorporated Mid-South Conference. Cotton Incorporated, Cary, NC, USA.

Logomasini, A., 2002, Chemical warfare: Ideological environmentalism's quixotic campaign against synthetic chemical. In: Bailey R. (ed.), Global Warming and Other Eco-Myths. Prima Publish-Random House, NY, pp. 149-177.

Losey, J. E., Rayor, L. S. and Carter, M. E., 1999, Transgenic pollen harms monarch larvae. Nature, **399**, 214.

Mabonga, S., 2003, African Countries Query EU's Move on GM Foods. Biosafety News, **38** February/March (Kenya) http://www.biosafetynews.com/febmarch03/bi.htm

Machuka, J., 2003, Developing food production systems in sub-Saharan Africa. In: Chrispeels M. J. and Sadava D. E. (eds), Plants, Genes, and Crop Biotechnology, 2ed., Jones and Bartlett Publishers, Inc., Sudbury, MA, pp. 100-123.

MacIntosh, S. C., Stone, T. B., Sims, S. R., Hunst, P. L., Greenplate, J. T., Marrone, P. G., Perlak, F. J., Fischhoff, D. A. and Fuchs, R. L., 1990, Specificity and efficacy of purified *Bacillus thuringiensis* proteins against agronomicly important insects. J. Invertebr. Pathol., **56**, 258-266.

Mascarenhas, V. J., Shotkoski, F. and Boykin, R., 2003, Field performance of Vip cotton against various lepidopteran cotton pests in the USA. Proceedings Beltwide Cotton Conferences, **5**, 1316-1322.

Matten, S. R., 2000, EPA regulation of transgenic pesticidal crops and insect resistant management of B. T. cotton. Proceedings of the Beltwide Cotton Conference, National Cotton Council, Memphis, TN, **1**, 71-76.

Matthews, G. A. and Turnstall, J. P., 1994, Insect Pests of Cotton, CAB International., Wallingford, UK.

McClintock, J. T., Schaffer, C. R. and Sjoblad, R. D., 1995, A comparative review of the mammalian toxicity of *Bacillus thuringiensis*-based pesticides. Pestic. Sci., **45**, 95-105.

Mengech, A. N., Saxena, K. N. and Gopatan, H. N. B., 1996, Integrated Pest Management in the Tropics: Current Status and Future Prospects, John Wiley & Sons, NY.

Metcalf, R. L. and Luckmann, W. H., 1994, Introduction to Insect Pest Management (3rd edition), John Wiley & Sons, NY.

Metcalf, R. L. and Metcalf, R. A., 1993, Destructive and Useful Insects: Their Habits and Control (5th edition), McGraw-Hill, NY.

Moellenbeck, D. J., Peters, M. L., Bing, J. W., Rouse, J. R., Higgins, L. S., Sims, L., Nevshemal, T., Marshall, L., Ellis, R. T., Bystrak, P. G., Lang, B. A., Stewart, J. L., Kouba, K., Sondag, B., Gustafson, V., Nour, K., Xu, D., Swenson, J., Zhang, J., Czapla, T., Schwab, G., Jayne, S., Stockhoff, B. A., Narva, K., Schnepf, H. E., Stelman, S. J., Poutre, C., Koziel, M. and Duck, N., 2001, Insecticidal proteins from *Bacillus thuringiensis* protected corn from corn rootworms. Nat. Biotech., **19**, 668-672.

Monsanto Company, 2002, Insect Efficacy Testing with Bollgard® *II Cotton*. Public Interest Document, submitted to EPA. Monsanto Co., St. Louis, MO, USA.

Monsanto Company, 2003, Personal communications with Todd DeGooyer, Walt Mullins and Eric Sachs. Monsanto Co., St. Louis, MO, USA.

Moore, P., 2003, Environmentalism for the 21st century. http://www.greenspirit.com/

Moore, G. C., Benedict, J. H., Fuchs, T. W. and Friesen, R. D., 1999, Bt Cotton Technology in Texas: A Practical View. *L-5169* (Revised). Tex. Agric. Ext. Serv. Texas A&M University, College Station, TX, USA.

Morris, J. and Bate, R., 1999, Fearing Foods, Butterworth, Oxford, UK.

MSUES, 2001, Managing insects attacking corn. Publication 899. Miss. State Univ. Ext. Serv., Starkville, MS, USA. http://msucares.com/pubs/publications/pub899.pdf

Munkvold, G. .P., Hellmich, R. L. and Rice, L. G., 1999, Comparison of fumonisin concentrations in kernels of transgenic Bt maize hybrids and non-transgenic hybrids. Plant Dis., **83**, 130-138.

National Research Council, 1989, Alternative Agriculture. National Academy Press, Washington, DC.

Nill, K., 2002, Let the Facts Speak for Themselves. The Contribution of Agricultural Crop Biotechnology to American Farming. Interim Report. American Soybean Association, St. Louis, MO, USA. http://www.tomorrowsbounty.org/library/prepubvs91502a.htm

Obrycki, J. J., Losey, J. J., Taylor, O. R. and Jesse, L. C. H., 2001, Transgenic insecticidal corn: Beyond insecticidal toxicity to ecological complexity. BioSci., **51**, 353-361.

Olsen, K. M. and Daly, J. C., 2000, Plant-toxin interactions in transgenic Bt cotton and their effect on mortality of *Helicoverpa armigera* (Lepidoptera: Noctuidae). J. Econ. Entomol., **93**, 1293-1299.

Ostlie, K. R., Hutchison, W. D. and Hellmich, R. L., 1997, Bt corn and European corn borer. *BU-07055-GO* Univ. of Minn. Ext. Serv., St. Paul, MN, USA. http://www.extension.umn.edu/distribution/cropsystems/DC7055.html#ch16

Pardey, G. and Wright, B. D., 2003, Agricultural R&D, productivity, and global food prospects. In: Chrispeels M. J. and Sadava D. E. (eds), Plants, Genes, and Crop Biotechnology, (2ed.), Jones and Bartlett Publishers Inc., Sudbury, MA, USA, pp. 22-51.

Park, S. H., Park, J. and Smith, R. H., 2001, Herbicide and insect resistant elite transgenic rice. J. Plant Physiol., **158**, 1221-1226.

Peferoen, M., 1997, Insect control with transgenic plants expressing *Bacillus thuringiensis* crystal proteins. In: Carozzi N., and Koziel M. (ed.), Advances in Insect Control: The Role of Transgenic Plants, Taylor & Francis, Bristol, PA, USA, pp. 21-48.

Penn, J. B., 2000, Biotechnology in the pipeline: Sparks companies' update. Proceedings Beltwide Cotton Conferences, National Cotton Council, Memphis, TN, USA.**1**, 51-75.

Perlak, F. J., Deaton, R. W., Armstrong, T. A., Fuchs, R. L., Sims, S. R., Greenplate, J. T. and Fischhoff, D. A., 1990, Insect resistant cotton plants. Bio/Tech., **8**, 939-943.

Perlak, F. J., Fuchs, R. L., Dean, D. A., McPherson, S. L and. Fischhoff, D. A., 1991, Modification of the coding sequence enhances plant expression of insect control protein genes. Proc. Natl. Acad. Sci. USA., **88**, 3324-3328.

Perlak, F. J., Oppenhuize, M., Gustafson, K., Voth, R., Sivasupramaniam, S., Herring, D., Carey, B., Ihrig, R. A. and Roberts, J. K. (2001) Development and commercial use of Bollgard® cotton in the USA—Early promises versus today's reality. The Plant J., 27, 489-501.

Persley, G. J., 1996, Biotechnology and Integrated Pest Management, CAB International, Oxon, UK.

Persley, G. J. and Siedow, J. N., 1999, Applications of biotechnology to crops: Benefits and risks. Issue Paper 12, Council for Agricultural Science and Technology, Ames, IA, USA. 8 pp. http://www.cast-science.org

Pfadt, R. E., 1985, Strategies of insect control. In: Pfadt R. E. (ed.), Fundamentals of Applied Entomology, Macmillan, NY, pp.179-202.

Pfeiffer, T. W., 2003, From classical plant breeding to modern crop improvement. In: Chrispeels M. J. and Sadava D. E. (eds.), Plants, Genes, and Crop Biotechnology, (2ed.), Jones and Bartlett Publishers, Sudbury, MA, USA, pp. 360-389.

Pray, C., Huang, J., Hu, R., and Rozzelle, S., 2002, Five years of Bt cotton in China—The benefits continue. The Plant J., **31**, 423-430.

Price, P. W., 1986, Ecological aspects of host plant resistance and biological control: Interactions among three trophic levels. In: Boethel D. J. and Eikenbary R. D. (eds), Interactions of Plant Resistance and Parasitoids and Predators of Insects, John Wiley & Sons, NY, pp. 12-30.

Price, P. W., 1997, Insect Ecology, John Wiley & Sons, NY.

Qaim, M. and Zilberman, D., 2003, Yield effects of genetically-modified crops in developing countries. Science, **299**, 900-902.

Rice, W. C. and Choo, H. Y., 2000, Rice pests. In: Lacey L. L. and Kaya H. K. (eds.), Field Manual of Techniques in Invertebrate Pathology, Kluwer Academic Publishers, NY, pp. 425-465.

Roush, R. T., 1999, Strategies for resistance management. In: Hall F. R. and Menn J. J. (eds), Biopesticides: Use and Delivery, Humana Press, Totowa, NJ, pp. 575-593.

Roush, R. T. and Tabashnik, B. E., 1991, Pesticide Resistance in Arthropods, Kluwer Academic Publishers, NY.

Sachs, E. S., Benedict, J. H., Stelly, D. M., Taylor, J. F., Altman, D. W., Berberich, S. A and Davis, S. K., 1998, Expression and segregation of genes encoding CryIA insecticidal proteins in cotton. Crop Sci., **38**, 1-11.

Sachs, E. S., Benedict, J. H., Taylor, J. F., Stelly, D. M., Davis, S. K. and Altman, D. W., 1996 Pyramiding CryIA(b) insecticidal protein and terpenoids in cotton to resist tobacco budworm (Lepidoptera: Noctuidae). Environ. Entomol., 25, 1257-1266.

Sadava, D.E., 2003, Human population growth: lessons from demography. In: Chrispeels M. J. and Sadava D. E. (eds), Plants, Genes, and Crop Biotechnology, (2ed.), Jones and Bartlett Publishers, Sudbury, MA, USA., pp. 1-21.

Sears, M. K., Hellmich, R. L., Stanley-Horn, D. E., Oberhauser, K. S., Pleasants, J. M., Mattila, H. R., Siegfried, B. D. and Dively, G. P., 2001, Impact of Bt corn pollen on monarch butterfly populations: A risk assessment. Proc. Natl. Acad. Sci USA., **98**, 11937-11942.

Sedlacek, J. D., Komaravalli, S. R., Hanley, A. M., Price, B. D. and Davis, P. M., 2001, Life history attributes of Indian meal moth (Lepidoptera: Pyralidae) and Angoumois grain moth (Lepidoptera: Gelechiidae) reared on transgenic corn kernels. J. Econ. Entomol., **94**, 586-592.

Shaver, J. M., 2003, Toward a greener agriculture. In: Chrispeels M. J. and Sadava D. E. (eds), Plants, Genes, and Crop Biotechnology, (2ed.), Jones and Bartlett Publishers, Sudbury, MA, USA., pp. 472-499.

Shelton, A. M., Tang, J. D., Roush, R. T., Metz, T. D. and Earle, E. D. (2000) Field tests on managing resistance to Bt-engineered plants. Nat. Biotech., **18**, 339-342.

Shelton, A. M., Zhao, J. Z., and Roush, R. T., 2002, Economic, ecological, food safety, and social consequences of the deployment of Bt transgenic plants. Annu. Rev. Entomol., **47**, 845-881.

Simmonds, N. W. and Smartt, J., 1999, Principles of Crop Improvement. Iowa State University Press, Blackwell Publishing, Ames, IA, USA.

Sims, M. A., Dennehy, T. J., Shriver, L., Holley, D., Carriere, Y., Tabashnik, B., 2002, Susceptibility of Arizona pink bollworm to *Cry*1Ac. Proceedings Beltwide Cotton Production Conferences, National Cotton Council, Memphis, TN, **3**, 312-314.

Smith, C. M., 1989, Plant Resistance to Insects: A Fundamental Approach, John Wiley & Sons, NY.

Sofos, J. N., 1998, Naturally-Occurring Antimicrobials in Food. Council for Agricultural Science and Technology, Ames, IA, USA. 103 pp. http://www.cast-science.org

Stern, V. M., Smith, R. F., van den Bosch, R. and Hagen, K. S., 1959, The integrated control concept. Hilgardia, 29, 81-101.

Stewart, Jr., C. N. and Wheaton, S. K., 2003, Urban myths and scientific facts about the biosafety of genetically modified (GM) crops. In: Chrispeels M. J. and Sadava D. E. (eds.), Plants, Genes, and Crop Biotechnology, (2nd ed.), Jones and Bartlett Publishers Inc., Sudbury, MA, USA, pp. 528-551.

Syngenta, 2003, Syngenta Plans to Introduce a New Choice for Transgenic Control of Worms in Cotton. News Release. http://www.syngentacropprotection-us.com/media/article.asp?article_id=303

Tabashnik, B. E., Liu, Y. -B., Malvar, T., Heckel, D.G., Masson, L. and Ferre, J., 1998, Insect resistance to *Bacillus thuringiensis*: uniform or diverse? Philos. Trans. R. Soc. Lond., **353**, 1751-1756.

TAES, 2003, Managing cotton insects in the southern, eastern and blackland areas of Texas. Publication E5. Texas A&M University Agricultural Extension Service, College Station, TX, USA. http://tcebookstore.org/tmppdfs/507787-E5.pdf

Thompson, J. A., 2002, Genes for Africa: Genetically Modified Crops in the Developing World. University of Cape Town Press, Cape Town, South Africa.

TIACT, 1998, Chemical Hormesis: Beneficial Effects at Low Exposures, Adverse Effects at High Exposures. INSIGHTS Report. Texas Institute for Advancement of Chemical Technology, Texas A&M University, College Station, TX, USA. http://www-chen.tamu.edu/TIACT/hormesisinsights.pdf

Trewavas, A., 2003, UK Organic Farming in Proper Perspective. AgBioWorld.org http://www.agbioworld.org/biotech_info/articles/orgfarmperspective.html

Tu, J., Zhang, G., Datta, K., Xu, C., He, Y., Zhang, Q., Khush, G. S. and Datta, S. K., 2000, Field performance of transgenic elite commercial hybrid rice expressing *Bacillus thuringiensis* delta-endotoxin. Nat. Biotechnol., **18**, 1101-1104.

UIES, 2003, Illinois Agricultural Pest Management Handbook. University of Illinois Extension Service, Urbana, IL, USA. http://www.ipm.uiuc.edu/pubs/iapmh/index.html

USDA, 2002, Oilseeds: World Markets and Trade. Report June-August 2002. US Department of Agriculture, Washington, DC.

USDA ARS, 2002, Butterflies and Bt Corn: Allowing Science to Guide Decisions. http://www.ars.usda.gov/sites/monarch/

US EPA, 1988, Guidance for the registration of pesticide products containing *Bacillus thuringiensis* as an active ingredient. NTIS PB 89-164198

US EPA, 1995, Pesticide fact sheet for *Bacillus thuringiensis* ssp. *kurstaki Cry*I(A)b delta-endotoxin and the genetic material necessary for the production (plasmid vector pCIB4431) in corn. EPA731-F-95-004. US EPA, Washington, DC.

US EPA, 1999, EPA and USDA position paper on insect resistance management in Bt crops. 5/27/99, minor revision, 7/12/99. US EPA Washington, DC.

US EPA, 2000, White paper on the possible presence of *Cry*9C protein in processed human foods made from food fractions produced through the wet milling of corn. US EPA, Washington, DC. http://www.epa.gov/oppbppd1/biopesticides/pips/wetmill18.pdf

US EPA, 2001a, Biopesticides registration action document: *Bacillus thuringiensis (Bt)* plant-incorporated protectants (BRAD). US EPA, Washington, DC. http://www.epa.gov/oppbppd1/biopesticides/pips/Bt_brad2/2-id_health.pdf

US EPA, 2001b, Biopesticide registration action document *Bacillus thuringiensis Cry*1F corn (BRAD). US EPA, Washington, DC. http://www.epa.gov/pesticides/biopesticides/ingredients/tech_docs/brad_006481.pdf

US EPA, 2003a, Currently registered section 3 PIP registrations. US EPA, Washington, DC. http://www.epa.gov/oppbppd1/biopesticides/pips/pip_list.htm

US EPA, 2003b, Current federal register notices of Experimental Use Permits granted for experimental Bt biopesticides. US EPA, Washington, DC. http://www.epa.gov/pesticides/biopesticides/regtools/frnotices2003.htm

US EPA, 2003c, *Bacillus thuringiensis Cry*3Bb1 Protein and the Genetic Material Necessary for its Production (Vector ZMIR13L) in Event MON863 Corn Fact Sheet. US EPA, Washington, DC. http://www.epa.gov/pesticides/biopesticides/ingredients/factsheets/factsheet_006484.htm

US House Agriculture Committee, 2002, The facts on U.S. farm policy. US Federal Government, Washington, DC. http://www.agriculture.house.gov/fbfocus.pdf

Vasquez, B. L., 1995, University of Florida Book of Insect Records. University of Florida, Gainesville, FL, USA. http://ufbir.ifas.ufl.edu/chap15.pdf

Wang, Z., Shu, Q., Ye, G., Cui, H., Wu, D., Altosaar, I. and Xia, Y., 2002, Genetic analysis of resistance of Bt rice to stripe stem borer (*Chilo suppressalis*). Euphytica, **123**, 379-386.

Warren, G. W., 1997, Vegetative insecticidal proteins: novel proteins for control of corn pests. In: Carozzi N. and Koziel M. (eds.), Advances in Insect Control. Taylor & Francis, London, UK, pp. 109-121.

Wiseman, B. R., Lynch, R. E., Plaisted, D. and Warnick, D., 1999, Evaluation of Bt transgenic sweet corn hybrids for resistance to corn earworm and fall armyworm (Lepidoptera: Noctuidae) using a meridic diet bioassay. J. Entomol. Sci., **34**, 415-425.

Williams, W. P., Buckley, P. M., Sagers, J. B. and Hanten, J. A., 1998, Evaluation of transgenic corn for resistance to corn earworm (Lepidoptera: Noctuidae), fall armyworm (Lepidoptera: Noctuidae), and southwestern corn borer (Lepidoptera: Crambidae) in a laboratory bioassay. J. Agric. Entomol., **15**, 105-112.

World Health Organization, 2000, Safety Aspects of Genetically Modified Foods of Plant Origin. Joint FAO/WHO Report, Expert Consultation on Foods Derived from Biotechnology, 29 May-2 June. http://www.who.int/fsf/GMfood/FAOWHO_Consultation_report_2000.pdf

Wraight, C., Zangeri, A. R., Carroll, M. J. and Berenbaum, M. R., 2000, Absence of toxicity of *Bacillus thuringiensis* pollen to black swallowtails under field conditions. Proc. Natl. Acad. Sci. USA, **97**, 7700-7703.

Ye, G. Y., Shu, Q. Y., Cui, H. R., Hu, C., Gao, M. W., Xia, Y. W., Cheng, X. and Altosaar, I., 2000, A leaf-section bioassay for evaluating rice stem borer resistance in transgenic rice containing a synthetic *cry1Ab* gene from *Bacillus thuringiensis* Berliner. Bull. Entomol. Res., **90**, 179-182.

Yu, C. G., Mullins, M., Warren, G. W., Koziel, M. G. and Estruch, J. J., 1997, The *Bacillus thuringiensis* vegetative insecticidal protein Vip3A lyses midgut epithelium cells of susceptible insects. Appl. Environ. Microbiol., **63**, 532-536.

Zhao, J.-Z., Li, Y. -X., Collins, H. L., Cao, J., Earle, E. D. and Shelton, A. M., 2001, Different cross-resistance patterns in diamondback moth (Lepidoptera: Plutellidae) resistant to *Bacillus thuringiensis* toxin *Cry*1C. J. Econ. Entomol., **94**, 1547-1552.

3

INSECT-RESISTANT TRANSGENIC CROPS EXPRESSING PLANT LECTINS

JESUSA CRISOSTOMO LEGASPI, B.C. LEGASPI, JR.,*
AND M. SÉTAMOU**

*USDA-ARS-CMAVE, Center for Biological Control, Florida A&M University Tallahassee, FL 32307, USA; *Department of Entomology Texas & A&M University College Station, TX 77843, USA*
**Texas Agricultural Experiment Station*
2415 East Hwy 83, Weslaco, TX 78596, USA

INTRODUCTION

Transgenic insecticidal crop cultivars are in the process of revolutionizing agriculture and are likely to become a major insect management tactic worldwide. Introducing novel resistance genes into economically-important crops can develop insect-resistant crops. This tactic has a potentially key role in integrated pest management of several important pests (Gatehouse and Gatehouse, 1998). Most of the research and development thus far has focused on the transfer of genes expressing toxins produced by the soil-borne bacterium, *Bacillus thuringiensis* (Berliner) (Bt) into food, grain and fiber plants. The insecticidal Bt d-endotoxins are a family of related proteins with different levels of activity against species of Lepidoptera, Coleoptera and Diptera (see Vaeck et al., 1987). Bt-transgenic cultivars have been produced for tobacco (*Nicotiana tabacum*), potato (*Solanum tuberosum*), cotton (*Gossypium hirsutum*), maize (*Zea mays*), rice (*Oryza sativa*), tomato (*Lycopersicon esculentum*), broccoli (*Brassica oleracea botrytis*), alfalfa (*Medicago sativa*) and other crops (Hilder and Boulter, 1999). In March 2000, Bt maize and Bt cotton each comprised 18% of the total acreages

---

*Contact through senior author

planted to these crops in the USA (USDA ERS, 2000). These acreages are expected to increase dramatically in the near future, both in developed and developing countries (Carozzi and Koziel, 1997; Gould, 1998). Field evaluations of Bt transgenic cultivars have produced mixed results. Effective pest suppression was reported for Bt cotton against cotton bollworm, *Helicoverpa zea* (Boddie) (Lepidoptera: Noctuidae) and pink bollworm, *Pectinophora gossypiella* (Saunders) (Lepidoptera: Gelechiidae), Bt potato against Colorado potato beetle, *Leptinotarsa decemlineata* (Say) (Coleoptera: Chrysomelidae), and Bt maize against European corn borer, *Ostrinia nubilalis* (Hübner) (Lepidoptera: Pyralidae) (Hilder and Boulter, 1999). However, the failure of the 800,000 ha. planted with Bt cotton against *H. zea* and *P. gossypiella* was widely reported in 1996. Furthermore, Bt cotton was also reported to have failed against *H. armigera* (Hübner) in Australia (Hilder and Boulter, 1999).

In addition to Bt endotoxins, genes expressing other insecticidal proteins have been transferred into crops of economic importance. These include: protease inhibitors, which disrupt amino acid metabolism; α-amylase inhibitors, which target carbohydrate metabolism; and, enzymes such as chitinase, which target the insect exoskeleton (Boulter, 1993; Estruch et al., 1996; Hilder and Boulter, 1999). In this chapter, we shall discuss the transfer of genes expressing secondary plant compounds called 'lectins' into crops of agricultural significance. This focus is primarily on *Galanthus nivalis* agglutinin (GNA) from the snowdrop lily, because it is an especially promising lectin with a great deal of research interest.

LECTIN TRANSGENICS: BACKGROUND AND DEVELOPMENT OF GNA-TRANSGENIC CROP CULTIVARS

Lectins are a large, heterogeneous family of carbohydrate-binding proteins. Plant lectins can be divided into different families based on protein structure, e.g. the cysteine-rich chitin-binding family (Zhu-Salzman et al., 1998). The precise functions of the different plant lectins are unclear but are believed to be important during interactions between the plant and other organisms. For example, the symbiotic relationship between *Rhizobium* and legumes is apparently facilitated by lectins (Bauchrowitz et al., 1996). Plant lectins, especially those in seeds, may play a role in defense against attack by bacteria and fungi, and predation by animals and insects (Chrispeels and Raikhel, 1991). Janzen et al. (1976) first proposed the theory that plant lectins acted as chemical defenses against insect seed predators in the case of the common bean (*Phaseolus vulgaris*). This lectin, phytohaemagluttinin (PHA), was shown to be toxic to cowpea bruchid, *Callosobruchus maculatus* (Fabricius), when incorporated into artificial diet (Gatehouse et al., 1984). A number of plant lectins exhibiting insecticidal characteristics are being evaluated as alternatives to Bt δ-endotoxins (Estruch et al., 1996). The precise mechanism of the action of insecticidal lectins is unknown. However, binding to specific carbohydrates and agglutination (fusion by adhesive substances) in the insect midgut has been clearly demonstrated. Specifically, lectins may interfere with development and structural integrity of the midgut peritrophic membrane (Harper et al., 1998); bind to glycosated targets in the insect midgut, thereby inhibiting nutrient absorption or cell disruption in the midgut; or bind or block the peritrophic membrane protecting the insect midgut surface (Fitches et al., 1997). Insecticidal activity of the lectins can be measured in two general ways (see also Boulter, 1993): (i) purified lectin or extracts from the transformed plant material is applied topically or incorporated into artificial diets and presented to the target insects; and (ii) whole plants or excized sections of the plants may be presented to insects, i.e. *in planta* methods, in order to evaluate transgenic plant lines for insect resistance.

Artificial Diet Bioassays

Evaluation of plant lectins by artificial diet bioassays is essential because development of transgenic plants is an expensive process. However, some difficulties inherent to this approach are that test insects may be unable to complete their development and the diet may be inadequate in simulating whole plant effects (Czapla, 1997). Some important lectins with demonstrated insecticidal effects are summarized in Table 3.1. To date, few lectins have been screened for insecticidal activity compared to bacterial isolates (Czapla, 1997). Possible reasons include paucity of source materials and growth media, difficulties in protein purification and amounts of purified protein required for bioassays. The majority of bioassays have been performed against the cowpea bruchid, *C. maculatus*, a key pest of cowpea, *Vigna unguiculata* (L.) (Walp.), which is an important source of dietary protein in sub-Saharan Africa (Machuka et al., 1999). Murdock et al. (1990) performed artificial diet bioassays using commercially-available plant lectins against cowpea bruchid, which remains a major storage pest of cowpea throughout the developing world. Of 17 lectins tested, the five produced a marked decline of insect development at 0.2 to 1.0% (w/w) with wheat germ (*Triticum vulgaris L.*) agglutinin (WGA) showing the greatest effect. Also working on cowpea pests, Omitogun et al. (1999) extracted insecticidal lectins from 20 insect-resistant species of African legumes, categorized as being *Vigna* or non-*Vigna* species. Lectin extracts were tested against three

Table 3.1 Lectins with demonstrated insecticidal effects against insect pests in artificial diet bioassays

Name	Plant source	Target insect	Reference(s)
1	2	3	4
Phytohemagglutinin (PHA)	Black bean (*Phaseolus vulgaris*)	Cowpea weevil (*Callosobruchus maculatus*) Potato leafhopper (*Empoasca fabae*) rice brown planthopper (*Nilaparvata lugens*) *Callosobruchus assimilis*[1]	Janzen et al. (1976) Gatehouse et al. (1984) Habibi et al. (1993) Hilder et al. (1993); Powell et al. (1993) Boulter et al. (1986)
Pea lectin	Pea (*Pisum sativum*)	Cowpea weevil Potato leafhopper	Boulter et al. (1986) Hepher et al. (1989) Habibi et al. (1993)
Maclura pomifera lectin (MPL)	Osage orange (*Maclura pomifera*)	Cowpea weevil Southern corn rootworm (*Diabrotica undecimpunctata howardi*)	Murdock et al. (1990) Czapla and Lang (1990)
	Peanut (*Arachis hypogaea*)	Cowpea weevil	Murdock et al. (1990)
Wheat germ agglutinin (WGA)	Wheat (*Triticum aesticum*)	Cowpea weevil European cornborer (*Ostrinia nubilalis*) Corn rootworm *Diabrotica sp.* rice brown planthopper Potato leafhopper rice leafhopper (*Nephotettix cincticeps*)	Huessing et al. (1991a) Czapla and Lang (1990) Czapla (1997) Hilder et al. (1993); Powell et al. (1993) Habibi et al. (1993) Hilder et al. (1993); Powell et al. (1993)

Table 3.1 contd.

[1]Possible contamination by α-amylase inhibitor

Table 3.1 contd.

1	2	3	4
Rice lectin	Rice (*Oryza sativa*)	Cabbage aphid (*Brevicoryne brassicae*)	Cole (1994)
		Cowpea weevil	Huessing et al. (1991a,b)
		European cornborer	Czapla (1997); Machuka et al. (1999)
Urtica dioica lectin (UDL)	Stinging nettle (*Urtica dioica*)	Cowpea weevil	Huessing et al. (1991b)
		Corn rootworm	Czapla (1997)
		Cabbage aphid	Cole (1994)
Allium sativum agglutinin (ASA)	Garlic (*Allium sativum*)	Cowpea weevil	Gatehouse et al. (1992)
		Rice brown planthopper	Powell et al. (1995)
Galanthus nivalis agglutinin (GNA)	Snowdrop (*Galanthus nivalis*)	Cowpea weevil	Gatehouse et al. (1992)
		Rice brown planthopper	Hilder et al. (1993); Powell et al. (1993)
		Rice leafhopper	Hilder et al. (1993); Powell et al. (1993)
		Green peach aphid (*Myzus persicae*)	Hilder et al. (1993, 1995); Sauvion et al. (1996)
		Spodoptera littoralis	Gatehouse et al. (1992)
		Maruca pod borer (*Maruca vitrata*)	Machuka et al. (1999)
		Glasshouse potato aphid (*Aulacorthum solani*)	Down et al. (1996)
		Mexican rice borer (*Eoreuma loftini*); Sugarcane borer (*Diatraea saccharalis*)	Legaspi and Mirkov (2000), Sétamou et al. (2002a)
Bauhinia purpurea agglutinin (BPA)	Camel's foot tree (*Bauhinia purpurea*)	European cornborer	Czapla and Lang (1990)

Table 3.1 contd.

Table 3.1 contd.

1	2	3	4
	Couch grass (*Agropyron repens*)	European cornborer	Czapla (1997); Machuka et al. (1999)
	Bandeiraea simplicifolia	Southern corn rootworm	Czapla and Lang (1990)
Jacalin	Jackfruit (*Artocarpus integrifolia*)	Southern corn rootworm Potato leafhopper	Czapla and Lang (1990) Czapla (1997) Habibi et al. (1993)
Codium fragile lectin (CFL)	Green marine algae (*Codium fragile*)	Southern corn rootworm	Czapla and Lang (1990) Czapla (1997)
Eranthis hyemalis lectin (EHL)	Winter aconite (*Eranthis hyemalis*) Scotch broom (*Cytisus scoparius*)	Corn rootworm (*Diabrotica sp.*) Corn rootworm	Czapla (1997); Kumar et al. (1993) Czapla (1997)
	Chickpea (*Cicer arientinum*)	Corn rootworm	Czapla (1997)
	Lentil (*Lens culinaris*)	Potato leafhopper Rice brown planthopper	Habibi et al. (1993) Hilder et al. (1993); Powell et al. (1993)
	Horse gram (*Dolichos biflorus*)	Potato leafhopper rice brown planthopper	Habibi et al. (1993) Hilder et al. (1993); Powell et al. (1993)

Table 3.1 contd.

Table 3.1 contd.

1	2	3	4
Concanavalin A	*Canavalia ensiformis*	Rice brown planthopper	Hilder et al. (1993); Powell et al. (1993) Rahbé and Febvay (1993)
		pea aphid (*Acyrthosiphon pisum*) Green peach aphid	Sauvion et al. (1996)
Brassica lectin (BL)	*Brassica* species	Cabbage aphid	Cole (1994)
Soybean lectin	Tepary bean (*Phaseolus acutifolius*)	*Manduca sexta* *Acanthoscelides obtectus*	Shukle and Murdock (1983) Pratt et al. (1990)
	Sweet hyacinth bean (*Lablab purpureus*)	*Maruca* pod borer, cowpea weevil, *Clavigralla tomentosicollis*	Omitogun et al. (1999)
	African yam bean (*Sphenostylis stenocarpa*)	*Maruca* pod borer, cowpea weevil, *Clavigralla tomentosicollis*	Omitogun et al. (1999)
	Winged bean (*Psophocarpus tetragonolobus*)	Cowpea weevil	Gatehouse et al. (1991)
	Pokeweed (*Phytolacca americana*)	Southern corn rootworm	Czapla and Lang (1990)
	Hairy vetch (*Vicia villosa*)	Southern corn rootworm	Czapla and Lang (1990)
Narcissus pseudonarcissus agglutinin (NPA)	Daffodil (*Narcissus pseudonarcissus*)	Rice brown planthopper Green peach aphid	Powell et al. (1995) Sauvion et al. (1996)
Listera ovata agglutinin (LOA)	*Listera ovata*	*Maruca* pod borer	Machuka et al. (1999)
	Jimson weed (*Datura stramonium*)	Cowpea weevil	Murdock et al. (1990)

important cowpea pests: *Maruca vitrata* (Fabricius), *C. maculatus*, and *Clavigralla tomentosicollis* Stäl. The lectins were presented to insects either in the form of artificial seeds (5% w/w), or by incorporation into an artificial diet (1% w/v). Extracts from the non-*Vigna* species, sweet hyacinth bean, *Lablab purpureus* (L.) and the edible African yam bean, *Sphenostylis stenocarpa* (Hoechst), produced greater agglutination than the wild *Vigna* species studied. The most toxic extracts caused >80% mortality, with that from *S. stenocarpa* selected for purification and possible genetic transformations. Gatehouse et al. (1991) also found the lectin from the winged bean, *Psophocarpus tetragonolobus* DC, toxic to *C. maculatus*.

Czapla and Lang (1990) performed artificial diet bioassays using 26 commercially-available lectins from six different plant families against neonate larvae of *O. nubilalis* and Southern corn rootworm, *Diabrotica undecimpunctata howardi* Barber (Coleoptera: Chrysomelidae). Topical applications of lectins (2%) from castor bean (*Ricinus communis* L.), camel's foot tree (*Bauhinia purpurea* L.) and WGA caused 100% mortality in *O. nubilalis* after seven days. Surviving larvae incurred a 50% weight loss relative to controls. Lectins from castor bean, pokeweed (*Phytolacca americana* L.), and green marine algae (*Codium fragile* [Suringar]) were toxic to *D. undecimpunctata howardi* larvae. Other lectins that inhibited larval growth by at least 40% included WGA and lectins from jackfruit (*Artocarpus integrifolia* Lamarck), hairy vetch (*Vicia villosa* Roth), osage orange (*Maclura pomifera* Rafinesque), and *Bandeiraea simplicifolia* (Baillon).

Two lectins (from rice and couch grass, *Agropyron repens*) screened against *O. nubilalis* demonstrated activity against this pest (Machuka et al., 1999). WGA and castor bean lectin caused >50% suppression of the larval weight increase at 0.1% topical application. Contrasting results were reported for soy bean (*Glycine max* L.) lectin, which increased larval weights of *O. nubilalis* (Czapla and Lang, 1990), yet reduced larval growth in the tomato hornworm, *Manduca sexta* Johannsen (Lepidoptera: Sphingidae) at 1% (w/v) (Shukle and Murdock, 1983). Habibi et al. (1993) claimed the first published demonstration of effectiveness of lectins against piercing-sucking homopteran pests. A total of 14 plant lectins from 13 plant species were compared for effects on survival of potato leafhopper, *Empoasca fabae* (Harris) (Homoptera: Cicadellidae). At 0.2 to 1.5% (w/w) diet concentrations, WGA and PHA, as well as lectins from jackfruit, pea (*Pisum sativum*), lentil (*Lens culinaris*), and horse gram (*Dolichos bifloris*) reduced survival of *E. fabae*. In artificial diet bioassays, 25 commercially-available proteins of different classes were evaluated for toxicity to pea aphid, *Acyrthosiphon pisum* (Harris) (Rahbé and

Febvay, 1993). The lectin concanavalin A (Con A) from *Canavalia ensiformis* (L.) DC displayed significant toxicity and growth inhibition, whereas others such as WGA did not.

The mannose-binding lectin from the snowdrop lily (*Galanthus nivalis* agglutinin; GNA), is perhaps the most promising insecticidal lectin identified to date because of its efficacy against sap-sucking Homopterans. In preliminary evaluations, GNA concentrations of 0.1% of artificial diet displayed antimetabolic effects against several insect species and were used as a base-line level for testing (Hilder et al., 1995). GNA was toxic to two representatives of different families of sap-sucking pests of rice, the rice green leafhopper, *Nephotettix cincticeps* (Uhler) (Powell et al., 1993) and the brown planthopper *Nilaparvata lugens* (Stål) (Hemiptera: Delphacidae). Several lectins were compared for efficacy against *N. lugens* in artificial diet bioassays (Powell et al., 1995). GNA was the most toxic of the mannose-binding lectins, causing 60% mortality in *N. lugens* at concentrations as low as 0.025% (w/v) (Powell et al., 1995). In contrast, daffodil (*Narcissus pseudonarcissus*) agglutinin (NPA) and garlic (*Allium sativum*) agglutinin (ASA) caused measurable, but lower levels of mortality. Of the chitin-binding lectins, WGA caused 78% mortality, whereas *Oryza sativa* agglutinin (OSA) and *Urtica dioica* agglutinin (UDA) did not show any significant antimetabolic effects against *N. lugens* at a concentration of 0.1% (w/v). Sauvion et al., (1996) compared the mannose-binding lectins, GNA, NPA and ASA for toxicity to peach-potato aphid, *Myzus persicae* (Sulzer) (Aphididae) in artificial diet bioassays. GNA was the most toxic, causing 42% nymphal mortality at a concentration of 30 mM. NPA and ASA caused no significant mortality, although growth was inhibited. When presented at a sublethal concentration of 4.3 mM, GNA and Con A reduced total aphid fecundity by 29 and 19%, respectively.

In other artificial diet bioassays, Machuka et al. (1999) screened 25 lectins (2% w/w) from 15 plant families for insecticidal effects against the *Maruca* pod borer, *Maruca vitrata* (Fabricius) (Lepidoptera: Pyralidae), an important pest of legumes in tropical Africa and Latin America. Biological parameters measured were larval survival and weight, feeding inhibition, pupation, adult emergence and fecundity. Sixteen of the lectins tested showed significant effects, although only *Listera ovata* agglutinin (LOA) and GNA affected all of the 6 variables recorded. Larval mortality and feeding inhibition was highest using LOA (60%). Down et al. (1996) incorporated GNA (0.1% w/v) into artificial diet fed to glasshouse potato aphid, *Aulacorthum solani* (Kaltenbach) (Aphididae). GNA in the diet resulted in declines of ≤ 10% in aphid survival and up to 65% in fecundity. Furthermore, the rate of increase in aphid body size decreased by up to 40%. Fitches et al. (1997) highlighted the variable toxicity of GNA against the different insect species tested. Variability was attributed to poorly-understood differences in the mode of action of lectins, particularly in relation to insect-feeding habits. For example, high toxicity of GNA to the rice brown planthopper, *N. lugens,* may be a consequence of its relatively monophagous diet and adaptation to a specific host plant, while the more polyphagous species, *Lacanobia oleracea* (Linnaeus) and *A. solani,* may possess more diverse detoxification mechanisms or resistance to antimetabolic factors for coping with a wider host plant range.

In Planta Bioassays

The transfer of a glucose/mannose-binding gene from pea (*Pisum sativum*) to tobacco provided the first demonstration of insecticidal qualities in a lectin-expressing transgenic plant (Boulter et al., 1990; Hilder and Boulter, 1999). This particular lectin was chosen because of its low mammalian toxicity when compared to the more toxic chitin-binding lectins WGA and PHA. The low mammalian toxicity of pea aphid lectin also resulted in low insect toxicity in transgenic plants although some insecticidal effects were observed against *Heliothis virescens* (Fabricius) in bioassays using transformed tobacco (Hilder and Boulter, 1999). The expression of even more toxic lectins has not always produced a desirable insect resistance. Three leading members of the lectin family (WGA, jacalin and rice lectin) were expressed in corn plants and tested against larvae of *O. nubilalis* and *D. undecimpunctata.* Of them, WGA-transgenic corn showed only modest larval growth inhibition and little mortality activity against the target pests (Maddock et al., 1991). However, concerns remain regarding the use of more toxic lectins such as WGA in transgenic crops produced for human consumption. Another potential difficulty in the development of commercial lectin-expressing transgenic crops in general is the relatively high levels of expression required for occurrence of insecticidal effects. When evaluated against target pests, lectin concentrations in artificial diets must be at least an order of magnitude greater than δ-endotoxins to be effective against target pests. (Estruch et al., 1996).

Transfer of the GNA lectin into tobacco was accomplished by introducing the target gene and cauliflower mosaic virus 35S (CaMV35S) gene promoter into the tobacco genome using the standard *Agrobacterium*-mediated leaf disc transformation method described by Horsch et al., (1985). Transgenic tobacco plants were produced, which expressed acceptable levels of GNA of upto 1% of total leaf protein (Gatehouse et al., 1994). Machuka et al. (1999) suggested that levels >2% were unlikely to be achieved in transgenic plants, although Hilder et al.

(1995) reported levels as high as ~ 2.5% of total tobacco leaf soluble protein. Variations in the levels of expression in different parts of the plant and possibly at different times during development must be considered while deploying these transgenic crops as components of pest management systems.

GNA-TRANSGENIC LECTIN CROPS

Tobacco

The insecticidal effects of GNA-tobacco were evaluated against *M. persicae*, a serious polyphagous pest worldwide, and a vector of >100 disease-causing viruses. In leaf disc bioassays, mean aphid numbers in the transgenic treatment were lower than in the control (8.3 versus 38.8 per disc, respectively) and doubling times were longer (2.47 d versus 1.62 d, treatment and control, respectively) (Hilder et al., 1995). Whole plant bioassays produced similar results, with doubling time of 4.8 d in the treatment and 4.4 d in the control.

Rice

GNA-transgenic rice was developed and evaluated against the brown planthopper, *N. lugens*, by Rao et al. (1998). *Nilaparvata lugens* is an important pest of rice throughout Southeast Asia. The planthopper feeds on plant phloem, causing a condition known as 'hopper burn', and is an important vector of viral diseases. Because of previous experiments showing efficacy of GNA against *N. lugens* when administered through artificial diet (Powell et al., 1995), GNA-transgenic rice was developed using constructs where a phloem-specific promoter drove expression (from the rice sucrose synthase RSs1 gene) and by a constitutive promoter (from the maize ubiquitin ubi1 gene). Successful transformation was confirmed by Western blot analyses that revealed expression levels of 2% of total protein. GNA-transgenic rice decreased feeding, development, survival and overall fecundity of *N. lugens* when evaluated through insect bioassays and feeding studies.

Using particle bombardment, Tingjuangjun et al. (2000) were the first to develop GNA-transgenic lines of two elite rice varieties, 'Khao Dawk Mali 105' (KDLM105) and 'Supanburi 60' (SP60), known to resist transformation. Over 30 transgenic lines were produced, with highest levels of expression = 0.25% of total soluble protein. Constitutive and phloem-specific expression of GNA reduced nymphal survival of *N. lugens* by 37 and 42%, respectively. Rao et al. (1998) maintain that GNA-rice is the first transgenic line to exhibit insecticidal activity against sap-sucking insects of an important cereal crop.

Potato

Transgenic potato (*Solanum tuberosum* L. cv. 'Desiree') was developed using the CaMV 35S promoter through *Agrobacterium*-mediated gene transfer (Fitches et al., 1997). GNA expression levels in transgenic leaves were 0.3-0.4% of total soluble protein (Down et al., 1996). The GNA-potatoes were evaluated for deleterious effects against *A. solani* through laboratory bioassays and large-scale greenhouse experiments. Laboratory bioassays showed reductions in fecundity on transgenic plants compared to non-transgenic control plants at levels similar to those observed in the artificial diet bioassays. In the greenhouse experiments, population increase on the transgenic plants was = 4-fold lower than on the control plants. Down et al. (1996) concluded that the insecticidal effects of GNA against *A. solani* as measured using artificial diet studies could be reproduced *in planta* under both laboratory bioassays and large scale glasshouse conditions.

Gatehouse et al. (1997) measured the insecticidal effects of genes expressing three plant-derived proteins against the tomato moth, *Lacanobia oleracea* (Linnaeus) [Lepidoptera: Noctuidae]). GNA, bean (*Phaseolus vulgaris*) chitinase inhibitor (BCH), and wheat a-amylase inhibitor (WAI), were studied and compared against cowpea trypsin inhibitor (CpTI). Transgenic potato lines were developed containing GNA, BCH and WAI singly and in pairwise combinations using the CaMV 35S promoter. GNA was measured at upto 2% of total soluble protein in the transformed plant tissue. The introductions of single foreign genes produced plants with higher levels of the protein expression than counterparts containing two foreign genes. When exposed to tomato moth larvae, GNA-transgenic tomato plants suffered 50% less leaf damage due to feeding compared to controls. Larval survival declined by 20%, and total insect biomass was reduced by 45-65%. GNA-transgenic potato showed the greatest insect resistance compared to other transformed lines. BCH-potato was not protected against *L. oleracea*; CpTI-transgenic potato showed similar effects as GNA-potato on biomass and survival of *L. oleracea*, but did not confer protection against feeding damage. Therefore, GNA-potato was the most promising among the transgenic potato lines developed by Gatehouse et al. (1997) against *L. oleracea*.

Also, using the tomato moth as the target pest, Fitches et al. (1997) studied insecticidal effects of GNA on insect development, consumption and survival, using three distinct bioassay methods. Larvae were reared on artificial diet containing GNA at 2% (w/w) dietary protein; on excized leaves of GNA-potato (expressing GNA at ~ 0.07% of the total soluble proteins); and on GNA-potato plants (~ 0.6% total soluble proteins) in greenhouse experiments. All of the three delivery methods produced significant reductions in larval growth.

Larval biomass was reduced by 32, 23 and 48% in artificial diet, excized leaves, and greenhouse trials, respectively. Prolonged larval development was observed in the artificial diet and excized leaf treatments. Larval consumption was reduced in the artificial diet and greenhouse bioassays. Prolonged feeding in the excized leaves resulted in 15% greater consumption than the control. Larval survival was not adversely affected in the artificial diet and the excized leaf bioassays. However, the greenhouse bioassay indicated a 40% reduction in survival. Reduced feeding in the artificial diet bioassays was the result of reduced larval size, rather than to a detrimental effect of GNA. Reduction in larval biomass, especially in the excized leaf experiments was attributed to reduced nutrient uptake caused by GNA. Fitches et al. (1997) concluded that GNA had significant insecticidal effects against the tomato moth, regardless of method of delivery.

Wheat

Stoger et al. (1999) developed GNA-transgenic wheat (*Triticum aestivum* L. cv 'Bobwhite') through particle bombardment using two promoters: the rice synthase sucrose gene (RSs1) and the maize ubiquitin 1 promoter. A total of 32 GNA-transgenic lines of wheat, exhibiting levels of expression from 0 to 0.2% total soluble protein were developed. Higher levels of expression were found in those lines developed using the RSs1 promoter. Seven lines were further evaluated against the grain aphid, *Sitobion avenae* (Fabricius) (Aphididae) in the greenhouse. GNA at levels > 0.04% total soluble protein decreased the fecundity, but not the survival of *S. avenae*. Insect survival over the experimental period was ~ 60% regardless of GNA expression level. However, expression levels did affect aphid fecundity. Low expressors did not significantly affect aphid fecundity, while high expressors caused fecundity reductions of 40 to 65%. Stoger et al. (1999) suggested that GNA-wheat might comprise a viable component of integrated pest management systems against pests of wheat.

GNA-TRANSGENIC SUGARCANE

Genetic Transformation of Sugarcane

At the Texas Agricultural Experiment Station in Weslaco, sugarcane was successfully transformed for resistance to herbicides (Gallo-Meagher and Irvine, 1996) and viruses (Ingelbrecht et al., 1999). Recently, research attention turned to the development of transgenic sugarcane expressing insecticidal characteristics, specifically developed for use against the key sugarcane pest in south Texas, the Mexican rice borer, *Eoreuma loftini*

(Dyar) (Lepidoptera: Pyralidae). Genes selected for transfer expressed GNA and CHA, another mannose-binding lectin from the *Cymbidium* orchid (see van Damme et al., 1994). However, attempts to incorporate CHA into the sugarcane genome proved to be unsuccessful. Transgenic sugarcane expressing GNA was developed using a modified Sanford gene gun technique (Irvine and Mirkov, 1997; Mirkov and Irvine, 1998). After testing several promoters, transformation was accomplished using the maize ubiquitin promoter (combination of *Ubi-nptII* and *Ubi-gna*) to induce GNA expression in two commercial sugarcane varieties 'CP65-357' and 'NCo 310', commonly grown in the Lower Rio Grande Valley of Texas. Modified DNA was coated onto tungsten pellets, which were then blasted by pressurized helium into masses of embryogenic sugarcane calluses using the gene gun. Untransformed cells were killed by the antibiotic geneticin contained in the growth medium on which the calluses developed into transgenic shoots. The derived transgenic shoots were transplanted in the fields, where they developed into whole transgenic plants. Based on the insertion level of the gene in the chromosome, several transgenic lines expressing different levels of lectin were obtained. Expression of GNA in the genetically-transformed plants was tested using Western and Southern blot analyses, and GNA concentrations of 0.5 to 0.7% of the total soluble protein were found in the leaf tissue, and 0.8-1.0% of total soluble protein in the stem tissue.

Field Evaluation of Transgenic Plants

Resistance of GNA-expressing sugarcane plants to damage caused by lepidopteran stalkborers, i.e. *E. loftini* and *Diatraea saccharalis* (F.) (Pyralidae), was evaluated in 1998 and 1999 under field conditions. In south Texas, *E. loftini* and *D. saccharalis* are the two borers damaging crops of sugarcane, but *E. loftini* constitutes > 95% of the collections and is by far the more important pest (Legaspi et al., 1997). The variety CP65-357 was selected for field evaluation based on preliminary greenhouse tests, which revealed high levels of GNA in the transgenic line 83 of this variety. In two out of three seasons, transgenic plants showed lower percentages of bored internodes compared to non-transgenic plants. These data support the hypothesis that expression of GNA at the levels found in this study confers some level of protection from stalkborer damage in the field (Legaspi and Mirkov, 2000). However, the results are not conclusive and require additional investigation. The biological mechanisms conferring such protection are unclear at present, especially in the light of the absence of significant deleterious effects in laboratory studies. One hypothesis suggested by recent studies is that transgenic sugarcane is not favored for oviposition by stalkborer adults. In preliminary greenhouse experiments, D. *saccharalis* adults oviposited >90% of their eggs on non-transgenic sugarcane, while the remainder were deposited on transgenic counterparts (Sétamou et al., 2002a). In the course of the field seasons, differences in oviposition rates, coupled with prolonged exposure to GNA, may have resulted in lower levels of damage.

Effects of Transgenic Sugarcane on Bionomics of Stalkborers in Laboratory Bioassays

The effects of transgenic sugarcane on the bionomics of *D. saccharalis* and *E. loftini* were studied for two successive generations. One-day-old neonate larvae of each borer species were offered meridic diet supplemented with leaf sheath tissue from either transgenic or non-transgenic sugarcane. Survival, growth, development and reproduction of each borer species were compared between the two types of diet at 30 ± 1 °C, 70 ± 2% R.H., and 12:12 (L: D) photoperiod regime. Transgenic sugarcane differentially affected the biological parameters of *D. saccharalis* and *E. loftini*. While no significant differences were observed between larval survival of *D. saccharalis* on transgenic and non-transgenic diet, significantly more *E. loftini* larvae survived on non-transgenic versus the transgenic diet. Consequently, the percentage of pupae formed was comparable in both diet treatments for the sugarcane borer, whereas more pupae were formed on non-transgenic versus transgenic diet for the Mexican rice borer, *E. loftini* (Table 3.2). Likewise, significantly fewer adults emerged from a transgenic diet relative to a non-transgenic diet for *E. loftini*, but no significant effect was evident in *D. saccharalis*. Significant generation effects on larval survival, pupal and adult emergence rates were not evident for both *D. saccharalis* and *E. loftini*.

Despite lower survival of *E. loftini* on transgenic versus non-transgenic diet, larval weights were comparable in both diet treatments. Likewise, male pupal weight was comparable in both diets, while female pupae had similar weights in the first generation but lower weight on transgenic diet in the second generation. In contrast, larvae of *D. saccharalis*-fed transgenic diet weighed significantly more than the larvae fed non-transgenic diet in both generations (Table 3.2). Consequently, both female and male pupae of *D. saccharalis* from transgenic diet weighed more than pupae from non-transgenic diet in the first generation, but this effect was not evident in the second generation (Table 3.2). The larval period of *D. saccharalis* males was longer on non-transgenic relative to transgenic diet in the first but not second generation, whereas that of females was similar among diets in both

Table 3.2 Survival and growth parameters of D. saccharalis and E. loftini on non-transgenic (NT) and transgenic (T) diets.

Generation	Treatment	Pupation (%)	Adult emergence (%)	Larval weight (mg)	Pupal weight (mg)	
					Males	Females
			D. saccharalis			
G1	NT	94.0±1.1a[1]	88.9±3.5a	117.5±3.3b	80.1±2.1b	118.9±2.7b
	T	93.0±1.9a	85.1±3.5a	135.8±4.8a	88.6±1.6a	135.8±2.6a
G2	NT	83.3±4.4a	81.7±4.4a	112.1±5.2b	82.9±2.2a	115.4±4.9a
	T	85.0±2.9a	73.3±9.2a	126.1±6.2a	85.2±2.3a	128.5±5.9a
			E. loftini			
G1	NT	86.7±5.8a	78.7±5.8a	61.8±3.6a	36.1±1.1a	64.5±2.3a
	T	76.0±4.0b	67.9±4.6b	62.2±4.5a	33.6±1.3a	65.2±1.8a
G2	NT	91.5±1.8a	89.8±0.2a	65.2±4.0a	37.3±0.8a	73.8±2.1a
	T	85.0±2.9b	76.7±1.7b	57.9±3.1b	35.8±1.2a	57.7±3.0b

[1]Means followed by the same letter within the same column and for each generation are not significantly different at P = 0.05 (Student Newman Keuls' test).

generations (Table 3.3). In contrast, significant differences were not observed in larval periods of both males and females of E. loftini. The larval diet did not significantly affect the pupal periods of both species and sexes, except in the first generation males of E. loftini where the pupal period was slightly shorter on transgenic diet (Table 3.3).

In general, diet had a significant effect on the longevity of adults. For E. loftini, adult longevity of both males and females fed on transgenic diet was shorter than on non-transgenic diet in the second generation, but the differences were not significant in the first generation (Table 3.3). D. saccharalis adult females lived significantly shorter on transgenic relative to non-transgenic diet in both generations, whereas longevity of adult males was shorter only in the second generation (Table 3.3). Although the proportion of males obtained for both species tended to be higher in the transgenic diet, no significant differences were observed between diets (Fig. 3.1). A generation effect was not observed on the sex ratio of both borer species.

The fecundity of D. saccharalis did not vary significantly with diet. In contrast, significantly fewer eggs were laid by females of E. loftini emerging from larvae-fed transgenic diet relative to a non-transgenic diet in the first generation. Although the same trend was observed in the second generation, the differences were not significant (Fig. 3.2).

Estimation of life table parameters for D. saccharalis and E. loftini revealed contrasting results. While the population growth parameters of D. saccharalis increased in the transgenic relative to the non-transgenic

Table 3.3 Immature developmental periods and adult longevity of *D. saccharalis* and *E. loftini* on non-transgenic (NT) and transgenic (T) diets

Generation	Treatment	Larval period (days)		Pupal Period (days)		Adult longevity (days)	
		Males	Females	Males	Females	Males	Females
		D. saccharalis					
G1	NT	24.1±0.4a[1]	26.1±0.3a	6.6±0.1a	6.4±0.1a	4.1±0.2a	4.4±0.2a
	T	21.7±0.2b	25.1±0.4a	6.9±0.1a	6.8±0.1a	4.5±0.2a	3.8±0.2b
G2	NT	21.9±0.4a	24.6±0.7a	6.8±0.2a	6.8±0.3a	4.5±0.2a	4.5±0.3a
	T	22.1±0.4a	25.5±0.7a	6.5±0.2a	6.5±0.2a	4.1±0.2a	3.9±0.2a
		E. loftini					
G1	NT	29.3±0.8a	30.7±0.7a	8.2±0.2a	8.4±0.2a	7.3±0.4a	7.7±0.4a
	T	28.6±0.8a	29.2±0.6a	7.8±0.3a	7.9±0.2a	6.7±0.5a	7.1±0.4a
G2	NT	26.1±0.5a	29.8±0.4a	7.7±0.1a	7.2±0.1a	9.4±0.4a	8.9±0.3a
	T	27.4±0.7a	30.0±0.7a	7.8±0.1a	7.5±0.1a	7.7±0.4b	6.3±0.4b

[1]Means followed by the same letter within the same column and for each generation are not significantly different at $P = 0.05$ (Student Newman Keuls' test).

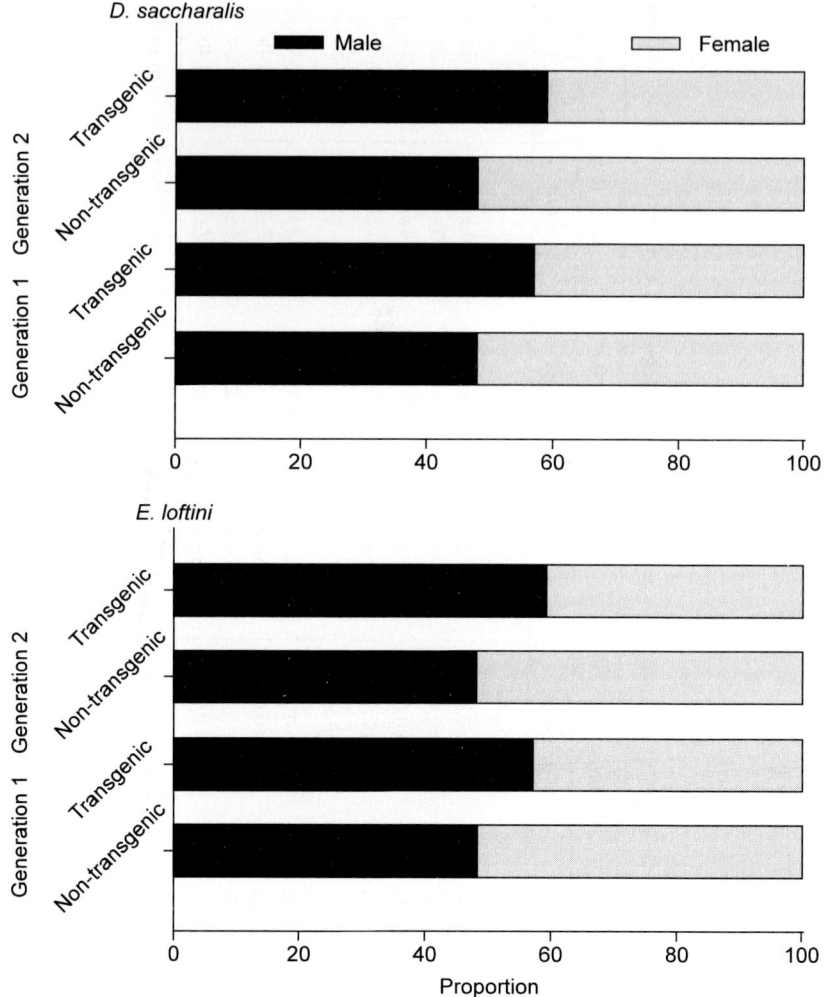

Fig. 3.1 Sex ratio of *D. saccharalis* and *E. loftini* pupae as affected by the type of their larval diet

diet, significant reductions were recorded in the net reproductive rate, intrinsic rate of increase, and total progeny of *E. loftini* in the transgenic diet. Additionally, *E. loftini* doubling time was highest on transgenic diet (Fig. 3.3), revealing the potential impact of transgenic sugarcane on *E. loftini* populations.

The observed increases in larval weights of *D. saccharalis* in the artificial diet bioassays (Sétamou et al., 2002a) were initially unexpected. However, other researchers found similar results using GNA diet

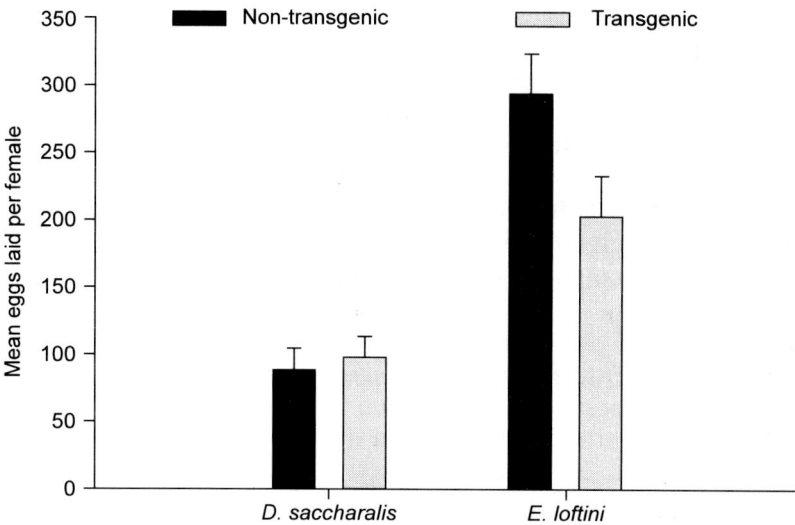

Fig. 3.2 Fecundity of D. *saccharalis* and E. *loftini* reared on non-transgenic versus transgenic diet

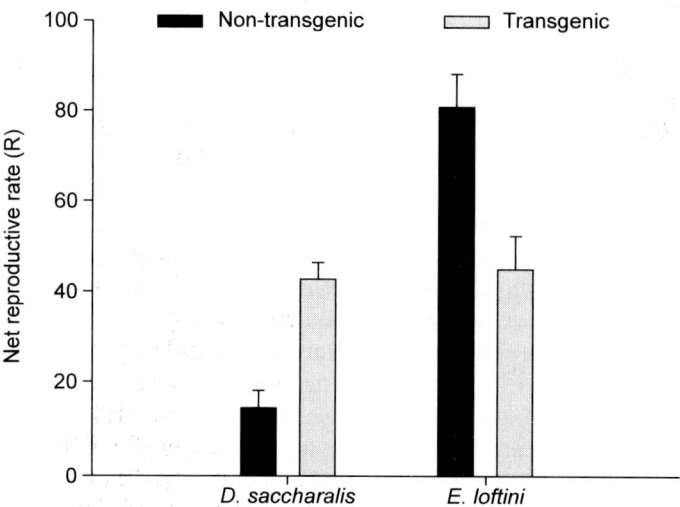

Fig. 3.3 Net reproductive rate (R_0) of D. *saccharalis* and E. *loftini* on non-transgenic and transgenic diets

bioassays. For example, G. *max* lectin increased larval weight by >25% (Czapla and Lang, 1990). The lectin from camel's foot tree increased the survival of E. *fabae* (Habibi et al., 1993). Sauvion et al. (1996) reported a slight but significant stimulation in the growth of M. *persicae* upon ingestion of the lectins GNA, ASA, and NPA (10 mg/ml), with increases of 26, 18 and 11%, respectively, over the controls. Two explanations have been proposed for these results: (i) Lectins can act as a source of proteins and induce weight gain in test insects (Machuka et al., 1999; Habibi et al., 1993); and (ii) sugars in the rearing medium may bind to active sites in the lectin, rendering them non-toxic (Trigueros et al., 2000). Similar effects may occur in transgenic plants due to plant sugars, especially those present in the sap.

In other evaluations of GNA-sugarcane, Nutt et al. (1999) reported the range of expression of the lectin as being from 0 to 0.05% of total soluble protein in different transgenic lines, with lines expressing levels of 0.05%, causing decreased larval weights of the canegrub, *Antitrogus consanguineus* (Blackburn) (Coleoptera: Scarabaeidae). Ineffectiveness of purified lectin evident in the artificial diet trials suggested significant *in planta* effects at concentrations 100 to 200-fold less than when incorporated into artificial diets (Nutt et al., 1999).

Effects of Transgenic Sugarcane on Fitness Parameters and Biological Performance of a Larval Parasitoid

Cotesia flavipes (Cameron) is the prime natural enemy responsible for maintaining *D. saccharalis* below economic levels. Thus, it was of great interest to determine whether *C. flavipes* would be negatively affected by the transgenic cultivars, and if so, whether biological control of *D. saccharalis* would be disrupted. *D. saccharalis* larvae reared on either transgenic or non-transgenic diets were exposed to *C. flavipes* females for two consecutive generations. Acceptability of *D. saccharalis* by females of *C. flavipes* for oviposition as well as their suitability for parasitoid development did not vary with generation in both diet treatments. Likewise, significant differences were not observed between the transgenic and non-transgenic diets in acceptability of *D. saccharalis* by *C. flavipes* in both generations. However, the suitability of *D. saccharalis* larvae for parasitoid development varied with diet treatment. Significantly fewer parasitized larvae yielded parasitoid adults in the transgenic diet compared to the non-transgenic diet ($P = 0.04$) (Table 3.4). Differences in total developmental periods between diet treatments were observed ($P = 0.29$). *C. flavipes* took on average ca. 15 days from egg to adult emergence. The total developmental period did not vary with generation ($P = 0.58$) (Table 3.4).

Brood size (number of parasitoids or cocoons emerging per host) of *C. flavipes* was not affected by any generation. However, the number of cocoons and adult parasitoids emerging per host were significantly lower in the transgenic diet relative to the non-transgenic diet in the first

Table 3.4 Parasitism and biological parameters of C. flavipes as affected by its host larval diet

Generation	Treatment[1]	Acceptability (%)	Parasitism (%)	Development period (days)	Cocoons/larva	Adults/larva	Adult emergence (%)
G1	NT	100a[2]	96.0a	15.3±0.3	48.1±2.1a	46.9±2.0a	92.7±1.2b
	T	96.1a	88.2b	15.2±0.2	40.4±2.5b	37.2±2.7b	83.3±3.3a
G2	NT	90a	75.8a	15.2±0.2	46.6±2.6a	43.2±3.1a	92.0±1.9b
	T	96.4a	78.6b	15.0±0.4	42.9±3.2a	41.8±2.9a	87.0±2.0a

[1]NT = non-transgenic and T = Transgenic
[2]Means followed by the same letter within the same column and for each generation are not significantly different at P = 0.05 (Student Newman Keuls' test).

generation ($P < 0.05$) but not in the second generation ($P > 0.05$) (Table 3.4). In addition, the proportion of cocoons yielding adult parasitoids was significantly lower in the transgenic diet compared to the non-transgenic diet in both generations. The sex ratio of C. flavipes was significantly female-biased in the two generations for both diet treatments ($P < 0.001$; Fig. 3.4). An increase in the proportion of males in the progeny ($P = 0.0006$) was observed in the second generation, relative to the first generation in both diet treatments, but the increase was greater for the transgenic diet. Consequently, the proportion of males in the transgenic diet was significantly higher than that in the non-transgenic diet in the second generation ($P < 0.01$). Differences were not observed in the sex ratio between transgenic and non-transgenic diet treatments in the first generation ($P = 0.89$). Neither the generation ($P = 0.74$) nor the diet treatment ($P = 0.47$) significantly affected the oviposition potential of C. flavipes. Mean egg loads of females emerging from hosts fed on transgenic and non-transgenic diets were comparable (Fig. 3.5).

Adult longevity of C. flavipes females averaged 36 to 52 h, depending on generation and diet treatment (Fig. 3.5). Longevity of adult females emerging from hosts reared on non-transgenic diet was similar in both generations, but decreased in the second generation relative to the first in hosts fed transgenic diet ($P < 0.0001$). Consequently, longevity was significantly shorter in the transgenic diet relative to the non-transgenic diet in the second generation ($P < 0.0001$). Significant differences were

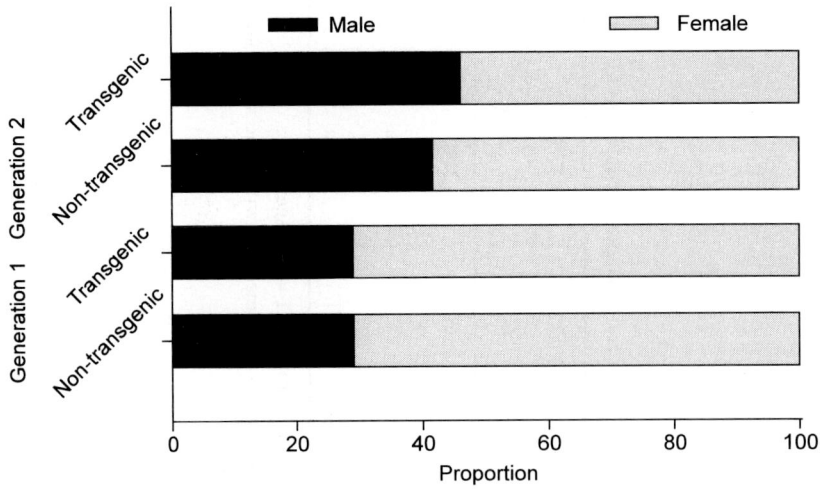

Fig. 3.4 Sex ratio of C. flavipes emerging from D. saccharalis reared on non-transgenic versus transgenic diet

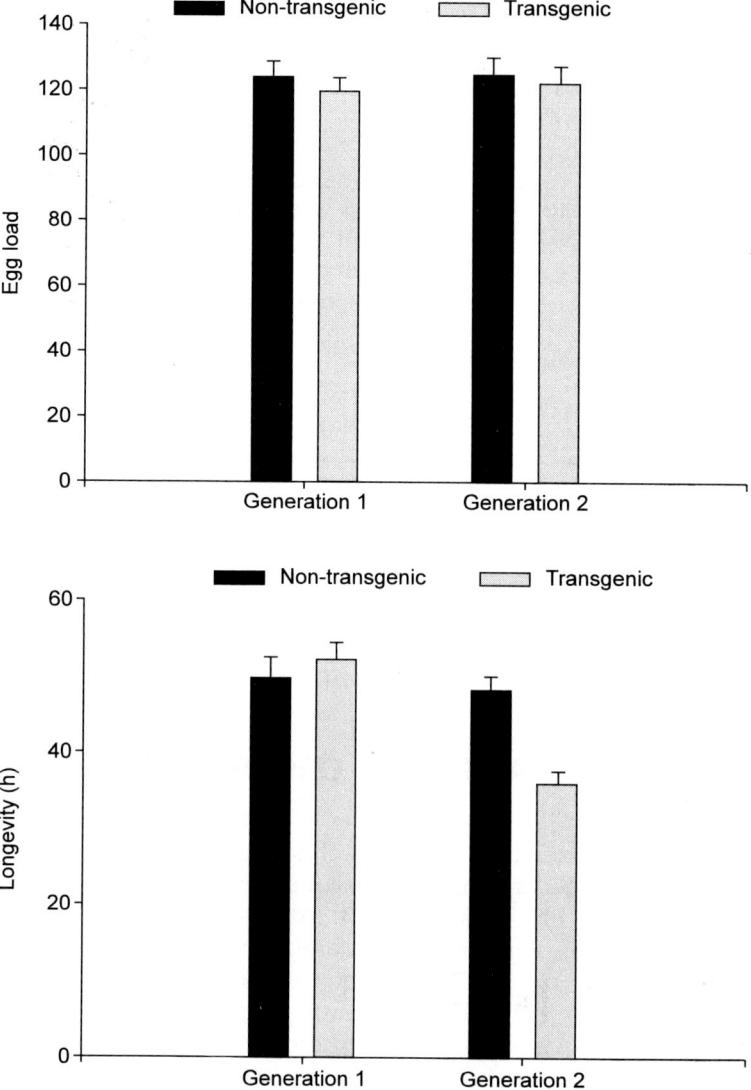

Fig. 3.5 Egg load and longevity of *C. flavipes* as affected by their host larval diet

not observed between the transgenic and the non-transgenic diet in the first generation (P = 0.37).

Covariance analyses showed that brood sizes based on the numbers of cocoons or adults emerging per host larva were significantly influenced by host size in both diets ($P < 0.0001$) (Table 3.5). However, the slopes of the regressions were comparable between the transgenic

Table 3.5 Parallel line analysis for the relationships between host larval size and brood size, and between adult female size and its brood size for *C. flavipes* emerging from *D. saccharalis* reared on non-transgenic (NT) and transgenic (T) diets.

Treatment	Regression equation	t-value for slope[1] comparison	t-value for intercept comparison
	Relationship between host larval size and brood size of *C. flavipes*		
NT	27.9 + 0.19x, $R^2 = 0.13$, $P = 0.001$	0.25[ns] (df = 132)	3.85** (df = 132)
T	16.9 + 0.24x, $R^2 = 0.19$, $P = 0.007$		
	Relationship between *C. flavipes* adult female hind tibia length and its egg load		
NT	−19.3 + 328.1x, $R^2 = 0.52$, $P < 0.0001$	0.70[ns] (df = 172)	2.90** (df = 172)
T	−36.6 + 345.8x, $R2 = 0.62$, $P < 0.0001$		

[1] ns = non-significant ($P > 0.05$) and ** = highly significant ($P < 0.01$).

and non-transgenic diet, indicating the absence of significant treatment-mediated effects on brood size. Similarly, the size of the adult female significantly affected its egg load. Significant differences were not observed between the regression slopes of the two diet treatments indicating that there were no treatment-mediated effects in the egg load of *C. flavipes*.

EFFECTS OF TRANSGENIC CROPS ON NON-TARGET INSECTS

As with other novel pest control technologies, consideration must be made for unintended ecological consequences, such as the development of resistance to lectins (Czapla, 1997), toxicity to humans (Pusztai et al., 1996) and non-target insects, especially beneficial species (Bell et al., 1999; Sétamou et al., 2002b). Genetic engineering of inherent insect resistance in economically-important plant species offers possible environment- and consumer-friendly alternatives for crop protection to meet the present (and future) demands of sustainable agriculture (Hilder and Boulter, 1999). However, an objective evaluation of the ecological risks involved in the development and deployment of transgenic crops is especially difficult due to the political and economic climate surrounding genetic engineering in general.

The issue of ecological implications of transgenic crops has largely been limited to that of Bt-transgenics (e.g. Hokkanen and Wearing, 1994) due to their current widespread deployment. Hilbeck et al. (1999) investigated the effects of Bt toxins on immature predatory lacewings,

Chrysoperla carnea (Stephens) (Neuroptera: Chrysopidae). An artificial diet containing Bt toxins was fed to the larvae of *Spodoptera littoralis* (Boisduval), which were presented as prey for larvae of *C. carnea*. Mortality of chrysopid larvae reared on Bt-fed prey was always significantly higher than in the control, although total development time was not consistently affected. Tritrophic effects on *C. carnea* were also investigated using Bt-corn (Hilbeck et al., 1998). Two different insect prey species—European corn borer, *O. nubilalis* and *S. littoralis* were allowed to feed on Bt-corn, and the corresponding non-Bt variety. The mean total immature mortality for chrysopid larvae raised on Bt-fed prey was 62%, compared with 37% on Bt-free prey. Differences in mortality were due to the presence or absence of Bt in the host plant, rather than in prey species. Hilbeck et al. (1999) concluded that tritrophic studies are necessary in order to assess the long-term compatibility of transgenic insecticidal crops with natural enemies.

Birch et al. (1999) studied the tritrophic effects of GNA-transgenic plants using a system comprising of GNA-potatoes, pest aphid and a predatory beetle. Peach potato aphids, *M. persicae*, were allowed to feed on GNA-potatoes. The aphids were then fed to adult 2-spot ladybirds, *Adalia bipunctata* (Linnaeus) (Coleoptera: Coccinellidae) for 12 d. Ladybird fecundity and egg viability declined significantly over the following 2-3 wk period. Ladybird longevity was reduced by up to 51%, although no acute toxicity was attributed to the transgenic plants. Adverse effects on ladybird reproduction caused by eating GNA-fed aphids were reversed after switching the ladybirds to a diet of pea aphids raised on non-transgenic beans. Although the deleterious effects on transgenic plants on beneficial insects were reversible, Birch et al. (1999) suggested that potential ecological risks in the field warrant further investigation.

In contrast to Bell et al. (1999), Down et al. (2000) suggested in a subsequent study that GNA had no direct adverse effects on *A. bipunctata*. Instead, the observed reductions in predator fitness could be attributed to their feeding on aphid prey rendered sub-optimal by consuming a diet containing GNA. Aphids reared on artificial diet containing GNA were smaller and weighed less than those fed the control diet. Protein extracts from GNA-fed aphids showed an accumulation of the lectin within 24 h of consumption of an artificial diet and the evidence of binding to the aphid body tissue. Upon consumption, aphids containing GNA transferred the lectin to *A. bipunctata* larvae. Immunohistological tests indicated the presence of GNA in ladybird larval guts, although damage to the epithelium was not observed (Down et al., 2000). However, predation on GNA-fed aphids caused two measurable deleterious effects compared to the control: ladybird body weights were less and development times were longer. Down et al. (2000) suggested that GNA did not exhibit direct antibiosis against *A. bipunctata* and that feeding on smaller aphids and thus, receiving less food caused the observed deleterious effects. The authors concluded that the transgenic plants caused no direct antibiosis in the predator and that any indirect reductions in predator fitness were lower than those induced by chemical insecticides.

Bell et al. (1999) studied the effects of GNA delivered by artificial diet and GNA-potato on parasitism of the tomato moth, *L. oleracea,* by the gregarious ectoparasitoid *Eulophus pennicornis* (Hymenoptera: Eulophidae). Third and fourth instar larvae of *L. oleracea* were fed corn-based and potato-based diets containing GNA, as well as the excized leaves of GNA-potato. The larvae were then exposed to mated adult parasitoid females for parasitism. Parasitoid size, longevity, egg load and fecundity were not adversely affected by GNA, regardless of the method of delivery. Bell et al. (1999) then concluded that the use of GNA crops should not adversely affect parasitism of the tomato moth by its eulophid parasitoid.

In a subsequent study, Bell et al. (2001) performed similar experiments using GNA-potato (expressing »1.0% total soluble protein) against the tomato moth in the greenhouse. The transgenic crop resulted in reduced levels of damage caused by *L. oleracea*. Furthermore, the use of *E. pennicornis* as a biological control agent resulted in significantly less damage in both the GNA-potato treatment and the non-transgenic control. In the transgenic plants, the presence of the parasitoid resulted in further reductions (»21%) in pest damage. The transgenic host plant did not affect the fecundity of female *E. pennicornis* adults, although the mean size and longevity of female parasitoids was reduced. The transgenic plants also did not affect the numbers of F2 progeny produced by the parasitoid. Therefore, it was suggested that biological control by parasitoids might be compatible and complementary to the transgenic host plants in an integrated pest management system (Bell et al., 2001).

CONCLUSIONS

Technologies with the potential to revolutionize their industries are often viewed with public fear and mistrust. During the infancy of the computer revolution, a common theme in the media portrayed computers as the product of an uncontrolled technology that would enslave mankind, or at the very least, unleash unforeseen dire consequences upon humanity. Today, few would argue that computerization has been of benefit to mankind, to the extent that modern society has actually become dependent upon it. Some problems that were foreseen have indeed been realized, e.g. loss of privacy and computer-related crimes. However, when viewed in perspective, the benefits of computerization clearly outweigh its costs. In fact, today's generation that has known computers and the Internet all their lives, would probably be amused to learn of the misconceptions expressed by many of their parents.

The advent of genetic engineering shares many parallels with that of computerization. We hold the belief that much of the public misgivings regarding this technology lie in the process of creating transgenic organisms, rather than with the transgenic products themselves. If a Bt-expressing crop could be created using conventional plant-breeding techniques, we doubt that much public outcry would exist regarding its development and widespread deployment. Transgenic crops represent one of the many tools available to agricultural scientists for use in integrated pest management programs. As such, the technology should be evaluated objectively, as is the case of any other pest control tactic. However, critical to this objective assessment is the acceptance of genetic engineering as an extension of conventional plant breeding, rather than being a radically new and untested technology. One encouraging example of this acceptance is the recently-reported adoption of transgenic tobacco by the Amish—a society traditionally known for its aversion to new technologies (Strawley, 2001).

In conclusion, we feel that transgenic crops have a great deal of potential to contribute to the development of integrated pest management (IPM) strategies for crops, because they can confer resistance against pests even in situations where resistance was previously unavailable. However, because of the commercial aspects of transgenic crops, especially those expressing Bt, the development, testing and deployment have proceeded at an accelerated pace, with little participation from IPM scientists. Thus, we are currently in a situation where the IPM, in some cases, has had to adapt to transgenic cultivars, rather than vice versa. Consequently, scientists are left to address issues that should have been addressed prior to deployment. As agricultural scientists, correcting this situation represents our greatest problem, and our greatest opportunity.

ACKNOWLEDGEMENTS

We thank J.S. Bernal (Texas A&M University) for helpful comments on the manuscript. T. E. Mirkov and Z. Yang (Texas Ag. Expt. Sta., Weslaco, Texas) performed the sugarcane transformation. E. van Damme and W.

Peumans (Katholieke Universiteit Leuven, Heverlee, Belgium) provided cDNA encoding GNA. Invaluable technical support was given by R. R. Saldaña, M. Garcia, S. Alvarez, J. Huerta, M. Diaz, E. Bustamante, D. Alejandro, A. Cardoza, and E. Hernandez (TAES, Weslaco). Research funding was provided by the American Sugarcane League, Inc., the Texas A&M University Research Enhancement Program and USDA-NRI-CGP Project Number 99-35316-7913. The Texas Agricultural Experiment Station sugarcane program received much support from the Rio Grande Valley Sugar Growers, Inc. (Sta. Rosa, Texas) and Hatch Project No. 8595. This article reports the results of research only. Mention of a commercial or proprietary product does not constitute an endorsement or recommendation for its use by the US Department of Agriculture, nor the Texas A&M University System. Approved for publication by the Director of the Texas Agricultural Experiment Station and US Department of Agriculture.

REFERENCES

Bauchrowitz, M. A., Barker, D. G. and Truchet, G., 1996, Lectin genes are expressed throughout root nodule development and during nitrogen-fixation in the *Rhizobium—Medicago* symbiosis. Plant J., **9**, 31-43.

Bell, H. A., Fitches, E. C., Down, R. E., Marris, G. C., Edwards, J. P., Gatehouse, J. A. and Gatehouse, A. M. R., 1999, The effect of snowdrop lectin (GNA) delivered via artificial diet and transgenic plants on *Eulophus pennicornis* (Hymenoptera: Eulophidae), a parasitoid of the tomato moth, *Lacanobia oleracea* (Lepidoptera: Noctuidae). J. Insect Physiol., **45**, 934-991.

Bell, H. A., Fitches, E. C., Marris, G. C., Bell, J., Edwards, J. P., Gatehouse, J. A. and Gatehouse, A. M. R., 2001, Transgenic GNA expressing potato plants augment the beneficial biocontrol of *Lacanobia oleracea* (Lepidoptera, Noctuidae) by the parasitoid *Eulophus pennicornis* (Hymenoptera: Eulophidae). Transgenic Res., **10**, 35-42.

Birch, A.N.E., Geoghegan, I. E., Majerus, M.E.N., McNicol, J. W., Hackett, C. A., Gatehouse, A. M. R. and Gatehouse, J. A., 1999, Tritrophic interactions involving pest aphids, predatory 2-spot ladybirds and transgenic potatoes expressing snowdrop lectin for aphid resistance. Mol. Breed., **5**, 75-83.

Boulter, D., 1993, Insect pest control by copying nature using genetically engineered crops. Phytochemistry, **34**, 1453-1466.

Boulter, D., Edwards, G. A., Gatehouse, A. M. R., Gatehouse, J. A. and Hilder, V. A., 1990, Additive protective effects of incorporating two different higher plant derived insect resistance genes in transgenic tobacco plants. Crop Prot., **9**, 351-354.

Boulter, D., Croy, R. R. D., Ellis, R. J., Evans, I. M., Gatehouse, A. M. R., Gatehouse, J. A., Shhirsat, A. and J. H.Yarwood, J. H., 1986, Isolation of genes involved in pest and disease resistance. In: Magnien E. (ed.), Report of the EEC Biomolecular Engineering Programme, Martinus Nijhoff/Junk, The Hague, pp. 715-725.

Carozzi, N. and Koziel, M., 1997, Advances in Insect Control: The Role of Transgenic Plants. Taylor & Francis, Bristol, PA.

Chrispeels, M. J. and Raikhel, N. V., 1991, Lectins, lectin genes and their role in plant defense. Plant Cell, 3, 1-9.

Cole, R. A., 1994, Isolation of a chitin-binding lectin, with insecticidal activity in chemically-defined synthetic diets, from two wild *Brassica* species with resistance to cabbage aphid *Brevicoryne brassicae*. Entomol. Exp. Appl. **72**, 181-187.

Czapla, T. H., 1997, Plant lectins as insect control proteins in transgenic plants. In: Carozzi N. and Koziel. M. (eds), Advances in Insect Control: The Role of Transgenic Plants, Taylor & Francis, Bristol, PA, pp. 123-138.

Czapla, T. H. and Lang, B. A., 1990, Effect of plant lectins on the larval development of European corn borer (Lepidoptera: Pyralidae) and Southern corn rootworm (Coleoptera: Chrysomelidae). J. Econ. Entomol., **83**, 2480-2485.

Damme, E. J. M. van, Smeets, K., Torrekens, S., van Leuven, F. and Peumans, W. J., 1994, Characterization and molecular cloning of mannose-binding lectins from the Orchidaceae species *Listera ovata, Epipactis helleborine* and *Cymbidium hybrid*. Eur. J. Biochem., **221**, 769-777.

Down, R. E., Gatehouse, A. M. R., Hamilton, W. D. O. and Gatehouse, J. A., 1996, Snowdrop lectin inhibits development and decreases fecundity of the glasshouse potato aphid (*Aulacorthum solani*) when administered in vitro and via transgenic plants in both laboratory and greenhouse trials. J. Insect Physiol., **42**, 1035-1045.

Down, R. E., Ford, L., Woodhouse, D., Raemaekers, J. J. M., Leitch, B., Gatehouse, J. A. and Gatehouse, A. M. R., 2000, Snowdrop lectin (GNA) has no acute toxic effects on a beneficial insect predator, the 2-spotted ladybird (*Adalia bipunctata* L.). J. Insect Physiol., **46**, 379-391.

Estruch, J. J., Carozzi, N. B., Desai, N., Duck, N. B., Warren, G. W. and Koziel, M., 1996, Transgenic plants: an emerging approach to pest control. Nature Biotech., **15**, 137-141.

Fitches, E., Gatehouse, A. M. R. and Gatehouse, J. A., 1997, Effects of snowdrop lectin (GNA) delivered via artificial diet and transgenic plants on the development of tomato moth (*Lacanobia oleracea*) larvae in laboratory and glasshouse trials. J. Insect Physiol., **43**, 727-739.

Gallo-Meagher, M. and Irvine, J. E., 1996, Herbicide resistant transgenic sugarcane plants containing the *bar* gene. Crop Sci., **36**, 1367-1374.

Gatehouse, A.M.R., and Gatehouse, J. A., 1998, Identifying proteins with insecticidal activity: use of encoding genes to produce insect-resistant transgenic crops. Pestic. Sci., **52**, 165-175.

Gatehouse, A. M. R., Dewey, F. M., Dove, J., Fenton, K. A. and Pusztai, A., 1984, Effect of seed lectin from *Phaseolus vulgaris* on the development of larvae of *Callosobruchus maculatus*: mechanism of toxicity. J. Sci. Food Agric., **35**, 373-380.

Gatehouse, A. M. R., Howe, D. S., Flemming, J. E., Hilder, V. A. and Gatehouse, J. A., 1991, Biochemical basis of insect resistance in winged bean, (*Psophocarpus tetragonolobus*) seeds. J. Sci. Food Agric, **55**, 63-74.

Gatehouse, A., Hilder, V., van Damme, E., Peumans, W., Newell, C. and Hamilton, W., 1992, Insecticidal Proteins, World Intellectual Patent Organization Application, No. WO 92/02139.

Gatehouse, A.M.R., Hilder, V. A., Powell, K. S., Wang, M., Davison, G. M., Gatehouse, L. N., Down, R. E., Edmonds, H. S., Boulter, D., Newell, C. A., Merryweather, A., Hamilton, W. D. O. and Gatehouse, J. A., 1994, Insect-resistant transgenic plants: choosing the gene to do the 'job'. Biochem. Soc. Trans., **22**, 944-948.

Gatehouse, A. M. R., Davison, G. M., Newell, C. A., Merryweather, A., Hamilton, W. D. O., Burgess, E. P. J., Gilbert, R. J. C. and Gatehouse, J. A., 1997, Transgenic potato plants enhanced resistance to the tomato moth, *Lacanobia oleracea*: growth room trials. Mol. Breed., **3**, 49-63.

Gould, F., 1998, Sustainability of transgenic insecticidal cultivars: integrating pest genetics and ecology. Annu. Rev. Entomol., **43**, 701-726.

Habibi. J., Backus, E. A. and Czapla, T. M., 1993, Plant lectins affect survival of the potato leafhopper (Homoptera: Cicadellidae). J. Econ. Entomol., **86**, 945-951.

Harper, M. S., Hopkins, T. L. and Czapla, T. H., 1998, Effect of wheat germ agglutinin on formation and structure of the peritrophic membrane in European corn borer (*Ostrinia nubilalis*) larvae. Tissue and Cell, **30**, 166-176.

Hepher, A., Edwards, G. and Gatehouse, J., 1989, Improvements relating to transgenic plants, European Patent Application No. 89201899.5.

Hilbeck, A., Baumgartner, M., Fried, P. M. and Bigler, F., 1998, Effects of transgenic *Bacillus thuringiensis* corn-fed prey on mortality and development time of immature *Chrysoperla carnea* (Neuroptera: Chrysopidae). Environ. Entomol., **27**, 480-487.

Hilbeck, A., Moar, W. J., Pusztai-Carey, M., Filippini, A. and Bigler, F., 1999, Prey-mediated effects of Cry1Ab toxin and protoxin and Cry2A protoxin on the predator *Chrysoperla carnea*. Entomol. Exp. Appl. 91, 305-316.

Hilder, V. A. and Boulter, D., 1999, Genetic engineering of crop plants for insect resistance—a critical review. Crop Prot., **18**, 177-191.

Hilder, V. A., Gatehouse, A. M. R., Powell, K. S. and Boulter, D., 1993, Proteins with insecticidal properties against homopteran insects and their use in plant protection. World Intellectual Patent Organization Application No. WO 93/04177.

Hilder, V. A., Powell, K. S., Gatehouse, A. M. R., Gatehouse, J. A., Gatehouse, L. N., Shi, Y., Hamilton, W. D. O., Merryweather, A., Newell, C. A., Timans, J. C., Peumans, W. J., van Damme, E. and Boulter, D., 1995, Expression of snowdrop lectin in transgenic tobacco plants results in added protection against aphids. Transgenic Res., **4**, 18-25.

Hokkanen, H. M. T. and Wearing, C. H., 1994, The safe and rational deployment of *Bacillus thuringiensis* genes in crop plants: conclusions and recommendations of OECD workshop on ecological implications of transgenic crops containing Bt toxin genes. Biocontr. Sci. Technol., **4**, 399-403.

Horsch, R. B., Fry, J. E., Hoffman, N. L., Wallroth, M., Eichholtz, D., Rogers, S. G. and Fraley, R. T., 1985, A simple and general method for transferring genes into plants. Science, **227**, 1229-1231.

Huessing, J. E., Shade, R. E., Chrispeels, M. J. and Murdock, L. L., 1991a, α-amylase inhibitor, not phytohemagglutinin, explains resistance of common bean seeds to cowpea weevil. Plant Physiol., **96**, 993-996.

Huessing, J. E., Murdock, L. L. and Shade, R. E., 1991b, Rice and stinging nettle lectins: insecticidal activity similar to wheat germ agglutinin. Phytochemistry, **30**, 3565-3568.

Ingelbrecht, I. L., Irvine, J. E. and Mirkov, T. E., 1999, Posttransscriptional gene silencing in transgenic sugarcane: dissection of homology-dependent virus resistance in a monocot that has a complex polyploid genome. Plant Physiol., **119**, 1187-1197.

Irvine, J. E. and Mirkov, T. E., 1997, The development of genetic transformation of sugarcane in Texas. Sugar J., **60**, 25-29.

Janzen, D. H., Juster, H. B. and Liener, I. E., 1976, Insecticidal action of the phytohemagglutinin in black bean on a bruchid beetle. Science, **192**, 795-796.

Kumar, M. A., Timm, D. E., Neet, K. E., Owen, W. G., Peumans, W. J. and Rao, G. A., 1993, Characterization of the lectin from the bulbs of *Eranthis hyemalis* (winter aconite) as an inhibitor of protein synthesis. J. Biol. Chem., **268**, 25176-25183.

Legaspi, J. C. and Mirkov, T. E., 2000, Evaluation of transgenic sugarcane against stalkborers. In: Allsopp P. G. and Suasa-ard W. (eds), Sugarcane Pest Management Strategies in the New Millennium. Intl. Soc. Sugarcane Tech. Brisbane, Australia, pp. 68-71.

Legaspi, J. C., Legaspi, B. C. Jr, King, E. G. and Saldaña, R. R., 1997, Mexican rice borer, *Eoreuma loftini* (Lepidoptera: Pyralidae) in the Lower Rio Grande Valley of Texas: its history and control. Subtrop. Plant Sci., **49**, 53-64.

Machuka, J., van Damme, E. J. M., Peumans, W. J. and Jackai, L. E. N., 1999, Effect of plant lectins on larval development of the legume pod borer, *Maruca vitrata*. Entomol. Exp. Appl., **93**, 179-187.

Maddock, S. E., Huffman, G., Isenhour, D. J., Roth, B. A., Raikhel, N. V., Howard, J. A. and Czapla, T. H., 1991, Expression in maize plants of wheat germ agglutinin, a novel source of insect resistance. 3rd Int. Cong. Plant Mol. Biol., Tucson, Arizona.

Mirkov, T. E. and Irvine, J. E., 1998, The development of genetic transformation of sugarcane in Texas. In: Rozeff N., Amador J. M. and Irvine J. E. (eds), South Texas Sugarcane Production Handbook, Texas A&M University and Rio Grande Valley Sugar Growers Inc., pp. 7.1-7.8.

Murdock, L. I., Huesing, J. E., Nielsen, S. S., Pratt, R. C. and Shade. R. E., 1990, Biological effects of plant lectins on the cowpea weevil. Phytochemistry, **29**, 85-89.

Nutt, K. A., Allsopp, P. G., McGhie, T. K., Sheperd, K. M., Joyce, P. A., Taylor, G. O., McQualter, R. B. and Smith, G. R., 1999, Transgenic sugarcane with increased resistance to canegrubs. Proc. Aust. Soc. Sugar Cane Technol., **21**, 171-176.

Omitogun, O. G., Jackai, L.E.N. and Thottappilly, G., 1999, Isolation of insecticidal lectin-enriched extracts from African yam bean (*Sphenostylis stenocarpa*) and other legume species. Entomol. Exp. Appl., **90**, 301-311.

Powell, K. S., Gatehouse, A. M. R., Hilder, V. A. and Gatehouse, J. A., 1993, Antimetabolic effects of plant lectins and plant fungal enzymes on the nymphal stages of two important rice pests, *Nilaparvata lugens* and *Nephotettix cinciteps*. Entomol. Exp. Appl., **66**, 119-126.

Powell, K. S., Gatehouse, A. M. R., Hilder, V. A., van Damme, E. J. M., Peumans, W. J., Boonjawat, J., Horsham, K. and Gatehouse, J. A., 1995, Different antimetabolic effects of related lectins towards nymphal stages of *Nilaparvata lugens*. Entomol. Exp. Appl., **75**, 61-65.

Pratt, R. C., Singh, N. K., Shade, R. E., Murdock, L. L. and Bressan, R. A., 1990, Isolation and partial characterization of a seed lectin from tepary bean that delays bruchid beetle development. Plant Physiol., **93**, 1453-1459.

Pusztai, A., Koninkx, J. , Hendriks, H., Kok, W., Hulscher, S., van Damme, E .J. M., Peumans, W. J., Grant, G. and Bardocz, S., 1996, Effect of the insecticidal *Galanthus nivalis* agglutinin on metabolism and the activities of brush border enzymes in the rat small intestine. J. Nutr. Biochem., **7**, 677-682.

Rahbé, Y. and Febvay, G., 1993, Protein toxicity to aphids: an in vitro test on *Acyrthosiphon pisum*. Entomol. Exp. Appl., **67**, 149-160.

Rao, K. V., Rathore, K. S., Hodges, T. K., Fu, X., Stoger, E., Sudhakar, D., Williams, S., Christou, P., Bharathi, M. and Brown, D. P., 1998, Expression of snowdrop lectin (GNA) in transgenic rice plants confers resistance to rice brown planthopper. Plant J., **15**, 469-477.

Sauvion, N., Rabhe, Y., Peumans, W. J., van Damme, E. J. M., Gatehouse, J. A. and Gatehouse, A. M. R., 1996, Effects of GNA and other mannose binding lectins on development and fecundity of the peach potato aphid *Myzus persicae*. Entomol. Exp. Appl., **79**, 285-293.

Sétamou, M., Bernal, J. S., Legaspi, J. C., Mirkov, T. E. and Legaspi, B., 2002a, Evaluation of lectin-expressing transgenic sugarcane against stalkborers (Lepidoptera: Pyralidae): effects on life history parameters. J. Econ. Entomol., **95**, 469-477.

Sétamou, M., Bernal, J. S., Legaspi, J. C. and Mirkov, T. E. 2002b, Effects of snowdrop lectin (*Galanthus nivalis* agglutinin) expressed in transgenic sugarcane on fitness of *Cotesia flavipes* (Hymenoptera: Braconidae), a parasitoid of the non-target pest *Diatraea saccharalis* (Lepidoptera: Crambidae) Ann. Entomol. Soc. Amer., **95**, 75-83.

Shukle, R. H. and Murdock, L. L., 1983, Lipoxygenase, trypsin inhibitor and lectin from soybeans: effects on larval growth of *Manduca sexta* (Lepidoptera: Sphingidae). Environ. Entomol., **12**, 787-791.

Strawley, G., 2001, Technology preserving rustic lifestyle of Amish. Associated Press, May 2, 2001.

Stoger, E., Williams, S., Christou, P., Down, R. E. and Gatehouse, J. A., 1999, Expression of the insecticidal lectin from snowdrop (*Galanthus nivalis* agglutinin; GNA) in transgenic wheat plants: effects on predation by the grain aphids *Sitobion avenae*. Mol. Breed., **5**, 65-73.

Tingjuangjun, P., Loc, N. T., Gatehouse, A. M. R., Gatehouse, J. A. and Christou, P., 2000, Enhanced insect resistance in Thai rice varieties generated by particle bombardment. Mol. Breed., **6**, 391-399.

Trigueros, V., Wang, M., Pere, D., Paquereau, L., Chavant, L. and Fournier. D, 2000, Modulation of a lectin insecticidal activity by carbohydrates. Arch. Insect Biochem Physiol., **45**, 175-179.

[USDA ERS] United States Department of Agriculture Economic Research Service, 2000, http://usda.mannlib.cornell.edu/reports/nassr/field/pcp-bbp/psp10300.txt

Vaeck, M., Reynaerts, A., Höfte, H., Jansens, S., de Beuckeleer, M., Dean, C. , Zabeau, M., von Montagu, M. and Leemans, J., 1987, Transgenic plants protected from insect attack. Nature, **328**, 33-37.

Zhu-Salzman, K., Shade, R. E., Koiwa, H., Salzman, R. A., Narasimhan, M., Bressan, R. A., Hasegawa, P. M. and Murdock, L. L., 1998, Carbohydrate binding and resistance to proteolysis control insecticidal activity of *Griffonia simplicifolia* lectin II. Proc. Natl. Acad. Sci., **95**, 15123–15128.

GENETICALLY-MODIFIED HERBICIDE-TOLERANT CROPS: A EUROPEAN PERSPECTIVE WITH A UNITED KINGDOM EMPHASIS

K. BERRY*, P.J.W. LUTMAN*, L.A.P. LOTZ** AND C. KEMPENAAR**

*Rothamsted Research, Harpenden, UK
**Plant Research International, Wageningen Ur, The Netherlands

INTRODUCTION

Since the early 1990s, the commercial development of genetically-modified herbicide-tolerant (GMHT) crops has proceeded swiftly across most of the world, especially in North and South America and China. By 2003, GM crops were being grown in 18 countries and for the first time, exceeded 65 million ha., of which 73% were herbicide tolerant. Six countries grew 99% of GM crops—USA (63%), Argentina (21%), Canada (6%), China (4%), Brazil (4%) and South Africa (1%) (James, 2003). However, GM crops have not been commercialized in Europe and the situation is more uncertain, with effectively a moratorium (at the time of writing this chapter) on commercialization.

The European farmer can look across the ocean and worry that the N. American farmers are deriving commercial advantage from producing crops more efficiently by using GM technology. European farmers are being denied access to these benefits. Conversely, the European consumer, fed a diet of sensationalist press reports, worries that this agrarian technology is a threat to human health and to the environment. The European regulator has to endeavour to plot a path through conflicting advice and attitudes and reach a conclusion as to whether to permit the commercialization of these crops. At present, good scientifically-based conclusions on the risks and benefits arising from commercialisation can only be based on information from commercial cropping elsewhere in the world, particularly from N. America, and from European-funded research projects (both by national governments and the EU). This chapter looks at the available data on GMHT crops from European eyes—especially from the United Kingdom—and discusses the perceived risks and benefits in relation to agriculture and the environment.

GMHT crops have the potential to induce a major impact on many aspects of European arable farming, as by utilizing this technology the application of broad-spectrum herbicides is no longer limited to the pre-emergence stage of the crop. The most likely GMHT crops to be grown in the UK and other areas of temperate Europe (e.g. France, Germany, The Netherlands) are oilseed rape, sugar beet and fodder maize that are tolerant to the herbicides glufosinate or glyphosate. As mentioned above, the possible commercialization of GM crops in Europe has raised many points for discussion. The debate centres on whether the potential benefits of the technology can outweigh concerns about its safety (Table 4.1). Supporters of GMHT crops refer to agricultural and environmental improvements ranging from a more convenient, efficient weed control system, to the increased use of minimum tillage, with its potential to reduce soil erosion and water loss. Conversely, opponents fear that their introduction and possible mismanagement would give rise to new HT weeds, HT volunteers and feral plants and a resultant loss of biodiversity. (Volunteers are those crop plants which appear in later crops from seed shed at harvest and so HT volunteers are from a HT crop. Feral plants are those present on roadside verges and field margins.) HT weeds can be of two types: those that have acquired herbicide tolerance via gene flow from HT crops, and those that have developed resistance through the selection pressure exerted by the continued use of the herbicide associated with the GMHT crop.

This chapter aims to look at the potential impact of herbicide tolerance on weed management, with reference to the present situation in Europe. Weed control, gene flow and associated side effects, and post-harvest management will all be discussed and related to the likely crops to be grown in Europe. The final section describes certain possible long-term weed management issues, which may affect the use of GMHT crops.

Table 4.1 Potential advantages and disadvantages of using GMHT crops in the UK

Parameter	Oilseed rape	Sugar beet	Fodder maize
Advantages			
More flexible herbicide timing	✓	✓	✓
Better control of resistant and troublesome weeds	✓	✓	✓
More convenient weed control	✓	✓	✓
Increased use of minimum tillage and other less intensive cultivations	✓	?	?
More environmentally-benign herbicides	✓	✓	✓
Possible decrease in herbicide use	?	✓	?
More reliable weed control	✓	✓	✓
Disadvantages			
Gene flow to wild relatives	✓	(✓)	✗
Gene flow to non-GM crops	✓	(✓)	✓
Gene stacking	✓	(✓)	✓
Seed persistence	✓	(✓)	✗
HT volunteers and feral populations	✓	✓	✗
Change in biodiversity/weed shifts	✓	✓	✓
Possible increase in herbicide use	✓	✓	✓
Loss of farmer control through 'Technology Use Agreements'	✓	✓	✓

(✓) Only of concern if GMHT beet 'bolts' and produces flowers (or is for seed production)

OILSEED RAPE (*Brassica napus* L.)

Oilseed rape (*B. napus*) and turnip rape (*B. rapa* L.) are important oil crops in China, Europe and Canada (Table 4.2), and were one of the first crops to be genetically modified, as the techniques used for transforming tobacco—one of the earliest GM species—were relatively easily adapted for oilseed rape. GM oilseed rape was first commercialized in Canada in 1995. In 2000, Canadian growers planted the largest area ever of GM oilseed rape, i.e. 69% of the total oilseed rape area (>3.0 million hectares), of which over 70% was GMHT Round-up Ready (Canola Council of Canada, 2001). In 2001, oilseed rape was farmed in Europe (EU states) on 3.0 million hectares. Thus, the management opportunities available for European farmers to benefit from GMHT oilseed rape are similar.

Weed Control

Weed control is one of the main concerns in management of oilseed rape. For example, 50% of Western Canadian farmers stated their main reason for growing GMHT *B. napus* was more convenient and efficient weed

Table 4.2 Production of sugar beet and oilseed rape in 2001 (FAOSTAT, 2002)

Country	Sugar beet (Mt)	Oilseed rape (Mt)
Canada	0.82	5.06
China	10.90	11.32
France	26.72	2.91
Germany	24.40	4.17
Poland	13.00	1.07
Russian Federation	14.54	0.11
UK	8.18	1.16
USA	23.36	0.91
Eastern Europe	25.60	2.64
European Union (15)	104.01	9.03
World	234.25	36.22

N.B. Turkey, Spain and Italy are among the top 10 producers of sugar beet; Austria, Czech Republic and Denmark are in the top 10 producers of oilseed rape.

control (Canola Council of Canada, 2001). GMHT technology offers a more flexible and convenient but equally effective weed control system. Amann (1998) found that depending on the weed infestation, glyphosate could be applied only once from autumn to early spring, while still sufficiently controlling the weeds present. Glufosinate has also exhibited high levels of weed control that are as effective as conventional systems (Read and Ball, 1999a; Gotz and Ammer, 2000).

Winter oilseed rape

Most oilseed rape grown in Europe is autumn-sown, which contrasts to the predominantly spring-sown Canadian (canola) crop. At present, herbicides used in conventional systems in the UK are primarily post-emergence but the emphasis in France, for example, is more on pre-emergence products. In the 2002 UK Pesticide Usage Survey (Garthwaite et al., 2003), the most popular herbicide against broad-leaved weeds was metazachlor, although clopyralid, trifluralin and propyzamide were also widely used, while cycloxydim and propaquizafop were applied against grass weeds. Elsewhere in Europe, the herbicide regime utilized in conventional systems can be very different. For example, France at present grows around one million hectares of oilseed rape each year. The main herbicide applied in France is trifluralin (30%), while others used include metazachlor (12%), and mixtures of tebutam plus clomazone (14%), or quinmerac plus metazachlor (8%) (Messean, pers.comm.). Thus, more than 50% of the weed control is based on pre-drilling or pre-emergence herbicides, which contrasts with the situation in the UK, where most products are applied post-emergence.

Table 4.3 shows the weed spectrum for the conventional herbicides used in winter rape in the UK, as compared to the GM herbicides. Glufosinate and glyphosate control most grasses and broad-leaved weeds as efficiently as conventional herbicides. The availability of GMHT cultivars would simplify the herbicide regime, as the farmer could apply one herbicide anytime in the autumn, or even in the spring, against most weeds, provided conditions were suitable for the application of the herbicide.

The broad-spectrum herbicides used in GMHT crops can control troublesome weeds, including crop volunteers, herbicide-tolerant weeds and weeds similar to the crop. In Canada, the use of GMHT oilseed rape cultivars has allowed the control of herbicide-resistant weed biotypes of common weeds like *Avena fatua* L. (wild-oats) and *Setaria viridis* (L.) Beauv. (green foxtail), as well as difficult-to-control weeds like *Galium aparine* L. (cleavers), *Chenopodium album* L. (fat hen), *Kochia scoparia* (L.) Schrad. (kochia) and perennial species like thistles (*Cirsium* spp. and *Sonchus* spp.) (Devine and Buth, 2001). Herbicide-tolerant varieties could have a similar benefit in the UK against resistant weeds like *Alopecurus myosuroides* Huds. (black-grass) and *A. fatua*, difficult weeds like *G. aparine* and weeds in the Brassicaceae.

Spring oilseed rape

Spring oilseed rape is more widely grown in Canada, though it is also cultivated in Europe, when autumn conditions are unsuitable for drilling, and where weather and cropping systems favour spring planting. Fewer herbicides have been registered for use in spring rape; so the farmer relies on good crop establishment and the competitiveness of the crop to outgrow the weeds, reducing herbicide inputs and costs. The main herbicides used in a conventional system are trifluralin against broad-leaved weeds and cycloxydim against grasses. The window of application is short due to fast crop growth; therefore, GMHT spring oilseed rape can increase the application window and allow control of weeds later in a crop which has established poorly.

Gene Flow

One possible negative consequence of growing GMHT crops, particularly oilseed rape, is that it could lead to gene flow. This can be defined as movement of a gene, via pollen or seed, followed by gene establishment in a new population. Pollen movement alone does not indicate the level of likely cross-fertilization as factors such as the size of the receptor pollen cloud and the viability of the incoming pollen must also be considered. Gene movement via pollen occurs in space and time and can result in gene flow to wild relatives, to other crops, (including other

Table 4.3 A selection of weed species controlled by conventional (CON) and GM herbicides in oilseed rape (OSR), sugar beet (SB) and fodder maize (FM) (based on company recommendations in product guides)
S—susceptible, MS—moderate susceptible, MR—moderate resistance, R—resistance.

Weed species	OSR CON herbicides		SB CON herbicides		FM CON herbicides			OSR/SB	GM herbicides	
	pre @ 2.5 l/ha metazachlor	pre @ 0.5 l/ha clopyralid	EC @1.7 kg/ha metamitron	EC @ 1.25 ml/ha phen + des + etho*	pre @3.0 l/ha atrazine	M @ 2 kg/ha pyridate	PT @ 2.0 l/ha bromoxynil	– / – graminicides	PT @ 3.0 l/ha glufosinate	PT @ 3.0 l/ha glyphosate
1	2	3	4	5	6	7	8	9	10	11
Grasses										
Agrostis spp.	–	–	–	–	MR	–	–	S	MS	S
Alopecurus myosuriodes	S	–	–	–	S	–	–	S	S	S
Avena sativa	–	–	–	–	–	–	–	S	S	S
Elymus repens	–	–	–	–	R	–	–	S	MS	S
Hordeum vulgare	R	–	–	–	–	–	–	S	–	S
Poa annua	–	–	–	–	S	–	–	R	S	S
Triticum aestivum	R	–	–	–	–	–	–	S	S	S
Broad-leaved weeds										
Brassica napus**	R	R	–	–	–	–	–	–	S	S
Beta sativa	–	–	R	R	–	–	–	–	S	S

Table 4.3 contd.

Table 4.3 contd.

1	2	3	4	5	6	7	8	9	10	11
Chenopodium album	–	–	S	–	S	S	S	–	S	S
Cirsium spp./ Sonchus spp.	–	S	–	–	S	–	–	–	S/MS	MS
Galium aparine	R***	–	R	S	MR	S	–	–	S	S
Lamium purpureum	–	–	–	S	S	–	S	–	S	S
Matricaria spp.	S	S	S	S	S	S	S	–	S	S
Papaver rhoeas	R	–	S	–	S	–	–	–	S	S
Polygonum spp.	MR	MS	S	S	MR	–	S	–	S	MS
Senecio vulgaris	S	S	S	S	S	–	S	–	S	S
Sinapis arvensis	–	–	S	S	S	–	–	–	S	S
Solanum nigrum	–	–	–	S	–	–	S	–	S	S
Stellaria media	S	–	S	S	S	S	–	–	S	S
Veronica spp.	S	–	S	S	MS	S	–	–	S	S
Viola arvensis	MR	–	S	S	MS	–	–	–	–	S

* phenmedipham + desmedipham + ethofumesate. ** In sugar beet, clopyralid (Dow Shield) is used against volunteer rape, Cirsium spp. and mayweeds; in fodder maize rimsulfuron (Titus) is used against volunteer oilseed rape. *** metazachlor + quinmerac (Katamaran) is used against cleavers. pre = pre-emergence application; EC = early cotyledon stage of weed; M = when maize is at 1 leaf stage; PT = post-emergence application

GMHT crops, i.e. gene stacking) and to feral populations. Conversely, gene survival via seeds occurs in time (i.e. seed persistence), leading to GMHT volunteers emerging in later years. There is evidence of hybridization to wild relatives in 44 of the world's crops, including banana, cassava, cotton, maize, millet, oats, potato, oilseed rape, rice, soybean and wheat. Ellstrand (2002) lists the 13 most important crops in the world and the wild relatives with which they are known to hybridize (Table 4.4). GM crops will probably behave in a similar manner to normal crops (Hokanson et al., 1997). Therefore, gene flow is liable to occur. This section will review the pollen dispersal for oilseed rape and consequent spatial gene flow, via pollen, to wild relatives and from crop to crop.

Table 4.4 Wild relatives which are known to hybridize with the 14 most important crop species.

Crop species		Wild relative
Wheat	Triticum aestivum	Wild T. aestivum, T. turgidum, Aegilops spp.
Rice	Oryza sativa	O. sativa f. spontanea
Maize	Zea mays	Teosintes, e.g. Zea mays ssp. mexicana
Soybean	Glycine max	G. soja, G. gracilis
Barley	Hordeum vulgare	H. spontaneum wild H. vulgare
Cotton seed	Gossypium hirsutum	Wild and feral G. hirsutum, G. barbadense
Sorghum	Sorghum bicolour	S. halepense and wild S. bicolor
Millet	Eleusine coracana	E. coracana ssp. africana
Beans	Phaseolus vulgaris	P. aborigineus, P. mexicanus
Rapeseed	Brassica napus	B. campestris/rapa, B. juncea
Groundnut	Arachis hypogaea	A. monticola
Sunflower seed	Helianthus annuus	Wild H. annuus
Sugar beet	Beta vulgaris	Wild annual B. vulgaris, B. vulgaris ssp. maritima
Sugarcane	Saccharum officinarum	Wild Saccharum spp.

Source: Ellstrand et al. (1999)

Pollen dispersal

Pollen movement and, therefore, the potential for gene flow, seems inevitable. Oilseed rape pollen can be dispersed by wind and by insects, e.g. honeybees (*Apis mellifera*). For spatial gene flow to occur, pollen must be disseminated beyond the crop perimeter, either to wild relatives or to other crops. Eastham and Sweet (2002) listed six main factors affecting pollen dispersal:

(i) size of pollen source and receptor;
(ii) pollination vectors;
(iii) environmental factors, i.e. weather, local environment and physical barriers like hedges;
(iv) pollen viability and competitive ability;
(v) crop outbreeding level; and
(vi) the synchrony of flowering times.

Rape pollen is viable for a period of 4-5 days in the field and during this time, it can be carried for several kilometres by wind or insects (Ramsey et al., 1999; Thompson et al., 1999). The likelihood of pollen movement and gene flow in oilseed rape is higher than for the other HT crops, as it is at least partly outbreeding, produces large amounts of pollen, is widely grown and has a number of European wild relatives.

It should be noted that the environmental and agronomic consequences of gene flow depend not only on the degree of pollen movement and frequency of cross fertilization but also on the impact of the gene(s) being transferred. If the gene(s) transferred have little environmental or agronomic impact, their effect will be small, however great the level of cross-fertilization. Thus, genes affecting seed quality will probably have less environmental impact than, for example, those influencing sensitivity to diseases or pests.

Gene flow to wild relatives

Gene flow to wild relatives has the potential to create problems including HT 'superweeds', loss of biodiversity and genetic pollution of native gene pools (Gray and Raybould, 1998). Oilseed rape is part of a genomic triangle of the diploid species *Brassica rapa* L. (*B. campestris*), *B. oleracea* L. and *B. nigra* (L.) Koch, and the tetraploids derived from them, *B. juncea* (L.) Czern. & Coss, *B. carinata* Braun and *B. napus* L. (Eastham and Sweet, 2002). Chevre et al. (1999) listed the following aspects that need to be studied in order to determine the probability of interspecific hybridization and introgression:

(i) the existence of close relatives growing in sympatry, with overlapping flowering times;
(ii) a mating system allowing pollen flow and allogamy;
(iii) the production and survival of fertile F_1 interspecific hybrids;
(iv) fertile plants produced in successive generations; and
(v) the possibility of chromosome recombination and spread of the gene in populations of the wild species. (This is affected by factors including chromosomal location, adaptation encoded for and wild population size (Darmency, 1994)).

In the 1990s, many gene flow studies on oilseed rape were undertaken and have shown variable out-crossing rates with wild relatives and subsequent hybrid persistence. A general rule seems to be that the closer the relationship between the parental genotypes, the greater the chance of hybridization. However, gene introgression to wild relatives will vary further, depending on the wild species population and the crop cultivar grown (Darmency, 1994; Baranger et al., 1995; Gueritaine and Darmency, 2001).

In Europe, various wild species have been investigated for potential hybridization to oilseed rape. The most studied species at present are *B. rapa*, *B. juncea* and *Raphanus raphanistrum* L. Table 4.5 lists the main wild relatives studied and their hybridization and back-crossing potential with oilseed rape.

B. rapa (wild turnip)

In Europe, *B. rapa* distribution varies with each country. In the UK, it is very localized, while in Denmark it is a common weed in agricultural systems, especially in oilseed rape fields. It is one of the most studied Brassicaceae, because oilseed rape is a tetraploid derivative of *B. rapa* and *B. oleracea*, and consequently they have a close genetic relationship. Oilseed rape can openly hybridize in the field or a laboratory with *B. rapa* to produce fertile, viable F_1 hybrids with either species as parent (Jorgensen et al., 1996; Mikkelsen et al., 1996a,b). Jorgensen et al. (1996) showed that a natural population of *B. rapa* in an oilseed rape field produced progeny, 60% of which were interspecific hybrids, although these hybrids produced fewer seeds per plant than the weedy parent. Later work by Metz et al. (1997) followed the transmission of the transgene into fertile F_1 hybrids, where it remained active.

Studies of fitness components of the hybrid plants produced showed that the hybrids were intermediate between their parents, for example in seed number/pod and pollen viability (Metz et al., 1997; Hauser et al., 1998b), while seed germination in the hybrids varied with some resembling one parent and some the other. For example, some seeds were dormant, like the weed and others non-dormant, like the crop (Landbo and Jorgensen, 1997). Once a hybrid is produced, the transgene can only become part of the wild population if the hybrid can back-cross (BC) with the wild relative in order to produce fertile BC generations. In Danish field studies, F_1 hybrids of *B. napus* and *B. rapa* were successfully back-crossed to the weed to form BC generations. In general, the BC_1 plants produced had a lower fitness than the parents for fitness components such as pollen viability and seed set (Hauser et al., 1998a).

Table 4.5 Hybridisation and back-crossing potential between oilseed rape (*Brassica napus*) and the listed wild relatives.

Wild relative			Hybridisation		Back-cross	Reference(s)
Latin name	Common name	crop ♂	crop ♀		to weed (BC$_1$)	
1	2	3	4		5	6
Brassica (g)						
Brassica rapa	Wild turnip, turnip rape	✓	✓		✓	Jorgensen et al. (1996), Mikkelsen et al., (1996a), Mikkelsen et al. (1996b), Metz et al. (1997), Landbo and Jorgensen (1997), Hauser et al. (1998a,b), Snow et al. (1999), Eastham and Sweet (2002)
B. juncea	Brown mustard	✓	✓		✓	Frello et al. (1995), Jorgensen et al. (1996)
B. oleracea	Wild cabbage	?✓	?✓		?✓*	Scheffler and Dale (1994), Wilkinson et al. (2000), Eastham and Sweet (2002)
B. carinata	Ethiopian mustard	?✓	?✓		?✓	Scheffler and Dale (1994)
B. nigra	Black mustard	?✓	?✓		?✓	Scheffler and Dale (1994), Eastham and Sweet (2002)
Other genera						
Raphanus raphanistrum	Wild radish, runch	✓	✓		✓	Baranger et al. (1995), Lefol et al. (1997), Chevre et al. (1997), Darmency et al. (1998).

Table 4.5 contd.

128 *Transgenic Crop Protection*

Table 4.5 contd.

1	2	3	4	5	6
					Chevre et al. (2000), Rieger et al. (2001), Gueritaine and Darmency (2001), Gueritaine et al. (2002), Eastham and Sweet (2002)
Sinapis arvensis	Charlock, wild mustard	?✓	?✓	?✓	Lefol et al. (1996), Chevre et al. (1996), Sweet et al., (1997), Moyes et al. (1999), Moyes et al. (2002)
Hirschfeldia incana	Hoary mustard	✓	✓	✓*	Lefol et al. (1995), Lefol et al. (1996a,b); Darmency and Fleury (2000)
Erucastrum gallicum	Dog mustard	?✓	?✓	?✓	Lefol et al. (1997)

✓ when a hybrid is produced by open pollination in field or glasshouse (even with low fertility)
?✓ when a hybrid is produced using controlled methods (including emasculation and hand pollination)
* BC hybrids became non-viable.
Source: Based on Scheffler and Dale (1994)

The BC_2 generation produced seeds which had pronounced seed dormancy and 42% of seeds still had the transgene present (Mikkelsen et al., 1996a). Further back-crosses between glufosinate-tolerant rape and *B. rapa* were performed by Snow et al. (1999) in the glasshouse. They obtained BC_3 plants (with *B. rapa*) which had 88-95% pollen fertility and ~50% chance of being HT. Therefore, there is potential for future transgene spread, and possible establishment, in the wild population both spatially and temporally (Hauser et al., 1998a). However, variation in fitness of BC generations may slow down the introgression of transgenes into these populations.

From these results, it seems clear that oilseed rape will cross and back-cross freely with *B. rapa* in the field (Jorgensen et al., 1996), and that gene transfer to *B. rapa* can potentially take place. Snow et al. (1999) showed that a transgene can introgress into *B. rapa* populations at no cost to the plants, i.e., BC_3 plants produced the same amount of seed as the wild weed. However, work in the UK by Scott and Wilkinson (1998) on two natural *B. rapa* populations that grew beside rape fields, showed that transgenes spread into the seed but were unlikely to form transgenic seedlings. Therefore, transgene spread is liable to be slow and uncertain unless the gene has a selective advantage. Other factors which will affect transgene spread into wild *B. rapa* populations include transgene location on the crop genome (whether the transgene is located on the ancestral *B. rapa*, *B. oleracea* or *B.nigra* part of the genome), the variation in F_1 and BC generations and selection pressures exerted on wild populations.

B. juncea (L.) Czern. & Coss (brown mustard)

This species is self compatible, but research has shown that upto 30% of the seeds are still obtained by out-crossing (Frello et al., 1995). Therefore, it is no surprise that hybrids do arise in the field by open pollination with oilseed rape in upto 3% of progeny (Frello et al., 1995; Jorgensen et al. 1996). However, pollen fertility is low. Production of hybrids using *B. juncea* as the pollinator have been less successful with low pollen fertility (Jorgensen, 1999). This is relatively uncommon in the UK and rare in and around arable fields.

Frello et al. (1995) conducted back-crosses of the F_1 hybrids to *B. juncea*, which produced BC_1 hybrids with improved pollen fertility, and these plants were able to produce seeds by open pollination in the glasshouse. The authors monitored for the presence of 20 markers from oilseed rape in the BC_1 generation. They found that some markers occurred at a higher frequency, which could be explained by gene copy number or the position of the gene on the chromosome. The transgene conferring resistance to the herbicide glyphosate occurred in 52% of the BC_1s.

Raphanus raphanistrum L. (wild radish, runch)

This is a common weed in France and the UK, and has been studied extensively by Chevre and colleagues in France. Its flowering period overlaps considerably with the flowering period of rape (Chevre et al., 2000). Comparison of formation of pollen tubes showed a directional preference, i.e. 34% of the weed pollen germinated on crop stigmas, whilst only 12% of crop pollen germinated on weed stigmas (Rieger et al., 2001). However, further work by Gueritaine and Darmency (2001) found that the amount of possible hybridization varied with respect to the individual plant phenotype, i.e. the ability of oilseed rape pollen to hybridize with *R. raphanistrum* varied with the cultivar and the weed population used. It may, therefore, be possible to breed for the varieties which are less likely to hybridize.

The formation of pollen tubes does not mean that hybrid seed will be produced but studies have shown that hybrid seeds can be produced, especially on the crop plant. In controlled crosses Lefol et al. (1997) produced vigorous but sterile hybrids on the crop plant, while Rieger et al. (2001) in the field obtained fertile hybrids from the crop only. Work in France using a varietal association (Synergy, 80% male sterile) (Chevre et al., 2000) or male sterile rape (Baranger et al., 1995) produced fertile hybrids (low pollen fertility) on the crop. Under normal field conditions, competition from crop pollen will limit the number of hybrids, which can be produced on the crop plant from weed pollen. The chances of obtaining an interspecific hybrid on the weed is increased when there is a lone *R. raphanistrum* plant, due to self-incompatibility and high pressure from crop pollen (Darmency et al., 1998). However, in the UK, a five-year programme monitoring gene flow from GM oilseed rape crops to *R. raphanistrum* found no evidence of hybridization (Eastham and Sweet, 2002).

As the likely gene flow will be towards the crop (weed pollen fertilising crop ovules rather than *vice versa*), chances of movement into the wild population appear to be low. Hybrids, which occur in the field, can be controlled by herbicides used to destroy HT volunteers. But, as for *B. rapa*, hybrid formation with *R. raphanistrum* is affected by crop cultivar, i.e. pure line or hybrid, which could lead to a significant amount of variation in hybrid production (Baranger et al., 1995).

From the above studies, formation of hybrid seeds on the weeds seems unlikely; however, if F_1 hybrids were produced, would they be able to back-cross into *R. raphanistrum*? Darmency et al. (1998) produced BC_1 hybrids which had a higher seed production than the F_1s. Chevre et al. (1997) produced fertile HT plants with chromosome numbers similar to the weed after four back-crosses to the weed. In this study, female fertility increased continuously with each back-cross generation. Gueritaine et al. (2002) found after five back-crosses with *R. raphanistrum*, followed by a final back-cross to oilseed rape, the progeny had 1% seed production of those whose final back-cross was to the weed, and a lower fitness which included lower seed germination, survival and biomass. The BC_6 plants, i.e. when the final back-cross was with the weed, could act as a stepping point for introgression into the wild radish population, because female fertility is gradually restored. However, in this case the transgene (glufosinate-ammonium resistance) had a negative effect on the fitness, as resistant plants had much lower fitness than the non-resistant plants. They had reduced flowering rates, pollen fertility and seed production. The authors suggested that this effect was due to the chromosomal positioning rather than the transgene itself, and may not be seen if chromosomal positioning was different.

Sinapis arvensis L. (charlock)

In the UK, this weed probably poses the greatest potential danger for hybridization. It is an obligate out-breeder and widespread in the UK, overlapping all these areas where oilseed rape is grown, thus increasing contact and likelihood of hybridization. However, *S. arvensis* is not thought to be compatible in the field with rape, as the only way of getting hybrids is by hand pollination, embryo rescue or ovule culture (Chevre et al., 1996; Lefol et al., 1996a; Sweet et al., 1997; Moyes et al., 1999). Recent work by Moyes et al. (2002) obtained a hybrid using hand pollination of emasculated buds with *S. arvensis* as the maternal parent. Crossing rates were always very low (<0.01%), especially when the weed was the maternal parent (~0.002%), and there was no variation in results due to cultivar or weed population. Hybrid fitness was also very low, as none of the hybrids formed were able to back-cross with the weed, and few seeds were produced when back-crossed with the crop (Lefol et al., 1996a; Moyes et al., 2002).

Hirschfeldia incana (L.) Lagrèze-Fossat (B. adpressa) (hoary mustard)

Several authors have studied this weed. *H. incana* has been found to be a self-incompatible weed, which is found in the mild oceanic borders of Europe, especially the Mediterranean. Lefol et al. (1996b), working in France, studied the fertilization potential of rape and weed ovules using foreign pollen. Seed set of hoary mustard ovules by oilseed rape pollen was 1.67% of that by hoary mustard pollen, while it was 0.13% in the reciprocal cross. Hybrids have been produced by open pollination using male sterile rape grown with the weed (Lefol et al., 1996b; Darmency and Fleury, 2000). Analysis of hybrid seed from rape showed intermediate vegetative characteristics, for example, leaf number and biomass. However, the characteristics changed, depending on environmental conditions (Lefol et al., 1995). Although the hybrids seemed more competitive as compared to the weed, they had a very low seed output, and when back-crossed to the weed, they became non-viable by the fifth generation (BC_5), when no viable seeds were produced (Darmency and Fleury, 2000). These results suggest that good weed management can control these hybrids. However, they would be likely to survive in the field the year in which they germinate.

Work on seed survival by Chadoeuf et al. (1998) in cultivated soil showed a lower survival of buried hybrid seeds when compared to the weedy parent, but a similar survival rate in comparison to the rape parent. Hybrid survival was intermediate between the parents in undisturbed soil, although like the rape parent no seeds germinated after 41 months.

Other wild relatives

Other relatives which have been looked at for their potential to hybridize with oilseed rape include its progenitor species *B. oleracea* (wild cabbage). This is a long-lived perennial and is rare in the UK, occurring only on a few well-known coastal sites (Wilkinson et al., 2000). Hybridization is possible and the production of BC fertile hybrids after three generations using *B. oleracea* as the pollinator (Chevre et al., 1998) has been achieved. However, there have been no reports of hybridization by open pollination (Scheffler and Dale, 1994; Eastham and Sweet, 2002). Other Brassica species that have been studied are *B. carinata* Al. Braun (Ethiopian mustard) and *B. nigra* (L.) Koch (black mustard). Hybrids with *B. carinata*. have been obtained, especially when rape was the female parent (Scheffler and Dale, 1994), although male fertility was low and few seeds were produced. Hybridization in the field with *B. nigra* seems unlikely as no hybrid has ever been identified in nature and even under controlled conditions, those produced had low fertility (Scheffler and Dale, 1994; Eastham and Sweet, 2002). Finally, *Erucastrum gallicum* (Willd.) O.E. Schulz (dog mustard), can cross with *B. napus* to produce a hybrid which was able to back-cross with *E. gallicum* so as to produce fertile offspring (Lefol et al., 1997). The hybrids grew vigorously to produce abundant seed if transplanted into a non-competitive environment but when grown within a rape field, they failed to mature enough to produce seed.

As it seems likely that transgene escape to wild relatives could occur, albeit rarely, an important question is: what affect will the gene have in the wild population? Will the presence of the transgene in the wild population have a selective advantage? The effect of the transgene is difficult to predict as the manner in which a gene will be expressed in a different genetic background, i.e. a wild population, is not clear (Eastham and Sweet, 2002). The presence of a HT transgene is unlikely to confer a fitness advantage or disadvantage to the weed unless the herbicide is used (Metz et al., 1997; Snow et al., 1999). However, a transgene against a biotic or an abiotic stress factor could be of selective advantage in a wild population (Hancock et al., 1996). This fitness advantage of a transgene will depend on the gene, the ecosystem and the selective pressures on the transgene (Jorgensen, 1999). Finally, it is important to note that gene flow is a potential problem for all HT crops— whether produced by GM technology or by other plant-breeding techniques.

Crop-to-crop gene flow

Crop-to-crop gene flow is the movement of genes from one crop to another, and is used here to refer to gene flow between the same crop species. The chances of gene flow to adjacent rape crops vary with the cultivars used, as hybrid cultivars along with those which are varietal associations have a greater susceptibility to external pollen than normal cultivars, as they have a proportion of male sterile individuals, that are as a consequence more vulnerable to incoming pollen.

The potential for crop-to-crop gene flow in oilseed rape directly relates to pollen movement. This raises questions about the need for isolation distances between adjacent crops. Present isolation distances used in the UK for GMHT rape field experiments (e.g. Farm Scale Evaluations (Firbank et al., 1999) see 'environmental effects' – below) are designed to reduce gene flow to below 1%, are:

    200m from organic crops;
    50m from conventional varieties and restored hybrids; and
    100m from varietal association and partially-restored hybrids.

(Agriculture and Environment Biotechnology Commission, 2001). These isolation distances are based on those currently used for production of certified seed of oilseed rape. The Soil Association, one of the organizations representing organic farmers in the UK, has suggested that a separation distance of 6000m (Treu and Emberlin, 2000) is needed in order to prevent organic crops being 'contaminated' by GM rape crops. The question is: 'what isolation distance could reduce gene flow to other crops to an acceptable level and what is this acceptable level'? Pollen movement of transgenic oilseed rape can be directly compared to conventional varieties and conventional pollen has been detected as far as 4 km from the nearest source (Thompson et al., 1999). This does not necessarily mean that fertile cross-pollinated seeds would have been produced, because the success of incoming pollen will be affected by its viability, the size of the receptor pollen cloud and the cultivar being grown. Studying gene flow on a regional scale is difficult and problematic as sources or receptors of pollen will include feral populations, which when combined with wild relatives, may act as possible escape routes for transgenes spatially and temporally (Squire et al., 1999; Thompson et al., 1999). Isolating all GMHT oilseed rape by 4 km or more from other oilseed rape crops is impractical and unworkable and would be a surrogate for banning the technology. Shorter isolation distances could still cause practical management problems and hence could prove to be expensive. Perry (2002) suggests that if only 1 in 200 pollination events are over long range (in excess of 200 m)—perhaps as a result of the activity of bees—then it could be necessary to have large separation distances, in excess of 1km, to keep cross-pollination thresholds to very low levels. It seems necessary to accept that 'zero tolerance' of gene flow from GMHT oilseed rape crops is not possible, and that if permission is given to commercialize the crop, some gene flow will occur. There is a need to choose a manageable isolation distance so as to keep the level within nationally-defined threshold levels and at an agreed percentage of 'contamination'/cross-fertilization.

If gene flow occurs, what level of cross-fertilization into non-HT oilseed rape crops will occur? Out-crossing rates are variable but are estimated at c.1% within 100m of the common field border. The rate decreases rapidly after 50m but is still present at 800m (0.07%) (Scheffler et al., 1995; Downey, 1999; Beckie, 2001; Rieger et al., 2002). Therefore, 100m could be an acceptable isolation distance if c.1% is a tolerable level of gene flow. However, this would probably be an unacceptable contamination for some organic growers and certified seed growers. The problem of cross-fertilization is greater for hybrid seed crops than for non-hybrid ones as the breeding system uses male sterile varieties in order to make hybrid seed. If the pollinating variety does not establish well, there is the opportunity for extensive 'foreign' pollen to cross-fertilize the crop. Consequently, isolation distances for hybrid certified seed crop production would have to be greater. At the moment, the threshold being considered in the EU is 0.9%, although there has been a suggestion of reducing acceptable contamination to 0.1%, which would mean an as yet unspecified increase in the separation distances.

Another aspect of crop-to-crop gene flow is the potential for gene stacking, which could be a major management concern for farmers. Gene stacking of HT genes could occur spatially if several HT oilseed rape crops are grown close together (Champolivier et al., 1999; Downey, 1999;

Kempenaar and Lotz, 1999; Beckie, 2001), or temporally, if rape crops with different herbicide tolerances are grown in sequence and volunteers appeared in the later GMHT crop. Gene stacking would produce volunteers or feral plants that are resistant to more than one herbicide, and these volunteers could be difficult to control if tolerant to an entire range of herbicides. Good control of multi-tolerant volunteers is vital to ensure that more seed is not produced to cause problems in following crops. In practice, the control of herbicide-resistant canola volunteers has not caused major problems in Canada, as alternative herbicides to glyphosate and glufosinate are available for their control—even those with stacked genes, in most other crops.

A widely reported incident, which led to gene stacking, occurred in 1997/98 in Canada. A field in Alberta was planted with both glufosinate- and imazethapyr-tolerant oilseed rape and separated by a 22m road from a glyphosate-tolerant rape crop. In 1998, glyphosate alone was applied to kill volunteers in the field. However, it did not kill all the volunteers, indicating that gene flow had occurred (Downey, 1999). Subsequently, seeds were collected from these volunteers and tested for herbicide resistance, of which two plants showed triple resistance (Hall et al., 2000). This report shows what could happen if adequate care is not taken to control volunteers and also emphasizes the need to avoid planting HT oilseed rape crops tolerant to different herbicides close to each other. It will be impossible to prevent very low levels of gene flow, but strategies can be put in place to bring it down to a minimal level.

Post-harvest Management

Transgenes can escape in time through the development of dormancy in the rape seed and, thus, its persistence in the soil seed-bank. The potential for a long-lived, large seed bank is present, as estimates of seed losses range from 5-10,000 seeds/m^2, which can persist for more than 8 years (Pessel et al., 2001). Post-harvest management affects the number of seeds that will become dormant, as both the timing and type of cultivation alter the secondary seed dormancy and therefore persistence. Pekrun et al., (1998) showed that seeds become dormant if they are kept dry and in the dark. Thus, if post-harvest cultivation is delayed, leaving the seeds in the light until after it rains, this causes a flush of seedlings and reduces the potential number of dormant seeds. It is important to minimize seed persistence as seed numbers in the seed bank reflect the number of volunteers in later crops, and similarly, good control of volunteers will lead to a rapid decrease in seed bank levels (Pekrun et al., 1999; Legere et al., 2001). If a volunteer is from a herbicide-tolerant crop, there are fewer chemical control options, especially if gene stacking has occurred. However, there is still a range of alternative herbicides, herbicide mixtures and non-chemical control methods that can be put into use. For example, in the UK, farmers may switch to a paraquat/diquat mixture for post-harvest weed management, or even use a hormone herbicide like mecoprop in order to enhance the control of rape volunteers. This may increase the cost of control.

It is likely that GM volunteers will occur after a GM crop. However, research by Linder (1998) suggested that in a controlled environment, for most test conditions, GM varieties with modified oil content have the same persistence rate as equivalent non-GM varieties. At present, there is no evidence that GM volunteers are more weedy or invasive compared to traditional varieties (Rasche and Gadsby, 1997). A study by Fredshavn et al. (1995) compared three transgenic lines of the cultivar Drakkar with three non-transgenic cultivars (Drakkar, Line and Topas). When grown in monocultures, or mixtures with barley, none of the transgenic lines were more competitive than the non-transgenic cultivars. These studies suggest that control of GMHT volunteers will be comparable to controlling non-GMHT volunteers.

Feral Rape

Oilseed rape forms feral populations on disturbed soils and in semi-natural habitats. These are often found along major roads and motorways due to loss of seed in transit. If the soil is not continually disturbed, there is a continual turnover of populations, with some plants reappearing after a successful control measure (Wilkinson et al., 1995). Further investigation of these populations using PCR analysis found evidence of cultivars that were commercially obsolete, suggesting self-sustaining feral populations or a long-lived viable seed bank. However, Crawley and Brown (1995) reported that most populations become extinct after 2-4 years due to competition from perennial grasses. GMHT feral rape populations are unlikely to be more invasive or have enhanced survival because of the HT transgene, especially when the herbicide is not present (Crawley et al., 1993; Norris et al., 1999; Beckie, 2001; Crawley et al., 2001; Pessel et al., 2001), although variability in persistence is likely with site (Hails et al., 1997). Crawley et al. (2001) monitored oilseed rape for 10 years at 12 sites in the UK, and found no difference in ecological competitiveness, i.e. persistence or invasiveness, between genetically-modified and conventional plants. However, it is important to note that the easiest way to control feral rape populations is to prevent them appearing, through careful harvesting, packaging and transportation.

SUGAR BEET (*Beta vulgaris* ssp *vulgaris* L.)

Sugar beet (*B. vulgaris* ssp *vulgaris* L.) was developed in Europe during the Napoleonic Wars. Original forms were fodder beets (Silesian beets) and had only 4% sugar, while modern varieties have upto 18% sugar. Sugar beet is an important crop in Europe, where over 50% of the world's yield is produced (Table 4.2). Sugar beet is sensitive to weed competition, especially for the first eight weeks after crop emergence and, traditionally, has had high herbicide inputs. Quantities used peaked in the 1970s and 1980s and since then, the use of repeat low dose herbicide strategies have reduced the amount of product used, especially for grasses, though the number of treatments has increased. GMHT sugar beet may have the potential to reduce further the herbicide load without risking decreased yield through weed competition.

Weed Control

In the UK, conventional herbicide programmes in sugar beet endeavour to ensure no yield reduction due to weed competition, with a grower spraying upto seven times with a variety of herbicide types (Pesticide Safety Directorate, 1998). Such multiple sequences are widespread in many other sugar beet growing countries in Europe, though with perhaps greater emphasis on mechanical control than is practiced in the UK. Initially, residual herbicides are sprayed pre-crop emergence followed by a series of post-emergence mixtures depending on the weed infestation (Table 4.6). The most common herbicides applied in the UK against broad-leaved weeds are: metamitron, phenmedipham, clopyralid, triflusulfuron-methyl and ethofumesate, while cycloxydim and propaquizafop are applied against grass weeds (Garthwaite and Thomas, 1998). This conventional programme is time-consuming, inconvenient, complicated and expensive, as well as potentially damaging to the crop (Wilson, 1994) and to the environment, as a large amount of active ingredient is applied. In The Netherlands, conventional weed control programmes in sugar beet are comparable to those in UK (Lotz et al., 2000). Such programmes generally use 2.4 kg active ingredient (a.i.) per ha (on average 3.5 pre- and post-emergence sprays). In addition, an integrated weed management programme was recently adopted by Dutch farmers and used only upto 1.8 kg a.i. per ha. (on average 3 post-emergence sprays and mechanical control) (Wevers, 1998; Lotz et al., 2000).

Arguably, one of the most damaging weeds in sugar beet in the UK is weed annual beet, which occurred on ~70% of sugar beet fields in 2001 (BBRO, 2002). Weed beet, at a density of 1 plant/m^2, can reduce yield by 11%. The exact origins of bolters and weed beet are varied and include

Table 4.6 Examples of herbicides and target weeds used in a conventional herbicide programme in sugar beet

Weed	Timing	Herbicide applied
Broad-leaved weeds	Pre-drilling	Glyphosate
	Pre-emergence	Metamitron
	Post-emergence	Mixtures of metamitron, phenmedipham, ethofumesate and desmedipham
Grasses	Pre-drilling	Glyphosate, especially against common couch
	Post-emergence	Graminicides like cycloxydim, propaquizafop or tri-allate if severe
Problem weeds	Pre-emergence	Chloridazon and lenacil against volunteers
	Post-emergence	Clopyralid against thistles and mayweeds

Source: Garthwaite and Thomas (1998)

sources from groundkeepers, feral seed banks, existing annual plants that hybridize with crop bolters and from vernalized crop plants. Weed beet between sugar beet rows can be controlled using tractor hoeing or application of trifluralin at crop 4-6 leaf stage. However, the control of weed beet within sugar beet rows is extremely difficult and can only be done after bolting has occurred. Some control of these plants can be achieved by hand roguing if the infestation is low; weed wiping with a non-selective herbicide (glyphosate) and cutting if the infestation is high. All methods of controlling weed beet are time consuming, repetitive and susceptible to failure if carried out incorrectly. Potato volunteers also constitute a particular threat to sugar beet crops when both potato and sugar beet are present in a crop rotation. Control of potato volunteers in sugar beet is also difficult and time consuming. Dutch farmers sometimes spent more than 10 hours per ha in controlling potato volunteers chemically (selective treatment with glyphosate or band spraying with clopyralid) or mechanically (hand weeding) (Kempenaar and Lotz, 1999).

GMHT sugar beet simplifies the weed control programme by combining a more flexible application timing along with reliable weed control. Such strategies allow easy control of troublesome weeds like oilseed rape and potato volunteers, herbicide-resistant weeds and weed beet, while reducing the herbicide load on the environment compared to conventional programmes. Research has shown that a two (2 x 3 l/ha.) or three (3 x 2 l/ha.) post-emergence spray programme of glyphosate or glufosinate would achieve similar weed control to the conventional programme (Madsen and Jensen, 1995; Moll, 1997; Read and Bush, 1998; Wevers, 1998). Finally, GMHT sugar beet could allow the use of new management strategies, which could benefit the environment. For example, Pidgeon et al., (2001) reported experiments where band spraying of glyphosate in GMHT beet was used initially only to control weeds within the rows. This caused an increase in beetle numbers (e.g. Carabidae) on the treated plots, which was apparently associated with the weeds remaining between the rows, which were not controlled until much later.

Although the weed control opportunities of GMHT crops are clear, these very same opportunities have the potential to alter the management of sugar beet. Glufosinate and glyphosate are not residual, and the lack of persistence could, in some cases, lead to a more active ingredient being applied. Delayed spraying, which is possible with these herbicides, as outlined above, could lead to yield loss if the herbicide is applied too late.

Gene Flow

This section on gene flow primarily applies to sugar beet plants that set seed. Most plants in commercial crops do not 'bolt' but if they are sown early or are exposed to a late frost, many plants can start flowering. Sugar beet can hybridize freely with the annual, weed beet (weedy *B. vulgaris* ssp *vulgaris*) and wild beet (*B. vulgaris* ssp *maritima* L.) on seacoasts of Europe and Asia. Weed beets will form flowering shoots in sugar beet, which hybridise with other 'natural' bolters, including any which may be transgenic, or with wild beet (Vigouroux et al., 1999). The main difficulties associated with gene flow are in areas of Europe such as southern France and Italy that produce seeds for sugar beet crops. There is good evidence that in these areas, hybridization between wild *Beta* spp. and sugar beet has occurred in the past and the risks of GM traits escaping into wild species is significant (Bartsch et al., 1999).

Since all 3 types of beet (cultivated, weed and wild) can exchange genes freely, the chance of gene escape is also appreciable in commercial beet crops, but only if the beet plants bolt and produce flowers. This stage needs to be prevented by strictly removing all bolters from GMHT beet crops, choosing varieties that don't bolt and producing seed for transgenic beet well away from wild beet populations. The present isolation distances used in the Farm Scale Evaluation experiments with glyphosate-resistant beets are 600m from seed or organic crops and 6m from other crops, because the beet is not permitted to flower in the experiments (Agriculture and Environment Biotechnology Commission, 2001). However, the UK Soil Association has suggested an isolation distance of 1000m (Treu and Emberlin, 2000). If hybridization does occur with a HT sugar beet, then research has suggested that hybrids are likely to be more competitive than both parents (Madsen et al., 1998). Further, if a GMHT crop does contain flowering bolters, the most likely candidate for hybridization is weed beet due to its proximity to the crop. In coastal areas, it is possible that hybridization with wild (sea) beet could also occur.

Fredshavn and Poulsen (1996) investigated the competitive ability of sugar beet compared to sea beet and an interspecific hybrid. In monocultures, there was no significant differences between the subspecies when comparing their above-ground biomass, while competitiveness of the subspecies—when in different density mixtures—showed that sugar beet and the hybrid were less influenced by competition than the wild sea beet. All beet lines were poor competitors when compared to other plant species (in this study, oilseed rape and barley). Therefore, it is plausible to conclude that only a dramatic change in the competitiveness of beet would change its distribution and invasiveness in natural habitats. Further experiments comparing eight transgenic sugar beet lines with a non-transgenic line produced no significant differences in biomass production.

Post-harvest Management

An important aspect of post-harvest management in sugar beet is to reduce the numbers of beet seeds returning to the seed bank to reduce future weed beet problems. Beet seed is known to become dormant and persist in the seed bank for >10 years, so it is important to delay cultivation after harvest in order to encourage seed germination and predation (BBRO, 2002). A shallow tine-cultivation also encourages seed germination, which can be controlled in a following cereal. Crawley et al. (2001) monitored conventional and transgenic sugar beet lines in natural habitats and all sugar beet lines were extinct by the end of Year 2. A further source of flowering beets is from re-grown crowns of sugar beet roots left in the field after harvest. These crowns should be buried or removed from the field after harvest to prevent regrowth. Regrowth of these groundkeepers and seedlings can be controlled in cereals using a sulfonyl urea herbicide (e.g. metsulfuron). An effective method to reduce weed beet seed is to grow a spring-sown crop, e.g. carrots or linseed.

Occasionally, a sugar beet crop cannot be harvested because of exceptional conditions, such as severe rains in the autumn of 1998 in The Netherlands. Farmers who were not able to harvest their sugar beet crops that year were recommended to treat the beets with glyphosate to prevent sugar beet growth in the next crop. Obviously, this recommendation would not apply to a GMHT crop.

MAIZE (*Zea mays* (L.))

Maize (or corn) is an important crop worldwide, especially in North and South America. In the USA, GMHT maize has not been as well received as other GMHT crops, e.g. soybean (Owen, 2000), although Bt-tolerant maize is grown quite widely. The conventional weed control system in maize is cheap and effective, although the use of glufosinate as part of this system could improve the control of some weeds (Culpepper and York, 1999).

Maize is widely grown in France, Germany and The Netherlands, but does not constitute a major crop in the UK. It is mostly grown in the south of England for fodder.

Weed Control

Maize is susceptible to weed competition for the first 8-10 weeks after emergence. In the UK early spring growth is slow due to low temperatures, so good weed control is important in order to maintain yield. In conventional systems, the herbicides which are regularly used in the UK include atrazine, bromoxynil, pyridate (against *Solanum nigrum* L (black nightshade) and *C. album*), rimsulfuron (against volunteer rape), and graminicides (Table 4.2). Atrazine is a residual herbicide that controls a wide variety of species and is the basis of maize weed control in many countries, including the UK. However, because of its persistence, it has become a serious pollutant of groundwater and, as a consequence, it has been banned in a number of EU countries. Problem weeds in maize include atrazine-tolerant *S. nigrum*, *C. album*, *Atriplex patula* L. (orache), *G. aparine* and *Polygonum* spp. as well as volunteers of potato and oilseed rape. In recent years, some new herbicides have been registered in maize in the EU, including dimethenamid, sulcotrione and nicosulfuron. The number of registered herbicides in maize is relatively high, offering farmers a wide range of options. Modern weed control programmes in The Netherlands follow integrated management approaches (as a result of the EU regulations on cross compliance) and use less than 1 kg a.i. per ha. (e.g. Lotz et al., 1999; Schans and Weide, 2001). This integrated system is now used on over 60% of the total Dutch maize acreage. Harrowing is part of this system.

Herbicide-tolerant maize varieties would have similar benefits to those already discussed for HT sugar beet and oilseed rape, as they can improve the control of volunteers such as potato and oilseed rape and problem weeds like *S. nigrum* and *G. aparine*. Read and Ball (1999b) showed that a dual application of 2 x 2 l/ha of glufosinate will achieve the same level of weed control as the present conventional system. An important benefit of GMHT maize could be the reduction in water contamination. For example, it is predicted that in the USA, herbicide run-off could be reduced to 20% of the present value for atrazine, by adopting GMHT maize (Wauchope et al., 2001). Glufosinate and glyphosate are more environmentally benign and less toxic so their use could benefit the environment.

Unlike atrazine, glyphosate and glufosinate are not residual herbicides. Correct timing is, therefore, an integral part of their use. American studies have shown that delayed glyphosate or glufosinate application will increase weed control, but yield can be reduced through weed competition, reducing the crop growth at the early growth stages. The lack of consistency in weed control using glufosinate and glyphosate has prompted many authors to recommend the use of a residual herbicide or mechanical control in conjunction with glufosinate and glyphosate applied early post-emergence (Tharp and Kells, 1999; Wychen van et al., 1999; Bradley et al., 2000; Hamill et al., 2000; Johnson et al., 2000). Consistent weed control seems to be achieved using GMHT maize as part of a weed management system rather than as a 'stand alone' solution (Culpepper and York, 1999). The use of glufosinate and glyphosate, alongside a residual herbicide, e.g. atrazine, may lead to more active ingredient being applied than in a conventional system, which would negate some of the environmental benefits cited for reasons to use GMHT crops.

Gene Flow

Pollen dispersal

Maize pollen is known to be viable for about 24 hours. However, the distances travelled by maize pollen in this period varies greatly depending on the weather conditions. Solar convection currents or wind turbulence can lift pollen high into the atmosphere, increasing its potential dispersal distance (Emberlin et al., 1999).

Gene flow to wild relatives

The problem of gene flow from maize to wild relatives is negligible in Europe as maize is not known to hybridize with any wild European species. The nearest relatives to maize are the teosintes (wild *Zea* species) occurring in Mexico and Central America. Gene flow is likely to occur between maize and these species, and hybrids are known to develop, but it is believed that they are less likely to persist because selection by man against seed dispersal in maize results in non-dispersed hybrid seeds (Doebley, 1990).

Crop-to-crop gene flow

In maize, the most likely occurrence of spatial gene flow will be to other maize crops. The current isolation distance for maize, used by the UK Farm Scale Evaluation project, is 200m from organic crops and 80m from conventional crops, (Agriculture and Environment Biotechnology Commission, 2001). Such isolation will reduce cross-fertilization to a very low level but will not achieve the 'zero' level advocated by some proponents of organic sweet corn (Treu and Emberlin, 2000).

Post-harvest Management

The problems of volunteers and seed persistence are not relevant for maize in northern Europe, as maize seeds and volunteers do not survive prolonged cold, reducing their survival rates. All available evidence from N. America suggests that maize does not persist, that maize seed cannot reproduce outside of cultivated fields and that it is non-invasive in nature. In a study conducted in the UK by Crawley et al. (2001) which monitored changes in maize populations in natural habitats, all their populations were extinct by the beginning of the second year. Maize seeds have little dormancy and no evidence of persistence, but can occur as volunteers in mild climates in the year following cultivation. This volunteer maize can be controlled using ACCase inhibitor herbicides (graminicides) like sethoxydim (Beckie, 2001).

LONG-TERM MANAGEMENT

As there is no commercial planting of GMHT crops in Europe, it is difficult to predict the likely outcome of commercialization—both for agriculture and the environment. However, it is important to consider the likely issues that may arise. Some of the consequences of the continuing use of GMHT crops should by now be apparent in N. America, where the crops have been grown extensively since 1995. Are effects such as weed shifts to more tolerant weeds and subsequent resistance, environmental damage, effects on the soil and environmentally-threatening changes in management practices, emerging?

Weed Shifts and Resistance

Weed shifts have occurred when man has changed the agricultural environment. Changes in production systems such as a switch from spring to autumn-sown crops can change the weed flora. Similarly, reliance on one herbicide or a group of similar products can cause changes in the weed flora by 'encouraging' those species that are intrinsically less sensitive to the herbicides being used. There are fears that the eventual outcome of the repeated use of a restricted number of herbicides in GMHT crops would be a change to more tolerant weed species. There are already examples of the repeated use of glufosinate and glyphosate causing a weed shift to species which are less susceptible and which, over a period of time, have become more prevalent, e.g. intensive use of glyphosate on plantations in SE Asia has altered the weed diversity selecting for *Amaranthus* spp., *Conyza canadensis* Cronq (Canadian fleabane), *K. scoparium* and *C. album* (Baylis, 2000). Protracted use of glufosinate has favoured the occurrence of *Abutilon theophrasti* Medic. (velvetleaf) in glasshouse trials (Tharp et al., 1999). From Table 4.3, it can be seen that some species are less sensitive to glyphosate or glufosinate, including thistles and *Polygonum* spp. These could become more significant with the repeated use of the two herbicides used in GMHT crops.

Secondly, an individual plant or group of plants of a normally sensitive species—which by chance are less susceptible to the herbicide—may become dominant as a result of continuous selection by the herbicide, and so the population will become resistant. This has already occurred widely with 'conventional' herbicide usage. At present, there are four weed species known to be resistant to glyphosate: rigid ryegrass (*Lolium rigidum* Gaud.) (Australia & N. California), Italian ryegrass (*L. multiflorum*) (Chile), goosegrass (*Elusine indica* (L.) Gaertn.) (Malaysia) and Canadian fleabane (*Conyza canadensis*) (Canada) (Ogg and Jackson, 2001; Heap, 2002). No species have yet developed resistance to glufosinate (Heap, 2002). The occurrence of widespread resistance to glyphosate or glufosinate could have a serious impact on the tools available to farmers for weed management. It would remove one of the benefits of using GMHT crops because one set of resistance problems would be replaced by another. Integration of GMHT crops into a weed management programme over a rotation rather than as 'stand alone' control will reduce the likelihood of resistant weeds, thus, increasing the effective 'life' of each individual variety.

So, will the repeated use of GMHT crops lead to a switch in weeds to those that are more tolerant? Will these species become resistant? What is the best method to prevent or control this occurrence? Canadian studies of weed composition within GMHT oilseed rape have shown that, at present, there are no long-term changes in weed populations (Derksen et al., 1999). However, the answers to these questions are uncertain as far as Europe is concerned, but present knowledge on reducing resistance occurring in weeds, through the rotation of crops and herbicides, or herbicide mixtures could defer, or even prevent, resistance development. The future of GMHT crops in agricultural systems may depend on monitoring for initial changes, which could indicate developing weed shifts or resistance. Such monitoring should be a part of the post-release monitoring required as a part of the approval for the growing on GMHT crops.

Environmental Effects

There is more concern about the indirect environmental effects of GMHT crops in Europe, especially the UK, than there has been in North America. In the UK, over 70% of the land surface is farmed and so any substantial change in agricultural practice inevitably has an effect on the overall ecosystem. Concern has already been expressed about the influence of crop production practices on arable ecosystems (Chamberlain et al., 2000) and it has been postulated by conservationists that the more intensive weed management associated with GMHT crops will have a further negative effect on biodiversity. It is not clear whether this is true and there is conflicting opinion as to the likely impact of GMHT crops. As a consequence, a series of trials in the 'Farm Scale Evaluation of GMHT crops' programme was started in the UK in 1999. This project has been evaluating the biodiversity effects of GMHT oilseed rape, beet and maize, compared to standard farm practice, for three years (Firbank et al., 1999). Information has been collected on plants and invertebrates, both in the fields and in surrounding areas. The report of the work has shown that weed control was higher in GMHT oilseed rape and beet but was lower in the GMHT maize, than in the conventional treatments (Heard, 2003). The differences in the surviving weed flora was mirrored by changes in the numbers of invertebrates.

In contrast, research has also been done to establish whether the greater flexibility in weed control possible with GMHT crops can be exploited in order to improve biodiversity in fields. Delayed weed control is practicable using this technology but is less so with conventional herbicide treatments, as greater crop sensitivity and non-susceptibility of weeds can be important with late applications of traditional herbicides. This delayed weed control can have environmental benefits. The potential for this has been explored in comparisons of GMHT and conventional crops of sugar beet, in particular (Pidgeon et al., 2001).

Kempenaar and Lotz (1999) and Lotz et al. (2000) have compared a number of environmental effects of conventional, integrated and GMHT systems for sugar beet, maize and potato. Existing methodologies (e.g. environmental yard stick of CLM, Utrecht, and Exposure Risk Index of PPO, Lelystad, The Netherlands) were used to assess the direct effects of herbicides on soil organisms, water organisms, quality of deep soil water and evaporation. It was concluded that GMHT will have a smaller adverse effect on the parameters mentioned when compared with conventional systems, but that this benefit is small to absent when compared with new integrated systems.

Soil Effects

Soil organisms are an important part of a healthy, productive soil. The term soil organisms include bacteria, fungi and nematodes, all of which are responsible for nutrient recycling in soils. Herbicides are degraded by soil organisms, so will the increased use of glyphosate and glufosinate have an effect on these communities? Haney et al. (2000) found that glyphosate was readily and quickly degraded by indigenous microorganisms, even at high application rates. A later study, looking at nine different soils, found that glyphosate increased microbial population size as well as their activity (Haney et al., 2002). However, this research indicated that soil organic carbon, soil pH and clay content do not necessarily indicate the ability of a soil to degrade glyphosate. This work suggests that the use of glyphosate may boost microbial activity and biomass. Termorshuizen and Lotz (2002) hypothesized from a literature survey that weed control by glyphosate and glufosinate may enhance opportunistic plant pathogens that multiply in the dying roots of weeds. This enhancement may be partly due to a reported suppressive effect of glyphosate on saprophytes (which results in an increase in dead weed roots) and a reported suppressive effect of glufosinate on antagonists of these root pathogens. These pathogens might eventually affect the HT crop itself when cropped on a large scale, especially if the timing of herbicide application is delayed in the HT crop to optimize weed control. The authors concluded that controlled field experiments on the interactions between crops, pathogens and weeds are required to test the hypothesis.

One potential benefit of HT crops is the use of no-tillage rather than inversion tillage systems. 'No-till' systems have been connected with beneficial effects on soil organisms, where the wetter, denser and cooler conditions result in additional organic matter and increased microbial activity/biomass (Young and Ritz, 2000). The dynamics of microbial populations are affected by the tillage system used. Young and Ritz (2000) suggest that bacteria, rather than fungi or earthworms, will dominate in tilled soils, for example, and inversion tillage will result in a reduction in soil-borne fungal diseases, which could be due to disruption of fungal mycelia.

Changes in Management Practices

GMHT crop technology is a new tool in the management of weeds. The commercial production of GMHT crops may require a change in present agriculture practices to fulfil their long-term potential benefits to farmers and possibly to the environment, whilst reducing the environmental risks. The sowing of GMHT crops in N. America has increased the use of minimum tillage, which has further reduced soil erosion and water contamination. However, as glyphosate is frequently used for stubble management, the presence of glyphosate-resistant volunteers derived from GMHT crops could jeopardize its effectiveness, unless changes are introduced in the manner it is applied. The development of GMHT wheat in N. America has been delayed as practitioners are concerned that the presence of glyphosate-resistant volunteer wheat will put at risk the success of minimum tillage techniques used to establish subsequent crops (van Acker, pers. comm.). Other changes are less obvious, for example, before a spring GMHT beet crop, cereal stubbles could be left untreated over the winter (Pidgeon et al., 2001), a practice not advocated in 'conventional' beet because of the risks of poor weed control.

It is suggested that the more reliable weed control possible in GMHT rape, sugar beet and maize will encourage farmers to take a more 'relaxed' approach to weed management in preceding crops. In the past, the difficulty of controlling some weeds in rape, beet and maize crops has meant that weed control has aimed at elimination of the problematic species in preceding crops. Such needs are not so acute where GMHT crops are included in the rotation.

Conversely, the management practices introduced to reduce gene flow, via pollen and seed, may be unacceptable to farmers (Moll, 1997). Similarly, it is possible that there will be an increase in herbicide use, or a change to more environmentally-damaging products, in order to combat HT volunteers, HT weeds and HT feral populations (Lotz et al., 1999; Orson, 2002). Experiences with research projects growing GMHT oilseed rape in rotation with wheat suggest that the increased use of broad-leaved weed herbicides may occur in subsequent cereal crops, to ensure that any GM rape volunteers do not survive and set seed. (Lutman and Berry, unpublished data). HT crops with stacked tolerances (e.g. crops that are tolerant to both glyphosate and glufosinate) may be possible in the future, but these carry particular risks, especially at the farm level (Kempenaar and Lotz, 1999).

All herbicides have weaknesses, which will have to be taken into account to ensure correct management of GMHT crops. For example, glyphosate has antagonism to hard water (Ca^{2+} ion) and many mixture partners, and has poor rainfastness, all of which may lead to higher application rates (Baylis, 2000). Glufosinate has failed to achieve full weed control when spraying has been delayed and as a consequence the weeds have become too big and at too advanced a growth stage (Harker et al., 2000). Glyphosate and glufosinate are non-residual foliar acting herbicides and are affected by environmental factors, e.g. less effective on drought-stressed plants or at low temperatures due to poor absorption (Wychen van et al., 1999). They also will have no effect on weeds emerging after the products have been applied. Consequently, different herbicide strategies may be needed, depending on the main weeds to be controlled (Bradley et al., 2000; Johnson *et al.*, 2000), for example, weeds which emerge in the spring or throughout the season are better controlled by delayed application of foliar-acting herbicides. Herbicides in the future may be needed to fill in the gaps left by glufosinate and glyphosate due to inadequate timing of control or a reduced weed spectrum (Hamill et al., 2000).

The zero tolerance attitude towards weed levels shown by some farmers and advisors may have to change to make the best use of GMHT crops within a sustainable agricultural system. For example, beet farmers prepared to leave weeds unsprayed for a period after crop and weed emergence, in a herbicide-tolerant crop treatment programme, may use less herbicide active ingredient than growers using conventional technologies. Herbicides in GMHT crops, which can be applied late post-emergence, may also give farmers the opportunity to use a weed threshold-based management strategies, which cannot be easily used where weed control is based around pre-emergence products. Farmers might be willing to make more use of mechanical weed control in a HT crop, knowing that when the efficacy of this treatment was low due to wet weather conditions, possibilities existed to still control the weeds with a broad spectrum herbicide in a later cropping stage. In such strategies, an improved knowledge of weed ecology will aid the management of weeds.

Finally, management of HT crops will be different as compared to conventional crops. There may be risks and problems associated with the use of the technology, as well as benefits. The farmers, along with those supplying the seeds and herbicides, will be jointly responsible for minimizing these risks. It is essential that good management guidelines for the crops concerned and sound strategies for post-release monitoring are prepared prior to commercialization.

CONCLUSIONS

The rapid uptake of the technology in N. America, in particular, shows that there are benefits to farmers. The main benefit seems to be the simplification of weed management rather than direct reductions in the cost of weed control. However, weed control can be cheaper and also tends to be more flexible and sometimes more effective in HT crops. As far as the three crops discussed in this chapter are concerned, one would predict appreciable reductions in the cost of weed control in sugar beet, some reduction in oilseed rape and only marginal differences in maize. The potential of the technology to provide more flexibility in the timing weed control provides the opportunity to exploit the technology to benefit the environment, both directly (e.g. using threshold-based weed management rather than prophylactic treatment) as well as indirectly (e.g. promoting more minimum tillage). Both glyphosate and glufosinate have relatively benign environmental profiles and so their use, replacing older 'conventional' products, will have environmental benefits.

Conversely, the introduction of GMHT crops may pose new management problems and challenges and may also have a negative impact on the environment. The broader spectrum weed control attainable with glyphosate and glufosinate may have a greater negative impact on the arable ecosystem than conventional practices. These aspects have been explored in the UK Farmscale Evaluations (Firbank et al., 1999; Heard et al., 2003). Desk studies from The Netherlands show that GMHT crops probably have a less adverse effect on selected environmental parameters (soil organisms, water organisms, quality of deep soil water and evaporation) when compared with conventional systems, but a comparable effect when compared to new integrated systems (Kempenaar and Lotz, 1999; Lotz et al., 2000). Other effects, such as gene flow, will vary from crop to crop. The issue to be borne in mind when considering the consequences of gene escape is the impact of the gene on the fitness of the plants affected. There is no evidence that the presence of herbicide-tolerant genes impacts on the fitness of plants, in the absence of the herbicide. This would not be true of other transgenes, such as those providing stress tolerances, and emphasizes that the impact of new GM crops needs to be assessed on a case-by-case basis.

Oilseed Rape—Risks

As there are a number of wild related species to oilseed rape in Europe and the crop is partially outbreeding, the potential of gene flow to wild relatives is greater than with either sugar beet or maize. *Brassica rapa* is the relative that seems most vulnerable to gene flow from the rape crop. *Raphanus raphanistrum* will also sometimes hybridize. Other Crucifer species are not sufficiently compatible with rape and will only rarely (or not at all) produce viable hybrid and backcross plants. Of greater concern is the risk of crop-to-crop gene flow, resulting in the presence of HT plants in adjacent non-HT crops. Good management will keep this to a minimum, but some gene flow will occur. The EU has to decide what level of mixture (0.9%, 0.1%) is acceptable. Once this is decided, management strategies and isolation requirements can be developed to meet that limit of cross fertilization. Good management can also make sure that HT volunteer plants do not pose a threat to weed control in subsequent years, and can ensure that the risks of the development of glyphosate and glufosinate resistant weeds is also minimized.

Sugar Beet—Risks

The risks with sugar beet are primarily associated with the avoidance of flowering. If the HT beet crops do not 'bolt' and flower, then there is minimal risk from growing the crop. If it does flower, the pollen will transfer the HT genes to wild and weed beets, which will jeopardize the value of the technology to manage these problem weeds.

Maize—Risks

As maize has no relatives in Europe, the only concern as far as gene flow is concerned is the potential to affect neighbouring non-HT maize crops. Some gene flow is likely to occur. As with rape, a decision has to be reached in the EU, as to what level of cross-fertilization is acceptable. With this information, workable isolation requirements can be devised.

Overall, the benefits to growers appear to be appreciable and in our view, the main risks are to the grower. Appropriate management is needed in order to maximize the benefits and minimize the risks. As herbicide-tolerant genes do not seem to affect other aspects of plant fitness, the impact of the technology on natural ecosystems seems to be negligible. Such a conclusion may not be appropriate for other GM crops. However, growers will need clear guidance on how to manage these crops and some post crop monitoring will be required to ensure that problems from such issues as gene stacking do not arise. The issue of the overall impact of the technology on arable ecosystems is still not fully resolved.

ACKNOWLEDGEMENTS

We would like to acknowledge the financial support of the UK government Department of Environment, Food and Rural Affairs, which has provided two of us (PJWL and KB) with resources to complete this review. Rothamsted Research receives grant aided financial support from the Biotechnology and Biological Sciences Research Council.

REFERENCES

Agriculture and Environment Biotechnology Commission, 2001, Crops on trial, Department of Trade and Industry, London.

Amann, A., 1998, Roundup Ready winter oilseed rape - three years field research experience in Europe. Z. Pflanzenkrank. Pflanzensch, **Sonderh. XVI**, 379-389.

Baranger, A., Chevre, A.M., Eber, F. and Renard, M., 1995, Effect of oilseed rape genotype on the spontaneous hybrisation rate with a weedy species - an assessment of transgene dispersal. Theor. Appl. Genet., **91**, 956-963.

Bartsch, D., Lehnen, M., Clegg, J., Pohl-Orf, M., Schuphan, I. and Ellstrand, N. C., 1999, Impact of gene flow from cultivated beet on genetic diversity of wild sea beet populations. Mol. Ecol., **8,** 1733-1741.

Baylis, A. D., 2000, Why glyphosate is a global herbicide: strengths, weaknesses and prospects. Pest Manage. Sci., **56,** 299-308.

BBRO, 2002, Sugar beet : A grower's guide. British Beet Research Organisation, website: www.bsonline.co.uk.

Beckie, H. J., 2001, Impact of herbicide-resistant crops as weeds in Canada. Brighton Crop Protection Conference—Weeds, **1,** 135-142.

Bradley, P. R., Johnson, W. G., Hart, S. E., Buesinger, M. L. and Massey, R. E., 2000, Economics of weed management in glufosinate-tolerant corn (*Zea mays* L.). Weed Technol., **14,** 495-501.

Canola Council of Canada, 2001, An Agronomic and Economic Assessment of Transgenic Canola, Canola Council of Canada, Winnipeg, Canada.

Chadoeuf, R., Darmency, H., Maillet, J. and Renard, M., 1998, Survival of buried seeds of interspecific hybrids between oilseed rape, hoary mustard and wild radish. Field Crops Res., **58,** 197-204.

Chamberlain, D. E., Fuller, R. J., Bunce, R. G. H., Duckworth, J. C. and Shrubb, M., 2000, Changes in the abundance of farmland birds in relation to the timing of agricultural intensification in England and Wales. J. Appl. Ecol., **37,** 771-788.

Champolivier, J., Gasquez, J., Messean, A. and Richard-Molard, M. (1999) Management of transgenic crops within the cropping system In: Lutman P. J. W. (ed.), Gene Flow and Agriculture - Relevance for Transgenic Crops, British Crop Protection Council, Farnham, Surrey, UK, pp. 233-240.

Chevre, A. M., Eber, F., Baranger, A., Kerlan, M. C., Barret, P., Festoc, G., Vallee, P. and Renard, M., 1996, Interspecific gene flow as a component of risk assessment for transgenic *Brassicas*. Acta Horticult., **407,** 169-179.

Chevre, A. M., Eber, F., Baranger, A. and Renard, M., 1997, Gene flow from transgenic crops. Nature, **389,** 924-924.

Chevre, A. M., Eber, F., Baranger, A., Renard, M., Boucherie, R., Broucqsault, L. M. and Bouchet, Y., 1998, Risk assessment on Crucifer species. Acta Horticult., **459,** 219-224.

Chevre, A. M., Eber, F., Darmency, H., Fleury, A., Picault, H., Letanneur, J. C. and Renard, M., 2000, Assessment of interspecific hybridization between transgenic oilseed rape and wild radish under normal agronomic conditions. Theor. Appl. Genet., **100,** 1233-1239.

Chevre, A. M., Eber, F., Renard, M. and Darmency, H. (1999) Gene flow from oilseed rape to weeds. In: Lutman P. J. W. (ed.), Gene Flow and Agriculture - Relevance for Transgenic Crops, British Crop Protection Council, Farnham, Surrey, UK, pp. 125-130.

Crawley, M. J. and Brown, S. L., 1995, Seed limitation and the dynamics of feral oilseed rape on the M25 motorway. Proc. Royal Acad. Lond., Series-B - Biol. Sci., **259,** 49-54.

Crawley, M. J., Hails, R. S., Rees, M., Kohn, D. and Buxton, J., 1993, Ecology of transgenic oilseed rape in natural habitats. Nature, **363,** 620-623.

Crawley, M. J., Brown, S. L., Hails, R. S., Kohn, D. D. and Rees, M., 2001, Transgenic crops in natural habitats. Nature, **409,** 682-683.

Culpepper, A. S. and York, A.C., 1999, Weed management in glufosinate-resistant corn (*Zea mays*). Weed Technol., **13,** 324-333.

Darmency, H., 1994, The impact of hybrids between genetically modified crop plants and their related species : introgression and weediness. Mol. Ecol., **3,** 37-40.

Darmency, H. and Fleury, A., 2000, Mating system in *Hirschfeldia incana* and hybridization to oilseed rape. Weed Res., **40**, 231-238.

Darmency, H., Lefol, E. and Fleury, A., 1998, Spontaneous hybridizations between oilseed rape and wild radish. Mol. Ecol., **7**, 1467-1473.

Derksen, D. A., Harker, K. N. and Blackshaw, R. E., 1999, Herbicide tolerant crops and weed population dynamics in western Canada. Brighton Crop Protection Conference - Weeds, **2**, 417-424.

Devine, M. D. and Buth, J. L., 2001, Advantages of genetically modified canola: a Canadian perspective. Brighton Crop Protection Conference - Weeds, **1**, 367-372.

Doebley, J., 1990, Molecular evidence for gene flow among *Zea* species. Bioscience, **40**, 443-448.

Downey, K., 1999, Gene flow and rape - the Canadian experience. In: Lutman P. J. W. (ed.), Gene Flow and Agriculture - Relevance for Transgenic Crops, British Crop Protection Council, Farnham, Surrey, UK, pp. 109-116.

Eastham, K. and Sweet, J. B., 2002, Genetically Modified Organisms (GMOs):The Significance of Gene Flow Through Pollen Transfer, European Environment Agency Report 28, Luxembourg.

Ellstrand, N. C., 2002, Gene flow from transgenic crops to wild relatives: What have we learned, what do we know and what do we need to know? In: Snow A. A. (ed.), Scientific Methods Workshop - Ecological and Agronomic Consequences of Gene Flow from Transgenic Crops to Wild Relatives, Ohio State University, Columbus, Ohio, pp. 39-46.

Ellstrand, N. C., Prentice, H. C. and Hancock, J. F., 1999, Gene flow and introgression from domesticated plants into their wild relatives. Ann. Rev. Ecol. Syst., **30**, 539-563.

Emberlin, J., Adams-Groom, B. and Tidmarsh, J., 1999, A report on the dispersal of maize pollen, Soil Association, Bristol, UK.

FAOSTAT, 2002, FAOSTAT, www.fao.org.

Firbank, L. G., Dewar, A. M., Hill, M. O., May, M. J., Perry, J. N., Rothery, P., Squire, G. R. and Woiwod, I. P., 1999, Farm-scale evaluation of GM crops explained. Nature, **399**, 727-728.

Fredshavn, J. R. and Poulsen, G. S., 1996, Growth behaviour and competitive ability of transgenic crops. Field Crops Res., **45**, 11-18.

Fredshavn, J. R., Poulsen, G. S., Huybrechts, I. and Rudelsheim, P., 1995, Competitiveness of transgenic oilseed rape. Transgenic Res., **4**, 142-148.

Frello, S., Hansen, K., Jensen, J. and Jorgensen, R., 1995, Inheritance of rapeseed (*Brassica napus*) specific RAPD markers and a transgene in the cross *B. juncea* x (*B. juncea* x *B. napus*). Theor. Appl. Genet., **91**, 236-241.

Garthwaite, D.G., Thomas, M.R., Dawson A and Stoddart H., 2003, Pesticide Usage Survey Report 187: Arable Farm Crops in Great Britain, Central Science Laboratory, York.

Gotz, R. and Ammer, F., 2000, Results of Liberty application in transgenic winter rape in Thuringia. Z. Pflanzenkrank. Pflanzensch., **Sonderh XVII**, 397-401.

Gray, A. J. and Raybould, A. F., 1998, Reducing transgene escape routes. Nature, **392**, 653-654.

Gueritaine, G. and Darmency, H., 2001, Polymorphism for interspecific hybridisation within a population of wild radish (*Raphanus raphanistrum*) pollinated by oilseed rape (*Brassica napus*). Sexual Plant Reprod., **14**, 169 -172.

Gueritaine, G., Sester, M., Eber, F., Chevre, A. M. and Darmency, H., 2002, Fitness of backcross six of hybrids between transgenic oilseed rape (*Brassica napus*) and wild radish (*Raphanus raphanistrum*). Mol. Ecol., **11**, 1419-1426.

Hails, R. S., Rees, M., Kohn, D. D. and Crawley, M. J., 1997, Burial and seed survival in *Brassica napus* subsp *oleifera* and *Sinapis arvensis* including a comparison of transgenic and non-transgenic lines of the crop. Proc. Royal Soc. Lond., Series-B - Biol. Sci., **264**, 1-7.

Hall, L., Topinka, K., Huffman, J., Davis, L. and Good, A., 2000, Pollen flow between herbicide-resistant *Brassica napus* is the cause of multiple-resistant *B. napus* volunteers. Weed Sci., **48**, 688-694.

Hamill, A. S., Knezevic, S. Z., Chandler, K., Sikkema, P. H., Tardif, F. J., Shrestha, A. and Swanton, C. J., 2000, Weed control in glufosinate-tolerant corn (*Zea mays*). Weed Technol., **14**, 578-585.

Hancock, J. F., Grumet, R. and Hokanson, S. C., 1996, The opportunity for escape of engineered genes from transgenic crops. Hort. Sci., **31**, 1080-1085.

Haney, R. L., Senseman, S. A. and Hons, F. M., 2002, Effect of Roundup Ultra on microbial activity and biomass from selected soils. J. Environ. Quality, **31**, 730-735.

Haney, R. L., Senseman, S. A., Hons, F. M. and Zuberer, D. A., 2000, Effect of glyphosate on soil microbial activity and biomass. Weed Sci., **48**, 89-93.

Harker, K. N., Blackshaw, R. E., Kirkland, K. J., Derksen, D. A. and Wall, D., 2000, Herbicide-tolerant canola : weed control and yield comparisons in western Canada. Can. J. Plant Sci., **80**, 647-654.

Hauser, T. P., Jorgensen, R. B. and Ostergard, H., 1998a, Fitness of backcross and F_2 hybrids between weedy *Brassica rapa* and oilseed rape (*B. napus*). Heredity, **81**, 436-443.

Hauser, T. P., Shaw, R. G. and Ostergard, H., 1998b, Fitness of F_1 hybrids between weedy *Brassica rapa* and oilseed rape (*B. napus*). Heredity, **81**, 429-435.

Heap, I., 2002, International survey of herbicide-tolerant weeds. website: www.weedscience.org.

Heard, M.S., Hawes, C., Champion, G.T., Clark, S.J., Firbank, L.G. et al. 2003, Weeds in fields with contrasting conventional and genetically modified herbicide-tolerant crops. 1. Effects on abundance and diversity Phil Trans Royal Soc: Biol Sci **358**: 1819-1832.

Hokanson, S. C., Hancock, J. F. and Grumet, R., 1997, Direct comparison of pollen-mediated movement of native and engineered genes. Euphytica, **96**, 397-402.

James, C., 2003, Global status of commercialised transgenic crops : 2003. ISAAA Briefs No. 30 ISAAA, New York, USA.

Johnson, W. G., Bradley, P. R., Hart, S. E., Buesinger, M. L. and Massey, R. E., 2000, Efficacy and economics of weed management in glyphosate-resistant corn (*Zea mays*). Weed Technol., **14**, 57-65.

Jorgensen, R. B., 1999, Gene flow from oilseed rape (*Brassica napus*) to related species. In: Lutman P. J. W. (ed.), Gene Flow and Agriculture - Relevance for Transgenic Crops, British Crop Protection Council, Farnham, Surrey, UK, pp. 117-124.

Jorgensen, R. B., Andersen, B., Landbo, L. and Mikkelsen, T. R., 1996, Spontaneous hybridisation between oilseed rape (*Brassica napus*) and weedy relatives. Acta Horticult., **407**, 193-200.

Kempenaar, C. and Lotz, L. A. P., 1999, Environmental risks of transgenic multiple herbicide resistance. Note 193, AB-DLO, Wageningen, The Netherlands, 30 pp.

Landbo, L. and Jorgensen, R. B., 1997, Seed germination in weedy *Brassica campestris* and its hybrids with *B. napus*: Implications for risk assessment of transgenic oilseed rape. Euphytica, **97**, 209-216.

Lefol, E., Danielou, V. and Darmency, H., 1996a, Predicting hybridization between transgenic oilseed rape and wild mustard. Field Crops Res., **45**, 153-161.

Lefol, E., Fleury, A. and Darmency, H., 1996b, Gene dispersal from transgenic crops. 2. Hybridization between oilseed rape and the wild hoary mustard. Sexual Plant Reprod., **9,** 189-196.

Lefol, E., Seguin-Swartz, G. and Downey, K., 1997, Sexual hybridisation of cultivated *Brassica* species with the crucifers *Erucastrum gallicum* and *Raphanus raphanistrum* : potential for gene introgression. Euphytica, **95,** 127-139.

Lefol, E., Danielou, V., Darmency, H., Boucher, F., Maillet, J. and Renard, M., 1995, Gene dispersal from transgenic crops. 1. Growth of interspecific hybrids between oilseed rape and the wild hoary mustard. J. Appl. Ecol., **32,** 803-808.

Legere, A., Simard, M. J., Thomas, J. E., Pageau, D., Lajeunesse, J., Warwick, S. I. and Derksen, D. A., 2001, Presence and persistence of volunteer canola in Canadian cropping systems. Brighton Crop Protection Conference - Weeds, **1,** 143-148.

Linder, C. R., 1998, Potential persistence of transgenes : seed performance of transgenic canola and wild x canola hybrids. Ecol. Appic., **8,** 1180-1195.

Lotz, L. A. P., Wevers, J. D. A. and Van der Weide, R. Y., 1999, My view. Weed Sci., **47,** 479-480.

Lotz, L. A. P, Brussaard, L., Gillissen, L. J. W. J., Gorissen, A., Kempenaar, C., van Loon, J. J. A., Noordam, M. Y., Termorshuizen., A. J. and van Vliet, P. C. J., 2000, Effecten van grootschalige toepassing van transgene herbicideresistente rassen. Report 2. Plant Research International (in Dutch), Wageningen, The Netherlands 110 pp.

Madsen, K. H. and Jensen, J. E., 1995, Weed control in glyphosate-resistant sugar beet (*Beta vulgaris* L.). Weed Res., **35,** 105-111.

Madsen, K. H., Poulsen, G. S., Fredshavn, J. R., Jensen, J., Steen, P. and Streibig, J. C., 1998, A method to study competitive ability of hybrids between seabeet (*Beta vulgaris* ssp *maritima*) and glyphosate tolerant sugarbeet (*B. vulgaris* ssp *vulgaris*). Acta Agric. Scandinavica Section B - Soil and Plant Sci., **48,** 170-174.

Metz, P. L. J., Jacobsen, E., Nap, J. P., Pereira, A. and Stiekema, W. J., 1997, The impact on biosafety of the phosphinothricin-tolerance transgene in inter-specific *B.rapa* x *B.napus* hybrids and their successive backcrosses. Theor. Appl. Genet., **95,** 442-450.

Mikkelsen, T. R., Andersen, B. and Jorgensen, R. B., 1996a, The risk of crop transgene spread. Nature, **380,** 31.

Mikkelsen, T. R., Jensen, J. and Jorgensen, R. B., 1996b, Inheritance of oilseed rape (*Brassica napus*) RAPD markers in a backcross progeny with *Brassica campestris*. Theor. Appl. Genet., **92,** 492-497.

Moll, S., 1997, Commercial experience and benefits of glyphosate tolerant crops. Brighton Crop Protection Conference - Weeds, **3,** 931-940.

Moyes, C. L., Lilley, J., Casais, C. and Dale, P. J., 1999, Gene flow from oilseed rape to *Sinapis arvensis* : variation at the population level. In: Lutman P. J. W. (ed.), Gene Flow and Agriculture - Relevance for Transgenic Crops, British Crop Protection Council, Farnham, Surrey, UK, pp. 143-148.

Moyes, C. L., Lilley, J. M., Casais, C. A., Cole, S. G., Haeger, P. D. and Dale, P. J., 2002, Barriers to gene flow from oilseed rape (*Brassica napus*) into populations of *Sinapis arvensis*. Mol. Ecol., **11,** 103-112.

Norris, C. E., Simpson, E. C., Sweet, J. B. and Thomas, J. E., 1999, Monitoring weediness and persistence of genetically modified oilseed rape (*Brassica napus*) in the UK. In: Lutman P. J. W. (ed.), Gene Flow and Agriculture - Relevance for Transgenic Crops, British Crop Protection Council, Farnham, Surrey, UK, pp. 255-260.

Ogg, A. G. and Jackson, P. J., 2001, Agronomic benefits and concerns for Roundup-Ready wheat. Proceedings of the Western Weed Science Society Meeting (2001), 80-90.

Orson, J., 2002, Gene stacking in herbicide tolerant oilseed rape : lessons from the North American experience. English Nature Research Reports, No.443, English Nature, Peterborough, UK, 17 pp.

Owen, M. D. K., 2000, Current use of transgenic herbicide-resistant soybean and corn in the USA. Crop Prot., **19**, 765-771.

Pekrun, C., Hewitt, J. D. J. and Lutman, P. J. W., 1998, Cultural control of volunteer oilseed rape (*Brassica napus*). J. Agric. Sci., **130**, 155-163.

Pekrun, C., Lane, P. W. and Lutman, P. J. W., 1999, Modelling the potential for gene escape in oilseed rape via the soil seedbank : its relevance for genetically modified cultivars In: Lutman P. J. W. (ed.), Gene Flow and Agriculture - Relevance for Transgenic Crops, British Crop Protection Council, Farnham, Surrey, UK, pp. 101-106.

Perry, J. N., 2002, Sensitive dependencies and separation distances for genetically modified herbicide-tolerant crops. Proc. Royal Acad.London, Series-B - Biol. Sci., **269**, 1173-1176.

Pessel, F. D., Lecomte, J., Emeriau, V., Krouti, M., Messean, A. and Gouyon, P. H., 2001, Persistence of oilseed rape (*Brassica napus* L.) outside of cultivated fields. Theor. Appl. Genet., **102**, 841-846.

Pesticides Safety Directorate, 1998, Scientific Review of the Impact of Herbicide Use on Genetically Modified Crops, Report, DEFRA, London, UK.

Pidgeon, J. D., Dewar, A. M. and May, M. J., 2001, Weed management for agricultural and environmental benefit in GMHT sugar beet. Brighton Crop Protection Conference—Weeds, **1**, 373-380.

Ramsey, G., Thompson, C. E., Neilson, S. and Mackay, G. R., 1999, Honeybees as vectors of GM oilseed rape pollen. In: Lutman P. J. W. (ed.), Gene Flow and Agriculture - Relevance for Transgenic Crops, British Crop Protection Council, Farnham, Surrey, UK, pp. 209-214.

Rasche, E. and Gadsby, M., 1997, Glufosinate-ammonium tolerant crops - international developments and experiences. Brighton Crop Protection Conference - Weeds, **3**, 941-946.

Read, M. A. and Ball, J. G., 1999a, Control of weeds in genetically modified crops of winter and spring oilseed rape with glufosinate-ammonium in the UK. Aspects of Applied Biology: Protection and Production of Combinable Break Crops, **56**, 27-33.

Read, M. A. and Ball, J. G., 1999b, The control of weeds with glufosinate-ammonium in genetically modified crops of forage maize in the UK. Brighton Crop Protection Conference - Weeds, 847-852.

Read, M. A. and Bush, M. N., 1998, Control of weeds in genetically modified sugar beet with glufosinate-ammonium in the UK. Aspects of Applied Biology: Protection and Production of Sugar Beet and Potatoes, **52**, 405-406.

Rieger, M. A., Potter, T. D., Preston, C. and Powles, S. B., 2001, Hybridization between *Brassica napus* L. and *Raphanus raphanistrum* L. under agronomic field conditions. Theor. Appl. Genet., **103**, 555-560.

Rieger, M. A., Lamond, M., Preston, C., Powles, S. B. and Roush, R. T., 2002, Pollen-mediated movement of herbicide resistance between commercial canola fields. Science, **296**, 2386-2388.

Schans, D. A. van, and van der Weide, R. Y., 2001, Schone mais met eg of schoffel en een beetje middel (in Dutch). PPO-Bullentin Akkerbouw **1**, 16-19.

Scheffler, J. A. and Dale, P. J., 1994, Opportunities for gene transfer from transgenic oilseed rape (*Brassica napus*) to related species. Transgenic Res., **3**, 263-278.

Scheffler, J. A., Parkinson, R. and Dale, P. J., 1995, Evaluating the effectiveness of isolation distances for field plots of oilseed tape (*Brassica napus*) using a herbicide-resistance transgene as a selectable marker. Plant Breeding, **114**, 317-321.

Scott, S. E. and Wilkinson, M. J., 1998, Transgene risk is low. Nature, **393**, 320.

Snow, A. A., Andersen, B. and Jorgensen, R. B., 1999, Costs of transgenic herbicide resistance introgressed from *Brassica napus* into weedy *B. rapa*. Mol. Ecol., **8**, 605-615.

Squire, G. R., Crawford, J. W., Ramsey, G. and Thompson, C., 1999, Gene flow at the landscape level. In: Lutman P. J. W. (ed.), Gene Flow and Agriculture—Relevance for Transgenic Crops, British Crop Protection Council, Farnham, Surrey, UK, pp. 57-64.

Sweet, J. B., Shepperson, R., Thomas, J. E. and Simpson, E. C., 1997, The impact of releases of genetically modified herbicide tolerant oilseed rape in the U.K. Brighton Crop Protection Conference - Weeds, **1**, 291-302.

Termoshuizen, A. J.and Lotz , L. A. P., 2002, Does large-scale cropping of herbicide-resistant cultivars increase the incidence of polyphagous soil-borne plant pathogens? Outlook Agric., **31**, 51-54

Tharp, B. E. and Kells, J. J., 1999, Influence of herbicide application rate, timing and interrow cultivation on weed control and corn (*Zea mays*) yield in glufosinate-resistant and glyphosate-resistant corn. Weed Technol., **13**, 807-813.

Tharp, B. E., Schabenberger, O. and Kells, J. J., 1999, Response of annual weed species to glufosinate and glyphosate. Weed Technol., **13**, 542-547.

Thompson, C. E., Squire, G. R., Mackay, G. R., Bradshaw, J. E., Crawford, J. W. and Ramsey, G., 1999, Regional patterns of gene flow and its consequences for GM oilseed rape. In: Lutman P. J. W. (ed.), Gene Flow and Agriculture - Relevance for Transgenic Crops, British Crop Protection Council, Farnham, Surrey, UK, pp. 95-100.

Treu, R. and Emberlin, J., 2000, Pollen dispersal in the crops maize (*Zea mays*), oilseed rape (*Brassica napus ssp oleifera*), potatoes (*Solanum tuberosum*), sugar beet (*Beta vulgaris ssp vulgaris*) and wheat (*Triticum aestivum*), Report from Soil Association, Bristol, UK.

Vigouroux, Y., Darmency, H., Gestat de Garambe, T. and Richard-Molard, M., 1999, Gene flow between sugar beet and weed beet. In: Lutman P. J. W. (ed.), Gene Flow and Agriculture—Relevance for Transgenic Crops, British Crop Protection Council, Farnham, Surrey, UK, pp. 83-88.

Wauchope, R. D., Estes, T. L., Allen, R., Baker, J. L., Hornsby, A. G., Jones, R. L., Richards, R. P. and Gustafson, D. I., 2001, Predicted impact of transgenic, herbicide-tolerant corn on drinking water quality in vulnerable watersheds of the mid-western USA. Pest Manage. Sci., **58**, 146-160.

Wevers, J. D. A., 1998, Agronomic and environmental aspects of herbicide-resistant sugar beet in the Netherlands. Aspects of Applied Biology: Protection and Production of Sugar Beet and Potatoes, **52**, 393-399

Wilkinson, M. J., Davenport, I. J., Charters, Y. M., Jones, A. E., Allainguillaume, J., Butler, H. T., Mason, D. C. and Raybould, A. F., 2000, A direct regional scale estimate of transgene movement from genetically modified oilseed rape to its wild progenitors. Mol. Ecol., **9**, 983-991.

Wilkinson, M. J., Timmons, A. M., Charters, Y., Dubbels, S., Robertson, A., Wilson, N., Scott, S., O'Brien, E. and Lawson, H. M., 1995, Problems of risk assessment with genetically modified oilseed rape. Brighton Crop Protection Conference - Weeds, **3**, 1035-1044.

Wilson, R. G., 1994, New herbicides for post-emergence application in sugar beet (*Beta vulgaris*). Weed Technol., **8**, 807-811.

Wychen van, L. R., Harvey, R. G., Vangessel, M. J., Rabaey, T. L. and Bach, D. J., 1999, Efficacy and crop response to glufosinate-based weed management in PAT-transformed sweet corn (*Zea mays*). Weed Technol., **13**, 104-111.

Young, I. M. and Ritz, K., 2000, Tillage, habitat space and function of soil microbes. Soil Tillage Res., **53**, 201-213.

5

TRANSGENIC RICE FOR DISEASE RESISTANCE

SWAPAN K. DATTA
Plant Breeding, Genetics, and Biotechnology Division
International Rice Research Institute
DAPO Box 7777, Metro Manila, Philippines

INTRODUCTION

Plant diseases cause billions of dollars in crop losses annually. Yield loss in rice alone is enormous. About 20% of total yield is lost because of biotic stresses, including severe diseases such as blast, sheath blight, bacterial blight, and tungro. Disease control is based on the principle of maintaining yield loss below an economic injury level. In most cases, agrochemicals such as fungicides/pesticides and biological control including, crop rotation, are used to control diseases. Developing varieties with disease resistance will most likely provide the best solution for disease control. This approach is inexpensive and environmentally friendly and management would be easier than before. The classic *R* gene, defined by plant breeders, is now isolated and characterized as a cloned gene and plant biotechnologists can transfer *R* genes along with pathogenesis-related genes into many crop plants, including rice (Van Loon, 1999; Datta and Muthukrishnan, 1999; Datta, 2002). *R* and *PR* genes are reported in Tables 5.1-5.3.

Plants use a variety of strategies to protect against pathogen attack (Dangl, 1998). Plant protection is manifested by a single gene or group of genes to make it work in coordination (Purkayastha, 1998). Resistance genes are regulatory in nature, whereas defense genes are functional.

Table 5.1 Classes of cloned plant disease-resistance genes

Class	R gene	Plant	Pathogen	Avr gene	Structure[a]	Reference(s)
1	Hm1	Maize	Cochliobolus carbonum (race 1)	None	HC-toxin reductase	Johal and Briggs (1992)
2	Pto	Tomato	Pseudomonas syringae pv. tomato	avrPto	Serine/threonine protein-kinase	Martin et al. (1993)
3a	RPS2	Arabidopsis	P. syringae pv. tomato	avrRpt2	LZ-NBS-LRR	Staskawicz et al. (1995)
	RPM1	Arabidopsis	P. syringae pv. maculicola	avrRpm1/avrB	LZ-NBS-LRR	Grant et al. (1995)
	Prf	Tomato	P. syringae pv. tomato	avrPto	LZ-NBS-LRR	Salmeron et al. (1994)
	I2	Tomato	Fusarium oxysporum f. sp. lycopersici	Unknown	LZ-NBS-LRR	Simons and Flur (1996)
3b	N	Tobacco	Tobacco mosaic virus	TMV replicase?	TIR-NBS-LRR	Whitham et al. (1994)
	L6	Flax	Melamsora lini	AL6	TIR-NBS-LRR	Lawrence et al. (1995)
	RPP5	Arabidopsis	Peronospora parasitica	AvrPp5	TIR-NBS-LRR	Parker et al. (1997)
3c	Xa1	Rice	Xanthomonas oryzae pv. oryzae	Unknown	NBS-LRR	Yoshimura et al. (1995)
3d	Mlo1	Barley	Erysiphe gramini f. sp hordei		NBS-LRR	Büschges et al. (1997)
4	Cf-9	Tomato	Cladosporium fulvum	Avr9	LRR-TM	Jones et al. (1994)
	Cf-2	Tomato	C. fulvum	Avr9	LRR-TM	Dixon et al. (1996)
5	Xa21	Rice	X. oryzae pv. oryzae	Unknown	LRR, protein kinase	Ikeda et al. (1990); Song et al. 1995
6	Stv-bi Rice	Rice	stripe virus	Unknown	Unknown	Hayano-Saito et al. (1998)

[a]LRR = leucine-rich repeat; LZ = leucine zipper, NBS = nucleotide binding site; TIR = Toll-1L-1R homology region; TM = transmembrane

However, R genes regulate the functions of defense genes. Defense genes are usually quiescent in healthy plants but become activated when a pathogen comes in contact with the plant, which then releases a signal (Vidhyasekaran, 1998).

Table 5.2 Pathogenesis-related protein families in plants (recognized and proposed)

Family	Representative plant	Molecular weight (kDa)	Biochemical properties	Gene symbol
PR-1	Tobacco	14, 15, 17	Antifungal, unknown	ypr1
PR-2	Tobacco	31, 33, 35	ß-1, 3-glucanase	ypr2, [gns2]
PR-3	Tobacco	27, 28, 32, 34	Chitinase	ypr3, chia
PR-4	Tobacco	13, 15, 20	Similar to potato proteins	ypr4, chid
PR-5	Tobacco Rice	24, 26 23	Thaumatin-like	ypr5
PR-6	Tomato	8, 13	Proteinase inhibitor	ypr6, pis ('pin')
PR-7	Tomato	69	Endo-proteinase	ypr
PR-8	Cucumber	28	Chitinase	ypr8, chib
PR-9	Tobacco	39, 40	Lignin-forming peroxidase	ypr9
PR-10	Parsley	17-19	Ribonuclease-like	ypr10
PR-11	Tobacco	41, 43	Class V chitinase	ypr11, chic
Others				
a. Thionins	Barley	5	Antimicrobial	
b. Plant defensins	Radish	5	Antifungal	

RESISTANCE, AVIRULENCE AND TRANSGENES

Every plant species possesses its own immune system, which signals a cellular response and leads to the death of the attacking pathogen. Plants recognize pathogen-encoded molecules through probable receptors encoded by disease-resistance (R) genes. A signal transduction study provides an excellent understanding of gene-for-gene resistance, which explains plant pathogen co-evolution in a given environment (Bent, 1996).

The gene-for-gene hypothesis as proposed (Flor, 1971) can also be explained by the defense response that is often activated by the action of a host-resistance (R) gene and a pathogen avirulence (Avr) gene (Gabriel and Rolfe, 1990). Resistance genes are regulatory in nature, whereas defense genes are functional. However, R genes regulate the functions of defense genes. Defense genes are usually quiescent in healthy plants but become activated when the pathogen comes in contact with the plant, which then releases a signal. When a plant R gene interacts with the corresponding avirulence gene (Avr) of the pathogen, this triggers a series of defense responses, as explained in Fig. 5.1a.

Table 5.3 Pathogenesis-related proteins and genes in rice

PR-protein family	Class/sub-family/enzyme activity	Name	Protein cDNA/gene	Tissue of expression	Induced by	Reference
PR-1	Acidic	16.5 kDaJIP	P	Roots	Jasmonic acid	Moons et al. (1997)
PR-2	Basic glucanase		P	Grain	Developmental	Akiyama et al. (1996)
PR-2	Acidic glucanase		P	Bran	Developmental	Akiyama et al. (1997)
PR-2	Glucanase		P	Leaves	Stress	Rakwal et al. (1999)
PR-3	Class Ib chitinase		P	Bran	Developmental	Anuratha et al. (1992)
PR-3	Class I chitinase		P	Leaves	Pathogen	Anuratha et al. (1996)
PR-3	Class II chitinase		P	Leaves	Pathogen	Anuratha et al. (1996)
PR-3	Class Ib chitinase	RCH-A, RCH-B	P	Suspension cells	Oligo (NAG)	Inui et al. (1996)
PR-3	Class III chitinase	RCH-C	P	Suspension cells	Oligo (NAG)	Inui et al. (1996)
PR-3	Class Ia chitinase	Chi11	G	Seeds		Huang et al. (1991)
PR-3	Class I chitinase	2-2W	C	Seeds	Developmental	Anuratha et al. (1992)
PR-3	Class I chitinase	RC-7	C	Leaves	Pathogen	Anuratha et al. (1996)
PR-5	TLP		P	Leaves	Jasmonic acid, Stress	Rakwal et al. (1999)
PR-5	TLP	pPIR2	C	Leaves	Pathogen	Dudler (1994)
PR-5	TLP	C22	C	Leaves	Pathogen	Velazhahan et al. (1998)
PR-5	TLP	D34	C	Leaves	Pathogen	Velazhahan et al. (1998)
PR-6	Cystatin	OC-1	P	Seeds	Developmental	Abe et al. (1987)
PR-6	Cystatin	OC-2	P	Seeds	Developmental	Abe et al. (1987)
PR-6	Cystatin	OC-26	C	Seeds	Developmental	Abe et al. (1987)
PR-6	Cystatin	OC-9b	C	Seeds	Developmental	Chen et al. (1992)
PR-9	Peroxidase	PO-C1	P	Seedlings	Pathogen	Reimers et al. (1992)
PR-9	Peroxidase		P	Roots	Jasmonic acid	Moons et al. (1997)
PR-10	RNAse	Osdrr	P	Roots	Jasmonic acid	Moons et al. (1997)

Fig. 5.1 Signal transduction pathway leads to control of plant defense
SAR = systematic acquired resistance; arrow mark shows the disease sympton in non-transformed control

The first cloned *R* gene, *Hm1* from maize, was obtained through transposon tagging. *Hm1* confers resistance to race 1 strains of the fungal pathogen *Cochliobolus carbonum*. *Hm1* encodes an NADPH-dependent reductase that inactivates the potent plant toxin produced by these fungal strains (Johal and Briggs, 1992). However, the *Hm1 R* gene does not involve pathogen *Avr* genes. Nevertheless, this work outlines the first natural or engineered plant disease resistance. Chromosome walking (positional cloning) and heterologous transposon tagging made it possible to clone several *R* genes (Table 5.1) and enhance plant resistance against the pathogenic fungus (Table 5.4). The *R/Avr* approach was reported in tomato (Martin et al., 1993). *Pto* cloned from tomato confers resistance against *Pseudomonas syringae* expressing the *Avr* gene. *Pto* encodes a protein with similarity to serine-threonine protein kinases. Further, a few *R* genes were cloned that encode proteins containing leucine-rich repeat (LRR) domains (Table 5.1). Barbara Valent's group has demonstrated the gene-for-gene hypothesis in rice (Jia et al., 2000). The *NPRI* gene cloned from *Arabidopsis* confers resistance to the pathogens *Pseudomonas syringae* and *Peronospora parasitica*. *NPRI* appears to succeed in enhancing plant immunity by over expressing regulatory induction of the SAR signaling pathway (Cao et al., 1998). Our work is in progress with transgenic rice with *NPRI* gene.

PR PROTEIN GENES AND THEIR ROLE IN PLANT DEFENSE

Pathogenesis-related proteins were first reported 30 years ago as new protein components induced by tobacco mosaic virus (TMV) in hypersensitively-reacting tobacco (Van Loon, 1999). Since then, a great deal of research has focussed on the isolation, characterization, and regulation of expression of this unique class of defense proteins in a variety of plants (Table 5.2) and selected details of PR proteins and genes of cereals (Table 5.3) (Datta, 1999). The major interest has focussed in recent years on the realization that several PR proteins had antimicrobial and insecticidal activity. Several studies led to the conclusion that over expression of PR proteins in transgenic crops can delay the progression of diseases caused by several pathogens belonging to diverse genera (Table 5.4). *PR* genes are designated as *ypr*, followed by the same suffix in accordance with the recommendations of the Commission for Plant Gene Nomenclature (Table 5.2) (Van Loon, 1999). It is necessary to gather information at both the nucleic acid and protein level when dealing with stress-related proteins. Newly-defined cDNAs may also be added to the existing families when they are shown to be induced by pathogens or specific elicitors. Defensins and thionins—both families of small, basic cysteine-rich polypeptides—qualify for inclusion as new families of PR proteins (Table 5.2). The monograph on PR proteins (Datta and Muthukrishnan, 1999) and chapters therein on individual classes of PR protein (PR-1 through PR-11) provide detailed information on the isolation, characterization, and function of individual PR proteins in plants, including the chapter on transgenic research (Datta et al., 1999a).

There is always some selectivity in the interaction between a PR protein and its intended target pathogens in the sense that PR proteins represent generalized plant defense responses for broad, albeit incomplete, protection against diverse pathogens (Table 5.4). Many transgenic plants have now been developed with constitutive inducible expression of PR proteins at effective levels and can be used as a tool to enhance or stabilize yield in areas where pathogens and pests are endemic (Tables 5.4, 5.5)

EVALUATION OF PR GENES FOR SHEATH BLIGHT RESISTANCE IN RICE

Sheath blight disease of rice causes significant yield losses every year and is widespread in all rice-growing countries. Resistance breeding for this disease is not feasible because resistant germplasm is not yet known. It now seems that genetic engineering to manage sheath blight is an attractive and powerful tool because it introduces PR genes and optimizes the over expression of PR proteins in transgenic plants. Two different types of PR genes—PR3-chitinases (Chi11 and RC7) and PR-5 thaumatin-like protein genes—have been introduced into rice (Lin et al., 1995; Datta, 1999; Datta et al., 1999a,b, 2001; Baisakh et al., 2001) (Tables 5.4, 5.5). Transformation was done with the biolistic, protoplast, and *Agrobacterium* systems described earlier (Datta et al., 1990, 1992, 1997).

Table 5.4 Increased fungal resistance in transgenic rice

Transgene(s)	Pathogen	Reference
Rice-*TLP*	*Rhizoctonia solani*	Datta et al. (1999)
Rice-*Chi11*	*R. solani*	Lin et al. (1995)
		Datta and Muthukrishnan (1999)
Rice-*Chi, RC7*	*R. solani*	Datta (2000)
Rice-*Chi11*	*R. solani*	Baisakh et al. (2001)
Rice-*PR3*	*Magnaporthe grisea*	Nishizawa et al. (1999)
Rice-*Pi-ta*	*M. grisea*	Jia et al. (2000)
Rice-*Pi1, Piz5, Xa21*	*M. grisea, X. oryzae*	Narayanan et al. (2002a,b)

Table 5.5 Status of transgenic rice plants at IRRI with pathogenesis-related genes

Cultivar	Gene of interest	Method used	No. of regenerated plants	Analysis (Southern+)	Fertility status (%)
IR72	Chi11	B	72	60	70
	RC7	B	20	15	75
IR64	RC7	B	3	1	Fertile
CBII	Chi11	P	56	30	90
	RC7	P	232	42*	90
	D34	P	141	30*	90
ML7	Chi11	B	20	14	85
IRRI-NPT	Chi11	B	133	48*	55
Basmati 122	Chi11	Agro	45	15*	50
Tulsi	Chi11	Agro	115	111	80
Vaidehi	Chi11	Agro	64	64	90
Dinorado	D34	B	54	5*	80
Swarna	Chi11	B			90
MH63	RC7	B	35	29	91

B = biolistic; P = protoplast; *Agro* = *Agrobacterium*; ML = maintainer line; IRRI-NPT = IRRI new plant type
*All the plants were not analyzed

Inheritance was studied by Southern blot analysis (for gene integration) and western blot analysis with a polyclonal antibody. The transformants synthesized high levels of PR proteins constitutively and exhibited enhanced resistance when challenged with the sheath blight pathogen (*Rhizoctonia solani*) (Fig. 5.1b). At least 10 rice cultivars have been transformed with several PR genes which are now at different stages of development, awaiting homozygous status with acceptable PR-protein expression (Table 5.5). A good phenotype with enhanced levels of antifungal activity is now being selected for future field-testing and breeding (Fig. 5.2).

BLAST RESISTANCE

Rice blast caused by *Magnaporthe grisea* is the most devastating plant disease in Asia, particularly where rice is irrigated or receives high amounts of rainfall and nitrogen fertilizer (Zeigler et al., 1994). Breeders have adopted three methods to suppress blast disease. First, they tried to use varieties with field resistance. Second, they introduced resistance genes into high-yielding cultivars from other varieties. However, shortly after their release, these cultivars became seriously susceptible to blast disease in many areas because of the appearance of new races of the blast fungus. Thus, this showed clearly that solving the blast problem was

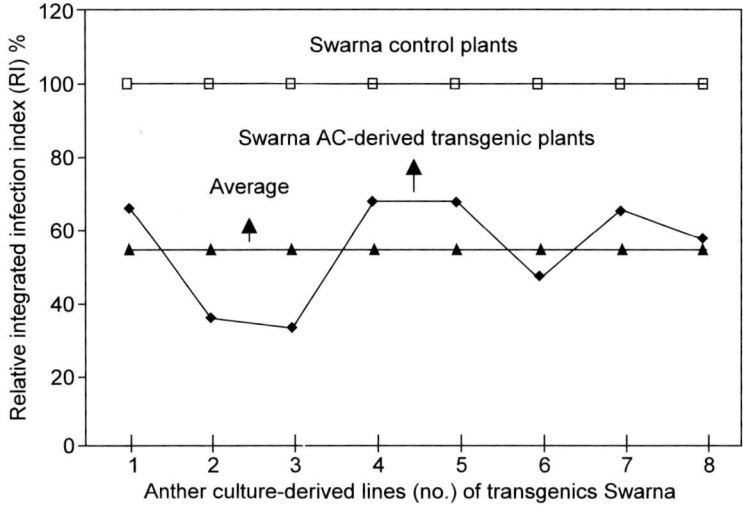

A) Bioassay results showing enhanced resistance of homozygous dihaploid transgenics of rice cultivar Swarna carrying chitinase transgene (*chi11*) to sheath blight fungus under greenhouse conditions (Baisakh et al., 2001)

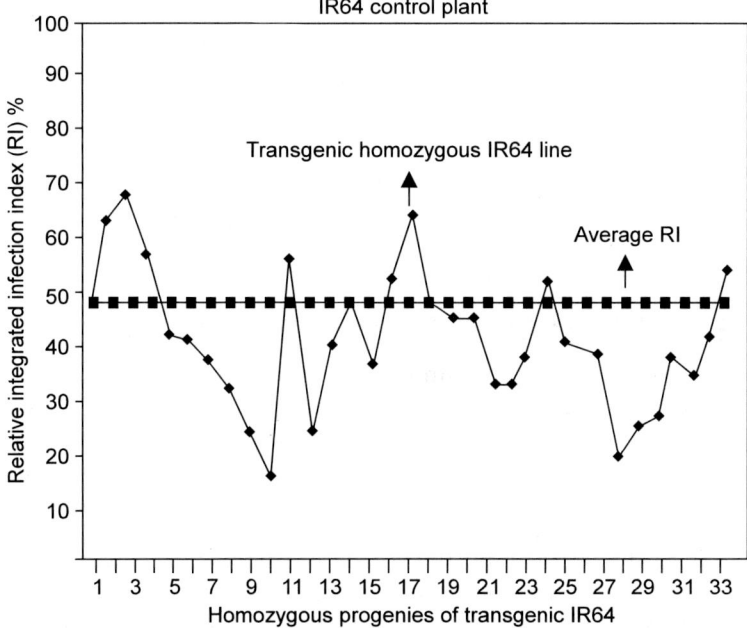

B) Bioassay results showing enhanced resistance of homozygous dihaploid transgenics of rice cultivar IR64 carrying chitinase transgene (*RC7*) to sheath blight fungus under greenhouse conditions (Datta et al., 2000)

Fig. 5.2 Transgenic rice with *Chitinase* gene showing enhanced resistance against sheath blight

more challenging than earlier thought. The third approach adopted was to develop varieties showing a high level of field resistance. Zeigler et al. (1994) proposed an alternative strategy called 'lineage exclusion' for using major genes to achieve durable resistance. Lavanya and Gnanamanickam (2000) found that two major blast resistance genes, *Pi-1* + *PiZ-5*, could exclude the entire population of *M. grisea*. Leister et al., (1999) mapped the *R*-gene candidates (*r2*, *r5*, and *r10*) in rice showing linkage to genes conferring race-specific resistance to blast (*Pi-k*, *Pi-f2*, and *Pi-1*). Genetic engineering allows us to shorten breeding time since the genes for a single trait would be transformed without altering the genetic makeup of the adopted cultivar. Two rice chitinase genes have been transferred into two japonica varieties of rice (Nipponbare and Koshihikari) by *Agrobacterium*-mediated transformation. The presence of transgenic *chitinase* genes was confirmed by PCR (polymerase chain reaction) and their expression in leaves was followed by northern blot analysis (Nishizawa et al., 1999). The product of the *cht-2* gene was intracellular, whereas *cht-3* was extracellular in accumulation. The constitutive expression of *cht-2* and *cht-3* chitinases exhibited enhanced resistance against the blast fungus, *M. grisea*, compared with that of the control plants. Instead of enhanced resistance, some transgenic lines showed reduced disease resistance, which might be due to the co-suppression of endogenous *PR* genes or gene silencing. A similar gene silencing was observed in transgenic rice (Chareonpornwattana et al., 1999) as well as in many other plants. Based on blast pathogen population dynamics and the lineage exclusion assay, two rice cultivars, CO39 and IR50, containing the *Pi-1* and *Piz-5* genes were transformed with the *Xa21* gene and showed resistance to blast and bacterial blight (Narayanan et al., 2002a,b).

VIRUS RESISTANCE IN TRANSGENIC PLANTS

Plant viruses cause severe damage to numerous crops, including rice. Considerable progress has been made in engineering crop plants with virus resistance. More than $1 billion in losses is reported yearly for rice in Southeast Asia (Herdt, 1991). The coat protein (CP) gene of tobacco mosaic virus (TMV) is the first report of virus-derived resistance in transgenic plants (Powell-Abel et al., 1986). Transgenic tobacco plants expressing high levels of TMV CP were more resistant to TMV virions than to TMV RNA inocula, suggesting that CP-mediated protection against TMV occurred through the inhibition of virion disassembly in the initially infected cells (Beachy et al., 1990).

Another approach of antisense RNA in homology-dependent resistance was hypothesized. The post-transcriptional gene silencing and pathogen-derived resistance to viruses were thought to be very effective, but they are yet to be well demonstrated in protecting crops against the viruses, particularly in cereals (Baulcombe, 1996).

TRANSGENIC RICE RESISTANCE TO YELLOW MOTTLE VIRUS: A CASE STUDY

Rice yellow mottle virus (RYMV) causes major yield losses in African rice production. Though endemic to Africa, RYMV is spreading in newly-established large-scale irrigated rice development schemes and experimental fields of Asian varieties. Controlling this disease is difficult because the virus is highly infectious and the epidemiology and role of vectors are not well understood. Natural resistance to RYMV is found in African land races of rice. However, the resistance is recessive and polygenic, and fertility barriers do not allow the introgression of this trait into cultivated rice. Genetic engineering based on pathogen-derived resistance was applied in this case to disrupt the pathogenesis. A transgene encoding the RNA-dependent RNA polymerase of RYMV, coupled to a 35S promoter and *hpt,* was transferred into susceptible rice cultivars. Fourteen fertile independent transgenic lines were produced that carried the transgenes. The transformed lines were resistant to RYMV strains from different African locations. One line completely suppressed virus multiplication. Resistance was stable over the last three generations. Further, in the most resistant line, transcription analysis indicated that the resistance derives from an RNA-based mechanism associated with post-transcriptional gene silencing (Pinto et al., 1999). Some examples also showed transgenic rice conferring resistance to rice stripe tenuvirus (Hayakawa et al., 1992) and rice dwarf virus (Zheng, 1997).

TRANSGENIC RICE DEVELOPED WITH BACTERIAL BLIGHT RESISTANCE

Bacterial blight (BB) caused by *Xanthomonas* pv. *oryzae* (*Xoo*) is one of the most destructive diseases of rice worldwide. Rice yield losses caused by BB in some areas of Asia can reach 50%. The use of resistant cultivars is the most economical and effective method to control this disease (Ogawa, 1993).

A dominant gene for resistance to BB was transferred from a wild species, *Oryza longistaminata*, to the cultivated variety IR24 (Khush et al., 1990). The resulting line with *Xa21* is called IRBB21. *Xa21* confers resistance to all the known races of *Xoo* in India and the Philippines (Ikeda et al., 1990). The molecular structure of *Xa21* represents an uncharacteristic class of plant disease resistance genes. From its deduced amino-acid sequence, the gene was found to be translated into a receptor kinase-like protein-carrying leucine-rich repeats in the putative extracellular domain, a single-pass transmembrane domain, and a serine theonine kinase intracellular domain. Further, *Xa21* supports a role for cellular signaling in plant disease resistance (Song et al., 1995).

Xa21 has been transferred to susceptible japonica rice T309, which showed resistance to BB (Wang et al., 1996). As T309 is not a commercial variety, we introduced the gene in elite breeding cultivars, such as IR72, MH63, IR51500, etc. Molecular analysis of transgenic plants revealed the presence of a 3.8-kb *Eco*RV-digested DNA fragment corresponding to most of the *Xa21*-coding region and its complete intron sequence, indicating the integration of *Xa21* in the genome of rice. Transgenic plants were challenged with two prevalent races (4 and 6) of *Xanthomonas oryzae*. T_0 and T_1 plants positive for the transgene were resistant to bacterial blight (Tu et al., 1998). We also observed that the level of resistance to race 4 of *Xoo* was higher because of pyramiding of *Xa21* in addition to *Xa4* already present in IR72. This is a very efficient way to improve BB resistance of rice without genetic dragging and it requires less than 2 years (Datta, 1999).

FIELD EVALUATION OF TRANSGENIC BB RESISTANCE IN IR72

Based on characterization of the resistance phenotype and molecular analysis, several homozygous lines carrying *Xa21* against the BB pathogen were obtained from previously-transformed indica rice IR72. The homozygous line, T103-10, with the best phenotype and seed-setting was tested repeatedly under normal field conditions in order to evaluate its resistance to the BB pathogen in Wuhan, China, in 1998 and 1999. The races of *Xoo* used in the experiments were PXO61, PXO79, PXO99, and PXO112 isolated from the Philippines, T2 isolated from Japan, and Zhe173 isolated from China. The results demonstrated that the transgenic homozygous line expressed the same resistance spectrum, but with a shorter lesion length for each inoculated race than the lesion length of the *Xa21* donor line IRBB21 (Table 5.6). The non-transformed control IR72 carrying *Xa4* was resistant to PXO61, PXO112, Zhe 173, and T2 but was susceptible to PXO79 and PXO99. The negative control variety IR24 was susceptible to all isolates under field conditions. The results demonstrated clearly that the *Xa21* transgene led to an excellent field performance of the induced bacterial blight resistance trait on the recipient plants (Tu et al.., 2000). The yield performance of transgenic

Table 5.6 Disease reaction of 90 plants of various lines to different races of *Xoo* under field conditions in 1999 (Huazhong Agricultural University, Wuhan, China).

Xoo race	Variety	Lesion length (cm)	Reaction
Philippines race 1 (PXO 61)	IR72	1.04 ± 0.12	S
	T103	0.31 ± 0.05	HR
	IR24	16.43 ± 1.32	HS
	IRBB21	0.97 ± 0.24	HR
Philippines race 3 (PXO79)	IR72	9.00 ± 0.86	S
	T103	0.61 ± 0.24	HR
	IR24	14.11 ± 1.46	HS
	IRBB21	0.82 ± 0.40	HR
Philippines race 5 (PXO112)	IR72	0.72 ± 0.13	HR
	T103	0.39 ± 0.09	HR
	IR24	7.60 ± 1.11	S
	IRBB21	1.02 ± 0.62	R
Philippines race 6 (PXO99)	IR72	9.37 ± 1.21	S
	T103	2.43 ± 0.53	R
	IR24	15.69 ± 1.24	HS
	IRBB21	8.00 ± 1.20	MS
Chinese race 4 (Zhe173)	IR72	1.41 ± 0.53	R
	T103	0.79 ± 0.31	HR
	IR24	11.16 ± 1.89	HS
	IRBB21	1.58 ± 0.73	R
Japanese race 2 (T2)	IR72	0.71 ± 0.45	HR
	T103	0.61 ± 0.31	HR
	IR24	20.01 ± 1.89	HS
	IRBB21	1.94 ± 0.911	R

HS = highly susceptible; S = susceptible; MS = moderately susceptible;
R = resistant; HR = highly resistant
Source: Tu et al. (2000)

homozygous line T103-10 is comparable with that of the control under field conditions (Table 5.7).

It was also noticed that an increased level of resistance to the BB pathogen persisted in transgenic plants through several generations, indicating its stable inheritance. The heritable increased level of resistance to the BB pathogen can, in turn, provide an advantage for genetic engineering over classical breeding in cases where the highest levels of resistance are desirable and can be achieved in a short time. It is also noteworthy that various national agricultural research systems in Asia are making efforts to incorporate the other *Xa* genes into popular cultivars through marker-aided selection.

The availability of various cultivars with different resistance genes could significantly decrease the yield loss. Assuming a minimum yield loss of 1% due to this disease, around $320.5 million could be saved over

Table 5.7 Agronomic traits of transgenic IR72 and IR72 control under field conditions (Huazhong Agricultural University, Wuhan, China, 1999)

Variety/line	Days to flowering	Plant height (cm)	Panicles/plant	Filled seeds/panicle	Empty grains/panicle	Total seeds/panicle	Seed-settting rate (%)	1000-seed weight (g)	Yield (t/ha)
IR72 transgenic	98	85.6	16.2	67.9	41.1	108.9	62.4	20.8	4.89
IR72 control	96	93.6	17.1	65.0	29.8	94.8	68.5	21.3	4.97

Source: Tu et al. (2000)

30 million ha. with an average yield of 5.5 t/ha. in China, whereas a yield loss of 0.75% covering 132.5 million ha. with an average yield of 3.6 t/ha in Asia translates into $715.5 million. Thus, transgenic rice with BB resistance will have a large economic impact.

This study shows that conventional and molecular breeding techniques could be a powerful combination in rice breeding. Genetic transformation is a one-step process of introducing novel genes into a desirable genetic background of important crops. As it is a fast and efficient gene integration tool, it could well be the answer to catching up with the pathogen's ability to mutate fast and render once-resistant plants susceptible. For instance, rice cultivars carrying the *Xa4* gene for resistance, which were widely deployed in the Philippines starting in the early 1970s, became susceptible to the predominant race of *Xoo* within 5 years (Mew et al., 1992). Transformation techniques could help in developing transgenic plants in less than 2 years in order to minimize the effects of a breakdown in resistance in the host plant. With the availability of resistance genes from other sources, the strategic deployment of transgenic rice with gene pyramiding may provide durable resistance in rice breeding.

CONCLUSIONS

Disease resistance will be an exciting and challenging research area in the coming decades. The cloning of many specific and broad-spectrum resistance genes can be anticipated in the near future. Functional analysis will allow the dissection of molecular specificity, leading to the *ex planta* generation of new transgenic plants with resistance-gene specificities. Disease-resistance genes control pathogens at a low phenotypic cost by inducing programmed cell death by hypersensitive reaction caused by the response of the pathogens. A signal transduction study now provides an excellent understanding of gene-for-gene resistance, which explains plant pathogen co-evolution in a given environment. The recent breakthrough in cloning several *R* genes such as *Xa21* shows that conventional and molecular breeding techniques could be a powerful combination in rice breeding. Genetic transformation has proved to be a one-step process of introducing novel genes into a desirable genetic background of important crops. Since this system is fast and efficient, it will create improved varieties faster than the pathogen can overcome the resistance. Over expression of *PR* genes and the combination of more than one gene will delay disease symptoms and protect plants in a sustainable manner (Datta et al., 2002). However, a disease is the outcome of interactions among the host, pathogen, and environment and, therefore, using all possible approaches, including genetic engineering in an adopted elite variety with enhanced disease resistance, is the only way to achieve durable resistance.

ACKNOWLEDGEMENTS

Financial support from BMZ (Germany) and the Rockefeller Foundation is gratefully acknowledged. Thanks are due to Dr Bill Hardy for comments and editorial assistance and to Dr Karabi Datta, Michelle Viray, Dr N.N. Narayanan, and Dr Niranjan Baisakh for artwork and general assistance.

REFERENCES

Abe, K., Emori, Y., Kondo, H., Suzuki, K. and Arai, S., 1987, Molecular cloning of a cysteine protease inhibitor of rice (*oryza cystatin*): homology with animal cystatins and transient expression in the ripening process of rice seeds. J. Biol. Chem., **262**, 16793-16797.

Akiyama, T., Kaku, H. and Shibuya, N., 1996, Purification and properties of a basic endol 1-3-b-glucanase from rice (*Oryza sativa* L.). Plant Cell Physiol., **37**, 702-705.

Akiyama, T., Shibuya, N., Hrmova, M. and Fincher, G. B., 1997, Purification and characterization of a (1-3)-beta-D-glucan endohydrolase from rice (*Oryza sativa*) bran. Carobhydr. Res., **14**, 365-374.

Anuratha, C. S., Huang, J. K., Pingali, A. and Muthukrishnan, S., 1992, Isolation and characterization of a chitinase and its cDNA clone from rice. J. Plant Biochem.Biotech., **1**, 5-10.

Anuratha, C. S., Zen, K. C., Cole, K. C., Mew, T. and Muthukrishnan, S., 1996, Induction of chitinases and β-glucanases in *Rhizoctonia solani*-infected rice plants:isolation of an infection-related chitinase cDNA clone. Physiol. Plant, **97**, 39-46.

Baisakh, N., Datta, K., Oliva, N., Ona, I., Rao, G. J. N., Mew, T. W. and Datta, S. K., 2001, Rapid development of homozygous transgenic rice using anther culture harboring rice *chitinase* gene for enhanced sheath blight resistance. Plant Biotechnol., **18**, 101-108.

Baulcombe, D. C., 1996, Mechanism of pathogen-derived resistance to viruses in transgenic plants. Plant Cell, **8**, 1833-1844.

Beachy, R. N., Loesch-Fries, S. and Tumer, N. E., 1990, Coat protein-mediated resistance against virus infection. Annu. Rev. Phytopathol. 24, 451.

Bent, A.F., 1996, Plant disease resistance genes: function meets structure. Plant Cell, **8**, 1757-1771.

Büschges, R., Hollricher,K., Panstruga, R., Simons, G., Wolter, M., Frijters, A., van Daelen, R., vander Lee, T., Diergaarde, P., Groenendijk, J., Töpsch, S., Vos, P., Salamini, F. and Schulze-Lefert, P., 1997, The barley *Mlo* gene: A novel control element of plant pathogen resistance. Cell, **88**, 695-705.

Cao, H., Li, X. and Dong, X., 1998, Generation of broad-spectrum disease resistance by over-expression of an essential regulatory gene in systematic acquired resistance. Proc. Natl. Acad. Sci. USA, **95**, 6531-6536.

Chareonpornwattana, S., Krishnarajapuran, V. T., Wang, L., Datta, S. K., Panbangred, W. and Muthukrishnan, S., 1999, Inheritance, expression, and silencing of a chitinase transgene in rice. Theor. Appl. Genet., **98**, 371-378.

Chen, M. J., Johnson, B., Wen, L., Muthukrishnan, S., Kramer, K. J., Morgan, T. D. and Reeck, G. R., 1992, Rice cystatin: bacterial expression, purification, cysteine proteinase inhibitory activity, and insect growth suppressing activity of truncated form of the protein. *Protein* Expt. Purif., **3**, 41-49

Dangl, J.L., 1998, Plants just say no to pathogens. Nature, **394**, 525-527.

Datta, K., Muthukrishnan, S. and Datta, S. K., 1999a, Expression and function of PR-protein genes in transgenic plants. In: Datta S.K. and Muthukrishnan S. (eds), Pathogenesis-Related Proteins in Plants, CRC Press, Boca Raton, FL, pp. 261-277.

Datta, K., Baisakh, N., Thet, K. M., Tu, J. and Datta, S. K., 2002, Pyramiding transgenes for multiple resistance in rice against bacterial blight, yellow stem borer and sheath blight. Theor. Appl. Genet. **106**, 1-8.

Datta, K., Tu, J., Oliva, N., Ona, I., Velazhahan, R., Mew, T. W., Muthukrishnan, S. and Datta, S. K., 2001, Enhanced resistance to sheath blight by constitutive expression of infection- related rice chitinase in transgenic elite indica rice cultivars. Plant Science, **160**, 405-414.

Datta, K., Koukolíková-Nicola, Z., Baisakh, N., Oliva, N. and Datta, S. K., 2000, *Agrobacterium*-mediated engineering for sheath blight resistance of indica rice cultivars from different ecosystems. Theor. Appl. Genet., **100**, 832-839.

Datta, K., Velazhahan, R., Oliva, N., Mew, T., Khush, G. S., Muthukrishnan, S. and Datta, S. K., 1999b, Overexpression of cloned rice thaumatin-like protein (PR-5) gene in transgenic rice plants enhances environmental friendly resistance to *Rhizoctonia solani* causing sheath blight disease. Theor. Appl. Genet., **98**, 1138-1145.

Datta, S. K., 1999, Transgenic cereals: *Oryza sativa* (rice). In: Vasil I. K. (ed.), Molecular Improvement of Cereal Crops. Kluwer Academic Publishers, The Netherlands, pp. 149-187.

Datta, S. K., 2000, Potential benefit of genetic engineering in plant breeding: Rice, a case study. Agric. Chem. Biotech., **43**, 197-206.

Datta, S. K., 2002, Transgenic rice breeding: protection against bacterial blight, sheath blight, and stem borer. In: Leong S. A., Teng P. S. and Cattlin N. (eds), A Colour Handbook of Pests, Diseases and Disorders of Rice, Manson Pub., London. (in press).

Datta, S. K. and Muthukrishnan, S., 1999, Pathogenesis-related Proteins in Plants. CRC Press, Boca Raton, FL.

Datta, S. K., Peterhans, A., Datta, K. and Potrykus, I., 1990, Genetically engineered fertile Indica-rice plants recovered from protoplasts. Bio/Technology, **8**, 736-740.

Datta, S. K., Datta, K., Soltanifar, N., Donn, G. and Potrykus, I., 1992, Herbicide-resistant Indica rice plants from IRRI breeding line IR72 after PEG-mediated transformation of protoplasts. Plant Mol. Biol., **20**, 619-629.

Datta, S. K., Torrizo, L., Tu, J., Oliva, N. and Datta, K., 1997, Production and Molecular Evaluation of Transgenic Rice Plants. IRRI Discussion Paper Series No. 21. International Rice Research Institute, Manila, Philippines.

Dudler, R., Mauch, F. and Reimmann, C. 1994, Thaumatin-like proteins. In: Witty M. and Higginbotham J.D. (ed), Thaumatin. CRC Press, Boca Raton, FL, pp. 193-199.

Dixon, M. S., Jones, D. A., Keddie, J. S., Thomas, C. M., Harrison, K. and Jones J. D. G., 1996, The tomato *Cf-2* disease resistance locus comprises two functional genes encoding leucine- rich repeat proteins. Cell, **84**, 451-459.

Flor, H., (1971), Current status of the gene-for-gene concept. Annu. Rev. Phytopathol., 9, 275-296.

Gabriel, D. and Rolfe, B., 1990, Working models of specific recognition in plant-microbe interactions. Annu. Rev. Phytopathol., **28**, 365-391.

Grant, M. R., Godlard, L., Straube, E., Ashfield, T., Lewald, J., Sattler, A., Inner, R. W. and Dangl, J. L., 1995, Structure of the *Arabidopsis RPM1* gene enabling dual specificity disease resistance. Science, **269**, 843-846.

Hayakawa, T., Zhu, Y., Itoh, K., Kimura, Y., Izawa, T., Shimamoto, K. and Toriyama, S., 1992, Genetically engineered rice resistant to rice stripe virus, an insect-transmitted virus. Proc. Natl. Acad. Sci. USA, **89,** 9865-9869.

Hayano-Saito, Y., Tsuji, T., Fujii, K., Saito, K., Iwasaki, M. and Saito, A., 1998, Localization of the rices tripe disease resistance gene, *Stv-b*, by graphical genotyping and linkage analyses with molecular markers. Theor. Appl. Genet., **96,** 1044-1049.

Herdt, R.D., 1991, Research priorities for rice biotechnology. In: Khush G.S. and Toenniessen. G. (eds), Rice Biotechnology, CAB International, Wallingford, UK, p 19.

Huang, J. K., Wen, L., Swegle, M., Tran, H. C., Thin, T. H., Naylor, H. M., Muthukrishnan, S. and Reeck, G. R., 1991, Nucleotide sequence of a rice genomic clone that encodes a class I endochitinase. Plant Mol. Biol., **16,** 479-480.

Ikeda, R., Khush, G. S. and Tabien, R., 1990, A new resistance gene to bacterial blight derived from *O. longistaminata*. Jpn. J. Breed., **40** (suppl 1), 280-281.

Inui, H., Yamaguchi, Y., Ishigami, Y., Kawaguchi, S., Yamada, T., Ihara, H. and Hirano, S., 1996, Three extracellular chitinases in suspension-cultured rice cells elicited by N-acetylchitooligosaccharides. Biosci. Biotechnol. Biochem., **60,** 1956-1961.

Jia, Y., McAdams, S. A., Bryan, G. T., Hershey, P. and Valent, B., 2000, Direct interaction of resistance gene and avirulence gene products confers rice blast resistance. EMBO J., **19,** 4004-4014.

Johal, G. S. and Briggs, S. P., 1992, Reductase activity encoded by the *Hm1* disease resistance gene in maize. Science, **258,** 985-987.

Jones, D. A., Thomas, C. M., Hammond-Kosack, K. E., Balint-Kurti, P. J. and Jones, J. D.G., 1994, Isolation of the tomato *Cf-9* gene for resistance to *Cladosporium fulvum* by transposon tagging. Science, **266,** 789-793.

Khush, G. S., Bacalangco, E. and Ogawa, T., 1990, A new gene for resistance to bacterial blight from *O. longistaminata*. Rice Genet. Newsl., **7,** 121-122.

Lavanya, B. and Gnanamanickam S. S., 2000, Molecular tools for characterization of rice blast pathogen (*Magnaporthe grisea*) population and molecular marker-assisted breeding for disease resistance. Curr. Sci., **78,** 248-257.

Lawrence, G. J., Finnegan, E. J., Ayliffe, M. A. and Ellis, J. G., 1995, The *L6* gene for flax rust resistance is related to the *Arabidopsis* bacterial resistance gene *RPS2* and the tobacco viral resistance gene, N. Plant Cell, **7,** 1195-1206.

Leister, D., Kurth, J., Laurie, D. A., Sasaki, T., Graner, A. and Schulze Lefert, P., 1999, RFLP- and physical mapping of resistance gene homologues in rice (*O. sativa*) and barley (*H. vulgare*). Theor. Appl. Genet., **98,** 509-520.

Lin, W., Anuratha, C. S., Datta, K., Potrykus, I., Muthukrishnan, S. and Datta, S. K., 1995, Genetic engineering of rice for resistance to sheath blight. Bio/Technology, **13,** 686-691.

Martin, G. B., Brommonschenkel, S. H., Chunwongse, J., Frary, A., Ganal, M. W., Spivey, R., Wu, T., Earle, E. D. and Tanksley, S. D., 1993, Map-based cloning of a protein kinase gene conferring disease resistance in tomato. Science, **262,** 1432-1436.

Mew, T. W., Vera Cruz, C. M. and Medalla, E. S., 1992, Changes in race frequency of *Xanthomonas oryzae* pv. *oryzae* in response to the planting of rice cultivars in the Philippines. Plant Dis., **76,** 1029-1032.

Moons, A., Prinsen, E., Bauw, G. and Van Montague, M., 1997, Antagonistic effects of abscisic acid and jasmonates on salt stress-inducible transcripts in rice roots. Plant Cell, **9,** 2243-2259.

Narayanan, N. N., Baisakh, N., Vera Cruz, C. M., Gnanamanickam, S. S., Datta, K. and Datta, S. K., 2002a, Molecular breeding for the development of blast and bacterial blight resistance in rice cv. IR50. Crop Sci. **42:** 2072-2079.

Narayanan, N. N., Baisakh, N., Oliva, N. P., Vera Cruz, C. M., Gnanamanickam, S., Datta, K. and Datta, S. K., 2002b, Gene stacking strategy for developing transgenic indica rice (cv. CO30) for blast and bacterial blight resistance. Mole Breed (in press).

Nishizawa, Y., Nishio, Z., Nakazono, K., Soma, M., Nakajima, E., Ugaki, M. and Hibi, T., 1999, Enhanced resistance to blast (*Magnaporthe grisea*) in transgenic japonica rice by constitutive expression of rice chitinase. Theor. Appl. Genet., **99**, 383-390.

Ogawa, T., 1993, Methods and strategy for monitoring race distribution and identification of resistance genes to bacterial leaf blight (*Xanthomonas campestris* pv. *oryzae*) in rice. JARQ, **27**, 71-80.

Parker, J., Coleman, M., Szabo, V., Daniels, J. and Jones, J. (1997) Resistance gene-dependent plant defense responses. In: K.E. Hammond-Kosack, J.D.G. Jones, eds. *Plant Cell*, **8**, 1771-1791.

Pinto, Y. M., Kok, R. A. and Baulcombe, D. C., 1999, Resistance to rice yellow mottle virus (RYMV) in cultivated African rice varieties containing RYMV transgenes. Nature Biotechnol., **17**, 702-706.

Powell-Abel, P., Nelson, R. S., De, B., Hoffmann, N., Rogers, S. G., Fraley, R. T. and Beachy, R. N., 1986, Delay of disease development in transgenic plants that express the tobacco mosaic virus coat protein gene. Science, **232**, 738-743.

Purkayastha, R. P., 1998, Disease resistance and induced immunity in plants. Indian Phytopathol., **51**, 211-221.

Rakwal, R., Agarwal, G. K. and Yonekura, M., 1999, Separation of proteins from stressed rice (*Oryza sativa* L.) leaf tissues by two-dimensional polyacrylamide gel electrophoresis: induction of pathogenesis-related and cellular protectant proteins by jasmonic acid, UV irradiation and copper chloride. Electrophoresis, **20**, 3472-3478.

Reimann, C. and Dudler, R. R., 1993, cDNA cloning and sequence analysis of a pathogen-induced thaumatin-like protein from rice (*Oryza sativa*). Plant Physiol., **101**, 1113-1114.

Reimers, P. J., Guo, A. and Leach, J. E., 1992, Increased activity of a cationic peroxidase associated with an incompatible interaction between *Xanthomonas oryzae* pv. *oryzae* and rice (*Oryza sativa*). Plant Physiol., **99**, 1044-1050.

Salmeron, J. M., Barker, S. J., Carland, F. M., Mehta, A. Y., Staskawicz, B. J., 1994 Tomato mutants altered in bacterial disease resistance provide evidence for a new locus controlling pathogen recognition. *Plant Cell*, **6**, 511-520.

Simons, G. and Flur, R., 1996, Resistance gene-dependent plant defense responses. Plant Cell, **8**, 1771-1791.

Song, W. Y., Wang, G. L., Chen, L. L., Kim, H. S., Pi, L. Y., Holsten, T., Gardner, J., Wang, B., Zhai, W., Zhu, L. H., Fauquet, C. and Ronald, P., 1995, A receptor kinase-like protein encoded by the rice disease resistance gene *Xa-21*. Science, **270**, 1804-1806.

Staskawicz, B. J., Ausubel, F. M., Baker, B. J., Ellis, J. G. and Jones, D. G., 1995, Molecular genetics of plant disease resistance. Science, **268**, 661-667.

Tu, J., Datta, K., Khush, G. S., Zhang, Q. and Datta, S. K., 2000, Field performance of *Xa21* transgenic indica rice (*Oryza sativa* L.), IR72. Theor. Appl. Genet., **101**, 15-20.

Tu, J., Ona, I., Zhang, Q., Mew, T. W., Khush, G. S. and Datta, S. K., 1998, Transgenic rice variety IR72 with *Xa21* is resistant to bacterial blight. Theor. Appl. Genet., **97**, 31-36.

Van Loon, L. C., 1999, Occurrence and properties of plant pathogenesis-related proteins. In: Datta, S. K., Muthukrishnan S. (eds), Pathogenesis-related Proteins in Plants, CRC Press, Boca Raton, FL, pp. 1-19.

Velazhahan, R., Cole, K. C., Anuratha, C. S. and Muthukrishnan, S., 1998, Induction of thaumatin-like proteins (TLPs) in *Rhizoctonia solani*-infected rice and characterization of two new cDNA clones. *Physiol. Plant.*, **102**, 21-28.

Vidhyasekaran, P., 1998, Molecular biology of pathogenesis and induced systematic resistance. Indian Phytopathol., **51**, 111-120.

Wang, G. L., Song, W. Y., Ruan, D. L., Sideris, S. and Ronald, P. C., 1996, The cloned gene *Xa21* confers resistance to multiple *Xanthomonas oryzae* pv. *oryzae* isolates in transgenic plants. Mol. Plant Microbe Interact., **9**, 850-855.

Whitham, S., Dinesh-Kumar, S. P., Choi, D., Hehl, R., Corr, C. and Baker, B., 1994, The product of the tobacco mosaic virus resistance gene N: similarity to Toll and the interleukin-1 receptor. Cell, **78**, 1011-1015.

Yoshimura, S., McCouch, S. R., Mew, T. W. and Nelson, R. J., 1995, Tagging and combining bacterial blight resistance genes in rice using RADP and RFLP markers. Mol. Breed., **1**, 375 387.

Zeigler, R. S., Leong, S. A. and Teng, P. S., 1994, Rice Blast Disease. CAB International. Wallingford, United Kingdom.

Zheng, H. H., 1997, Recovery of transgenic rice plants expressing the rice dwarf virus outer coat protein gene (*S8*). Theor. Appl. Genet., **94**, 522-527.

6

DEVELOPMENT OF RESISTANCE IN PESTS TO TRANSGENIC PLANTS: MECHANISMS AND MANAGEMENT STRATEGIES

CHARLES F. CHILCUTT* AND MARSHALL W. JOHNSON**
*Department of Entomology, Texas A&M University System
10345 Agnes St. Corpus Christi, TX 78406-9704, USA
**Department of Entomology
University of California at Riverside
Riverside, California 92521, USA

INTRODUCTION

Despite human efforts to control crop pests, every year, approximately 37 to 42% of the world's potential crop production is lost to pests, including 12% to plant pathogens, 14% to insects, and the remainder to weeds (Gatehouse and Gatehouse, 1998; FAO, 1993; Oerke et al., 1994). Control measures employed include pesticides, host plant resistance, cultural controls (e.g. as rotations or tillage types), behavioral modification, and biological control. Oerke et al. (1994) project that without pesticides, losses of wheat, soybeans, and potatoes would be more than 50%, and losses of corn and rice greater than 80%. Pesticides, in addition to saving about 10% of the world food supply, can also cause serious environmental and public health problems (Blair, 1989; Lefferts, 1989; Hotchkiss, 1992; Younes and Galal-Gorchev, 2000).

Genetic engineering has increased the potential for creating crops that are resistant to these pests as well as the herbicides used to control weeds. Resistance is a genetically-based decrease in the susceptibility of a population over a period of time. It may be in the form of a decrease in a plant's susceptibility to an insect pest or even a decrease in an insect's susceptibility to an insecticide. Thus, we are creating plants that are resistant to insects and pathogens while trying to prevent these pests from countering this resistance. In the case of weeds, the threat is not the weeds evolving resistance to the transgenic plants, but rather, an increased use of herbicides increasing the rate of resistance in weeds or in the transference of resistance genes through outcrossing between transgenic crops and related weed species.

Genetic engineering of pest- or herbicide-resistant plants offers a number of advantages over conventional breeding methods, including an increase in the number of useful genes and a reduction in the time needed to introduce desirable characters into a crop's genetic background (Hilder and Boulter, 1999).

As of October 2000, 15 insect-protected transgenic crop varieties, six virus-protected varieties, and 23 herbicide-resistant varieties had been cleared for commercial use in the US (USDA APHIS, 2000). All of the 15 insect-protected varieties were engineered with toxins from the bacterium *Bacillus thuringiensis* Berliner (Bt), including one tomato, two cotton, five potato, and seven corn varieties. The virus-protected plants included one papaya variety resistant to papaya ringspot virus (PRSV), two squash varieties resistant to watermelon mosaic virus 2 (WMV2) and zucchini yellow mosaic virus (ZYMV), with one of the varieties also resistant to cucumber mosaic virus (CMV), and two potato varieties resistant to potato leafroll virus (PLRV) and one resistant to potato virus Y (PVY). Herbicide-resistant transgenic plants included one beet, one flax, one rice, three rapeseed, four soybean, four cotton, and eight corn varieties resistant to glufosinate-ammonium, glyphosate, bromoxynil, or sulfonyl urea.

In this chapter, we shall review the mechanisms and management strategies of pest resistance to transgenic plants or herbicides used in combination with herbicide-resistant transgenic plants. Understanding the mechanisms and how best to manage resistance is essential for the continued use of transgenic plants. The most obvious reason for this is that many pests have already evolved resistance to a number of pesticides, including insects that are resistant to the Bt insecticidal proteins now common in insect-resistant transgenic plants (Tabashnik, 1994). Although transgenic resistance could be used against any type of arthropod, to date, most have concentrated on insects (mainly Lepidoptera and Coleoptera) and, therefore, we will confine our discussion to insect-resistant transgenic plants. Similarly, pathogen-resistant transgenic plants could potentially be created to defend against any kind of pathogen, but the primary focus of work has been viruses; so we will confine our discussions to these topics.

Many reasons exist for studying the mechanisms by which pests become resistant to transgenic plants. For one, understanding the mechanism of resistance is the first step in developing methods to monitor and detect resistance evolution and, eventually, to develop a successful management program. Also, if the mechanism of resistance is known, then a method of overcoming the resistance, such as the use of a synergist or the use of a transgenic plant with a different mechanism of killing the pest can be devised.

More important than understanding the mechanisms of resistance of pests is being able to retard or prevent the resistance mechanisms from evolving. Despite the slow progress in elucidating the many mechanisms of pest resistance to pesticides, there have been many theories and suggestions on how best to manage resistance. While this may seem premature, the value of pesticides necessitates that management strategies proceed before resistance mechanisms are fully understood. If not, we will lose environmentally-benign products such as Bt Already, more than 500 insect species and 145 weed species are resistant to one or more pesticides (Georghiou and Lagunes-Tejeda, 1991; Weedscience.com, 2001).

Two main strategies have been employed to genetically engineer plants with resistance to insect herbivores (Lal and Lal, 1993; Hilder and Boulter, 1999). By far, the most important is the insertion of insecticidal protein genes from *B. thuringiensis* (Hilder and Boulter, 1999; Morton et al., 2000). The second is the use of plant-derived compounds found in high levels in many plants and believed to be for defense against insect attack, especially leguminous seeds and potatoes (Jouanin et al., 1998). These compounds, including lectins and enzyme inhibitors, generally induce starvation or reduce nutrient uptake and growth in herbivores by inhibiting the active sites of digestive enzymes or binding to midgut membranes (Gatehouse and Gatehouse, 1998; Hilder and Boulter, 1999).

Although potentially, any number of herbicide-resistance mechanisms can be engineered into plants, resistance to glyphosphate, gluphosinate-ammonium, bromoxynil, and sulfonyl urea has been the most important, and, therefore, we mainly discuss resistance to herbicides within the same classes (Watrud et al., 1996). Similarly, although several strategies for virus resistance have been used in transgenic plants, the most important to date are plants engineered with viral coat protein resistance (Powell-Abel et al., 1986; Beachy et al., 1990). For this reason, we limit our discussion of plant pathogen-resistant transgenic plants to the use of viral coat protein genes. Also, because of the lack of discussion of viruses evolving resistance to transgenic plants, we discuss the more commonly-voiced concern regarding the release of these transgenic plants; the possibility of creating new viral pathogens through transencapsidation (de Zoeten, 1991). Along a similar vein, probably the most widespread concern about the introduction of transgenic plants is that it might create new weed species or increase the aggressiveness of existing weeds (Colwell et al., 1985; Tiedje et al., 1989). Therefore, we have included an additional section on gene flow in which we discuss the possibility of creating new weed species or plants that cause some type of environmental harm.

Our discussion of management strategies is relevant to any type of pest- or herbicide-resistant transgenic plant. Although these strategies may differ in some ways—depending on whether one looks at pathogen resistance, insect resistance, or herbicide resistance—the overall strategies can be useful for all.

RESISTANCE MECHANISMS

Insects

Factors that influence the evolution of resistance in insects include the initial allele frequency and dominance of alleles that confer resistance, the immigration rate of susceptible individuals, available refugia, dose, and the frequency and timing of applications (Georghiou and Taylor, 1977; Taylor and Georghiou, 1979; Caprio and Tabashnik, 1992). In theory, insecticides are only applied after a pest population level has reached or is expected to reach a certain threshold level. Therefore, unless toxin expression within transgenic plants can be turned on only when economically-significant damage has occurred or is only expressed in important areas of the plant, there will be a constant selection pressure, something that generally does not occur with modern pesticide applications. Also, transgenic plants tend to express toxins at much higher levels than typically achieved through routine treatments.

Bacillus thuringiensis endotoxins

Bt is a bacterial pathogen of several economically-important crop pests (Peferoen, 1997). During sporulation, it produces crystalline inclusions composed of proteins termed *Cry* toxins (Gill et al., 1992). These proteins are extremely toxic to certain insects, yet cause little or no harm to other animals (Croft, 1990; Peferoen, 1997). Bt had been applied in France since 1938 and in the US for more than two decades before resistance was first reported in field populations of the diamondback moth, *Plutella xylostella* (Linnaeus), in Hawaii (Tabashnik et al., 1990).

In laboratory selections with Bt, seven species of herbivorous insects have developed at least 10-fold resistance, including the lepidopterous species *P. xylostella* (Tabashnik et al., 1991), tobacco budworm, *Heliothis virescens* (Fabricius) (Stone et al., 1989; Sims and Stone, 1991; Gould et al., 1992), beet armyworm, *Spodoptera exigua* (Hubner) (Moar, 1993), cabbage looper, *Trichoplusia ni* (Hubner) (Estada and Ferré, 1992), Indian meal moth, *Plodia interpunctella* (Hubner) (McGaughey and Beeman, 1988; McGaughey and Johnson, 1992), and the beetles, Colorado potato beetle, *Leptinotarsa decemlineata* (Say) (Whalon et al., 1993) and cottonwood leaf beetle, *Crysomela scripta* Fabricius (see Tabashnik, 1994). At least five other lepidopterous species have exhibited significant increases in resistance to Bt toxins in laboratory selection experiments (Tabashnik, 1994): cotton leafworm, *Spodoptera littoralis* (Boisduval) (Salama and Matter, 1991; Sneh and Schuster, 1983), almond moth, *Cadra cautella* (Walker) (McGaughey and Beeman, 1988), spruce budworm, *Choristoneura fumiferana* (Clemens) (Van Frankenhuyzen and Milne, 1993), pink bollworm, *Pectinophora gossypiella* (Saunders) (Bartlett et al., 1996), and sunflower moth, *Homoeosoma electellum* Hulst (Brewer, 1991).

Despite possible problems with non-target lepidopterans being affected by pollen from Bt transgenic plants (Jesse and Obrycki 2000), transgenic plants come into contact with less non-target organisms and, therefore, potentially affect less non-target organisms than do conventional pesticide applications. While both conventional insecticides and insecticidal transgenic plants can affect any herbivore that feeds on the plants and natural enemies that feed on the herbivores, conventional pesticides can also affect any organism in a planting that is being treated and organisms in nearby fields where drift may occur. However, because of the high concentrations of Bt toxins in transgenic plants (Perlak et al., 1991), their increased persistence in the environment compared to Bt sprays (Gould, 1998) and, in some cases the presence of activated toxins (Peferoen, 1992), there is a much greater chance they will affect natural enemies of targeted pests and possibly non-targeted soil organisms. Also, because of the high concentrations and prolonged duration of the expression of these toxins (Benedict and Altman, 2001), the rate of herbivore resistance to these toxins is likely to increase.

For Bt to kill an insect, the crystal protein must first be ingested and then solubilized in the insect midgut (at an alkaline pH) in order to release the protoxins (or delta-endotoxins). The protoxins must be activated by cleavage to a truncated form by midgut proteases (Gill et al., 1992). These activated protoxins must then bind to the brush border membrane of the midgut epithelium, creating pores in the cell membrane, disturbing osmotic balance, and eventually causing cells to swell and lyse through a process called 'colloid-osmotic-lysis' (Knowles and Ellar, 1987). Finally, gut cell damage must occur faster than it can be repaired, permitting gut juices and bacteria to enter the insect haemocoel, followed by death due to starvation or septicemia. In the case of Bt transgenic plants, Bt is present in the plant in the form of protoxins and solubilization of the crystal is unnecessary. Also, in many Bt gene constructs, the protoxin is expressed in the activated form (Peferoen, 1992).

Behavioral detection and avoidance of Bt toxins have been reported for Lepidoptera (Gould and Anderson, 1991; Gould et al., 1991a). This could be important in those cases where the use of spatial toxin refuges for resistance management is considered or if the toxins are expressed at lower levels in some plants or plant parts (Halcomb et al. 2000). The type and size of the refuge would be important because mixtures or small refuges might act as sites for partially-resistant heterozygotes to feed after moving from toxic plants. Behavioral avoidance would also be a problem if toxins were only expressed in certain tissues. In this case, it would be relatively easy for any insect to search for plant parts that do not express toxins. On the other hand, *P. xylostella*, the only insect species documented as resistant to Bt toxins in the field, does not discriminate between plants treated with Bt and untreated plants (Schwartz et al., 1991).

Several physiological mechanisms exist that could be genetically inherited by insects and could confer resistance to Bt (Tabashnik, 1994; Gould, 1998). Two possibilities are decreased solubilization of Bt crystals and decreased cleavage of the Bt toxin into the active fragment (Tabashnik et al. 1992; Forcada et al. 1996; Michaud 1997; Gould, 1998). Changes in gut enzymes causing differences in the dissolution and activation of the protein crystal could be important, but would be overcome by the expression of activated protoxins in the plant. Oppert et al. (1997) found that resistant *P. interpunctella* may exhibit reduced rates of cleavage of full length Bt-toxins to their activated form, possibly due to a decrease in their protease function. This decrease could be due either to a loss of protease function or to an increase in levels of protease inhibitors.

A third possible physiological resistance mechanism is increased proteolytic digestion of the active toxin fragment (Gould, 1998). Although this resistance mechanism has not been documented for Bt, several cases exist where other mechanisms (such as midgut binding discussed below) have been ruled out and the possibility of proteolysis of the toxic fragment still exists (MacIntosh et al., 1991; Tabashnik, 1994; Ferré et al., 1995; Gould et al., 1992; Oppert et al., 1997; Tabashnik et al., 1999a). If proteolytic detoxification of Bt does evolve as a resistance mechanism in pest populations, and a protease that cleaves a conservative region of the Bt-toxin molecule is involved, it is likely to result in broad cross-resistance to Bt toxins (Heckel, 1994).

Two final possibilities for physiological mechanisms of resistance are: changes in the gut cell membrane that interfere with binding of the toxin, and changes that affect cell sensitivity to pore formation or to recovery from the toxin (Tabashnik, 1994; Gould, 1998). The best-known mechanism of resistance to Bt is reduced binding to the midgut membrane (Tabashnik, 1994). However, this is not the only existing mechanism, as evidenced by the lack of reduced binding found in several resistance cases (Gould et al., 1992; Tabashnik, 1994; Escriche et al., 1995; Ferré et al., 1995; Escriche et al., 1997; Oppert et al.,1997; Wright et al., 1997; Tabashnik et al., 1999a). Studies by Van Rie et al. (1990), Ferré et al. (1991) and Bravo et al. (1992a, 1992b) have provided convincing evidence that in *P. interpunctella* and *P. xylostella*, resistance is partially due to changes in the binding affinity of receptors on the brush border membrane of the insect midgut. Studies of *P. interpunctella* with radioactively-labeled CryIAb showed that a 50-fold reduction in binding to *Cry*IAb was correlated with a 100-fold reduction in toxicity (Van Rie et al., 1990). Recent studies implicate the binding site receptor to be aminopeptidase-N in diamondback moth and in gypsy moth, *Lymantria dispar* (Linnaeus), and N-acetylgalactosamine in tobacco hornworm, *Manduca sexta* (Johannsen) (Lee et al., 1995b; de Maagd et al., 1999).

Ferré et al. (1991) found that diamondback moth field populations were susceptible to CryIAb, a crystal found in insecticide formulations. The susceptibility occurred because the crystal did not bind to the brush-border membrane of the epithelial cells either due to a change in the receptor site or its absence. They found that the same populations of diamondback moth were susceptible to two other proteins, *Cry*IB and CryIC, not found in insecticidal formulations, because they bound with high affinity. Tang et al. (1996) produced similar results with different populations of diamondback moth. Lee et al. (1995a) found that the relative binding-site concentrations of CryIAa, CryIAb, and CryIAc to brush border membrane vesicles of resistant *H. virescens* were directly correlated with toxicity. They further suggested that alterations in binding proteins shared by all three proteins were a major factor in resistance. Granero et al. (1996) believed that a change in a shared midgut receptor of diamondback moth was responsible for cross-resistance to CryIA and CryIFa proteins.

Liu et al. (2000), using midgut brush border membrane vesicles, found that there was no difference in binding of radioactively-labeled CryIC between susceptible and resistant individuals. They also found that CryIC protoxin was significantly less toxic to resistant larvae than CryIC toxin, although there was no difference in mortality to susceptibles, between the protoxin and toxin. These results suggest that reduced conversion of protoxin to toxin is a minor mechanism of resistance to CryIC, and that neither reduced conversion nor reduced binding are major resistance mechanisms. Escriche et al. (1995) found that low levels of binding can occur without harmful effects, which suggests that other mechanisms are necessary for resistance evolution.

Midgut proteases of susceptible and resistant *P. interpunctella* were similar and pH was not correlated with susceptibility in 12 populations (Johnson et al., 1990; Kinsinger and McGaughey, 1979). Similarly, a CryIAb-resistant strain of *P. xylostella* from the Philippines (>200-fold resistance compared to a susceptible strain) showed little or no binding to the midgut membrane. MacIntosh et al. (1991) and Gould et al. (1992) found two strains of Bt-resistant *H. virescens* that had no differences in binding compared to susceptibles. The strain found by Gould et al. (1992) was broadly cross-resistant to other Bt-toxins. A number of other studies have confirmed that in some cases, resistant insects may have the same binding affinity for toxins as susceptible insects, suggesting mechanisms of resistance other than changes in binding receptors for diamondback moth (Masson et al., 1995; Luo et al., 1997).

Plant Derived Genes

Natural selection has resulted in the evolution of plant species that produce a variety of compounds capable of discouraging attack by fungi, insects, and vertebrates (Jongsma and Bolter, 1997). All plants contain varying degrees of these compounds that are regularly ingested without causing harm to most animals and people (Lal and Lal, 1993). However, these chemicals are present in relatively high concentrations in some plants and negatively impact many species of insects by interfering with the growth and development of insect larvae and even causing death (Gatehouse and Gatehouse, 1998; Hilder and Boulter, 1999). For example, lectins, protease inhibitors, and alpha-amylase inhibitors generally starve insect herbivores by inhibiting the active sites of digestion. Thus, these lectins and enzyme inhibitors are part of the natural defense mechanisms of plants that apparently evolved against herbivores (Lal and Lal, 1993).

Protease inhibitors

Many insect species contain proteases (and protease inhibitors) in their alimentary canals for digestion of proteins (Wolfson and Murdoch, 1990). Previous studies of dietary protease inhibitors demonstrated that they have deleterious effects on insect gut proteases that are detrimental to the growth and development of a wide range of insect herbivores (e.g. corn rootworm, *Diabroticus* spp.) (Ryan, 1990; Edmonds et al., 1996). They can decrease growth rate, prolong larval development, and even increase mortality. Two classes of protease inhibitors abundant in plant seeds and storage tissues are cysteines and serines (Reek et al., 1997). Their primary site of action is thought to be digestive proteases in the insect gut, leading to a disruption of amino acid metabolism. The target of protease inhibitors is the active region of proteolytic sites (Broadway 1994). Evidently, they cause a critical shortage of amino acids, leading to decreases in growth and/or death by starvation (Broadway and Duffey, 1986).

Increasing evidence suggests that insects can adapt to the ingestion of protease inhibitors (Broadway, 1994; Jongsma and Bolter, 1997). This should have been expected because insect species that feed on plants naturally exhibiting a specific inhibitor are generally not susceptible to that particular inhibitor (Steffens et al., 1978). Generally, the use of plant defensive compounds, such as protease inhibitors works on the principle that pests that have adapted to counter the defensive measures of their own plant hosts can still be susceptible to related defense mechanisms from non-host plants. For this reason, polyphagous species are likely to adapt to these defensive compounds faster than specialist species, making it important to recognize this possibility when deciding on targets for genetic engineering. Also important is understanding the fact that protease inhibitor effectiveness is dependent on the nutritional background in which it is presented; protease inhibitors can be overcome by adding amino acids to the insects' diet (Gatehouse and Boulter, 1983; Broadway and Duffey, 1986). This can make it difficult to determine whether a pest has evolved resistance to a protease inhibitor or if some arbitrary variation in its diet has disabled the protease inhibitor.

Jongsma et al. (1995a) found that of lepidopteran larvae fed on transgenic plants expressing the potato inhibitor II protease, only 18% of the protease activity of their gut abstracts was sensitive to inhibition by inhibitor II from the transgenic plants. In contrast, 78% of gut abstracts was sensitive in larvae fed on control plants. They showed that increased activity of a protease insensitive to inhibitor II compensated for the decline in inhibitor II sensitive activity. Jongsma et al. (1995b) found that insects become resistant to proteinase inhibitors by induction of proteinase inhibitor-insensitive protease activity that allows the insects to digest proteins independent of the level of proteinase inhibitors. Jouanin et al. (1998) stated that Lepidoptera and Coleoptera species can overexpress the existing gut proteases or induce the production of new types that are insensitive to the introduced protease inhibitors.

The ability of some insect species to compensate for protease inhibition by switching to an alternative proteolytic activity or even overproducing the existing protease, limits the application of their use in pest control. This also leads to the possibility of susceptible pests evolving resistance to protease inhibitors either by activating genes to produce alternative proteases, or by activating genes to increase the levels of existing proteases and overcome protease inhibitors by saturation (Broadway and Duffey, 1986; Jouanin et al., 1998). Several other possibilities for resistance evolution include the increased use of minor proteases already present in the insect, improving the affinity of target proteases, or inactivating the protease inhibitors before they can reach their target enzymes (Broadway, 1994; Jongsma and Bolter, 1997).

Just as pest protein metabolism has been targeted with genes encoding for protease inhibitors, so has their carbohydrate metabolism been targeted with α-amylase inhibitors (Hilder and Boulter, 1999). These inhibitors inhibit the α-amylase in the digestive tracts of mammals and Coleoptera (Powers and Whitaker, 1977; Powers and Culbertson, 1983). In both plants and animals, starch is broken down by amylases in order to yield glucose. Animals and fungi break down the plant starch to obtain glucose for their own metabolism. Amylase inhibitors are produced by plants, usually in the seeds and tubers (starch storage organs), possibly to ward off bacterial and fungal infections and animal herbivory (Gatehouse et al., 1997). Although their exact mechanism of action is unknown, it is believed that they affect insect larval development by inhibiting larval digestive amylases (Morton et al., 2000). In effect, amylase inhibitors act as an anti-nutrient by obstructing the availability of starch, and depriving insect larvae of starches needed for optimal growth and maturation.

α-Amylase inhibitors

Two α-amylase inhibitors, α-AI-1 and α-AI-2, from the common bean (*Phaseolus vulgaris* L.) were inserted into pea (*Pisum sativum* L.), which lacks α-amylase inhibitor activity. Morton et al. (2000) found that alpha-AI-1 was effective against the pea weevil, *Bruchus pisorum* (Linnaeus), whose larvae mature in developing peas while feasting on seed starches. They also found that α-AI-2 only weakly inhibited pea weevil alpha-amylase at pH 5.5 to 6.5, but substantially inhibited it at pH 4 to 4.5. Probably like other insects, pea weevils possess multiple α-amylase enzymes and not all are inhibited by α-AI-2 (Morton et al., 2000). This suggests several methods by which pea weevils could evolve resistance to α-amylase inhibitors. One way would be to activate genes that produce alternative α-amylase enzymes. Another way would be to increase levels of currently-expressed α-amylases so as to overcome the inhibitors. Storage pests of wheat (*Triticum aestivum*) have higher levels of α-amylase in their digestive tracts than insects that do not feed on wheat so that they are able to overcome the high levels of α-amylase inhibitors found in wheat and other cereals (Ishimoto and Chrispeels, 1996).

Many insects have a gut pH level that inhibits the optimal activity of amylase inhibitors (Jongsma and Bolter, 1997; Morton et al., 2000). Additionally, insects produce a variety of amylases, any of which may not be affected by different amylase inhibitors. For these reasons, it is important to both carefully select target insects and crops targeted for resistance enhancement by gene transfer and to determine the possibilities of the insect target evolving resistance to the α-amylase inhibitors.

Lectins

Lectins are a heterogeneous group of sugar-binding proteins that have detrimental effects on the survival and development of species of Lepidoptera, Coleoptera, Hemiptera, and Homoptera (Shukle and Murdoch, 1983; Czapla and Lang, 1990; Habibi et al., 1992; Powell et al., 1993; Hilder et al., 1995). Lectins recognize and bind to glycoproteins in insects and other organisms. As glycoproteins constitute a major part of insect digestive tract membranes, it is possible that the insect gut contains specific ligand-bonding molecules that are targets of plant defense lectins (Zhu-Salsman et al., 1998). Lectins probably affect insects by binding to the chitin in the peritrophic membrane or binding to the midgut epithelium (Fitches and Gatehouse, 1998; Zhu-Salsman et al., 1998). Probably as a result of this binding, nutrient uptake is reduced, causing reduced growth and/or starvation.

Gatehouse et al. (1989) found that lectins from common bean bind to the midgut epithelium of the beetle *Callosobruchus maculatus* (Fabricius) and are highly toxic to the beetle. These same lectins do not bind to the midgut of the bean weevil, *Acanthoscelides obtectus* (Say), and are not toxic to it. Harper et al. (1995) found that lectins that caused significant mortality to European corn borer, *Ostrina nubilalis* (Hubner), bound strongly to its brush border membrane proteins, but not all lectins that bound strongly caused mortality. They believed that the peritrophic membrane lining the midgut may act as particle size barrier, preventing lectins from reaching the epithelial cells or the lectin could become bound to glycoproteins. According to Gatehouse et al. (1989) the mechanism of tolerance in *A. obtectus* is the inability of intact lectin molecules to pass through the pores of the peritrophic membrane and bind to the midgut epithelium. Insect resistance to lectins could evolve through changes in the target receptor sites, such as a change in their structure so that lectins could not bind. Another possibility is that changes in the peritrophic membrane could prevent lectins from reaching the midgut epithelium. Finally, resistance to lectins could possibly evolve through increases in their number of target receptor sites as well as increases or activation of enzymes that could inactivate the lectins so they would not bind.

Weeds

Weeds are undesirable plants that either interfere with human activities or invade natural habitats due to human activities. About 10 to 15% of global crop production is lost to weeds every year (Oerke et al., 1994) despite annual expenditures of the 6 to 7 billion dollars on weed control (Kirschner, 1994; Hilder and Boulter, 1999). Three mechanisms of resistance have been engineered into plants: (i) the capability to inactivate herbicides metabolically; (ii) reduction in the ability of herbicide-sensitive proteins to bind to a herbicide; and (iii) overproduction of herbicide-target proteins in order to insure that after completely binding a herbicide, adequate proteins remain for cell functioning (Sherman et al., 1996). As of 2001, crops genetically engineered for herbicide resistance to glycines (glyphosate), glutamine synthesis inhibitors (glufosinate), nitriles (bromoxynil), and sulfonyl ureas are being commercially grown (USDA APHIS, 2000). Below we briefly review the mode of action of each of these classes of herbicides, resistance to herbicides in general, and the mechanisms of resistance in weeds to each herbicide for which crops have been genetically engineered with resistance. We shall also discuss the possibility of resistance spreading by gene flow from transgenic plants to wild relatives. For more information on herbicide-resistant transgenic plants see Berry et al. (this volume).

As of 2000, there were 249 cases of herbicide resistance in 145 species of weeds worldwide (Weedscience.com, 2001). Most cases include resistance to one or more of seven classes of herbicides: AACase inhibitors (25), ALS inhibitors (65), triazines (63), urea/amides (20), bipyradiliums (26), dinitroanilines (10), and synthetic auxins (20). This includes 96 species that are resistant to one single class of herbicide, and 49 species that are resistant to multiple classes of herbicides. Of the latter, there are 14 species resistant to three classes of herbicides, two each resistant to four, five, and, six classes. Barnyard grass, *Echinochloa crusgalli* (L.), and rigid ryegrass, *Lolium rigidum* Gaudin, are resistant to seven and 10 classes of herbicides, respectively.

For resistance to the herbicide glyphosate, the target-site method has been used by inserting both plant and microbe genes into plants (Padgette et al., 1996). In both the cases, the genes encode modified forms of the enzyme 5-enolpyruvate 3-phosphoshikimate phosphate (EPSP) synthase. The modified enzymes have a reduced affinity for binding glyphosate, creating glyphosate-resistant crops such as soybeans, corn, cotton, beets, and oilseed rape. For resistance to sulfonyl ureas, a target-site approach was used to insert herbicide-tolerant genes for acetolactate synthase from tobacco into cotton and tomato (Lee et al., 1988). For bromoxynil, the metabolic deactivation method was used (Stalker et al., 1996). A gene from the soil microbe *Klebsiella* sp. was cloned into cotton and canola in order to confer resistance to bromoxynil by coding the enzyme nitrilase that degrades the herbicide (Stalker et al., 1988). Similarly, the bar gene for phosphinothricin acetyl transferase was cloned from a soil species of Acetomycetes to confer resistance to the herbicide glufosinate to soybeans, corn, rice, beets, and rapeseed (Vasil, 1996).

Weedscience.com (2001) lists 68 weed species that are resistant to ALS insecticides, with most being resistant to sulfonyl ureas. Sulfonyl urea-resistant weed populations were first discovered in Australia in 1982 where rigid ryegrass populations were found with resistance to seven herbicide classes (Heap and Knight, 1986; Powles and Howat, 1990). Since then, sulfonyl urea-resistant weeds have been found in at least 20 other countries, including New Zealand, Japan, Malaysia, South Korea, Sweden, Denmark, the UK, Belgium, France, Poland, Portugal, Spain, Greece, Israel, South Africa, Brazil, Paraguay, Costa Rica, the USA., and Canada (Weedscience.com, 2001). These sulfonyl urea-resistant weeds have been found in fields of rice, wheat, winter wheat, soybean, canola, corn, barley, and other cereals as well as in pastures, forests, yards, railways, industrial sites, and on roadsides.

At least three weed species resistant to glyphosate have been found within the past five years (Weedscience.com, 2001). Glyphosate-resistant populations of rigid ryegrass were found in grain sorghum and wheat fields in Victoria, Australia, in 1996 in apple orchards in New South Wales, Australia (Lorraine-Colwill et al., 1999); and in almond orchards in California, USA (Weedscience.com, 2001). Glyphosate-resistant populations of goosegrass, *Eleusine indica* (L.), which were also found to be resistant to ACCase inhibitors, were found in orchards in Malaysia in 1997 (Weedscience.com, 2001). Also, glyphosate-resistant populations of horseweed, *Conyza canadensis* (L.), which infests soybean fields, were found in Delaware, USA. (Weedscience.com, 2001).

Common groundsel (*Senecio vulgaris* L.) is the only weed documented with resistance to bromoxynil; this was discovered in mint fields in Oregon, USA, in 1995 (Mallory-Smith, 1998). Populations of weeds with resistance to glufosinate have yet to be discovered.

The possibility of weeds developing herbicide resistance is much different from either arthropods or plant diseases developing resistance to compounds used for their control. In this case, the targets of the transgenes are the pesticides themselves, whereas for transgenes used against insects or plant pathogens, the pest itself is the target. Therefore, the possibility of weed species evolving resistance to the pesticide is only indirectly related to the use of transgenic crops. If transgenic plants increase the use of herbicides, then this could increase the possibility of weeds developing resistance. However, it is also possible that transgenic plants could decrease the use of herbicides, thereby slowing the evolution of herbicide resistance in weeds.

Despite all the facts about resistance development to date, little reason exists to believe that herbicide-resistant transgenic plants would increase the rate of resistance evolution by increasing the use of herbicides. On the contrary, over the last four years, as herbicide-resistant transgenic plant use has increased the overall use of herbicides has decreased (USDA ERS, 2000). While this may seem similar to the case of insect-resistant transgenic plants, it is different because if insecticide use halted, insect pests would still eventually evolve resistance to the toxins expressed (such as Bt proteins). With herbicide-resistant transgenic plants, if herbicide use was stopped, then there is no possibility of weeds evolving resistance, except through gene transfer. A factor that has been overlooked to this point, namely a similarity between engineered resistance to herbicides, insecticides and plant pathogens, is the possibility of gene flow to wild relatives. Along with the possible escape of crop plants to potentially become pests (because of their increased ability to resist herbicides, insects, or plant pathogens), these problems are of a much greater concern than weeds evolving resistance from herbicide oversue. These issues will be discussed under the section pertaining to gene flow.

Plant Pathogens

For many years, the only way to reduce viral diseases of plants was by breeding for natural resistance (Salomon, 1995). With the advent of genetic engineering, the scenario underwent a change. Now, coat protein (CP) genes from the virus itself, non-structural viral genes, and defective interfering RNA, to name a few, can be inserted into the genome of a crop species for protection against a virus.

Studies of transgenic plants that express genes from one of more than 15 different viruses indicate that coat protein genes have the broadest applicability in the development of resistant plants (Sela, 1996). Also, virus-resistant transgenic plants are the only commercially-available crops genetically engineered with resistance to plant pathogens (USDA APHIS, 2000). For these reasons, the remainder of our discussion on mechanisms of resistance and pathogen-resistant transgenic plants will center on virus coat protein-mediated resistance.

The mechanisms of CP-mediated resistance are poorly known (Lal and Lal, 1993). Although it seems pretty clear that CP molecules are responsible for the resistance, the question of how it confers resistance remains unanswered. CP-mediated resistance has several characteristics in common with classical cross protection (inoculating plants with a mild virus strain), suggesting they have similar resistance mechanisms. All types of viral resistance in plants usually can be overcome by inoculating with increasing concentrations of virus (Lal and Lal, 1993). As with classical cross protection, with coat protein-mediated resistance, increasing the viral inoculum concentration decreases the amount of time until symptoms develop and increases the proportion of plants infected (Powell-Abel et al., 1986). Also, both methods are much less effective when viral RNA is used.

Two main theories exist for the mechanisms of CP-mediated resistance. The first is that the endogenous CP subunits bind to RNA as soon as viral CP units are removed, preventing viral uncoating (Sturtevant and Beachy, 1993). The second theory is that endogenous CP binds to cellular receptors for viral entry or reassembly, thereby preventing the incoming virus from entering the cell or forming disassembly complexes (Sturtevant and Beachy, 1993).

Given our lack of knowledge about mechanisms of CP-mediated resistance, understanding mechanisms of resistance development by viruses to CP-mediated protection is impossible at this time. However, in the context of virus protected transgenic plants, virus resistance to the protection mechanism is generally not discussed as a potential drawback to the use of transgenic plants, as evidenced by the lack of such discussion in reviews (Lal and Lal, 1993; Lomonossoff, 1993; Sturtevant and Beachy, 1993; Sela, 1996). Instead, researchers usually discuss the possibility of creating new viral pathogens through processes such as transencapsidation or RNA recombination (de Zoeten, 1991). Transencapsidation can occur when a second virus infects a plant that has already been infected by a different virus (de Zoeten, 1991). In this case, when both viruses mature, both protein coats are present and the resulting virus particles may have a coat of mixed elements (Levin and Israeli, 1996). de Zoeten (1991) first suggested the possibility that new virus strains could arise from releasing virus-protected transgenic plants in the environment and exposing them to virus inoculation. This is common in nature and can give the viruses the ability to infect a different range of plants (Rochow, 1970; Osbourne et al., 1990). Also, both template switching or heterologous recombination and changes in the role of RNA helper-dependent complexes could occur naturally by way of mixed infections and could increase the range of hosts by increasing the transmissibility by vectors (Atreya et al., 1990; Gal-On et al., 1992). Although the few examples cited here pose questions about the use of viral CPs in pathogen resistance, much more information is needed, on a case-by-case basis, to ascertain the extent to which encapsidation occurs under field settings and the risks that it may cause.

Gene Flow: Insect, Weed, and Plant Pathogen Resistant Genes

Recent developments in plant genetic engineering may heighten the potential for development of invasive crop plants or invasive wild crop relatives through hybridization and gene flow (Colwell et al., 1985; Tiedje et al., 1989; Hoffman et al.,1992; Kareiva et al., 1994; Darmency, 1994; Rissler and Mellon, 1994). Many examples exist of gene flow from crops to wild relatives that go on to create new or more aggressive weeds (Levin and Israeli, 1996; Ellstrand, 1999). Several important examples given by Levin and Israeli (1996) are: pearl millet to wild millet, sorghum to Johnson grass, corn to teosinate, rice to wild rice, cultivated carrot to wild carrot, kayseri alfalfa to weedy relatives, durum wheat to wild emmer wheat, and foxtail millet to wild green foxtail (Levin and Israeli, 1996). Although gene flow to wild relatives may not create economically-important weeds, it could have a negative environmental impact on the wild relative or on herbivores of the wild relative (Wearing and Hokkanen, 1994). For example, cotton can hybridize with its wild relative, found in Hawaii, but this plant is usually not considered a weed. Of course, one possibility is that the resulting hybrid will become a weed, but it is more likely that the transfer of genes coding insect toxins will negatively impact endangered insect herbivores or will result in the displacement of endangered plant species (Ellstrand, 1988; Regal, 1988).

For the transfer of genes from genetically-modified crops to their wild relatives, there must first be wild relatives grown in the same region and secondly, hybridization must occur. Also, for the gene transfer to be relevant to pest control, the wild relative must change its fitness or change the fitness of other pests or beneficial organsims. For gene transfer or outcrossing to increase the weediness of the plant, it must increase its fitness. de Wet and Harlan (1975) list the wild races of crops, including many vegetable crops such as cabbages, watermelon and beets. Almost all crops have wild relatives at some level, but more important is whether their distributions overlap. However, most important genetically-modified food crops, such as rice, corn, soybean, potato and canola, are grown in many regions of the world, increasing the opportunities for transgenic crops to interact with their wild relatives (Wiersema and Leon, 1999).

Cultivated rice (*Oryza sativa* L.), grown in tropical, subtropical, and warm temperate regions throughout the world, has several wild weedy relatives (Wiersema and Leon, 1999). These include *Oryza barthii* A. Chev., *O. longistaminata* A. Chev. & Roehr., and *O. punctata* Kotschy ex Steud. in Africa, and *O. rufipogon* Griff. in China, East Asia, Southeast Asia, India, Malaysia, and Australia. All, except *O. punctata*, hybridize with wild rice, including *O. rufipogon*, which is found throughout important rice-growing regions (Wiersema and Leon, 1999).

Corn (*Zea mays* L.) has several wild relatives, known as teosinte, in central and northern Mexico and Mesoamerica. Since corn can hybridize with its teosinte relatives, gene flow from transgenic corn is possible in areas of Mesoamerica where it is grown. Currently, teosinte is not considered to be a weedy species. Thus, negative changes occurring from the gene would have to increase teosinte impacts on the environment or make teosinte highly competitive with other plants (making it a weed).

Potato (*Solanum tuberosum* L.), originated in Chile and is now cultivated worldwide (Wiersema and Leon, 1999). It has about 100 wild relatives, with which most it can hybridize. Several of these species are already weeds. Weedy relatives include *Solanum americanun* Mill. and *Solanum erianthum* D. Don located in the US, Mexico, South America, and widely distributed throughout the tropics; *Solanum nigrum* L.and *Solanum dulcamara* L. found in Africa, Asia, India, and Europe; *Solanum carolinense* L., *Solanum dimidiatum* Raf., *Solanum ptycanthum* Dunal, *Solanum rostratum* Dunal, *Solanum torvum* Sw., and *Solanum triflorum* Nutt. found in North America; *Solanum mauratianum* Scop., *Solanum sarrachoides* Sendtn., *Solanum sisymbriifolium* Lam., and *Solanum viarum* Dunal from southern South America; and *Solanum marginatum* L.f. and *Solanum linnaeanum* Hepper & P. Jaeger from Africa (Wiersema and Leon, 1999).

In North America, a number of crops grow sympatrically with their wild relatives; thus natural hybridization is always possible. Some examples include: sorghum and Johnson grass; radish and wild radish; turnip and wild relatives; rape and field mustard; cultivated *Amaranthus* spp. and wild *Amaranthus* spp.; carrot and wild carrot; sunflower and wild morphs; squash, pumpkin, and wild marrow; rye and wild rye types; lettuce and wild lettuce; oat and wild oat; and artichoke and its wild types (Simmonds, 1979).

Possible transfer of transgenic traits, for increased fitness to wild relatives of engineered crop plants, is difficult to assess in field or greenhouse studies. This is due to the regulatory status of transgenic organisms and the potential risks associated with the introduction of these genes into wild populations. Investigators have used simulations to address some of the problems of gene transfer (Klinger et al., 1991; Warwick and Black, 1993; Klinger and Ellstrand, 1994; Morris et al., 1994; Wang et al., 1997), but results of these simulations vary. Despite this lack of knowledge, many examples exist of gene transfer from non-transgenic crop plants to wild relatives, including those listed earlier. One example comes from India, where plant breeders selected for increased anthocyanin production in cultivated rice (resulting in purple leaves to facilitate field workers' ability to discriminate between wild and cultivated rice). Within several generations, the genes transferred to wild rice, colored the leaves of wild rice purple, and rendered the gene useless (Parker and Dean, 1976). Tucker and Sauer (1958) found many amaranth hybrids in California that had resulted from crosses between amaranth crops and wild relatives. It appeared to them that under cultivation, the hybrids could out-compete their weed parents because they had acquired traits for robust stature and high fecundity. Also, in California, hybridization between rye and its wild relatives was so extensive that it caused the farmers to stop growing rye for human consumption (Jain, 1977). de Wet and Harlan (1975) believed that within its native range, Johnson grass, *Sorghum halapense* (L) was not a weedy plant, but that it became so through gene transfer from cultivated sorghum, *Sorghum bicolor* (L).

More convincing evidence for gene flow to wild relatives has been found recently using molecular genetics. Second (1982) found that African rice (*Oryza barthii*) contained more isozymic variation than cultivated rice (*Oryza glaberrima* Steud.), probably through gene transfer between cultivated and wild rice. Doebley (1984) found two alleles from corn that had not been documented previously in the wild species *Zea diploperennis* Iltis *et al*. In Texas, Decker and Wilson (1987) found that alleles common to a squash cultivar (*Cucurbita pepo* L.) occur in a wild species, *Cucurbita texana* (Scheele), enhancing its weediness by making it more difficult to distinguish from the crop.

MANAGEMENT STRATEGIES

Over the past 40 years, a number of strategies for managing insecticide resistance have been proposed. These strategies include high doses (Taylor and Georghiou, 1979), refuges (temporal and spatial) (Comins, 1977; Taylor and Georghiou, 1979), mixtures (Tabashnik, 1994), rotations (Brown, 1977), mosaics (Tabashnik, 1994), synergists (Matsamura, 1975), and integrated pest management (IPM). Resistance management seeks to slow, prevent, or reverse the development of resistance in pests (Tabashnik, 1990). Strategies for managing insect resistance to toxins expressed in transgenic plants are comparable to those used for managing resistance to conventional insecticides, with only a few small exceptions (e.g. refugia or mosaics could be implemented on a much smaller scale through seed mixtures) (McGaughey and Whalon, 1992; Daly, 1994; Tabashnik, 1994; Gould, 1998; Roush, 1999). There are far fewer discussions of resistance management for weeds or plant pathogens, but generally they will fall into several of the categories described for insects. We shall now discuss the assumptions necessary for each of these strategies to work and the problems that are foreseen for each along with the results from models and field trials where relevant.

High Expression Level

High expression level is commonly known as the high dose strategy in pesticide resistance research. The premise is that toxin concentration is so high that all individuals of the target pest species are killed. Some traditional problems with the high dose strategy are that it is expensive and difficult to implement, affects natural enemies negatively, and the residues resulting from decaying toxins are essentially low doses (Taylor and Georghiou, 1979; Tabashnik and Croft, 1985; Gould, 1998). However, with transgenic plants, toxins can now be expressed at constantly high levels at no extra cost, few direct affects on natural enemies exist, and pesticide residues or very low doses will not pose a problem (Gould, 1998; Benedict and Altman, 2001).

A different problem for transgenics is that toxin expression must be consistently high and stable over a period of time or among different plant tissues. For example, a plant genetically engineered to produce Bt or protease inhibitors at a high dose must express this dose continually, even in the case of tissue-specific expression. Otherwise, the majority of feeding insects may fail to ingest the high dose, potentially increasing the rate of resistance development if resistant heterozygotes are able to survive better than susceptible homozygotes (Tabashnik and Croft, 1982). Taylor and Georghiou (1982) stated that the slower the rate of pesticide decay, the longer the period over which it exerts selection pressure against susceptible genotypes and in favor of resistant genotypes. In the case of transgenic plants, if the toxin expression is functionally lower (low enough to allow the survival of partially-resistant heterozygotes) at any time when the insects are feeding, then resistance might develop more rapidly as it would towards more persistent pesticides. Evidence of this occurring in field populations receiving conventional insecticide applications supports this reasoning (Georghiou, 1972). Resistance to slowly-decaying cyclodienes and DDT has arisen more rapidly than to quickly-decaying organophosphates and carbamates (Georghiou, 1972). Also, Greenplate (1999) found that as Bt transgenic cotton plants matured Bt-toxin concentrations decreased, increasing the chances that pests such as *H. virescens* will encounter sublethal or low doses of Bt toxins, throwing into question the use of the high dose strategy with Bt transgenic plants. Greenplate (1999) also found different toxin levels in various plant parts, also increasing the chances that target pests will receive a lower dose.

Another factor to consider is that even ultra-high doses may never kill highly-resistant pests. For example, Tabashnik et al. (1993) found that two *P. xylostella* strains were resistant to concentrations of Bt greater than 4,000 times the concentration used in the field, and at least 1,000 times more resistant than a susceptible strain. Similarly, if wild host plants of herbivorous pests, or non-transgenic host plants of the same or different species are located in close proximity to the transgenic plants, then pests can move to these non-toxic plants before they receive a lethal dose (Ramachandran et al., 1998; Dirie et al., 2000). Either of these cases could actually increase the rate of resistance by allowing the survival of partially-resistant heterozygotes.

For protease inhibitors, amylase inhibitors and lectins, the doses that can be engineered into plants are generally insufficient to control a target pest (Jongsma, 1996). For these compounds, the high dose strategy cannot be implemented. Although some of these substances may be engineered at high doses in the future, many of these substances (particularly at high doses) are harmful to people, other mammals, and birds (Liener, 1986).

Similarly, even at high levels, CP-mediated resistance usually does not completely control disease symptoms even when the CPs are expressed at the highest levels possible (Lal and Lal, 1993). Unfortunately, the higher the level of CP expression achieved, the greater the chance that transencapsidation with other viruses will occur.

For herbicides, the high dose strategy is unnecessary. If a low herbicide dose is used and kills most of the weeds present in a crop, then the few remaining weeds can be controlled by cultural methods. Weed destruction would have to be accomplished before the weeds reproduce, otherwise locating and destroying pollen and seeds would be necessary. This is possible for herbicides because, unlike insects or viruses, weeds are stationary, relatively large, and easier to locate in the field.

Refuges

Elimination or minimization of selection pressure has always been recognized as the best way to delay or avoid the evolution of resistance (Georghiou, 1983). One way to achieve reduced selection pressure is to use spatial or temporal refuges. The goal of the refuge strategy is to allow susceptible genotypes to survive and mate with resistant individuals in order to limit the expression of resistance traits, thus slowing the evolution of resistance in a pest population. For recessive resistance genes, migration from untreated refuge populations into treated populations can delay the rate of resistance development by producing a temporary equilibrium at a low gene frequency (Comins, 1977). Spatial refuges include mixtures of transgenic and non-transgenic plants. The mixture may be achieved by mixing seeds before planting or by planting a smaller proportion of the field with the non-transgenic strain. Another type of spatial refuge results from the expression of toxins only in critical areas of the plant. A third type of spatial refuge, plantings of Bt-transgenic crops with non-Bt transgenic crops, such as Bt-transgenic corn with non-transgenic cotton, could be applied for resistance management with polyphagous pests such as *H. zea* (Wigley and Chilcott, 1992). This would have to be used either as a decision-making tool for deciding which crops should be engineered with the gene (Wigley and Chilcott, 1992), or in conjunction with refuges within a single crop itself. If they were not used in this manner, the rate of resistance development in monophagous pests such as the cotton feeding *P. gossypiella* would not be slowed. Of the spatial refuges proposed, the seed mixture is generally preferred because it is easy to adopt and much easier to regulate (Hokkanen and Wearing, 1995). Temporal refuges involve alternating transgenic plants with non-transgenic plants over growing seasons. Also, temporal control of toxin expression in a plant can act as a refuge.

Georghiou and Taylor (1977) found that refugia might be an important factor in delaying the evolution of resistance, especially when coupled with a fitness disadvantage to resistance. For the refuge strategy to work, either the resistant genotype must have a fitness disadvantage in the absence of the toxin or the refuge must be used at very low resistance allele frequencies (Tabashnik and Croft, 1982; Caprio, 1994). If no fitness advantage to resistance exists in the absence of the toxin, then a refuge may still slow resistance, but will not stop it. Luckily, loss of resistance to Bt in the absence of selection pressure has been found in *P. xylostella* (Groeters et al., 1994; Tabashnik et al., 1994), *L. decemlineata* (Trisyono and Whalon, 1997), and *O. nubilalis* (Bolin et al., 1999). However, McGaughey and Beeman (1988) found that resistance to Bt by colonies of *P. interpunctella* remained stable after 5 generations without selection.

Another aspect of resistance management with refuges is that there must be random mating between individuals from the refuge and the transgenic plants, which means that the refuge must be located close to the transgenic planting (Caprio, 1998; Tabashnik et al., 1999b). At the same time, for individual pests that tend to feed on (or infect) several plants rather than just one, the refuge and toxic plant plots must be located far enough apart so that susceptible individuals do not move from non-toxic to toxic plants (Mallet and Porter, 1992; Gould, 1998). Similarly, if refuge fields are less attractive than transgenic plantings to female insects for oviposition, the effective refuge size would be decreased (Alstad and Andow, 1995; Ives, 1996). Finally, pest mortality caused by other factors such as chemical sprays or natural enemies must either be equal in refuges and transgenic plantings or less in the refuge. For example, if natural enemies tend to congregate and cause greater pest mortality in the refuge plantings, favoring them over transgenic plantings, then the effectiveness of the refuge would be decreased proportional to the increase in mortality in the refuge compared to the transgenic planting.

For species where individuals mate separately within refuges and transgenic plantings rather than randomly mating across these plantings, the use of temporal refuges or promoters for time-specific or site-specific toxin expression within the plants would overcome this problem. However, site-specific toxin expression in itself would not be useful as a refuge if the pest moves from non-toxic to toxic areas of the plant. Also, promoters may not prevent lowered toxin expression in some tissues and could potentially increase the rate of resistance (McGaughey and Whalon, 1992). Finally, for weed species or other pest species that reproduce asexually, such as aphids, fungi, or viruses, spatial refuges would be of limited use.

Several recent papers discuss research on refuges used in combination with Bt transgenic plants. Dirie et al. (2000) found that two rice stem borers, *Chilo suppressalis* (Walker) and *Scirpophaga incertulas* (Walker), moved among Bt transgenic and non-transgenic rice plants during larval development. They concluded that the refuges created by seed mixtures of Bt and non-Bt rice would have limited effectiveness in resistance management and that Bt and non-Bt plantings would have to be separated so as to ensure survival of susceptible insects. However, they also discussed problems with mandating and enforcing refuge requirements in Asian rice farms (>250 million rice farms) without having seed released only as mixtures (Hossain, 1998; Dirie et al., 2000). Similarly, Ramachandran et al. (1998) found that *P. xylostella* larvae moved between Bt transgenic and non-transgenic canola plants. They also concluded that this movement would limit the use of seed mixtures as refuges, because susceptible insects were more likely to feed on toxic plants and they added that the rate of resistance could actually increase if potentially-resistant insects moved from transgenic to non-transgenic plants before obtaining a lethal dose. Halcomb et al., (2000) and Parker and Luttrell (1999) observed similar results with *H. virescens* and *H. zea*; both pests moved between transgenic and non-transgenic cotton plants during larval development.

In 2000, all Bt field cotton and corn products were required to have mandatory structured refuge requirements (Benedict and Altman, 2001; USEPA, 2001). In case of Bt-cotton, there was a choice of a 4% unsprayed refuge or a 20% refuge that could be sprayed with certain insecticides other than Bt For Bt-corn grown outside cotton-growing areas, growers had to have a minimum of 20% non-Bt corn refuge (treatable with other pest control products). In cotton-growing regions, a minimum of 50% non-Bt corn refuge had to be planted. These requirements are only for transgenic Bt crops; we know of no such requirements for management of herbicide or plant pathogen resistance.

Synergists

The use of synergists has been proposed to manage resistance to pesticides by increasing the toxicity of a pesticide, although by itself a synergist has little toxicity (Matsamura, 1975). In the case of insect- or pathogen-resistant transgenic plants, this would mean incorporating a second gene into the plant. On the other hand, for herbicide-resistant transgenic plants, this would require a chemical combined with the herbicide, just the same as with conventional crop plants. To slow the development of resistance, a synergist would have to increase the mortality of resistant heterozygotes and homozygotes as with the high dose strategy. For example, in Australia, when *H. armigera* developed pyrethroid resistance, piperonyl butoxide was recommended to block the resistance mechanism (oxidative metabolism) so that resistant insects could be killed (Forrester, 1988).

Serine protease inhibitors synergize the effects of Bt on four lepidopteran species and the coleopteran *L. decemlineata*, but not susceptible or resistant *P. xylostella* (MacIntosh et al., 1990; Tabashnik et al., 1992). However, by definition, synergists are not toxic alone, so that combinations of toxic protease inhibitors and Bt are referred to as mixtures (discussed below). According to Tabashnik (1994), no data is available on synergists and resistance to Bt Similarly, no data exists on herbicide resistance and synergists or on the use of synergists with viral coat protein defense.

Mixtures

Insecticide mixtures are applied so that organisms are simultaneously exposed to two or more toxins. With transgenic plants, this is equivalent to a single crop variety expressing several different toxins, known as pyramiding or gene stacking. This strategy assumes that resistance to each toxin is monogenic, there is no cross-resistance among toxins used, resistant pest individuals are rare so that no one individual is resistant to both toxins, and the toxins have equal persistence (Mani, 1985; Curtis, 1985; Comins, 1986). If this is achieved, then pests that survive one of the toxins will be killed by the other toxin (Georghiou, 1983). In the case of weeds, mixtures of herbicide could be used against crop species that have been engineered with resistance to two or more herbicides.

Georghiou (1983) comments that the use of mixtures and rotations as countermeasures of resistance are discussed frequently in the literature with no consensus as to their usefulness and few conclusions from actual research. Early work on mixtures for resistance management indicated that they were not effective (Brown, 1977).

Currently, all commercially-available cultivars of insect-resistant and herbicide-resistant transgenic plants contain only one Bt-toxin or herbicide resistant gene (USEPA, 2001). Roush (1999) believes that pyramids can increase the time for resistance to develop while using smaller and more acceptable refuges. However, Tabashnik et al. (1991) found that field populations of *P. xylostella* were capable of evolving resistance to mixtures of Bt toxins.

Researchers believe that mixtures of plant protease inhibitors may be the best way to prevent insect resistance to these compounds (Jongsma and Bolter, 1997). Plants naturally improve the efficacy of protease inhibitors by exhibiting multidomain or multimeric variants. These tactics are common in plants and a large number of combinations exist (Jongsma and Bolter, 1997). Feeding studies have confirmed that inhibitor combinations act synergistically to reduce insect herbivore growth. Although multiple protease inhibitors may prevent each other's breakdown, they will be more effective in preventing the digestion of dietary protein (Oppert et al., 1993; Orr et al., 1994; Marwick et al., 1995). Since genes for different proteases have been found in nature and engineered into crop plants, the success rates for the use of protease and amylase inhibitors in host plant resistance to insects can be greatly improved, especially through pyramiding genes that produce a variety of these compounds.

Rotations

In the case of transgenic plants, rotations would be the alternation over time of two or more varieties containing different toxins. Most resistance researchers assume that resistance of individuals to one toxin must decline during the use of the second toxin in order for rotations to be effective. However, assuming no occurrence of cross-resistance between toxins and that resistance does not increase in the absence of the toxins, then alternating two toxins over a period of time would actually at least double (if the increase were linear) the time to resistance, and alternating three toxins would at least triple the time to resistance, and so on. However, if a fitness cost to resistance or negative cross-resistance exists, rotations can be particularly effective.

One of the few field tests of rotations, conducted by MacDonald et al. (1983), suggested that rotations of permethrin and dichlorvos did not slow resistance development by the housefly, *Musca domestica* Linnaeus. A number of models have also suggested that rotations will not substantially suppress resistance (see Tabashnik, 1990), although Roush (1989) believes rotations may be more effective than other multiple toxin tactics (such as mixtures or mosaics). For insect-resistant transgenic plants, there are limited possibilities for the use of rotations. Bt toxins have similar modes of action and cross-resistance is a problem, thus rotating between different toxins from year to year is unlikely to slow resistance. So far, other insect toxins such as enzyme inhibitors and lectins do not give adequate protection when used alone, and, therefore, would not be useful in rotations, although they may be used in mixtures with Bt toxins.

For herbicide-resistant transgenic plants, the uses of rotations along with cultural control measures are the best methods for controlling weed resistance. In practice, many crops are rotated on a yearly basis, and generally different herbicides are used for different crops. Also, where crops are rotated, each crop could be engineered with resistance to a different type of herbicide, ensuring that herbicide rotations would be carried out.

For viral CP-resistant transgenic plants, the use of rotations would probably not decrease the probability of transencapsidation or RNA recombination occurring. This is because, unlike resistance evolution, which takes many generations to occur, transencapsidation can occur in one generation. Also, because coat protein resistance is currently the only effective type of resistance to viruses, there are really no other measures with which it can be rotated.

Mosaics

Mosaics are a spatial patchwork of two or more varieties of the same crop species expressing different toxins at the same time. For mosaics to delay resistance, it is assumed that there is no cross-resistance among the different toxins, so that individuals resistant to one toxin are susceptible to the other toxin(s). Thus, susceptible individuals are maintained and are able to move among the different patches. Resistance management strategies for transgenic plants differ from conventional insecticides in that mixtures of seeds for plants expressing different toxins can be used to create a plant-to-plant mosaic—a much smaller scale than is practical for conventional insecticides. Also, because of problems with drift when applying conventional insecticides, transgenic plants can make it easier to define the size of and distance between patches within a mosaic.

To date, there is little experimental evidence related to this tactic. In possibly the only report of a test of mosaics, Pimentel and Bellotti (1976) found that resistance by the housefly to a mixture of six toxic salts did not evolve in 32 generations, although resistance to each salt tested separately evolved within 10 generations. However, results from models indicate that mosaics would not be useful for slowing the development of resistance (Comins, 1986; Curtis, 1985). Also, the evolution of resistance to pesticides by numerous insects, weeds, and plant pathogens does not bode well for the use of mosaics, at least on a large scale. Although experiments may be lacking, the use of pesticides has always been a large-scale mosaic, since few if any growers actually coordinate with their neighbors to use the same insecticides, herbicides, or plant pathogen control techniques.

The use of rotations with viral coat protein-resistant transgenic plants would definitely not decrease the probability of transencapsidation or RNA recombination occurring; it would only confine it to a smaller area. Also, because coat protein resistance is the only effective type of resistance to viruses, there are really no other measures for use in a mosaic of differently treated fields.

Isolation or Barriers to Gene Flow

Although generally not considered in case of insect or plant pathogen management, isolation of transgenic plants from their wild relatives would be the best way to prevent the dissemination of genetic material (Lieberman et al., 1996). If resistance genes for herbicides or insect toxins are transferred to wild relatives, it could increase the evolution of resistance to herbicides by weeds or insects in weeds by increasing the selection pressure. This method of containing transgenic plants would only be plausible for certain types of crops, such as greenhouse crops that can be physically separated from hybridizing with their wild relatives.

To date, most transgenic crops are field crops such as corn, soybean, and cotton, grown on a broad scale. Distances that they must be separated from their non-transgenic counterparts to prevent hybridization are not well known. During field tests, regulatory agencies such as the U.S. Environmental Protection Agency are mainly interested in separating transgenic and non-transgenic crops rather than preventing hybridization with wild relatives. Despite the regulations, researchers have recently found genes from one Bt transgenic corn hybrid in seeds of hybrids that were not supposed to contain the genes (USEPA, 2001). Given the fact that this occurred when the transgenic corn hybrid was supposedly being grown under strict control trials to keep it separated from other corn hybrids and that it had not been cleared for commercial release, this finding leaves little doubt that prevention of gene flow in field crops is impossible at present.

Integrated Pest Management and Natural Enemies

Last, but not the least, the most important way to manage resistance to transgenic plants is to diversify pest management techniques as well as resistant management techniques. Georghiou (1972) discussed IPM as the most important enemy of pest resistance and vice versa. For an IPM program to remain viable, it must ensure that resistance does not evolve, because changes of insecticides are usually so disruptive to biological control that new control programs must be formulated. In the case of insecticidal transgenic crops, in particular those expressing Bt toxins they can significantly reduce the use of synthetic chemical insecticides, as much as 50 to 60% in cotton, increasing the abundance of natural enemies and thereby pest control (Fitt, 1994; Roush and Shelton, 1997). This increased control by natural enemies could be followed by increased refuge size without sacrificing yield, which would in turn slow the evolution of resistance to the plant toxin (Tabashnik, 1994).

The combined effect of partial plant resistance and pest-regulating natural enemies offers a more diversified and potentially more durable approach to crop protection with IPM (Van Emden, 1991). Evidence from the entomological literature demonstrates that resistance in crops that causes a low level of pest mortality, but alters the behavior or slows the growth of immatures can enhance the effectiveness of natural enemies (Boethel and Eikenbary, 1986; Hare and Luck, 1991; Johnson and Gould, 1992; Bottrell et al., 1997). This could slow the evolution of resistance if resistance were partially recessive and natural enemies selectively killed heterozygotes whose growth or behavior was altered, as shown with models by Gould et al. (1991b). However, if for this same reason natural enemies kill disproportionately higher numbers of susceptible than resistant individuals, they could speed the evolution of resistance. On the other hand, if natural enemies discriminate against sick or unhealthy individuals, they may be more likely to kill resistant individuals, in particular partially-resistant heterozygotes and slow the rate of resistance evolution.

Several laboratory studies indicate how natural enemies may interact with plant resistance. Lu et al. (1996) found that the predator *Coleomegilla maculata* (DeGeer) caused higher mortality to Colorado potato beetle genotypes that were more susceptible to their tomato hosts than those that were resistant. Similarly, the fungus *Nomuraea rileyi* caused higher mortality to susceptible *Helicoverpa virescens* genotypes than those resistant to Bt toxins in its transgenic tobacco host (Johnson et al., 1997). On the other hand, *Campoletis sonorensis* (Cameron) caused higher mortality to the Bt-resistant *H. virescens* genotype. A third possibility is that natural enemies will have no affect on the development of herbivore resistance to transgenic plants. As an example, the parasitoid *Cotesia plutellae* Kurdjumov showed no difference in choice between resistant and susceptible hosts treated with *Bacillus thuringiensis*, implying that they would not affect resistance development (Chilcutt and Tabashnik, 1999). Another example is parasitism rates of *Diadegma insulare* Cresson on *P. xylostella* and incidence of the four most abundant *P. xylostella* were unaffected by toxic versus non-toxic collard plants (Riggin-Bucci and Gould, 1997).

With transgenic plants that express toxins on a wound-induced basis, natural enemies could slow the development of resistance by keeping the pest population below the economic levels that would trigger the toxins. One advantage for resistance management of tissue-specific expression of toxins (refuge within a plant) over spatial refuges within a whole field is that surviving susceptible insects theoretically would be distributed throughout the field. As refuges are currently used for Bt transgenics, they consist of a large plot of non-Bt plants within a field of Bt plants. For natural enemies, this means that they have a choice of a large area with few and widely distributed prey (or hosts) or a smaller area with many concentrated prey. In this scenario, it is not only more likely that natural enemies will be concentrated within the refuge, but because there are more prey in the refuge, those natural enemies that possess a functional response to prey density will actually kill a higher proportion of pests. The use of tissue-specific expression of toxins could overcome this problem by ensuring that surviving insects would be distributed throughout the field.

Hokkanen and Wearing (1994) suggested multiple tactics for Bt resistance management in oilseed that could be applied to any transgenic plant: (i) provide refuge for susceptibles; (ii) do not use pesticides that kill susceptibles; (iii) do not use pesticides that kill natural enemies; (iv) enhance natural control through crop rotation or tillage; and (v) rotate between susceptible and resistant crop genotypes over large areas. Current U. S. EPA (US Environmental Protection Agency, 2001) recommendations for resistance management (i.e. a 4% untreated refuge or a 20% non-Bt refuge) only include item #1 from Hokkanen and Wearing's (1994) list. Currently-used strategies for resistance management do not consider IPM or integrated resistance management; instead they use the simplest strategy with the least potential for crop damage over the short run. If more emphasis is not placed on the use of IPM (e.g. integrating natural enemies with transgenic plants), the growers are destined to relive the control failures and problems associated with relying solely on synthetic insecticides (Johnson, 2001).

As with arthropod resistance management, weed resistance management must be an integrated effort combining different weed control practices to reduce dependence on one type of control. In contrast to arthropod or pathogen management, herbicide-resistant transgenic plants have the potential to act as a management tactic that will help to slow the evolution of herbicide resistance by weeds.

REFERENCES

Alstad, D. N. and Andow, D. A., 1995, Managing the evolution of insect resistance to transgenic plants. Science, **268**, 1894-1896.

Atreya, C. D., Raccah, B. and Pirone, P. T., 1990, A point mutation in the coat protein abolishes aphid transmissibility of a potyvirus. Virology, **178**, 161-165.

Ballester, V., Granero, F., de Maagd, R. A., Bosch, D., Mensua, J. L. and Ferré, J., 1999, Role of *Bacillus thuringiensis* toxin domains in toxicity and receptor binding in the diamondback moth. Appl. Environ. Microbiol., **65**, 1900-1903.

Ballester, V., Granero, F., Tabashnik, B. E., Malvar, T. and Ferré, J., 1999, Integrative model for binding of *Bacillus thuringiensis* toxins in susceptible and resistant larvae of the diamondback moth (*Plutella xylostella*). Appl. Environ. Microbiol., **65**, 1413-1419.

Bartlett, A. C., Dennehy, T. J. and Antilla, L., 1996, An evaluation of resistance to Bt toxins in native populations of the pink bollworm. Proc. 1996 Beltwide Cotton Res. Con., National Council of America, Memphis, Tennessee.

Beachy, R. N, Loesch-Fries, S. and Tumer, E., 1990, Coat protein-mediated resistance against virus infection. Annu. Rev. Phytopath., **28**, 451-474.

Benedict, J. H. and Altman, D. W., 2001, Commercialization of transgenic cotton expressing insecticidal crystal protein. In: Jenkins J. N. and Saha S. (eds), Genetic Improvement of Cotton: Emerging Technologies, Science Publishers, Inc., Enfield, NH, USA, pp. 137-201.

Birch, A. N. E., Geoghegan, I. E., Majerus, M. E. N., McNicol, J. W., Hackett, C. A., Gatehouse, A. M. R. and Gatehouse, J. A., 1999, Tri-trophic interactions involving pest aphids, predatory 2-spot ladybirds and transgenic potatoes expressing snowdrop lectin for aphid resistance. Mol. Breed., 5, 75-83.

Blair, D., 1989, Uncertainties in pesticide risk estimation and consumer concern. Nutr. Today, 24, 13-19.

Boethel, D. J. and Eikenbary, R. D., 1986, Interactions of Plant Resistance and Parasitoids and Predators of Insects, Ellis Horwood Ltd., Chichester, England.

Bolin, P. C., Hutchison, W. D. and Andow, D. A., 1999, Long-term selection for resistance to *Bacillus thuringiensis* Cry1Ac endotoxin in a Minnesota population of European corn borer (Lepidoptera: Crambidae). J. Econ. Entomol., 92, 1021-1030.

Bottrell, D. G., Barbosa, P. and Gould, F., 1997, Manipulating natural enemies by plant variety selection and modification: a realistic strategy? Annu. Rev. Entomol., 43, 347-367.

Bravo, A., Hendrickx, K., Jansens, S. and Peferoen, M., 1992a, Immunocytochemical analysis of specific binding of *Bacillus thuringiensis* insecticidal crystal proteins to lepidopteran and coleopteran midgut membranes. J. Invertebr. Pathol., 60, 247-253.

Bravo, A., Jansens, S. and Peferoen, M., 1992b, Immunocytochemical localization of *Bacillus thuringiensis* insecticidal crystal proteins in intoxicated insects. J. Invertebr. Pathol., 60, 237-246.

Brewer, G. J., 1991, Resistance to *Bacillus thuringiensis* subsp. *kurstaki* in the sunflower moth (Lepidoptera: Pyralidae). Environ. Entomol., 20, 316-322.

Broadway, R. M., 1994, Are insects resistant to plant proteinase inhibitors? J. Insect Physiol., 41, 107-116.

Broadway, R. M. and Duffey, S. S., 1986, Plant proteinase inhibitors: mechanisms of action and effect on the growth and physiology of larval *Heliothis zea* and *Spodoptera exigua*. J. Insect Physiol., 32, 827-833.

Brown, A. W. A., 1977, Epilogue: resistance as a factor in pesticide management. Proceedings, XV International Congress of Entomology, Entomological Society of America, College Park, Maryland.

Caprio, M. A., 1994, *Bacillus thuringiensis* gene deployment and resistance management in single- and multi-tactic environments. Biocontr. Sci. Technol., 4, 487-497.

Caprio, M. A., 1998, Evaluating resistance strategies for multiple toxins in the presence of external refuges. J. Econ. Entomol., 91, 1021-1031.

Caprio, M. A. and Tabashnik, B. E., 1992, Gene flow accelerates local adaptation among finite populations: simulating the evolution of insecticide resistance. J. Econ. Entomol., 85, 611-620.

Chilcutt, C. F. and Tabashnik, B. E., 1999, Effects of *Bacillus thuringiensis* on adults of *Cotesia plutellae* (Hymenoptera: Braconidae), a parasitoid of the diamondback moth, *Plutella xylostella* (Lepidoptera: Plutellidae). Biocontr. Sci. Technol., 9, 435-440.

Colwell, R. R., Norse, E. A., Pimentel, D., Sharples, F. E. and Simberloff, D., 1985, Genetic engineering in agriculture. Science, 229, 111-112.

Comins, H. N., 1977, The development of insecticide resistance in the presence of migration. J. Theor. Biol., 64, 177-197.

Comins, H. N., 1986, Tactics for resistance management using multiple pesticides. Agric. Ecosyst. Environ., **3**, 129-148.

Croft, B. A., 1990, *Arthropod Biological Control Agents and Pesticides*, John Wiley & Sons, New York.

Curtis, C. F., 1985, Theoretical models of the use of insecticide mixtures for the management of resistance. Bull. Entomol. Res., **75**, 259-265.

Czapla, T. H. and Lang, B. A., 1990, Effect of plant lectins on the larval development of European corn borer (Lepidoptera: Pyralidae) and southern corn rootworm (Coleoptera: Chrysomelidae). J. Econ. Entomol., **83**, 2480-2485.

Daly, J. C., 1994, Ecology and resistance management for *Bacillus thuringiensis* transgenic plants. Biocontr. Sci. Technol., **4**, 563-571.

Darmency, H., 1994, The impact of hybrids between genetically modified crop plants and their related species: introgression and weediness. Mol. Ecol., **3**, 37.

Decker, D. and Wilson, H. D., 1987, Allozyme variation in the *Cucurbita pepo* complex, *C. pepo* var *ovifera* vs. *C. texana*. Syst. Bot., **12**, 263-273.

de Maagd, R. A., Bakker, P. L., Masson, L., Adang, M. J., Sangadala, S., Stiekema, W. and Bosch, D., 1999, Domian III of the Bacillus thuringiensis delta-endotoxin CryIAc involved in binding to *Manduca sexta* brush border membranes and to its purified aminopeptidase N. Mol. Microbiol., **31**, 463-471.

de Wet, J. M. J. and Harlan, J. R., 1975, Weeds and domesticates: evolution in the man-made habitat. Econ. Bot., **29**, 99-107.

de Zoeten, G. A., 1991, Risk assessment: do we let history repeat itself? Am. Phytopath. Soc. Monogr., 81, 585.

Dirie, A. M., Cohen, M. B. and Gould, F., 2000, Larval dispersal and survival of *Scirpophaga incertulas* (Lepidoptera: Pyralidae) and *Chilo suppressalis* (Lepidoptera: Crambidae) on *cry1Ab*-transformed and non-transgenic rice. Environ. Entomol., **29**, 972-978.

Dirie, A. M., Cohen, M. B. and Gould, F., 2000, Larval dispersal and survival of *Scirpophaga incertulas* (Lepidoptera: Pyralidae) and *Chilo suppressalis* (Lepidoptera: Crambidae) on cry1Ab-transformed and non-transgenic rice. Environ. Entomol., **29**, 972-978.

Doebley, J. F., 1984, Maize introgression into teosinte-a reappraisal. Ann. Mo. Bot. Gard., **71**, 1100-1113.

Edmonds, H. S., Gatehouse, L. N., Hilder, V. A. and Gatehouse, J. A., 1996, The antimetabolic effects of oryzacystatin on larvae of the southern corn rootworm (*Diabrotica undecimpunctata howardi*); use of bacterial expression system for oryzacystatin. Entomol. Exp. Appl., **78**, 83-94.

Ellstrand, N. C., 1988, Pollen as a vehicle for the escape of engineered genes? Trends Biotech., **6**, S30-S32.

Ellstrand, N. C., Prentice, H. C. and Hancock, J. F., 1999, Gene flow and introgression from domesticated plants into their wild relatives. Annu. Rev. Ecol. Syst., **30**, 39-63.

Escriche, B., Ferre, J. and Silva, F. J., 1997, Occurrence of a common binding site in *Mamestra brassicae*, *Phthorimaea operculella*, and *Spodoptera exigua* for the insecticidal crystal proteins CryIA from *Bacillus thuringiensis*. Insect Biochem. Mol. Biol., **27**, 651-656.

Escriche, B., Tabashnik, B., Finson, N. and Ferre, J., 1995, Immunohistochemical detection of binding of CRYIA crystal proteins of *Bacillus thuringiensis* in highly resistant strains of *Plutella xylostella* (L.) from Hawaii. Biochem. Biophys. Res. Commun., **212**, 388-395.

Estada, U. and Ferré, J., 1992, Laboratory selection of *Trichoplusia n*i for resistance to *Bacillus thuringiensis* CryIA(b) delta-endotoxin. Presented at Annu. Meet. Soc. Invertebr. Pathol. 25th, Heidelberg, Germany.

FAO, 1993, Production Year Book, Food and Agriculture Organization, Rome.

Ferré, J., Real, M. D., Van Rie, J., Jansens, S., and Peferoen, M., 1991, Resistance to *Bacillus thuringiensis* bioinsecticide in a field population of *Plutella xylostella* is due to a change in a midgut membrane receptor. Proc. Natl. Acad. Sci. USA, **88**, 5119-5123.

Ferré, J., Escriche, B., Bel, Y. and Van Rie, J., 1995, Biochemistry and genetics of insect resistance to *Bacillus thuringiensis* insecticidal crystal proteins. FEMS Microbiol. Lett., **132**, 1-7.

Ferré, J., Real, M.D., Van Rie, J., Jansens, S. and Peferoen, M., 1991, Resistance to the *Bacillus thuringiensis* bioinsecticide in a field population of *Plutella xylostella* is due to a change in a midgut membrane receptor. Proc. Natl. Acad. Sci. USA, **88**, 5119-5123.

Fitches, E. and Gatehouse, J.A., 1998, A comparison of the short and long term effects of insecticidal lectins on the activities of soluble and brush border enzymes of tomato moth larvae (*Laconobia oleracea*). J. Insect Physiol., **44**, 1213-1224.

Fitt, G. P., 1994, Cotton pest management. 3. Australian perspective. Annu. Rev. Entomol., **39**, 543-562.

Forcada, C., Alcacer E., Garcera, D. and Martinez, R., 1996, Differences in the midgut proteolytic activity of two *Heliothis virescens* strains, one susceptible and one resistant to *Bacillus thuringiensis* toxins. Arch. Insect Biochem. Physiol., **31**, 257-272.

Forrester, N W., 1988, Field selection for pyrethroid resistance genes. Aust. Cotton Grower, **9**, 48-51.

Gal-On, A., Antignus, Y., Rosner, A. and Raccah, B., 1992, A zucchini yellow mosaic virus coat protein gene mutation restores aphid transmissibility but has no effect on multiplication. J. Gen. Virol., **73**, 2183-2187.

Gatehouse, A. M. R. and Boulter, D., 1983, Assessment of antimetabolic effects of trypsin inhibitor from cowpea (*Vigna unguiculata*) and other legumes on development of the bruchid beetle *Callosobruchus maculatus*. J. Sci. Food Agric., **34**, 345-350.

Gatehouse, A. M. R., Davison, G. M., Newell, C. A., Merryweather, A., Hamilton, W. D. O., Burgess, E. P. J., Gilbert, R. J. C. and Gatehouse, J. A., 1997, Transgenic potato plants with enhanced resistance to the tomato moth, *Lacanobia oleracea*: growth room trials. Mol. Breed., **3**, 49-63.

Gatehouse, A. M.R. and Gatehouse, J. A., 1998, Identifying proteins with insecticidal activity: use of encoding genes to produce insect-resistant transgenic crops. Pestic. Sci., **52**, 165-175.

Gatehouse, A. M. R., Shackley, S. J., Fenton, K. A., Bryden, J. and Pusztai, A., 1989, Mechanism of seed lectin tolerance by a major insect storage pest of *Phaseolus vulgaris*, *Acanthoscelides obtectus*. J. Sci. Food Agric., **47**, 269-280.

Georghiou, G. P., 1972, The evolution of resistance to pesticides. Annu. Rev. Ecol. Syst., **3**, 133-168.

Georghiou, G. P., 1983, Management of resistance in arthropods. In: Georghiou G. P. and Saito T. (eds), Pest Resistance to Pesticides, Plenum, New York, pp. 769-792.

Georghiou, G. P. and Taylor, C. E., 1977, Operational influences in the evolution of insecticide resistance. J. Econ. Entomol., **70**, 653-658.

Georghiou, G. P. and Lagunes-Tejeda, A., 1991, The Occurrence of Resistance to Pesticides in Arthropods. Food and Agriculture Organization of the United Nations, Rome. 318 pp.

Gill, S. S., Cowles, E. A. and Pietrantonio, P. V., 1992, The mode of action *of Bacillus thuringiensis* endotoxins. Annu. Rev. Entomol., **37**, 615-636.

Gould, F., 1998, Sustainability of transgenic insecticidal cultivars: integrating pest genetics and ecology. Annu. Rev. Entomol., **43**, 701-726.

Gould, F. and Anderson, A., 1991, Effects of *Bacillus thuringiensis* and HD-73 delta-endotoxin on growth, behavior, and fitness of susceptible and toxin-adapted strains of *Heliothis virescens* (Lepidoptera: Noctuidae). Environ. Entomol., **20**, 30-38.

Gould, F., Anderson, A., Landis, D. and Van Mellaert, H., 1991a, Feeding behavior and growth of *Heliothis virescens* larvae on diets containing *Bacillus thuringiensis* formulations or endotoxins. Entomol. Exp. Appl., **58**, 199-210.

Gould, F., Kennedy, G. G. and Johnson, M. T., 1991b, Effects of natural enemies on the rate of herbivore adaptation to resistant host plants. Entomol. Exp. Appl., **58**, 1-14.

Gould, F., Martinez-Ramirez, A., Anderson, A., Ferre, J., Silva, F. J. and Moar, W. J., 1992, Broad-spectrum resistance to *Bacillus thuringiensis* toxins in *Heliothis virescens*. Proc. Natl. Acad. Sci. USA, **89**, 701-726.

Granero, F., Ballester, V. and Ferre, J., 1996, *Bacillus thuringiensis* crystal proteins Cry1Ab and Cry1Fa share a high affinity binding site in *Plutella xylostella* (L.). Biochem. Biophys. Res. Commun., **224**, 779-783.

Greenplate, J. T., 1999, Quantification of *Bacillus thuringiensis* insect control protein Cry1Ac over time in Bollgard cotton fruit and terminals. J. Econ. Entomol., **92**, 1377-1383.

Groeters, F. R., Tabashnik, B. E., Finson, N. and Johnson, M. W., 1994, Fitness costs to resistance to *Bacillus thuringiensis* in the diamondback moth (*Plutella xylostella*). Evolution, **48**, 197-201.

Habibi, J., Backus, E. A. and Czapla, T. H., 1992, Effect of plant lectins on survival of potato leafhopper. Proc. XIX Int. Congress of Entomology, Beijing, p. 373.

Halcomb, J. L., Benedict, J. H., Cook, B. and Ring, D. R., 1996, Survival and growth of bollworm and tobacco budworm on nontransgenic and transgenic cotton expressing a CryIA insecticidal protein (Lepidoptera: Noctuidae). Environ. Entomol., **25**, 250-255.

Halcomb, J. L., Benedict, J. H., Cook, B., Ring, D. R. and Correa, J.C., 2000, Feeding behaviour of bollworm and tobacco budworm (Lepidoptera: Noctuidae) larvae in mixed stands of nontransgenic and transgenic cotton expressing an insecticidal protein. J. Econ. Entomol., **93**, 1300-1307.

Hare, J. D. and Luck, R. F., 1991, Indirect effects of citrus cultivars on life history parameters of a parasitic wasp. Ecology, **72**, 1576-1585.

Harper, S. M., Crenshaw, R. W., Mullins, M. A. and Privalle, L. S., 1995, Lectin binding to insect brush border membranes. J. Econ. Entomol., **88**, 1197-1202.

Heap, I. and Knight, R., 1986, The occurrence of herbicide cross-resistance in a population of annual ryegrass, *Lolium rigidum*, resistant to diclofop-methyl. Aust. J. Agric. Res., **37**, 149-156.

Heckel, D. G., 1994, The complex genetic basis of resistance to *Bacillus thuringiensis* toxin in insects. Biocontr. Sci. Technol., **4**, 405-417.

Hilder, V. A. and Boulter, D., 1999, Genetic engineering of crop plants for insect resistance- a critical review. Crop Prot., **18**, 177-191.

Hilder, V. A., Powell, K. S., Gatehouse, A. M. R., Gatehouse, J. A., Gatehouse, L. N., Shi, Y., Hamilton, W. D. O., Merryweather, A., Newell, C. A., Timans, J. C., Peumans, W. J., Van Damme, E. and Boulter, D., 1995, Expression of snowdrop lectin in transgenic tobacco plants results in added protection against aphids. *Transgenic Res.*, **4**, 18-25.

Hoffman, T., Golz, C. and Schieder, O., 1992, Preliminary evidence for horizontal gene transfer between higher plants and *Aspergillus niger*. In: Casper R. and Landsmann J. (eds), Proceedings of the Second International Symposium on the Biosafety Results of Field Tests of Genetically-Modified Plants and Microorganisms, May 11-14, Biologische Bundesanstalt fur Land- und Forstwirtschaft, Braunshwieg, Germany, p. 247.

Hokkanen, H. M. T. and Wearing, C. H., 1994, The safe and rational deployment of *Bacillus thuringiensis* genes in crop plants: conclusions and recommendations of OECD workshop on ecological implications of transgenic crops containing Bt toxin genes. Biocontr. Sci. Technol., **4**, 339-403.

Hokkanen, H. M. T. and Wearing, C. H., 1995, Assessing the risk of pest resistance evolution to *Bacillus thuringiensis* engineered into crop plants: a case study of oilseed rape. Field Crops Res., **45**, 171-179.

Hossain, M., 1998, Sustaining food security in Asia: economic, social, and political aspects. In: Dowling N. G., Greenfield S. M. and Fischer K. S. (eds), Sustainability of Rice in the Global Food System, Pacific Basin Study Center and International Rice Research Institute, Manila, pp. 19-44.

Hotchkiss, J. H., 1992, Pesticide residue controls to ensure food safety. Crit. Rev. Food Sci. Nutr., **3**, 191-203.

Ishimoto, M. and Chrispeels, M. J., 1996, Protective mechanism of the Mexican bean weevil against high levels of alpha-amylase inhibitor in the common bean. Plant Physiol., **111**, 393-401.

Ives, A. R., 1996, Evolution of insect resistance to *Bacillus thuringiensis*-transformed plants. Science, **273**, 1412-1413.

Jain, S. K., 1977, Genetic diversity of weedy rye populations in California. Crop Sci., **17**, 480-482.

Jesse, L. C. H. and Obrycki, J. J., 2000, Field deposition of Bt transgenic corn pollen: lethal effects on the monarch butterfly. Oecologia, **125**, 241-248.

Johnson, D. E., Brookhart, G. L., Kramer, K. J., Barnett, B. D. and McGaughey, W. H., 1990, Resistance to *Bacillus thuringiensis* by the Indian meal moth, *Plodia interpunctella*: comparison of midgut proteinases from susceptible and resistant larvae. J. Invertebr. Pathol., **55**, 235-243.

Johnson, M. T. and Gould, F., 1992, Interaction of genetically engineered host-plant resistance and natural enemies of *Heliothis virescens* (Lepidoptera: Noctuidae) in tobacco. Environ. Entomol., **21**, 586-597.

Johnson, M. T., Gould, F. and Kennedy, G. G., 1997, Effect of natural enemies on relative fitness of *Heliothis virescens* genotypes adapted and not adapted to resistant host plants. Entomol. Exp. Appl., **82**, 219-230.

Johnson, M. W., 2001, Our war with the insects: Analysis of lost battles. In: Heinz K. M., Frisbie R. and Bográn C., (eds), Challenges within Entomology: A Celebration of the Past 100 Years and A Look to the Next Century, Texas A&M University Press, College Station, TX, USA, pp.

Jongsma, M. A., 1996, Combatting inhibitor-insensitive proteases of insect pests. Trends Biotechnol., **14**, 331-333.

Jongsma, M. A., Bakker, P. L., Peters, J., Bosch, D. and Stiekema, W. J., 1995a, Adaptation of *Spodoptera exigua* to plant proteinase inhibitors by induction of proteinase activity insensitive to inhibition. Proc. Natl. Acad. Sci. *USA*, **92**, 8041-8045.

Jongsma, M. A., Bakker, P. L., Stiekema, W. J., and Bosch, D., 1995b, Phage display of a double-headed proteinase inhibitor: analysis of the binding domains of potato proteinase inhibitor II. Mol. Breed., **1**, 181-191.

Jongsma, M. A. and Bolter, C., 1997, The adaptation of insects to plant protease inhibitors. J. Insect Physiol., **43**, 885-895.

Jouanin, L., Bonade-Bottino, M., Girard, C., Morrot, G. and Giband, M., 1998, Transgenic plants for insect resistance. Plant Sci., **131**, 1-11.

Kareiva, P., Morris, W. and Jacobi, C. M., 1994, Studying and managing the risk of cross fertilization between transgenic crops and wild relatives. Mol. Ecol., **3**, 15-23.

Kinsinger, R. A. and McGaughey, W. H., 1979, Susceptibility of populations of Indianmeal moth and almond moth to *Bacillus thuringiensis*. J. Econ. Entomol., **72**, 346-349.

Kirschner, E. M., 1994, Agricultural chemical producers rebound from floods of 1993. Chem. Eng. News, **72**, 13.

Klinger, T., Elam, D. R. and Ellstrand, N. C., 1991, Radish as a model system for the study of engineered gene escape rates via crop-weed mating. Conserv. Biol. J. Soc. Conserv. Biol., **5**, 531-535.

Klinger, T. and Ellstrand, N. C., 1994, Engineered genes in wild populations: fitness of weed-crop hybrids of *Raphanus sativus*. Ecol. Appl., **4**, 117-120.

Knowles, B. and Ellar, D. J., 1987, Colloid-osmotic lysis is a general feature of the mechanisms of actions of *Bacillus thuringiensis* delta-endotoxins with different insect specificity. Biochem. Biophys. Acta, **924**, 509-518.

Lal, R. and Lal, S., 1993, Genetic Engineering of Plants for Crop Improvement, CRC Press, Boca Raton, Florida, USA. Lee, K. Y., Townsend, J., Tepperman, J., Black, M., Chui, C. F., Mazun, B., Dunsmuir, P. and Bedbrook, J., 1988, The basis of sulfonyl urea tolerance in tobacco. EMBO J., **7**, 1241.

Lee, M. K. and Dean, D. H., 1996, Inconsistencies in determining *Bacillus thuringiensis* toxin binding sites relationship by comparing competition assays with ligand blotting. Biochem. Biophys. Res. Commun., **220**, 575-580.

Lee, M. K., Rajamohan, F., Gould, F. and Dean, D. H., 1995a, Resistance to *Bacillus thuringiensis* CryIA delta-endotoxins in a laboratory-selected *Heliothis virescens* strain is related to receptor alteration. Appl. Environ. Microbiol., **61**, 3836-3842.

Lee, M. K., Young, B. A. and Dean D. H., 1995b, Domain III exchanges of *Bacillus thuringiensis* CryIA toxins affect binding to different gypsy moth midgut receptors. Biochem. Biophys. Res. Comms., **216**, 306-312.

Lefferts, L. Y., 1989, Risky business. Nutr. Action Health Lett., **16**, 1-7.

Levin, M. A. and Israeli, E., 1996, General overview of releases to date. In: Levin M. A. and Israeli E. (eds) Engineered Organisms in Environmental Settings: Biotechnological and Agricultural Applications, CRC Press, Boca Raton, pp. 13-40.

Lieberman, D. F., Wolfe, L., Fink, R. and Gilman, G., 1996, Biological safety considerations for release of transgenic organisms and plants. In: Levin M. A. and Israeli E. (eds), Engineered Organisms in Environmental Settings: Biotechnological and Agricultural Applications, CRC Press, Boca Raton, Florida, U.S.A., pp. 41-64.

Liener, I. E., 1986, Nutritional significance of lectins in the diet. In: Liener I. E., Sharon N. and Goldstein I. J., (eds), The Lectins, Academic Press, London, pp. 527-552.

Liu, Y-B. and Tabashnik, B. E., 1997, Experimental evidence that refuges delay insect adaptation to *Bacillus thuringiensis*. Proc. Roy. Soc. Lond. Ser. B, **264**, 605-610.

Liu, Y-B., Tabashnik, B. E., Masson, L., Escriche, B. and Ferré, J., 2000, Binding and toxicity of *Bacillus thuringiensis* protein Cry1C to susceptible and resistant diamondback moth (Lepidoptera: Plutellidae). J. Econ. Entomol., **93**, 1-6.

Lomonossoff, G. P., 1993, Virus resistance mediated by a nonstructural viral gene sequence. In: Hiatt A. (ed.), Transgenic Plants: Fundamentals and Applications, Marcel Dekker Inc., New York, pp. 93-114.

Lorraine-Colwill, D. F., Hawkes T. R., Williams P. H., Warner S.A.J., Sutton P. B., Powles S. B. and Preston, C., 1999, Resistance to glyphosate in *Lolium rigidum*. Pestic. Sci., **55**, 489-491.

Lu, W., Kennedy, G. G. and Gould, F., 1996, Differential predation by *Coleomegilla maculata* on Colorado potato beetle strains that vary in growth on tomato. Entomol. Exp. Appl., **81**, 7-14.

Luo, K., Tabashnik, B. E. and Adang, M. J., 1997, Binding of *Bacillus thuringiensis* Cry1Ac toxin to aminopeptidase in susceptible and resistant diamondback moths (*Plutella xylostella*). Appl. Environ. Microbiol., **63**, 1024-1027.

MacDonald, R. S., Surgeoner, G. A., Solomon, K. R. and Harris, C. R., 1983, Development of resistance to permethrin and dichlorvos by the house fly (Diptera: Muscidae) following continuous and alternating insecticide use on four farms. Can. Entomol., **15**, 1555-1561.

MacIntosh, S. C., Kishore, G. M., Perlak, F. J., Marrone, P. G. and Stone, T. B., 1990, Potentiation of *Bacillus thuringiensis* insecticidal activity by serine protease inhibitors. J. Agric. Food Chem., **38**, 1145-1152.

MacIntosh, S. C., Stone, T. B., Jokerst, R. S. and Fuchs, R. L., 1991, Binding of *Bacillus thuringiensis* proteins to a laboratory-selected line of *Heliothis virescens*. Proc. Natl. Acad. Sci. USA, **88**, 8930-8933.

Mallet, J. and Porter, P., 1992, Preventing insect adaptation to insect-resistant crops: are seed mixtures or refugia the best strategy? Proc. R. Soc. Lond., **250**, 165-169.

Mallory-Smith, C., 1998, Bromoxynil-resistant common groundsel (*Senecio vulgaris*). Weed Technol., **12**, 322-324.

Mani, G. S., 1985, Evolution of resistance in the presence of two insecticides. Genetics, **109**, 761-783.

Masson, L., Mazza, A., Brousseau, R. and Tabashnik, B., 1995, Kinetics of *Bacillus thuringiensis* toxin binding with brush border membrane vesicles from susceptible and resistant larvae of *Plutella xylostella*. J. Biol. Chem., **270**, 11887-11896.

Markwick, N. P., Reid, S. J., Lang, W. A. and Christeller, J. T., 1995, Effect of dietary protein and protease inhibitors on codling moth (Lepidoptera: Tortricidae). J. Econ. Entomol., **88**, 33-39.

Matsamura, F., 1975, Toxicology of Insecticides, Plenum Press, New York.

McGaughey, W. H. and Beeman, R. W., 1988, Resistance to *Bacillus thuringiensis* in colonies of Indianmeal moth and almond moth (Lepidoptera: Pyralidae). J. Econ. Entomol., **81**, 28-33.

McGaughey, W. H. and Johnson, D. E., 1992, Indianmeal moth (Lepidoptera: Pyralidae) resistance to different strains and mixtures of *Bacillus thuringiensis*. J. Econ. Entomol., **85**, 1594-1600.

McGaughey, W. H. and Whalon, M., 1992, Managing resistance to *Bacillus thuringiensis* toxins. Science, **258**, 1451-1455.

Michaud, D., 1997, Avoiding protease-mediated resistance in herbivorous pests. Trends Biotechnol., **15**, 4-6.

Moar, W. J., 1993, Development of *Bacillus thuringienis* CryIC resistance by *Spodoptera exigua*. 26th Annu. Meet. Soc. Invertebr. Pathol., USA.

Morris, W. F., Kareiva, P. M. and Raymer, P. L., 1994, Do barren zones and pollen traps reduce gene escape from transgenic crops? Ecol. Appl., **4**, 157-165.

Morton, R. L., Schroeder, H. E., Bateman, K. S., Chrispeels, M. J., Armstrong, E. and Higgins, T. J. V. (2000) Bean alpha-amylase inhibitor 1 in transgenic peas (*Pisum sativum*) provides complete protection from pea weevil (*Bruchus pisorum*) under field conditions. Proc. Natl. Acad. Sci. USA, **97**, 3820-3825.

Oerke, E. -C., Dehne, H. –W., Schonbeck, F., and Weber, A., 1994, Crop Production and Crop Protection: Estimated Losses in Major Food and Cash Crops. Elsevier, Amsterdam.

Oppert, B., Morgan, T. D., Culbertson, C. and Kramer, K., 1993, Dietary mixtures of cysteine and serine proteinase inhibitors exhibit synergistic toxicity towards the red flour beetle. Comp. Biochem. Physiol., **105**, 379-385.

Oppert, B., Kramer, K., Beeman, R. W., Johnson, D. and McGaughey, W. H., 1997, Proteinase-mediated insect resistance to Bacillus thuringiensis toxins. J. Biol. Chem., **272**, 23473-23476.

Orr, G. R., Strickland, G. A. and Walsh, T. A., 1994, Inhibition of Diabrotica larval growth by multicystatin from potato tubers. J. Insect Physiol., **40**, 893-900.

Osbourne, J. K., Sarkar, S. and Wilson, T. M. A., 1990, Complementation of coat proteindefective TMV mutants in transgenic tobacco plants expressing TMV coat protein. Virology, **179**, 921-925.

Padgette, S. R., Re, D. B., Berry, G. F., Eichholtz, D. E., Delannay, X., Fuchs, R. L., Kishore, G. and Fraley, R. T., 1996, New weed control opportunities: development of soybeans with a Roundup ready^{TM} gene. In S. O. Duke (ed.), Herbicide Resistant Crops: Agricultural, Environmental, Economic, Regulatory, and Technical Aspects, CRC Press, Inc., Boca Raton, Florida, USA, pp. 53-84.

Parker, C. and Dean, M. L., 1976, Control of wild rice in rice. Pestic. Sci., **7**, 403-416.

Parker, C. D. and Luttrell, R. G., 1999, Interplant movement of Heliothis virescens (Lepidoptera: Noctuidae) larvae in pure and mixed plantings of cotton with and without expression of the Cry1Ac delta-endotoxin protein of Bacillus thuringiensis Berliner. J. Econ. Entomol., **92**, 837-845.

Peferoen, M., 1992, Engineering of insect-resistant plants with Bacillus thuringiensis crystal protein genes. In: Gatehouse A. M. R., Hilder V. A., and Boulter D. (eds), Plant Genetic Manipulation for Crop Protection, CAB International, Wallingford, UK, pp. 135-153.

Peferoen, M., 1997, Insect control with transgenic plants expressing Bacillus thuringiensis crystal proteins. In: Carozzi N. and Koziel M. (eds), Advances in Insect Control: The Role of Transgenic Plants, Taylor & Francis, London, pp. 21-48.

Perlak, F. J., Fuchs, R. L., Dean, D. A., McPherson, S. L. and Fischhoff, D. A., 1991, Modification of the coding sequence enhances plant expression of insecticidal insect control genes. Proc. Natl. Acad. Sci. USA, **88**, 3324-3328.

Pimentel, D. and Bellotti, A. C., 1976, Parasite-host population systems and genetic stability. Am. Nat., **95**, 65-79.

Powell-Abel, P., Nelson, R. S., De, B., Hoffman, N., Rogers, S. G., Fraley, R. T. and Beachy, R. N., 1986, Delay of disease development in transgenic plants that express the tobacco mosaic virus coat protein gene. Science, **232**, 738-743.

Powell, K. S., Gatehouse, A. M. R., Hilder, V. A. and Gatehouse, J. A., 1993, Antimetabolic affects of plant lectins and fungal enzymes on the nymphal stages of two important rice pests, Nilaparvata lugens and Nephotettix cinciteps. Entomol. Exp. Appl., **66**, 119-126.

Powers, J. R. and Culbertson, J. D., 1983, Interaction of a purified bean (Phaseolus vulgaris) glycoprotein with an insect amylase. Cereal Chem., **60**, 427-429.

Powers, J. R. and Whitaker, J. R., 1977, Effect of several experimental parameters on combination of red kidney bean (Phaseolus vulgaris) alpha amylase inhibitor with porcine pancreatic alpha amylase. J. Food Biochem., **1**, 239-260.

Powles, S. B. and Howat, P. D., 1990, Herbicide-resistant weeds in Australia. Weed Technol., **4**, 178-185.

Ramachandran, S., Buntin, G. D., All, J. N., Raymer, P. L. and Stewart, C. N., Jr., 1998, Movement and survival of diamondback moth (Lepidoptera: Plutellidae) larvae in mixtures of nontransgenic and transgenic canola containing a *Cry*IA(c) gene of *Bacillus thuringiensis*. Environ. Entomol., **27**, 649-656.

Reek, G. R., Kramer, K. J., Baker, J. E., Kanost, M. R., Fabrick, J. A. and Behnke, C. A., 1997, Proteinase inhibitors and resistance of transgenic plants to insects. In: Carozzi N. and Koziel M. (eds), Advances in Insect Control: The Role of Transgenic Plants, Taylor & Francis, London, pp. 157-183.

Regal, P. J., 1988, The adaptive potential of genetically engineered organisms in nature. Trends Biotechnol., **6**, S36-S38.

Riggin-Bucci, T. M. and Gould, F., 1997, Impact of intraplot mixtures of toxic and nontoxic plants on population dynamics of diamondback moth (Lepidoptera: Plutellidae) and its natural enemies. J. Econ. Entomol., **90**, 241-251.

Rissler, J. and Mellon, M., 1994, No commercial gene-altered crop approvals until fed gov't assesses the ecological risks. Genetic Engin. News, **14**, 4-12.

Rochow, W. F, 1970, Barley yellow dwarf virus: phenotypic mixing and vector specificity. Science, **167**, 875-878.

Roush, R. T., 1989, Designing resistance management programs: how can you choose? Pestic. Sci., **26**, 423-441.

Roush, R. T., 1999, Two-toxin strategies for management of insecticidal transgenic crops: can pyramiding succeed where pesticide mixtures have not? In: Denholm I., Pickett J. A., and Devonshire A. L., (eds), Insecticide Resistance: From Mechanisms to Management, CABI Publishing, Wallingford, UK, pp. 75-80.

Roush, R. T. and Shelton, A. M., 1997, Assessing the odds; the emergence of resistance to Bt transgenic plants. Nature Biotechnol., **15**, 816-817.

Ryan, C. A., 1990, Proteinase inhibitors in plants: genes for improving defenses against insects and pathogens. Annu. Rev. Phytopathol., **28**, 939-943.

Salama, H. S. and Matter, M. M. (1991) Tolerance level to *Bacillus thuringiensis* Berliner in the cotton leafworm *Spodoptera littoralis* Boisduval (Lepidoptera: Noctuidae). J. Appl. Entomol., **111**, 225-230.

Salomon, R., 1995, Interference in plant viral vector transmission by proteolytic modifications of the coat protein. In: Reuveni R. (ed.), Novel Approaches to Integrated Pest Management, Lewis Publishers, Boca Raton, Florida, USA, pp. 129-138.

Schwartz, J. M., Tabashnik, B. E. and Johnson, M. W., 1991, Behavioral and physiological responses of susceptible and resistant diamondback moth larvae to *Bacillus thuringiensis*. Entomol. Exp. Appl., **61**, 179-187.

Second, G., 1982, Origin of the genetic diversity of cultivated rice (*Oryza* spp.): study of the polymorphism scored at 40 isozyme loci. Jpn. J. Genet., **57**, 25-57.

Sela, I., 1996, Engineered viruses in agriculture. In: Levin M. A. and Israeli E. (eds), Engineered Organisms in Environmental Settings: Biotechnological and Agricultural Applications, CRC Press, Boca Raton, Florida, USA, pp. 107-148.

Sherman, T. D., Vaughn, K. C. and Duke, S. O., 1996, Mechanisms of action and resistance to herbicides. In: Duke S. O. (ed.), Herbicide Resistant Crops: Agricultural, Environmental, Economic, Regulatory, and Technical Aspects, CRC Press, Inc., Boca Raton, Florida, USA, pp. 13-36.

Shukle, R. H. and Murdoch, L. L., 1983, Lipoxygenase, trypsin inhibitor and lectin from soybeans: effects on larval growth of *Manduca sexta* (Lepidoptera: Sphingidae). Environ. Entomol., **12**, 787-791.

Simmonds, N. W., 1979, Evolution of Crop Plants, Longman, New York.

Sims, S. R. and Stone, T. B., 1991, Genetic basis of tobacco budworm resistance to an engineered *Pseudomonas fluorescens* expressing the delta-endotoxin of of *Bacillus thuringiensis kurstaki*. J. Invertebr. Pathol., **57**, 206-210.

Sneh, B. and Schuster, S., 1983, Effect of exposure to sublethal concentration of *Bacillus thuringiensis* Berliner ssp. *entomocidus* on the susceptibility to the endotoxin of subsequent generations of the Egyptian cotton leafworm *Spodoptera littoralis* Boisd. (Lep.: Noctuidae). Z. Angew. Entomol., **96**, 425-428.

Stalker, D. M., Kiser, J. A., Baldwin, G., Coulombe, B. and Houck, C. M., 1996, Cotton weed control using the BXN system. In: Duke S. O. (ed.), Herbicide Resistant Crops: Agricultural, Environmental, Economic, Regulatory, and Technical Aspects, CRC Press, Inc., Boca Raton, Florida, USA, pp. 93-106.

Stalker, D. M., McBride, K. E. and Malyj, L. D., 1988, Herbicide resistance in transgenic plants expressing a bacterial detoxification gene. Science, **242**, 419-423.

Steffens, R., Fox, F. R. and Kassel, B., 1978, Effect of trypsin inhibitors on growth and metamorphosis of corn borer larvae *Ostrinia nubilalis* (Hubner). J. Agric. Food Chem., **26**, 170-174.

Stone, T. B., Sims, S. R. and Marrone, P. G., 1989, Selection of tobacco budworm for resistance to a genetically engineered *Pseudomonas fluorescens* containing the delta-endotoxin of *Bacillus thuringiensis* subsp. *kurstaki*. J. Invertebr. Pathol., **53**, 228-234.

Sturtevant, A. P. and Beachy, R. N., 1993, Virus resistance in transgenic plants: coat protein-mediated resistance. In: Hiatt A. (ed.), Transgenic Plants: Fundamentals and Applications, Marcel Dekker, Inc., New York, pp. 93-114.

Tabashnik, B. E., 1990, Modeling and evaluation of resistance management tactics. In: Roush R. T. and Tabashnik R. T. (eds), Pesticide Resistance in Arthropods, Chapman and Hall, New York, pp.153-182.

Tabashnik, B. E., 1994, Evolution of resistance to *Bacillus thuringiensis*. Annu. Rev. Entomol., **39**, 47-79.

Tabashnik, B. E. and Croft, B. A., 1982, Managing pesticide resistance in crop-arthropod complexes: interactions between biological and operational factors. Environ. Entomol., **11**, 1137-1144.

Tabashnik, B. E. and Croft, B. A., 1985, Evolution of pesticide resistance in apple pests and their natural enemies. Entomophaga, **30**, 37-49.

Tabashnik, B. E., Cushing, N. L., Finson, N. and Johnson, M. W., 1990, Field development of resistance to *Bacillus thuringiensis* in diamondback moth (Lepidoptera: Plutellidae). J. Econ. Entomol., **83**, 1671-1676.

Tabashnik, B. E., Finson, N., Groeters, F. R., Moar, W. J., Johnson, M. W., Luo, K. and Adang, M. J., 1994, Reversal of resistance to *Bacillus thuringiensis* in *Plutella xylostella*. Proc. Natl. Acad. Sci. USA, **91**, 4120-4124.

Tabashnik, B. E., Finson, N., and Johnson, M. W., 1991, Managing resistance to *Bacillus thuringiensis*: lessons from the diamondback moth (Lepidoptera: Plutellidae). J. Econ. Entomol., **84**, 49-55.

Tabashnik, B. E., Finson, N., and Johnson, M. W., 1992, Two protease inhibitors fail to synergize *Bacillus thuringiensis* in diamondback moth (Lepidoptera: Plutellidae). J. Econ. Entomol., **85**, 2082-2087.

Tabashnik, B. E., Finson, N., Johnson, M. W. and Moar, W. J., 1993, Resistance to toxins from *Bacillus thuringiensis* subsp. *kurstaki* causes minimal cross-resistance to *B. thuringiensis* subsp. *aizawai* in diamondback moth (Lepidoptera: Plutellidae). Appl. Environ. Microbiol., **59**, 1332-1335.

Tabashnik, B. E., Liu, Y-B, Malvar, T., Heckel, D. G., Maaon, L. and Ferré, J., 1999a, Insect resistance to *Bacillus thuringiensis*: uniform or diverse? In: Denholm I., Pickett J. A.

and Devonshire A. L. (eds), Insecticide Resistance: From Mechanisms to Management, CABI Publishing, Wallingford, UK, pp. 75-80.

Tabashnik, B. E., Patin, A. L., Dennehy, T. J., Liu, Y.-B., Miller, E. and Staten, R. T., 1999b Dispersal of pink bollworm (Lepidoptera: Gelechiidae) males in transgenic cotton that produces a *Bacillus thuringiensis* toxin. J. Econ. Entomol., **92**, 772-780.

Tang, J. D., Shelton, A. M., Van Rie, J., de Roeck, S., Moar, W. J., Roush, R. T. and Peferoen, M., 1996, Toxicity of *Bacillus thuringiensis* spore and crystal protein to resistant diamondback moth (*Plutella xylostella*). Appl. Environ. Microbiol., **62**, 564-569.

Taylor, C. E. and Georghiou, G. P., 1979, Suppression of insecticide resistance by alteration of gene dominance and migration. J. Econ. Entomol., **72**, 105-109.

Taylor, C. E. and Georghiou, G. P., 1982, Influence of pesticide persistence in evolution of resistance. Environ. Entomol., **11**, 746-750.

Tiedje, J. M., Colwell, R. K., Grossman, Y. L., Hodson, R. E., Lenski, R. E., Mack, R. N. and Regal, P. J., 1989, The planned introduction of genetically engineered organisms: ecological considerations and recommendations. Ecology, **70**, 298.

Trisyono, A. and Whalon, M. E., 1997, Fitness costs of resistance to *Bacillus thuringiensis* in Colorado potato beetle (Coleoptera: Chrysomelidae). J. Econ. Entomol., **90**, 267-271.

Tucker, J. M. and Sauer, J. D., 1958, Aberrant *Amaranthus* populations of Sacramento-San Joaquin delta, California. Madrono, **14**, 252-261.

U.S. Department of Agriculture, Animal and Plant Health Inspection Service, 2000, Petitions of nonregulated status granted by APHIS as of 10-12-2000. http://www.aphis.usda.gov/biotech/not_reg.html

U.S. Department of Agriculture, Economic Research Service, 2000, Agricultural outlook: resources & environment.

U. S. Environmental Protection Agency, 2001, Office of pesticide prgrams: biopesticides. http://www.epa.gov/pesticides/biopesticides.

Van Emden, H. F., 1991, The role of host plant resistance in insect pest mismanagement. Bull. Entomol. Res., **81**, 123-126.

Van Frankenhuyzen, K. and Milne, R. E., 1993, Application of Bt for control of spruce budworm in Canada: is there a risk of resistance? Presented at: Pacific Entomology Conference, Honolulu.

Van Rie, J., McGaughey, W. H., Johnson, D. E., Barnett, B. D. and Van Mallaert, H., 1990, Mechanism of insect resistance to the microbial insecticide *Bacillus thuringiensis*. Science, **247**, 72-74.

Vasil, I. K., 1996, Phosphinothricin-resistant crops. In: Duke S. O. (ed.), Herbicide Resistant Crops: Agricultural, Environmental, Economic, Regulatory, and Technical Aspects, CRC Press, Inc., Boca Raton, Florida, pp. 85-92.

Wang, T.Y., Chen, H. B., Reboud, X. and Darmency, H., 1997, Pollen-mediated gene flow in an autogamous crop: foxtail millet (*Setaria italica*). Plant Breed., **116**, 579-583.

Warwick, S. I. and Black, L. D., 1993, The biology of Canadian weeds. 61. *Sorghu halepense* (L.) PERS. Can. J. Bot., **62**, 1781-1790.

Watrud, L. S., Metz, S. G. and Fischhoff, D. A., 1996, Engineered plants in the environment. In: Levin M. A. and Israeli E. (eds), Engineered Organisms in Environmental Settings: Biotechnological and Agricultural Applications, CRC Press, Boca Raton, Florida, USA, pp. 165-190.

Wearing, C. H. and Hokkanen, H. M. T., 1994, Pest resistance to *Bacillus thuringiensis*: ecological crop assessment for *Bt* gene incorporation and strategies for management. In: Hokkanen H. M. T. and Lynch J. M. (eds), Biological Control: Benefits and Risks, Cambridge University Press, Cambridge, UK, pp. 236-252.

Weedscience.com, 2001, Herbicide resistant weeds. http://www.weedscience.org/in.asp.
Whalon, M. E., Miller, D. L., Hollingworth, R. M., Grafius, E. J. and Miller, J. R., 1993, Selection of a Colorado potato beetle (Coleoptera: Chrysomelidae) strain resistant to *Bacillus thuringiensis*. J. Econ. Entomol., **86**, 226-233.
Wiersema, J. H. and Leon, B., 1999, World Economic Plants: A Standard Reference. CRC Press, Boca Raton, Florida, USA.
Wigley, P. W. and Chilcott, C. N., 1992, Present use of and problems with, *Bacillus thuringiensis* in New Zealand. In: Milner R. and Chandler C. (eds), Proceeding of a Workshop on *Bacillus thuringiensis*, Canberra, Australia, pp. 34-37.
Wolfson, J. L. and Murdoch, L. L., 1990, Diversity in digestive proteinase activity among insects. J. Chem. Ecol., **16**, 1089-1102.
Wright, D. J., Iqbal, M., Granero, F. and Ferré, J., 1997, A change in single midgut receptor in the diamondback moth (*Plutella xylostella*) is only in part responsible for field resistance to *Bacillus thuringiensis* subsp. *kurstaki* and *B. thuringiensis* subsp. *aizawai*. Appl. Environ. Microbiol., **63**, 1814-1819.
Younes, M., Galal-Gorchev, H., 2000, Pesticides in drinking water—a case study. Food Chem. Toxicol., **38**, S87-S90.
Zhu-Salsman, K., Shade, R. E., Koiwa, H., Salzman, R. A., Narasimhan, M. Bressan, R. A., Hasegawa, P. M. and Murdoch, L. L., 1998, Carbohydrate binding and resistance to proteolysis control insecticidal activity of *Griffonia simplicifolia* lectin II. Proc. Natl. Acad. Sci. USA, **95**, 15123-15128.

7

GENE PYRAMIDING: A TRANSGENIC APPROACH TO ENHANCING RESISTANCE DURABILITY IN PLANTS

K.R. RAJYASHRI AND MADAN MOHAN*
Plant Resistance Group
International Centre for Genetic Engineering and Biotechnology
Aruna Asaf Ali Marg, New Delhi-110 067, India

INTRODUCTION

The world population is increasing at the rate of 160 people per minute, with about 90% of them residing in developing countries. Feeding these growing numbers will require a substantial increase in food production. However, since the early 1990s, there has not only been a reduction in the rate of growth of food production, but the area under cultivation has also shrunk, making it imperative to explore new avenues of increasing crop yield (Khush, 1999). While a lot of work is already underway to address the problem of increasing the yield potential, the best way of realizing this potential lies in reducing the loss of crop due to various factors. The crop yields obtained are greatly affected by various biotic and abiotic factors, resulting in a huge gap between the yield potential and the actual yield, especially in the developing countries with limited resources to combat these problems. Strategies aimed at developing crop cultivars with greater tolerance to abiotic stress like drought and salinity and resistance to pests and pathogens are the only viable means of improving the crop yields in these regions.

There is a substantial loss in the yield of crop worldwide due to various bacterial, viral and fungal diseases, infestation by different

---
*Corresponding author

nematodes and pests and due to competition from weeds. About 12% of the world crop production is lost annually to plant diseases caused by bacteria, fungi and viruses (Rommens and Kishore, 2000), while 14% of the total agricultural produce is lost to pests (Oerke et al., 1994). Fungi not only reduce the crop yield, but also lower the crop quality by producing toxins, such as aflatoxins, that affect human health negatively. Likewise, in addition to causing a direct damage, pests also act as vectors for various pathogens, thereby adding to the loss of crop due to pathogens. In India 20-30% of the crop productivity is lost due to damage by insects pests (Tuli et al., 2000). Nematodes, especially the plant parasitic nematodes *Meliodogyne* (root knot nematode), Hetrerodera and *Globodera* spp. are responsible for a loss of US $ 100 billion annually (Lilley et al., 1999). Viruses are yet other agents causing considerable loss of crop, especially in developing countries in the tropical and subtropical regions (Buck, 1991). In Africa alone, losses of cassava crops due to infection by African cassava mosaic virus have been valued at more than $ 300 million per annum (Harrison et al., 1987). Crop yields are also reduced by weeds, which compete for water, light and nutrients. Further, the presence of the competing weeds reduces the crop quality, simultaneously adding to the expense of production in terms of the labour needed to weed out the unwanted plants. Thus the total loss of crop due to biotic stress averages 30-40% every year. Any reduction in this loss would make a significant contribution to the crop yield.

CHEMICAL CONTROL OF BIOTIC STRESSES

Control of Pests and Pathogens

Chemical pesticides have been used on a very wide scale to control pests since the beginning of the twentieth century. According to a report, more than $ 33.5 billion are spent annually on agro-chemicals worldwide (www.fao.org; www.pmfai.org). However, several problems have been generated by their widespread use. Often, the pesticides not only kill the pests they are directed against, but are also toxic to non-target organisms, including pollinating insects and natural predators that would normally keep the pest population under control. The high selective pressure imposed on the target population by the toxicity and heavy application of the insecticides results in a build up of populations of pest species resistant to one or more insecticides within a few generations (Gatehouse et al., 1991). The indiscriminate use of pesticides has resulted in the elimination of a wide range of predatory species along with the primary pests; this has, in some cases, resulted in secondary pests transforming into primary pests with even more devastating effect. A case in point has been the brown planthopper, a minor pest of rice in the Asian countries, which after indiscriminate use of the pesticide DDT (dichloro diphenyl trichloroethane), has now emerged as a major pest in these countries (Heinrichs and Mochida, 1984).

Pesticides are also often used far in excess of requirements, leaving huge amounts of residue in food and resulting in environmental pollution. It has been estimated that 99.9% of the applied pesticides are wasted, thereby contributing to environmental pollution and toxicity to humans (Gatehouse et al., 1991). In some crops, like rice, the expenses incurred in pesticide use are already higher than the benefits (Rola and Pingali, 1993). Difficulties encountered in controlling pests affecting cotton, tomatoes and melons using synthetic pesticides have forced farmers in some countries to abandon growing these crops completely (www.fao.org/WAICENT/FAOINFO/AGRICULT/ags/agse).

Nematodes are another class of pests that cause a considerable loss of revenue to the farmers. Current control of plant parasitic nematodes relies very heavily on highly toxic and environmentally-harmful nematicides, the use of which is being increasingly restricted (Lilley et al., 1999). Various chemical fungicides and antibiotics—generally in the form of sprays—are also used to control fungal and bacterial diseases. However, most of them are either expensive or strong chemicals that are corrosive and harmful on ingestion by humans and animals.

Thus, the cost of controlling pests and pathogens affecting the crops is high, both in cost of input and their effect on the environment. In most countries, there is a growing concern about the known and unknown consequences of these chemicals on the environment and human health, and a consequent increase in strong public pressure to reduce their use. Hence, safer modes of controlling the pests and pathogens are being seriously explored.

An alternate to the chemical pesticides is the use of certain bacterial toxins with insecticidal properties. *Bacillus thuringiensis* (Bt) is a soil bacterium producing intracellular crystals that are toxic to certain insects, but not to humans and other mammals. These Bt toxins in various formulations—usually as whole sporulated bacteria—have been employed as insecticidal crop sprays for more than 30 years (Dulmage, 1981). More than 2000 tonnes of protein crystal/spore preparations of Bt are used as commercial preparations every year. The Bt toxins most commonly used are from three major sources: *B. thuringiensis* var. *kurstaki* which are effective against lepidopteran pests; *B. thuringiensis* var. *berliner*, aimed at the wax moth and *B. thuringiensis* var. *israelensis*, which acts as a control agent for dipteran vectors of human disease. However, the high costs of production and instability of the crystal protein under field spraying conditions limit their use (Gatehouse et al., 1991).

Viral Control

Viral diseases of cultivated plants cause substantial loss in food, forage and fiber crops throughout the world. At least 700 plant viruses causing various diseases have been recognized so far (Matthews, 1981). Crop plants are regularly exposed to one or more of these viral pathogens which cause variable loss of yield, depending on a variety of environmental and vector related factors that control viral epidemiology. No methods of curing the plants exist once they have been infected with a virus. Traditional methods to prevent the spread of viral diseases rely entirely on the prevention of infection through use of cultivars resistant to the viruses (Buck, 1991).

Weed Control

Weeds compare strongly with crops for water, light and nutrients and their presence in a field of crop greatly reduces the quality and quantity of the crop yield. The most widely-used modern method of controlling the weeds is through the use of chemical herbicides. Herbicides act selectively on weeds, mainly as a consequence of a differential uptake or metabolism of the herbicide, or by precise localization of the herbicide applied. The best way to control weeds would be to confer resistance against broad-spectrum compounds to crops. This would permit the use of broad-spectrum herbicides, as well as enable the farmer to select the most suitable and environmentally-friendly herbicides from a range of compounds. However, attempts to produce crops resistant to herbicides by conventional breeding methods have not been very successful.

Thus, although agro-chemicals have been routinely used to control various pests and pathogens, they are not effective against all the agents that affect the plants. Agro-chemicals are totally ineffective against viruses and only partly effective against many plant pathogens. Moreover, the use of these chemicals not only increases the cost of production of the crop, but also has undesirable consequences on the environment. The most cost-effective and environmentally friendly method would be to deploy varieties that have been developed for resistance to the various agents causing biotic stress. Although complete elimination of chemical control agents is not realistic, a judicious use of pesticides, crop rotation, field sanitation, pest-free seeds and exploitation of inherently-resistant plant varieties (host plant resistance) is the best method of controlling these agents.

RESPONSE TO BIOTIC STRESSES

A plant responds to infection by a pathogen with two types of resistance responses—a race/cultivar-specific host resistance response and a general non-host resistance response.

According to the gene-for-gene hypothesis suggested by Flor (1971), the product of a pest or pathogen avirulence (*avr*) gene, or a factor produced by it is recognized by a corresponding plant disease resistance (R) gene product, thereby triggering the race-specific host-resistance response. Subsequent studies have confirmed that the resistance response involves two steps: (i) The recognition of the gene product of an avirulence (*avr*) gene specific to the particular pathogen, by the product of the resistance (R) gene of the host; and (ii) triggering of downstream genes, leading to the resistance (Staskawicz et al., 1995). It is postulated that the pathogen or pest has an avirulence gene, which encodes a protein molecule that is recognized by a matching receptor in a particular plant genotype. The formation of the receptor-protein complex initiates a plethora of defense-related genes involved in hypersensitive response (HR) (see review DeWit, 1997), and subsequently establishes the systemic acquired resistance (SAR) response. Multiple signal transduction pathways, of which salicylic acid (SA), jasmonic acid (JA), ethylene (ET) and absiscic acid (ABA) are key components, mediate the defense response. R genes conferring resistance to bacterial, viral, and fungal pathogens appear to use multiple signaling pathways, which converge upon common downstream effectors. Manipulation of salicylic acid, jasmonic acid, ethylene and absiscic acid levels seems a promising way of evoking the defense pathways in response to insect attack. Howe et al., (1996) observed that mutation in genes involved in the synthesis of jasmonic acid compromised the ability of the plant to defend itself from insect attack, thereby suggesting that an increase in JA levels would confer greater resistance against the insects. Thus, the manipulation of R genes and their signaling pathways by transgenic expression appears to be a promising strategy to improve disease resistance in plants (Martin, 1999).

Non-host resistance or systemic acquired resistance (SAR) a broad-spectrum resistance is exhibited by all plant species that respond to all potential pathogens in a non-specific manner. It is independent of the pest and the cultivar affected. SA is a key component in triggering the SAR which involves upgradation of a battery of pathogenesis-related genes (PR). Verberne et al. (2000) observed that transformation of tobacco chloroplasts with two bacterial genes coding for enzymes that convert chorismate into SA by a two-step process led to a 500- to 1000-

fold increase in the accumulation of SA and SA glucoside compared to control plants. They observed that defense genes, particularly those encoding acidic pathogenesis-related (PR) proteins, were constitutively expressed in these transgenic plants. Over-expression of SA did not affect the plant phenotype, but these plants exhibited an increased resistance to viral and fungal infection resembling SAR in nontransgenic plants. Thus, a broad-spectrum resistance to pathogens can be obtained by overexpression of SA in plants.

As such, the manipulation of the signaling molecules and downstream genes involved in disease resistance should be able to confer broad-spectrum resistance to a range of pests and pathogens, while the specific resistance genes should be able to provide resistance to the specific pest or pathogen. Although most of the focus so far has been on the manipulation of the specific resistance genes for developing disease resistant crops, more and more interest is now being shown in manipulating the other genes of the signaling pathway.

RESISTANCE GENES IN CONTROL OF BIOTIC STRESSES

Natural resistance to a majority of the pests and pathogens that affect a crop is invariably available in the germplasm. These resistant varieties have been traditionally used for breeding the resistance into desired varieties. Some of the resistant varieties thus bred have been very successful. For example, the Ry gene of *Solanum stoloniferum* is a single major dominant resistance gene giving extreme resistance to all strains of potato virus Y (PVY). The Ry gene has been incorporated into several potato cultivars and is being used extensively in breeding programmes as a source of resistance to PVY (Ross, 1986). However, there are two major problems with this approach of introducing resistance into crops. Firstly, there may not be a source conferring resistance to the particular pest or pathogen; most viruses do not have corresponding resistance genes in the germplasm. Secondly, even where resistance genes are found in wild relatives, lack of sexual compatibility makes the resistance gene inaccessible. Moreover, even where there is a source for resistance, lengthy screening and breeding procedures are required before a homozygous resistance genotype is achieved (Mohan et al., 1997a).

Herbicides, used for weed control, are generally designed to affect the photosynthetic and metabolic processes common to both crop plants and weeds. As mentioned earlier, selectivity is based on the differential uptake between the crop and weed and detoxification of the herbicide by the crop plant. Though certain genes that provide resistance to herbicides have been identified in crops, attempts to produce crops resistant to herbicides by conventional breeding methods have not been successful so far.

Thus, while genes that provide resistance to different pests and pathogens can be exploited to breed for resistant plants, breeding for viral and herbicide resistance through the use of naturally-occurring resistance genes is generally not available. Alternate methods of introducing resistance into crops need to be explored and exploited.

ADVANTAGES OF USING TRANSFORMATION VERSUS BREEDING FOR INTROGRESSION OF RESISTANCE GENES

Selective breeding has been the most commonly-used method for introducing desired traits into new plant varieties. However, the production of hybrids requires that the two parents be sexually compatible. This not only limits the diversity of the genetic material available for crossing but also results in uncontrolled combination of thousands of uncharacterized genes. The constraints imposed by sexual incompatibility deny plant breeders access to a diverse range of genetic material, severely limiting the ability to improve crops through traditional means. Moreover, the introgression of R genes into elite cultivars via traditional breeding is a time consuming and laborious process, which can take upto 10-20 years. Although this process can be reduced by marker-assisted selection (see review Mohan et al.,1997a), it still involves making a large number of crosses between a range of parents selected for desirable traits and, subsequently, repeated rounds of crossing and back crossing over several generations in order to produce the desired combination of traits. The problems are compounded if the resistance genes are tightly linked to undesirable traits or are difficult to score for phenotype. For example, the soybean *rhg1* gene for control of cyst nematodes behaves as a quantitative trait locus and is very difficult to separate genetically from a reduced yield phenotype (Riggs and Wrather, 1992).

Under the circumstances, the best way of circumventing these difficulties is through the use of transformation, a tool through which specific genes can be introduced into elite cultivars in a fairly controlled manner as compared to hybridization by plant breeding. Transformation allows the introduction of several desirable genes in a single event through particle bombardment, *Agrobacterium*-mediated transformation or through electroporation. Details of these procedures can be obtained from many reviews (Christou, 1997; Komari et al., 1998; Bates, 1999; Finer et al., 1999; Maenpaa et al., 1999; Bogorad, 2000; Heifetz, 2000). The main advantage of the technology is that it enables the plant breeder to cross the species barriers, allowing genes from non-related plants and other organisms to be introduced into the desired crop plants. Precise genetic changes can be produced by well-characterized genes selected carefully from one organism and introduced into another. Novel molecules can be generated and used, the levels and patterns of gene expression can be controlled and different genes can be tried out in a short time span. The plants' own genome may also be modified to control, turn on or off the specific functions introduced into the plant. Using transformation technology, usually only one or two generations are needed to complete the gene transfer, thus reducing the time taken to introgress novel genes into an elite background. Transformation is being increasingly used to move genes among plants that cannot be hybridized using traditional methods, and to control specific functions within a plant. The greater precision of these techniques reduces the time and money required for generating crops carrying the desirable traits. The last few years have seen an enormous increase in the number of genes that have been transferred using this new technology, some of which are described in the subsequent sections.

DEPLOYMENT OF SINGLE VERSUS MULTIPLE RESISTANCE GENES WITHIN A CROP

Only a few R genes have so far proved able to control any pest or pathogen for an extended length of time (see review Rommens and Kishore, 2000). Most single resistance genes for agronomically-important diseases—including those for stem rust and rice blast, as well as most insect resistance genes—have proved to be of limited durability (Crute, 1985). Pests and pathogens have also been seen to evolve quite rapidly in response to the deployment of single resistance genes targeted against them. For example, Phalguna is a variety of rice carrying a single resistance gene, the *Gm2* – which confers resistance to the biotypes 1, 2 and 5 of gallmidge (Mohan et al., 1994; Nair et al., 1995; Sardesai et al., 2001). Extensive deployment of this variety for 5-6 years in the coastal regions of Andhra Pradesh in India led to a complete breakdown of resistance in the year 1986, leading to the loss of rice crop to a tune of $ 500 million in just one season (Sardesai et al., 2001).

Therefore, in order to overcome the very limited durability of single resistance (R) genes and to delay the evolution of new biotypes of pests and pathogens, it is necessary to deploy more than one resistance gene in the same plant. The deployment of resistance genes with different modes of action in the same plant will prove to be the most effective way of preventing the evolution of resistance to these genes in the pest or pathogen. Genome sequencing and mapping experiments have recently demonstrated that resistance genes are generally organized in tightly linked clusters (see review Michelmore, 2000; Michelmore and Meyers, 1998). Parniske et al. (1997) have also shown that tightly-linked resistance genes have the ability to act synergistically. Since transformation allows the simultaneous transfer of many genes (Hadi et al., 1996) and of large DNA fragments, many resistance genes present either in a single block or present on different plasmids can be transferred simultaneously (Hamilton, 1997), thereby enhancing the durability of the resistance.

Some of the genes that could be used for generating resistance to pathogens, pests, viruses and weeds have been enumerated below.

RESISTANCE TO PATHOGENS

Bacterial and fungal diseases are responsible for great losses in crop plants, with the highest losses occurring in cereals, vegetables and fruits. Although many of these diseases can be controlled through application of protective agro-chemicals, these chemicals are both expensive and environmentally hazardous. With recent improvements in transformation as a technique, agricultural scientists are increasingly turning to genetic engineering for rapid development of resistant varieties (Mourgues et al., 1998).

The two approaches being used for generating resistance to bacteria and fungi are

- The introduction of the resistance R gene/s effective against the pathogen; and
- The expression of genes that encode proteins with an inhibitory effect on the pathogen.

Expression of Cloned Resistance Genes

Since the past decade, the molecular basis of plant-pathogen interactions is being worked out in great detail and a number of resistance R genes that confer resistance against pathogens have been cloned. A majority of the R genes isolated encode proteins with common motifs—a carboxy terminal leucine rich repeat (LRR) region, which is probably involved in protein—protein interactions and an amino-terminal nucleotide-binding site (NBS). These R genes are now being transferred to susceptible crops and varieties. Some of these genes have proven their specificity on transfer not only within, but also across the species (Bent et al., 1994; Rommens et al., 1995; Tai et al., 1999).

Bs2 (Tai et al., 1999), a resistance gene from pepper, is one such durable resistance gene, which confers broad-spectrum resistance against different pathovars of the bacterial pathogen *Xanthomonas campestris*. The *Bs2* gene specifically recognizes and confers resistance to strains of *X. campestris* pv. *vesicatoria* that contain the corresponding bacterial avirulance gene, *avrBs2*. *avrBs2* is involved in pathogen fitness and is prevalent in many *X. campestris* pathovars, thereby rendering the *Bs2* gene durable in the field and providing resistance when introduced into other plant species. Introduction of *Bs2* into tomato resulted, as expected, in resistance against the bacterial spot disease caused by *X. campestris* pv. *vesicatoria* in tomato. This gene specifically controlled the hypersensitive response when transiently expressed in susceptible pepper and tomato lines, as well as in a non-host species, *Nicotiana benthamiana*. Thus, the expression of *Bs2* in stable transgenic tomatoes and tobacco suggests that it can be used as a source of resistance in other plant species as well (Tai et al., 1999).

Xa-21 is another such gene that confers resistance against bacterial blight caused by *X. oryzae* pv. *oryzae* in rice. Song et al., (1995) cloned the *Xa-21* gene; subsequently, this gene has been transformed into the improved rice variety IR72 in order to develop a variety of rice resistant to bacterial blight (Datta, 2000). This variety is now being deployed in China.

Production of Proteins with a Direct Inhibitory Effect on the Pathogen

The constitutive expression of genes encoding proteins which inhibit fungi and/or bacteria would be very effective in conferring resistance to the pathogen in the transgenic plant. Transformation of plants with genes encoding several of the inhibitory proteins has been attempted.

Chitinases and glucanases are hydrolic enzymes, which degrade the major cell-wall constituents of most filamentous fungi. Expression of these enzymes in transgenic plants has led to an enhanced resistance to fungi in the plants. Expression of plant or bacterial chitinase in transgenic tobacco have been effective in conferring resistance to *Rhizoctonia solani*, an endemic, chitinous soil-borne fungus that infects numerous plant species (Broglie et al., 1991). Small anti-fungal peptides, called defensins, also provide broad-spectrum resistance to fungal pathogens. The constitutive expression of *Rs-AFP*—a small plant defensin from radish— in tobacco conferred a high level of resistance to the pathogen *Alternaria longipes* (Terras et al., 1995) in the plants, which expressed high levels of the transgene.

Similarly, the expression of cloned anti-bacterial proteins of diverse origin has also been seen to confer resistance to bacterial diseases in plants (see review Mourgues et al., 1998). These proteins include certain lytic peptides, including the cecropins and attacins (isolated from the silk moth), which form pores in bacterial membranes thereby causing cell lysis and death. Synthetic analogues of these peptides have also been used to confer resistance to pathogens in crop plants. Introduction of a stable analog of cecropin, the *MB39*, into tobacco provides resistance to bacterial wilt and the transgenic plants show no necrosis even after infection by the *Pseudomonas syringae* pv. *tabaci*, the casual agent for bacterial wilts (Florack et al., 1995). The attacin E, when introduced into apple plants, reduces the plants' susceptibility to *Erwina amylovora*, the casual agent for fire blight (Norelli et al., 1994).

Lysozymes of diverse origin—from hen egg white, humans and T_4-bacteriophage with specific hydrolytic activity against the bacterial peptidoglycan (see review Mourgues et al., 1998) —have also been introduced into plants. Transgenic tobacco expressing hen-egg-lysozyme inhibits the growth of several species of bacteria, while potato plants expressing T_4-bacteriophage lysozyme show partial resistance to *E. carotovora* pv. *atrospetica*. Other antibacterial peptides including lactoferrin, tachylepsin, etc., have also been expressed in plants and have been seen to have a distinct anti-bacterial activity in the transgenes expressing them (Mitra and Zhang, 1994).

It has been observed that plants respond to infection by pathogens with a localized, transient, but massive generation of reactive oxygen species. This oxidative burst is accompanied by the accumulation of hydrogen peroxide, which triggers localized hypersensitive cell death and has a direct antimicrobial activity against the pathogens. At the same time, the hydrogen peroxide also acts as a diffusible signal for induction of defense genes in the surrounding tissues. This phenomenon has been used to induce resistance to bacterial infection in the potato plant. Transformation of potato plant with glucose oxidase gene from *Aspergillus niger* has resulted in the production of large amounts of hydrogen peroxide in the transgenic plants. The transgenic plants responded to *Erwinia carotovora* infection with increased expression of hydrogen peroxide and a concomitant increase in resistance to the bacterium (Wu et al., 1995).

As a rule, however, the transfer of individual defense genes confers only partial resistance to the transgenic plant. This is probably because, in a plant, the battery of defense reactions works synergistically against a pathogen. Consequently, the expression of a combination of heterologous, defense-related genes should be more promising to enhance the resistance levels in transgenic plants.

RESISTANCE TO PESTS

Naturally Occurring Resistance Genes

Gene-for-gene interactions have been well documented in the interaction of insect pests and their hosts—between the biotypes of raspberry aphid *Amphorophora idaei* (Born) and its host the raspberry (Briggs, 1965); the hessian fly, *Mayetiola destructor* (Say) and its wheat host (Hatchett and Gallun, 1970; Gallun, 1978); the gall midge, *Orseolia oryzae* (Wood-Mason) and its host, the rice plant (Bentur et al., 1992; Mohan et al., 1994, 1997b; Nair et al., 1995, 1996; Sardesai et al., 2001). It has been hypothesized that the product of the avirulence gene of the pest interacts with the resistance gene product of the host and thereby triggers the resistance response.

Although a large number of insect-resistance genes have been identified, only a few of them have been cloned to date. The only gene that confers resistance to insects and which has been isolated from plants so far is the tomato *Mi* gene. The *Mi* gene confers resistance to the potato aphid *Mi* gene *Macrosiphum euphorbiae* (Thomas), as well as the root-knot nematode in tomato. Introduction of the *Mi* gene into a susceptible tomato variety confers resistance to the potato aphid as well as to the root-knot nematode in the transgenic tomato plants (Rossi et al., 1998).

However, most of the plant resistance genes are highly specific not only to the particular pest, but also to specific biotypes of the pest. For example, certain varieties of indica rice show resistance to the gall midge, an insect pest that causes extensive economic damage in many parts of Southeast Asia and Africa. At least seven different non-allelic gall midge resistance genes have been identified so far, while 13 biotypes of the gall midge are known to exist (Sardesai et al., 2001). Each of these resistance genes provides resistance against a different set of biotypes of the insect. While the gall midge resistance gene *Gm1* provides resistance against the biotypes 1, 3 and 5, *Gm2* shows resistance to the biotypes 1, 2 and 5 and *gm3* to biotypes 1, 2, 3 and 4. The *Gm4* gene provides resistance to biotypes 1, 2, 3, 4 and *Gm5* to biotypes 1, 2 and 5. The *Gm6* gene from the cultivar Duokang # 1 confers resistance to biotypes of gall midge identified from China (Tan et al., 1993). Thus, the different gall midge resistance genes provide resistance against different sets of biotypes of the insect (Bentur and Amudhan, 1996). Deployment of only one of these genes would provide resistance only to the biotypes it is specific to, while the other biotypes would still be capable of damaging the crop. Hence, it would obviously be more prudent to deploy either more than one gene or deploy genes that provide a broad spectrum of resistance for a wider range of pests.

The most widely-used genes for the engineering of insect-resistant plants are those of the Bt toxins—which are insecticidal proteins from the bacteria *B. thuringiensis* (Bt) and *B. sphaericus* (Bs) (Charles et al., 1996; Gill et al., 1992). Other proteins like protease inhibitors, plant lectins, secondary plant metabolites; vegetative insecticidal proteins from Bt and related species, etc.—either alone or in combination—are also being used to generate insect-resistant plants (Hilder and Boulter, 1999). Many classes of plant proteins and secondary plant substances have also been shown to have a toxic or anti-metabolic effect on insects and have been proposed as possible candidates for engineering insect-resistant plants. Since plant derived gene products attack different sites in insects, they may be deployed in combination with exotic genes and insecticides, thereby enhancing the range and durability of the resistance genes.

Bacillus thuringiensis Toxin Genes

The bacterium Bt is a common spore-forming Gram-positive soil bacterium that produces a crystalline parasporal inclusion body. These crystals are responsible for the insecticidal activity of Bt (Hannay, 1953) and have been used to generate insect-resistant transgenic plants. The Bt produces several type of toxins, of which the δ-endotoxins are the most important. The δ-endotoxins are protoxins that are inactive until they are solubulized by the gut proteases (Tojo and Aizawa, 1983; Gill et al., 1992; Milne and Kaplan, 1993). Under the alkaline conditions of the insect midgut, the protoxins are proteolytically cleaved, releasing proteins of 65,000 to 1,60,000 KDa. Gut proteases further cleave these fragments to smaller toxic fragments that bind to specific receptors located on the apical brush border membrane of the columnar epithelial cells. The toxin molecule inserts itself irreversibly into the plasma membrane of the cells, thereby forming an ion channel, which allows a rapid flux of ions across the membrane. This results in midgut necrosis and death of the insect. The extent of toxicity to the larva correlates with the receptor number on the cells (Van Rie et al., 1989). The δ-endotoxins are not toxic to humans and other mammals, since the conditions necessary for the production of the active toxin fragments are not present in the guts of these animals (Gatehouse et al., 1991).

Genes for most of the crystal proteins have been cloned and their DNA sequences determined. About 140 genes coding for the δ-endotoxins have been described (Crickmore et al., 1998). Each of these genes encodes for proteins with a different specificity to the insects they are effective against. Studies on the mode-of-action of the Bt crystal proteins reveals that different crystal protein classes bind to different receptors present on the membrane of the midgut epithelial cells (Hofmann et al., 1988). The ability of the crystal (*Cry*) endotoxins to bind to the receptors determines the species sensitivity to the various toxins.

The first transgenic plants, produced in 1987, were tobacco plants carrying the Bt gene *Cry1A*, under the control of constitutive promoters. Their level of expression was quite low and resulted in 20% mortality of the tobacco hornworm, *Manduca sexta* (Johannsen) larvae (Vaeck et al., 1987). Many different Bt genes have been experimented with since then. Plants transformed with constructs containing the entire protoxin coding sequences were found to express Bt toxin at very low levels. Levels of Bt toxin expressed were higher in plants transformed with constructs containing truncated Bt toxin coding sequences. Subsequently, improved version of Bt genes were developed by replacing the bacterial codons with plant preferred codons in order to improve its translation (Perlak et al., 1991).

Bt toxins are now cloned in constructs containing promoters that allow expression of the toxin in high quantities in specific tissues (Koziel et al., 1993). Cotton plants transformed with *Cry1Ab* or *Cry1Ac* genes under the control of powerful promoters—CaMV35S (cauliflower mosaic virus 35S promoter) with duplicated enhancers—and sequence modifications in the areas of the gene with mRNA secondary structures showed a hundredfold increase in the levels of expression. The Bt toxin constituted upto 0.1% of the soluble proteins in these plants and the plants showed a very high degree of resistance to the cotton bollworm and pink bollworm (Perlak et al., 1990; Wilson et al., 1992).

Transformation of chloroplasts is another way of achieving high levels of expression of the toxins. Wong et al. (1992) transformed tobacco plants with *Cry1Ac* placed under the control of *Arabidopsis thaliana* small subunit promoter with its chloroplast transit peptide sequence. This increased the expression levels to 10-20 fold that of the gene placed under the CaMV35S promoter with the duplicated enhancer region. Bt toxin production was nearly 1% of the total leaf protein in the transgenic tobacco plants and these plants showed complete resistance to lepidopteran pests. Transgenic tobacco leaves expressing the *Cry2Aa2* protoxin in chloroplasts, when fed to susceptible, *Cry1A* resistant and *Cry2Aa2* resistant tobacco budworm, *Heliothis virescens* (Fabricius); cotton bollworm, *Helicoverpa zea* (Boddie), and the beet armyworm, *Spodoptera exigua* (Hubner), were able to elicit 100% mortality in all these insect species and strains (Kota et al., 1999). Transformed tobacco leaves express *Cry2Aa2* protoxin at levels between 2 and 3% of the total soluble protein 20-30 fold higher levels than the current commercial nuclear transgenic plants. These results suggest that plants expressing high levels of a nonhomologous Bt protein should be able to overcome, or at the very least, significantly delay broad-spectrum Bt resistance development in the field. The expression of Bt in chloroplasts also circumvents the need to modify the Bt genes since chloroplast transcriptional and translational apparatus are prokaryotic and it is possible to have many copies of the Bt gene in the cells. Moreover, chloroplasts being maternally inherited, there is no risk of transfer of the Bt genes through pollen to the related plants.

Tissue specific regulation of Bt *Cry1Ab* has also been achieved in leaves and pollen grains of maize using promoters from PEPC (Phospho Enol Pyruvate Carboxylase) that controls the expression of the gene in green tissue (Hudspeth and Grula, 1989); and the Calcium Dependent protein Kinase (CDPK) gene in pollens (Estruch et al., 1994). A combination of green tissue specific PEPC and pollen specific CDPK tissue promoters provides high *Cry1Ab* gene expression in leaves and pollen, where it is the most effective in controlling the European corn borer.

Rice transformed with the Bt toxins has also been developed (Fujimoto et al., 1993; Nayak et al., 1997). The transformed rice produce 0.05% toxin of the total soluble leaf protein and shows resistance to the rice leaf folder and yellow stem borer. These represent only a few examples of transgenics generated using the *Cry* genes. A large number of crops transformed with a variety of *Cry* genes have now been generated. Tobacco and potato plants transformed with synthetic *Cry III* genes show resistance to the Colorado potato beetle *Leptinotarsa decemlineata* (Say) (Perlak et al.,1993); tobacco and tomato plants expressing *Cry1Ab* and *Cry1Ac* genes show resistance to lepidopteran pests (van der Salm et al., 1994). *Cry 1Ac* introduced in groundnut provides resistance to corn stalkborer (Singsit et al., 1997); chickpea cultivars transformed with *Cry 1A(c)* shows resistance to *Helicoverpa armigera* (Hubner); potato transformed with *Cry V* provides protection against the potato tuber moth (Mohammed et al., 2000).

However, insects are known to develop resistance to the Bt toxins within a few generations of exposure to them. The Bt cotton showed very high levels of resistance against the bollworm larvae. But with each generation of larvae selected on Bt cotton, the resistance of the plant to the bollworm reduced, indicating a potential problem of development of the resistance in insects to Bt cotton. Laboratory experiments in which bollworm larvae were fed on Bt cotton flowers and bolls showed a 7.1 fold increase in resistance to Bt in the bollworm after 17 generations of selection on the transgenic plant tissues.

A number of Bt proteins do possess some insecticidal activity against specific insects, but the quantities needed for effective control are too high to be attained in transgenic plants (Estruch et al., 1996). Moreover, some economically-important insect pests such as northern and western corn rootworm, the black cutworm and boll weevil are not effectively controlled by any of the known Bt endotoxins. Under these circumstances, only other insecticidal proteins can provide resistance.

Vegetative Insecticidal Proteins

Bacillus thuringiensis and *B. cereus* release certain proteins that can be collected from the supernatant of their cultures (Estruch et al., 1996, 1997). Unlike the Bt endotoxins which are produced during sporulation, these proteins are produced and secreted during the vegetative stages and continue to be produced in sporulating cultures. Since they are produced in the vegetative phase and are also highly toxic to certain insects, they are called the vegetative insecticidal proteins or *Vip* proteins. The *Vip* proteins bind to the columnar epithelial cells of the insect midgut and bring about a complete lysis of the gut epithelial cells. Gut paralysis and larval death follows. The level of insecticidal activity exhibited by the *Vip* proteins is comparable to that exerted by the δ-endotoxins.

A binary system of two vegetative insecticidal proteins, *Vip1* and *Vip2*, isolated from certain *B. cereus* isolates, possesses acute activity against northern and western corn rootworms against which the Bt toxins are ineffective (Warren, 1996). Similarly, certain proteins in the supernatant of some of the Bt cultures have potent insecticidal properties against the black cutworm and fall armyworm, against which the Bt toxins have been proved to be ineffective. *Vip3A(a)* isolated from Bt strain AB88 and *Vip3(b)* from the strain AB424 exhibit a wide spectrum of insecticidal activity against lepidopteran insects, particularly black cutworm, fall armyworm, beet armyworm, tobacco budworm and corn earworm (Estruch et al., 1996), which are resistant to the Bt toxins.

Thus, the *Vip* proteins are effective insecticides and are especially useful for the control of many of the insects that are resistant to the Bt toxins. These proteins could be used to complement and extend the use of the Bt toxins.

Protease Inhibitors

One of the most effective natural defense mechanisms against herbivorous insects is the synthesis of protease inhibitors, which have an antimetabolic activity against a wide range of insects. Certain insects like the lepidopterans depend on serine proteases—trypsin, chymotrypsin, elastase, while others rely on thiol proteases for the digestion of proteins. Inhibition of proteases by serine protease inhibitors (SPIs) or thiol protease inhibitors (TPIs) is an effective method of controlling these insects by disrupting their ability to digest proteins. The introduction of protease inhibitors into plants by transformation has been an effective alternative approach for obtaining crops that are resistant to insect attack (Sharma et al., 2000).

Expression of SPIs in transgenic plants has been used as a means of controlling certain insects (Wasmann et al., 1994). Thomas et al. (1995) observed that transgenic cotton plants expressing anti-trypsin, anti-chymotrypsin and anti-elastase protease inhibitor (PI) genes under the CaMV35S promoter provide protection against the sweet potato whitefly, *Bemisia tabaci* (Gennadius). The cowpea trypsin inhibitors (*CpTI*) provide resistance to the bruchid beetle *Callosobruchus maculatus* (Fabricius) in cowpeas and have toxicity to a range of insect genera, including *Heliothis, Spodoptera, Diabrotica* and *Tribolium*, all of which cause significant economic losses (Sharma et al., 2000). Tobacco transformed with the *CpTI* were found to be highly resistant to lepidopteran tobacco budworm, *H. virescens*, causing only limited damage to the plants, while untransformed plants were almost completely devoured by these larvae (Hilder et al., 1987). The Soybean Kunitz trypsin inhibitor also confers resistance to brown planthopper, *Chilo suppressalis* (walker) (Lee et al., 1999), while the *CpTI* confers resistance to *C. suppressalis* and *Sesamia inferens* (Walker) (Xu et al., 1996). Expression of the potato trypsin inhibitor gene in rice confers resistance to the pink stem borer, *S. inferens* in the transgenic plants (Duan et al., 1996). TPIs have also been effective in controlling the maize rootworm, against which Bt proteins are not effective (Edmonds et al., 1996). Vain et al. (1998) transformed rice with a gene coding for an engineered cysteine protease inhibitor oryzacystatin-I delta D86, *OC-I* and found a significant level of reduction (55%) in the egg production by *Meliodogyne incognita*. These experiments suggest that the protease inhibitors are effective against the insects in the transgenic crop expressing them.

Protease inhibitors are also effective against a wide range of insects, including those that are resistant to the Bt endotoxin. But very high levels of the inhibitor are required for killing the insects. A major advantage of using protease inhibitors is that even large amounts of the inhibitors are easily inactivated during cooking, thereby preventing them from being harmful to humans.

Although several PI genes have been introduced into different crops, it has been observed that many species compensate for protease inhibitors by switching on to alternative proteolytic activity or by overproducing the protease. Cloutier et al. (2000) transformed a potato cultivar with Oryzacystatin I (*OC-I*), which confers resistance to the Colorado potato beetle. However, they observed that the potato beetle fed on the transgenic leaves expressing *OC-I* reacted rapidly to the presence of *OC-I* by producing *OC-I* insensitive protease and ingesting 2.4-2.5 times more *OC-I* rich foliage, apparently, as a compensatory response for nutritional stress due to the protease inhibitor in their diet. The poplar cultivar Jean Purlet, transformed with Kunitz proteinase inhibitor (*Kpi3*) gene, also did not show any increase in resistance to *Lymantria dispar* and *Clostera anastomosis* (Confalonieri et al., 1998) as compared to the untransformed controls. Moreover, many *SPIs* are toxic to beneficial insects such as honeybees (Malone et al., 1995; Burgess et al., 1996). A prudent use of *PIs* in combination with other insecticidal genes would probably give the best protection against a range of insects.

α-Amylase Inhibitors

Another method for controlling the insects could be by targeting their carbohydrate metabolism through the use of amylase inhibitors, which inhibit the larval digestive amylases. Carbonero et al. (1993) reported that transgenic tobacco expressing amylase inhibitors derived from wheat (Wheat α-Amylase Inhibitor—WAAI) increased the mortality of lepidopteran larvae by 30-40%. Similarly, transgenic pea seeds expressing α-amylase inhibitor derived from common beans were found to exhibit increased resistance to bruchid beetles and pea weevil (Shade et al., 1994; Schroeder et al., 1995). Ishimoto et al. (1999) observed enhanced levels of resistance to the cowpea weevil and adzuki bean weevil in transgenic adzuki beans expressing the α-amylase inhibitor of common bean. While even low levels of the amylase inhibitor was sufficient for providing resistance to the adzuki bean weevil, higher levels of the protein make the seeds resistant to the cowpea weevil, *C. maculatus* and the pea weevil, *C. chinensis* as well. Morton et al. (2000) have recently demonstrated that transgenic peas containing a bean α-amylase, α-AI-I showed complete protection from the pea weevil under field conditions.

Chitinases

Chitinases are enzymes that digest the chitin present in structures like the peritrophic membrane, which protects the delicate midgut cells of the insect gut lumen. It has long been thought that the chitinases could be used effectively against insects. Transgenic tobacco plants expressing chitinase have been shown to increase resistance to lepidopteran insects (Gatehouse, 1995) and to the tobacco budworm (Ding et al., 1998). Insects, however, continuously regenerate this membrane and hence, very large amounts of chitinase are required for an effective use as an insecticide.

Plant Lectins

Plant lectins are a group of carbohydrate-binding protenis that bind to the glycosylated proteins of the insect midgut, thereby affecting the survival and development of insect pests of different orders (Shukle and Murdock, 1983; Czapla and Lang, 1990; Habibi et al., 1993; Powell et al., 1993,1995; Gatehouse, 1995). Lectins have a low level of toxicity to mammals, rendering them safe in controlling insects. However, they afford good insecticidal properties only when the insects are exposed to microgram levels of the lectin in their diet. Lectins from snowdrop, pea, wheat, rice, castor, soybean, mungbean, garlic, sweet-potato, tobacco, chickpea and groundnut have all been isolated and characterized and some of them have been used for generating resistance to insects in plants. The sap-sucking Hemiptera are especially susceptible to certain lectins. The snowdrop lectin *GNA* has been the most frequently-used lectin for the generation of transgenics resistant to sap-sucking insects. Fitches et al. (1997) observed that the mean larval biomass of the tomato moth *Lacanobia oleracea* (Linnaeus) was reduced by 23-32% on being fed on leaves of transgenic tomato plants expressing *GNA* for three weeks. Zhao et al. (1998) generated transgenic tobacco expressing snowdrop lectin and observed upto 90% reduction in population of the peach-aphid infesting these transgenic plants. Rao et al. (1998) similarly observed, for the first time in a cereal crop, that transgenic rice plants expressing *GNA* exhibited a significant resistance to infestation by the rice brown plant hopper. *GNA* decreased the rate of survival of the insects and retarded their development. Transgenic sugarcane plants engineered to express the snowdrop lectin gene also had a significant detrimental effect on the sugarcane grubs, *Antitrogus consanguineus*, feeding on it (Nutt et al., 1999). Thus, *GNA* can be used quite effectively for the development of insect-resistant crops.

Concanavalin A (*Con A*) is another lectin, which is effective against insects. Transgenic potato plants expressing *Con A* inhibit the development of tomato and peach potato aphid (Cao et al., 1999). Ingestion of *Con A* retards larval development and decreases the larval weight but does not show any significant effect on the survival. Similarly, transgenic cotton expressing *Con A* retards the growth of the cotton budworm larvae, but does not result in larval mortality (Satyendra et al., 1998). Thus, *Con A* appears to have a larvistatic, but not larvicidal effect and may not be as effective as *GNA* in its effect on insects.

Cholesterol Oxidases

The boll weevil is a major cotton pest against which conventional insecticides have no effect (Estruch et al., 1996). A random screen of filtrates from microbial fermentation uncovered an association between sterol oxidases and insects. A protein present in two *Streptomyces* culture filtrates showed acute toxicity to the cotton boll weevil. This protein was found to be a cholesterol oxidase (*CO*) with a highly toxic effect on the first and second instar larvae of the boll weevil, comparable to that of Bt toxins on Bt susceptible insects (Purcell et al., 1993). Cholesterol is essential for the integrity and function of all cell membranes and any interference in its availability for incorporation into and maintenance of cell integrity would lead to toxicity. *CO* catalyses the oxidation of cholesterol in order to produce ketosteroids and hydrogen peroxide, thereby making the cholesterol unavailable for incorporation into and maintenance of cellular integrity. Thus, the *CO* compromises the integrity of the insect membranes, thereby, leading to cell lysis and death.

Secondary Metabolites

A number of secondary metabolites, called phytoalexins, are produced in response to insect feeding, infection by pathogens and abiotic stress factors (Sharma and Agarwal, 1983; Sharma and Norris, 1991; Sharma et al., 2000). Alkaloids, steroids, foliar phenolic esters (rutin, chlorogenic acid, etc.), terpenoids, cyanogenic glycosides, glucosinolates, saponins, flavonoids, pyrethrins and *non-protein* amino acids are some *secondary* metabolites which act as potent protective chemicals and are produced by plants to ward off insects and other pests. Manipulation of the key genes involved in the synthesis of secondary metabolites could lead to the enhanced production of many insecticides. However, Vrieling et al. (1991) observed that manipulation of components of secondary metabolic pathways reduces the yield of the host plants. The kind of effect observed is unlike that of the natural protection mechanisms based directly on protective proteins (Brown, 1988) where expression of large amounts of foreign proteins like *CpTI* does not impose a cost in the yield in the transgenics (Hilder and Gatehouse, 1991) and most transgenics are phenotypically normal. The reduced yields are a major constraint in the utilization of secondary metabolites to provide protection against the insects and other pests.

Thus, the choice of insecticidal genes that could be used to control insects are quite wide and are increasing daily as new classes of insecticidal genes are being discovered and exploited. Genes conferring resistance to insects have been inserted in a wide range of crops including maize, cotton, potato, tobacco, rice, broccoli, lettuce, walnuts, apples, alfalfa and soybean. However, as has been reiterated in the earlier sections, insects and other pathogens have a high capacity to develop resistance. A combination of different approaches is needed to combat the development of resistance to the genes deployed. Transgenic plants producing very high levels of insecticidal proteins are needed in order to eliminate the heterozygotes carrying a resistance allele. Plants producing additional insecticidal proteins with either different modes of action, different targets or both would be ideal to prevent the occurrence of resistance in insects. Since Bt toxins have already been shown to be effective in providing resistance to insects, they would probably continue to be a part of the transgenic approach to combat pests. In addition, proteins of the *Vip* family, protease inhibitors, amylase inhibitors, cholesterol oxidases, lectins, etc., could be used to complement these Bt toxins. Multiple resistance genes with differing modes of action could simultaneously be transferred into plants, thereby, enhancing the durability of resistance. The observation (Parniske et al., 1997) that tightly-linked *R* genes could act synergistically could be exploited and clusters of *R* genes that provide resistance against insects could be introduced in a single block.

GENE MANIPULATION FOR VIRAL CONTROL

In spite of the fact that the viral diseases of plants constitute a major economic problem, there are no effective chemical methods to control viral diseases in plants. Prevention of viral disease through use of resistant cultivars is the only method in this endeavour.

Transformation with Viral Resistance Genes

Certain crop plants have a natural resistance to some of the viruses, while there are no resistance genes available in the germplasm against most viruses. The isolation and subsequent transfer of the viral resistance genes from resistant to susceptible plant species is an effective means of controlling important viral diseases. For example, the *N* gene of tobacco confers resistance to the viral pathogen, tobacco mosaic virus (TMV) in a classical gene-for-gene interaction. It confers resistance to TMV by mediating defense responses that function to limit viral replication and movement. Whitman et al. (1996) generated transgenic tomato plants bearing the *N* gene and demonstrated that the *N* gene confers a hypersensitive response and effectively localizes TMV to sites of inoculation in transgenic tomato, as it does in tobacco. The ability to reconstruct the *N*-mediated resistance response to TMV in tomato demonstrates the utility of using isolated resistance genes from one crop for protecting other crop plants from the same virus/es (Erickson et al., 1999).

However, since most viruses do not have corresponding resistance genes in the germplasm, other methods of protecting the plants against viruses have been developed. Three strategies have been attempted for protection against the viruses, viz.—(i) pathogen-derived resistance; (ii) use of antisense RNA; and (iii) satellite RNA-mediated resistance.

Pathogen-Derived Resistance in Transgenic Plants

Introduction of pathogen (in this case, virus) derived proteins or nucleic acids into the host plants elicits a resistance response in the host. Viral-pathogen derived resistance (PDR) could be invoked either through the expression of viral proteins or through the accumulation of viral nucleic acid sequence in the transgenic plant. Genes that confer viral PDR include those for coat proteins, replicases. movement proteins, defective interfering RNAs and DNA, and non-translated RNAs. The viral-derived proteins generally provide high levels of resistance to a broad range of viral strains and viruses, while the viral nucleic acids provide very high levels of resistance to a specific virus strain (Beachy, 1997).

Coat protein (*CP*) genes of a virus, when introduced into a plant by way of transformation have been observed to provide resistance to the virus. *CP* mediated resistance (CPMR) probably acts by interfering with the disassembly of the virus, thereby preventing infection (Clark et al., 1995). There is a direct correlation between the amount of *CP* expressed and the level of resistance obtained (Powell et al., 1990). Expression of a single virus *CP* may also provide resistance to other viruses, the extent of resistance conferred depending upon the degree of similarity in the *CP* sequence of the two viruses. The *CP* of the TMV not only provides resistance to TMV itself (Powell et al., 1986, 1990), but also provides effective levels of resistance to closely-related strains of TMV and decreasing levels of resistance to tobamoviruses that share less CP sequence similarity (Nejidat and Beachy, 1990). Transgenic plants expressing Alfalfa Mosaic Virus (AIMV) CP are not only resistant to AIMV, but are also resistant to low concentrations of Potato Virus X (PVX) and Cucumber Mosaic Virus (CMV) (Anderson et al., 1989). Coat protein genes isolated from different RNA plant viruses have been transformed into various crop plants, including tobacco, tomato, rice (Hayakawa et al., 1992), potato, citrus spp, papaya, etc. (Horsch et al., 1985; Powell et al., 1990). For reviews on CPMR, see Miller and Hemenway (1998) and Beachy (1999).

Genes that encode complete or partial replicase proteins also confer a very high degree of immunity to viral infection. This immunity is, however, generally limited to the virus strain from which the gene sequence is obtained (Beachy, 1997). Golemboski et al. (1990) observed that transgenic tobacco plants containing a sequence that encodes a 54 kDa fragment of the TMV replicase were resistant to TMV even at an inoculum concentration of 100 mg/ml. These plants, however, did not show any resistance either to the CMV or to the related strains of TMV. The truncated mutant of replicase derived from a CMV subgroup I virus confers high levels of resistance in tobacco plants to all subgroup I CMV strains, but not to subgroup II strains or other viruses (Zaitlin et al., 1994). Thus, the replicase-mediated resistance is highly specific—with only a few exceptions—to the strain from which the replicase is derived. However, transformation of rice with fragments of the Rice Yellow Mottled Virus (RYMV) genome, which codes for the RNA dependent RNA polymerase of RYMV, conferred resistance to a number of strains of RYMV (Pinto et al., 1999).

Movement proteins are encoded by plant viruses and enable infections to spread between adjacent cells as well as through the plasmodesmata channels that traverse plant cell walls and provide a continuity between cells and tissues (Carrington et al., 1996). Transgenic plants carrying defective mutants of viral movement proteins inhibit the spread of the virus. Transgenic plants that contain the defective movement protein from TMV show resistance to several tobamoviruses as well as to AIMV, CaMV and other viruses (Lapidot et al., 1993; Cooper et al., 1995). Similarly, a mutation that disrupts one of the three movement proteins of white clover mottle virus confers different degrees of resistance to several potexviruses (Beck et al., 1994). They include the white clover mottle virus itself, Potato virus X, Narcissus mosaic virus, and the Carlavirus potato virus S, but not to tobacco mosaic virus of the tobamovirus group.

Transgenic potato plants expressing the mutant alleles for the movement protein of the Potato Leaf Roll Virus (PLRV) provide resistance not only to this luteovirus, but also provide a broad-spectrum resistance to the unrelated potyviruses PVY and PVX (Tacke et al., 1996). Transgenic tobacco plants expressing RNA sequences of the tomato spotted wilt virus *NSm* gene, which encodes the putative viral movement protein, were found to be highly resistant to infection with the virus (Prins et al., 1997). Thus, viral resistance can clearly be engineered into transgenic plants by expression of mutant forms of the movement proteins.

Use of Antisense RNA

The second approach to inhibit the infection of plants by viruses involves the use of antisense RNA. Antisense RNA binds to sense RNA and prevents its translation. Virus replication, packaging and transmission are all inhibited in the transgenic plants encoding an antisense strand to a specific viral sequence. The antisense RNA could be directed against different parts of the virus. It could also be directed against the mRNA for the viral coat protein RNA, the replicase or the movement proteins. One of the most severe diseases of cultivated tomato worldwide is caused by tomato yellow leaf curl virus (TYLCV), a geminivirus transmitted by the whitefly *B. tabaci*. Trasngenic *Nicotiana benthamiana* plants expressing antisense RNA for *C1*, a gene encoding the Rep protein of TYLCV, were seen to resist infection by TYLCV. Some of the resistant lines were symptomless, and the replication of challenging TYLCV almost completely suppressed in these lines (Bendahmane and Gronenborn, 1997). Similarly, the introduction of a 660bp sequence, including the C-terminal portion of the bean yellow mosaic potyvirus (BYMV) coat protein gene to produce antisense RNA in transgenic *Nicotiana benthamiana*, provided effective resistance to BYMV infection in transgenic plants even at the highest inoculum concentration (100 mg/ml). However, these plants were not resistant to other potyviruses such as pepper mottle or turnip mosaic potyvirus (Hammond and Kamo, 1995). Thus, antisense RNAs directed against many different viruses have been incorporated and expressed in different plants, each of them providing a high degree of resistance against the specific virus, but differing degrees of resistance against related viruses.

Satellite RNA Mediated Viral Resistance

Some plant RNA viruses harbour certain parasitic RNA sequences, which are replicated and packaged normally, bear no sequence homology to the main genomic RNA and are not required for viral replication and spread. These sequences are termed as satellite RNAs. These RNAs are dependent on the factors encoded by the viral genome for their replication. If the cDNA corresponding to satellite RNA is introduced into a plant, infection by the host virus of which the satellite RNA is a parasite evokes protection against the virus in the host plant and inhibits symptom information. The strategy, therefore, is to introduce DNA copies of satellite RNA into susceptible plants, which on transcription, inhibit symptom formation. It has been observed that even when the satellite RNA is under the control of CaMV35S promoter, satellite RNA replication is induced in response to infection by its host virus (Harrison et al., 1987). Gerlach et al. (1987) observed that transcription of the tobacco ringspot virus (TobRV) satellite RNA introduced into the tobacco plants was induced dramatically by infection with the TobRV, provided the cDNA comprised a trimer of the monomeric satellite. They observed a reduction in the level of replication of the TobRV with a concomitant reduction of ringspot symptoms in the transgenic plants. However, a problem with the satellite RNA approach is that the sequences that are protective in one species can be virulent in others or can even mutate to a form that is virulent in the originally-protected species.

Thus, a variety of strategies have been used to develop viral resistance in crop plants, many of which have been tested under field conditions for eventual commercial release. Of these, Coat Protein Mediated Resistance has been the most widely-used strategy for providing viral resistance to plants. Genetically-engineered resistance to virus infection by expression of viral capsid protein in transgenic plants has been demonstrated for TMV (Beachy, 1999), AIMV (Tumer et al., 1987; vanDun and Bol, 1988; vanDun et al., 1988), PVX (Hemenway et al., 1988; Hoekema, 1989), TSV (vanDun et al., 1988) and TRV (vanDun and Bol, 1988; Angenent et al., 1990). However, since all the methods used for generating viral resistance are, most often than not, specific to one or two viruses, resistance to more viruses requires manipulation for resistance to each virus. Potato plants resistant to both Potato Virus X and Potato Virus Y have been produced by transformation with the *CP* of PVX and PVY within the same plant (Kaniewski et al., 1990; Lawson et al., 1990).

GENE MANIPULATION FOR HERBICIDE RESISTANCE

Herbicides have become an indispensable tool of modern agriculture and are now widely used in weed control. They generally act by inactivating the target proteins essential for vital functions such as photosynthetic or plant biosynthetic pathways, most often in a non-selective manner. The best use of herbicides could be made if the crops were to be conferred resistance to broad-spectrum herbicides so that it would become possible to select the most suitable herbicide from a range of compounds available. Resistance to herbicides is generally conferred in three ways:

Over-Expression of Target Proteins

Over-expression of the protein, which is the target for the herbicide, ensures that a portion of the total amount of target protein synthesized remains unaffected by the herbicide, thereby making the plant tolerant to the herbicide. This principle has been used in the development of many herbicide-resistant crops. The best example is the production of transgenic plants with resistance to glyphosate. Glyphosate (Roundup from Monsanto) is a very popular and potent broad-spectrum non-selective herbicide. It works by competitive inhibition of the enzyme 5-enol-pyruvyl shikmate-3-phosphate synthase (EPSPS), which catalyzes an essential step in the aromatic amino acid biosynthetic pathway. Shah et al. (1986) transformed petunia plants by transformation with a cDNA for EPSP synthase under the control of the powerful CaMV35S promoter. The transgenic plants expressing EPSP synthase were tolerant to glyphosate and could tolerate upto 0.8 lb/acre of roundup herbicide, this being 2-4 times the level required to kill wild type plants. Subsequently, Daniell et al., (1998) introduced the petunia *EPSPS* gene into the tobacco chloroplast genome, in such a manner that 5000-10,000 copies of the gene were present per cell of the transgenic plants. While control plants were extremely sensitive to glyphosate, transgenic plants survived sprays of high concentration of glyphosate. Similarly, the over-expression of the *Arabidopsis* protoporphyrinogen oxidase, introduced into the tobacco chloroplast genome under the control of the CaMV35S promoter, neutralizes the action of the herbicide acifluorfen which acts to inhibit the enzyme (Lermontova and Grimm, 2000).

Transformation with Mutant Target Proteins

The second method of achieving resistance to any herbicide is to mutate the target protein in such a way as to reduce its binding to the inhibitor, without affecting its function. Genetically-engineered resistance to several herbicides has been achieved using genes coding for altered proteins, which substitute for their inactivated counterpart in the transgene. Mutant EPSP synthases with increased resistance to glyphosate introduced into tobacco plants confer a high degree of resistance to glyphosate in the transgenic plants (Comai et al., 1983; Della-Cioppa et al., 1987). Acetohydroxy acid synthase (AHAS) is an essential enzyme that catlyzes the first step in the biosynthesis of the branched-chain amino acids valine, isoleucine, and leucine. Several classes of herbicides including sulfonylureas, imidazolinones and triazolopyrimidines that are believed to bind to a common quinone-binding site inhibit it. A mutant AHAS gene with a single amino acid change within a conserved region of AHAS alters the binding-site of the herbicides, and when expressed in transgenic plants, confers a strong resistance to all three classes of herbicides (Hattori et al., 1995).

Detoxification of Herbicides

The introduction of herbicide-detoxifying enzymes—whose product converts the herbicide to a non-phytotoxic form—to produce herbicide-resistant plants has now become a very popular method of generating resistance to herbicides. The first use of detoxifying gene was for the development of bialophos-resistant plants. Bialophos is a non-selective herbicide that consists of phosphoinothricin (PPT) and two L-alanine residues. DeBlock et al. (1987) developed-bialophos resistant plants by the transfer of the *bar* gene coding for phosphoinothricin acetyl transferase (PAT) which converts PPT into a non-herbicidal acetylated form. The resultant transgenics were highly resistant to bialophos. The *bar* gene has, since, been introduced into a wide variety of crop plants and has been seen to provide resistance to high doses of PPT and bialophos in all these varieties. In fact, the *bar* gene is now being used as a selectable marker in routine transformation experiments, wherein transformants are selected on a medium containing bialophos.

Thus, herbicide resistance can be engineered into plants through over-expression of the target protein, expression of mutant proteins or through the introduction of detoxification genes. Herbicide-resistance genes are now routinely being used as a selectable marker during transformation, along with other genes of interest, effectively pyramiding herbicide resistance along with genes for other traits.

RESISTANCE GENES PYRAMIDED BY TRANSFORMATION

Many of the genes used to generate resistant crops have been either too specific to the targeted stressing agent or are not very effective against them. Moreover, pests and pathogens against which they are deployed develop resistance to the genes within a few generations of exposure to them. In order to increase the potency of the resistance in the transgenes, scientists have attempted to pyramid two or more resistance genes into the same plant. Hadi et al. (1996) have already proved the feasibility of transformation with multiple plasmids carrying a number of different genes. They have been successful in transforming soybean with twelve different plasmids, all of which were integrated into at least some of the lines generated, thus proving the possibility of pyramiding a number of genes through transformation.

Several attempts have been made to pyramid insect-resistance genes with differing modes of action. Zhao et al. (1998) generated transgenic tobacco plants expressing the Bt gene *CryIA* and the cowpea trypsin inhibitor gene *CpTI*. Comparisons of these transgenic plants to ones containing Bt alone and non-transgenic controls for their insecticidal activity against *H. armigera* larvae showed that the insecticidal action of the transgenic plants expressing both genes was significantly higher than that of the plants expressing the Bt gene alone. Moreover, even after 11 generations of selection of *H. armigera* for resistance to *Cry*1A and *Cp*TI, there was significantly less resistance to the insecticidal proteins in the larvae selected on the double transgenics than those selected on Bt plants.

MacIntosh et al. (1990) were the first to show that the activity of Bt in transgenic plants could be enhanced by serine proteases. Tobacco plants expressing *Bt* var *kurstaki* HD-73 δ-endotoxin or cowpea trypsin inhibitor (*CpTI*) have been shown to be much more effective against *Helicoverpa zea* rather than transgenic plants containing Bt gene alone (Hoffmann et al., 1992). Cornu et al. (1996) reported that transgenic poplars expressing proteinase inhibitor and *CryIIIA* genes exhibited reduced larval growth, altered development and increased mortality as compared to the controls.

There have also been successful attempts of pyramiding different Bt toxin genes in order to increase the durability and range of resistance. Rice plants of nine different varieties have been transformed with plant codon optimized versions of two synthetic *CryIAb* and *CryIAc* coding sequences from Bt (Cheng et al., 1998). These sequences have been placed under the control of the maize ubiquitin promoter, the CaMV35S and the *Brassica Bp10* gene promoter so as to achieve high tissue-specific expression of the lepidopteran-specific δ-endotoxins. Accumulations of high levels (upto 3% of soluble protein) of *Cry*IAb and *Cry*IAc proteins have been detected in the transgenic plants. Bioassays with transgenic plants indicate that the transgenic plants are highly toxic to two major rice insect pests—striped stem borer, *C. suppressalis* and yellow stem borer, *S. incertulas*, with mortalities of 97-100% within 5 days of infestation, thus offering a potential for effective insect resistance in transgenic rice plants.

Transgenic potato plants containing *CryIAb* gene *Bt884* and a truncated *CryIAb6* against potato tuber moth, *Pthorimaea operculella* (Zeller) resulted in increased resistance in the potato plants. While the effectiveness of *Bt884* gene alone reduced with time, there was no decrease in resistance in transgenics carrying both genes (Jansens et al., 1995).

Tobacco and tomato plants transformed with *CryIAb* or *CryIAc* genes showed a very high degree of resistance to the cotton bollworm and pink bollworm. The expression of *CryIAb—CryIAc* genes provides protection against *S. exigua*, *M. sexta* and *H. virescens*. Hence, it is feasible to increase the levels of resistance in transgenic plants and also employ different genes for managing development of resistance to Bt in insect populations. Evolving levels of insect resistance to Bt can be dramatically reduced through the genetic engineering of chloroplasts in plants or by inserting more than one gene in the same plant. Transgenic tobacco leaves expressing the *CryIA2* protoxin in chloroplasts when fed to susceptible insects, *CryIA*-resistant (20-40,000 fold resistant) and *Cry2A2* resistant (330-393-fold resistant) insects (*H. virescens, H. zea* and *S. exigua*), brought about 100% mortality in all the insect species and strains (Kota et al., 1999). These results suggest that the expression of high levels of a non-homologous Bt protein may be able to delay or overcome the development of broad-spectrum resistance to Bt in the field.

Multiple genes that confer resistance to different viruses have also been introduced into crop plants. A major commercial cultivar of potato, Russet Burbank, has been transformed with the coat protein genes of PVX and PVY simultaneously. Transgenic plants that expressed both CP genes were resistant to infection by PVX as well as PVY. Thus, the CP protection proved effective against mixed infection by two different viruses (Lawson et al., 1990).

Multiple genes conferring resistance to insects and viruses can also be engineered into crops very effectively. Transgenic potato carrying the Bt gene *CryV* (toxic to Lepidoptera and Coleoptera) and a potato Y potyvirus Y coat protein gene (PVY) were most resistant to the potato tuber moth and Potato virus Y infection rather than non-transgenic lines. The double transgenics were as lethal to the potato tuber moth as a *Cry V* transgenic line (Li et al., 1999).

Our laboratory at IGGEB is also in the process of pyramiding two insect resistance genes—*Gm2* and *Gm4*—which provide resistance to gall midge biotypes 1, 2 and 4 and 1, 2, 3 and 4, respectively (Mohan et al., 1994, 1997b; Nair et al., 1995). The resulting plants would be effective against the most prevalent biotypes of gall midge in India.

Genome sequencing and genetic mapping experiments have demonstrated that the *R* genes are generally organized in tightly-linked clusters (Michelmore and Meyers, 1998). These tightly-linked *R* genes are known to act synergistically (Parniske et al., 1997). The *R* gene clusters could be transferred in a block to elite varieties by transformation and thus the durability of the resistance could be enhanced.

Thus, genes providing resistance to pests, pathogens and viruses as well as to herbicides can all be pyramided into the same crop. Optimizing the activity of isolated resistance genes could further refine disease control programmes by replacement with plant specific codons, before reintroducing them into the plant. Incorporation of novel and powerful promoters, new and tighter tissue specific promoters as well as a better understanding of the principles governing plant gene expression would further enhance the effect of the resistance genes introduced. For example, the activity of the wild type tomato *Pto* gene is limited to certain races of *Pseudomonas syringae* pv. *tomato* (*Pst*) that contain the *avr* gene *avrPto*. Replacement of the weak endogenous promoter of *Pto* with the strong promoter of cauliflower mosaic virus resulted in not only a further increased resistance to *Pst* (*avrPto*) but also a partial control of unrelated pathogens, such as *Xanthomonas campestris* and *Cladosporium fulvum* (Tang et al., 1999). Chloroplast transformation also provides containment of foreign genes since plastid transgenes are not transmitted by the pollen. The escape of foreign genes via pollen is a serious environmental concern in nuclear transgenic plants, especially for the generation of herbicide-resistance crops, because of the high rates of gene flow from crops to wild weedy relatives.

THE RISK AND DRAWBACKS OF DEVELOPING TRANSGENIC CROPS

Presently, transgenic crops are grown commercially on several million hectares, principally in North America (see details in chapters 1 and 2 in this volume). There have been field trials conducted on transgenics from at least 52 species, including all the major field crops, vegetables, and several herbaceous and woody species in different parts of the world (Dunwell, 2000).

Although transgenic plants have been grown around the world for about 10 years now, exaggerated fears about genetically-modified crops remain. These fears have become stronger during the last couple of years in many countries around the world, especially in Europe. The US government recently appointed the committee on science to look into the matter. The latter, after a thorough study of the existing literature on the effect of GMOs on the environment, released a report in April 2000 assessing the benefits and risks of genetically-modified plants and plant-derived foods. The report concludes that there is no significant difference between plant varieties created using agricultural biotechnology and similar plants created using traditional cross breeding. In fact, it even recommends that regulation of US Department of Agriculture and the Environmental Protection Agency targeting biotechnology products be changed to focus on the characteristic of the plant, not the process used to develop it (www.house.gov/science). The report also says 'there is no evidence that transferring genes from unrelated organisms to plants poses any kind of unique risks. The risks associated with plant varieties developed using agricultural biotechnology are the same as those for similar varieties developed using the classical breeding methods. As the new methods are more precise and allow for better characterization of the changes being made, plant developers and food producers are in better position to assess the safety than when using classical breeding methods'. This report also refutes the charges of threat to wildlife posed by transgenic crops and says that 'widespread use of pest-resistant crop varieties developed using agricultural biotechnology is unlikely to accelerate the emergence of pesticide resistant insect strains and may actually be more effective in preventing their emergence when compared to spray applications of similar pesticides. Wraight et al. (2000) after a thorough study of the effects of pollen from Bt corn transgenic for the *CryIA(b)* endotoxin on black swallowtails under field conditions, came to the conclusion that the pollen had no effect on wild populations of black swallowtails under the highest pollen density tests (10,000 pollens/cm^2). Feeding chicken with Bt corn had no adverse effect on the chicken either (Brake and Vlachos, 1998). Thus, the deployment of transgenic crops does not appear to pose any added risks to humans, livestock or environment and may actually prove to be beneficial to all the three.

In spite of the tremendous advantages derived from using the transgene technology to produce plants with resistance to a wide range of pests and pathogens, the technique does have its own limitations as well. The generation of transgenics requires high levels of technical expertise, and is labour intensive as also expensive. Pests and pathogens could, with time, develop resistance to the genes deployed. Insect migration could reduce the effectiveness of the insect resistance genes deployed. Since secondary pests are not controlled through the use of transgenics, in the absence of sprays for the major pests, the secondary pests might begin increasing in number and take over as the major pest. On the other hand, control of secondary pests through chemical sprays would kill the pests' natural enemies and reduce the advantage of transgenics.

Saxena et al. (1999) found that Bt maize, which is resistant to corn borer pests, releases the Bt toxin through its roots into the soil. The toxin, not easily able to break down, remains insecticidal during 25 days of the plant growth. This could help both in controlling the pests as well as speed up development of insect-resistant pests. However, the insecticidal specificity of each protein is defined and Bt toxins have no adverse effects on non-target beneficial insects. According to Monsanto, the Bt toxin in each transgenic plant to date has degraded comparably or more rapidly than the microbial Bt products. They actually claim that there has been a significant increase in the beneficial insects, thereby establishing clear environmental and ecological benefits as compared to other insect control alternatives (www.agbiotechnet.com).

Another major worry regarding transgenics is the effect of expression of exogenous genes on the quality and quantity of the yield obtained. Purrington and Bergelson (1997) conducted a field experiment using mutant and transgenic *Arabidopsis thaliana* that differ by a single gene conferring resistance to either the herbicide chlorsulfuron or the antibiotic kanamycin. They found that herbicide-resistant individuals produced 26% fewer seeds than susceptible counterparts. It seems likely that, as in the case of bacteria carrying plasmids for antibiotic resistance, which are seen to have slower growth than those without, the expression of exogenous genes in the transgenic plants would cause metabolic drain, thereby reducing the yield of the plant.

However, an objective assessment of the risk: benefit ratio of deploying transgenic plants with multiple resistance suggests that the benefits that accrue would far outweigh all the risks. It is almost certain that in the not too distant future, a large proportion of the crops farmed would be of the transgenic kind. Perhaps the day is not far off when super-crops showing resistance to all the important pests, pathogens and viruses and at the same time carrying genes for resistance for broad-spectrum herbicides are generated and deployed routinely.

ACKNOWLEDGEMENTS

This work was supported in part by a grant from the Rockefeller Foundation. We would like to thank Ms Meenakshi Sharma for helping with the references.

REFERENCES

Anderson, E.J., Stark, D.M., Nelson, R.S., Powell, P.A., Tumer, N.E. and Beachy, R.N., 1989, Transgenic plants that express the coat protein genes of TMV and AIMV interfere with disease development of some non-related viruses. Phytopathology, **79**, 1284-1290.

Angenent, G.C., Van den Ouweland, J.M.W. and Bol, J.F., 1990, Susceptibility to virus infection of transgenic plants expressing structural and nonstructural genes of tobacco rattle virus. Virology, **175**, 191-198.

Bates, G.W., 1999, Plant transformation via protoplast electroporation. Methods Mol. Biol., **111**, 359-366.

Beachy, R.N., 1997, Mechanisms and applications of pathogen-derived resistance in transgenic plants. Curr. Opin. Biotechnol., **8**, 215-220.

Beachy, R.N., 1999, Coat protein-mediated resistance to tobacco mosaic virus: discovery, mechanisms and exploitation. Philos. Trans. R. Soc. Lond. B. Biol. Sci., **354**, 659-664.

Beck, D.L., Van Dolleweerd, C.J., Lough, T.J., Balmori, E., Voot, D.M., Andersen, M.T., O'Brien, E.W. and Forster, L.S., 1994, Disruption of virus movement confers broad-spectrum resistance against systemic infection by plant viruses with a triple gene block. Proc. Natl. Acad. Sci. (USA), **91**, 10310-10314.

Bendahmane, M. and Gronenborn, B., 1997, Engineering resistance against tomato yellow leaf curl virus (TYLCV) using antisense RNA. Plant Mol. Biol., **33**, 351-357.

Bent, A.F., Kunkel, B.N., Dahlbeck, D., Brown, K.L., Schmidt, R., Giraudat, J., Leung, J. and Staskawicz, B.J., 1994, *RPS2* of *Arabidopsis thaliana*: a leucine-rich repeat class of plant disease resistance genes. Science, **265**, 1856-1860.

Bentur, J.S. and Amudhan, S., 1996, Reaction of differentials of different populations of Asian rice gall midge (*Orseolia oryzae*) under green house condition. Indian J. Agric. Sci., **66**, 197-199.

Bentur, J.S., Pasalu, I.C. and Kalode, M.B., 1992, Inheritence of virulence in rice gall midge (*Orseolia oryzae*). Indian J. Agric. Sci., **62**, 490-493.

Bogorad, L., 2000, Engineering chloroplasts: an alternative site for foreign genes, proteins, reactions and products. Trends Biotechnol., **18**, 257-263.

Brake, J. and Vlachos, D., 1998, Evaluation of transgenic event 176 *'Bt'* corn in broiler chickens. Poult. Sci., **77**, 648-653.

Briggs, J.B., 1965, The distribution, abundance and genetic relationships of four strains of the rubus aphid, *Amphorophora rubi* (Kalt.) in relation to raspberry breeding. J. Hortic. Sci., **40**, 109-117.

Broglie, K., Chet, I., Holliday, M., Cressman, R., Biddle, P., Knowlton, S., Mauvais, C.J. and Broglie, R., 1991, Transgenic plants with enhanced resistance to the fungal pathogen *Rhizoctonia solani*. Science, **254**, 1194-1197.

Brown, D.G., 1988, The cost of plant defense: An experimental analysis with inducible proteinase inhibitors in tomato. Oecologia, **76**, 467-470.

Buck, K.W., 1991, Virus resistant plants. In: Grierson D. (ed.), Plant Biotechnology, Vol. 1, Plant Genetic Engineering, Chapman & Hall, London, pp. 136-178.

Burgess, E.P.J., Malone, L.A. and Christeller, J.T., 1996, Effects of two proteinase inhibitors on the digestive enzymes and survival of honeybees (*Apis mellifera*). J. Insect Physiol., **42**, 823-828.

Cao, J., Tang, J.D., Strizhov, N., Shelton, A.M. and Earle, E.D., 1999, Transgenic broccoli with high levels of *Bacillus thuringiensis Cry1C* protein controls diamond back moth larvae resistant to *Cry1A* or *Cry1C*. Mol. Breed., **5**, 131-141.

Carbonero, P., Royo, J., Diaz, I., Garcia-Maroto, F., Gonzalez-Hidalgo, E., Gutierez, C. and Casanera, P., 1993, Cereal inhibitors of insect hydrolases (α-amylases and trypsin): genetic control, transgenic expression and insect pests. In: Bruening G.J. Garcia-Olmedo F. and Ponz F.J. (eds.), Workshop on Engineering Plants Against Pests and Pathogens, Instituto Juan March de Estudios Investigaciones, Madrid, Spain, pp. 1-13.

Carrington, J.C., Kasschau, K.D.. Mahajan, S.K. and Schaad, M.C., 1996, Cell-to-cell and long distance transport of viruses in plants. Plant Cell, **8**, 1669-1681.

Charles, J.F., Nielsen-LeRoux, C. and Delecluse, A., 1996, *Bacillus sphaericus* toxins: Molecular biology and mode of action. Ann. Rev. Entomol., **41**, 451-472.

Cheng, X.G., Sardana, R., Kaplan, H. and Altosaar, I., 1998, Agrobacterium-transformed rice plants expressing synthetic *CryIA(b)* and *CryIA(c)* genes are highly toxic to striped stem borer and yellow stem borer. Proc. Natl Acad. Sci. (USA), **95**, 2767-2772.

Christou, P., 1997, Rice transformation: bombardment. Plant Mol. Biol., **35**, 197-203.

Clark, W.G., Fitchen, J.H. and Beachy, R.N., 1995, Studies on coat-protein mediated resistance to TMV using mutant CPI. The *PM2* assembly defective mutant. Virology, **208**, 485-491.

Cloutier, C., Jean, C., Fournier, M., Yelle, S. and Michaud, D., 2000, Adult Colorado potato beetles, *Leptinotarsa decemlineata* compensate for nutritional stress on oryzacystatin I-transgenic potato plants by hypertrophic behaviour and over-production of insensitive proteases. Arch. Insect Biochem. Physiol., **44**, 69-81.

Comai, L., Sen, L.C. and Stalker, D.M., 1983, An altered *aroA* gene product confers resistance to the herbicide glyphosate. Science, **221**, 370-371.

Confalonieri, M., Allegro, G., Balestrazzi, A., Fogher, C. and Delledonne, M., 1998, Regeneration of *Populus nigra* transgenic plants expressing a Kunitz proteinase inhibitor (*KTi3*) gene. Mol. Breed., **4**, 137-145.

Cooper, B., Lapidot, M., Heick, J.A., Dodda, J.A. and Beachy, R.N., 1995, Multivirus resistance in transgenic tobacco mosaic virus. Virology, **206**, 307-313.

Cornu, D., Leple, J.C., Bonade-Bottino, M., Ross, A., Augustin, S., Delplanque, A., Jouanin, L., Pilate, G. and Ahuja, M.R., 1996, Expression of a proteinase inhibitor and a *Bacillus thuringiensis* d-endotoxin in transgenic poplars. In: Boerjan W. and Neale D.B. (eds), Somatic Cell Genetics and Molecular Genetics of Trees, Kluwer Academic Publishers, Dordrecht, Netherlands, pp. 131-136.

Crickmore, N., Ziegler, D.R., Fietelson, J., Schnepf, E., Van Rie, J., Lereclus, D., Baum, J. and Dean, D.H., 1998, Revision of the nomenclature for *Bacillus thuringiensis* pesticidal crystal proteins. Microbiol. Mol. Biol., **62**, 807-813.

Crute, I.R., 1985, The genetic bases of relationship between microbial parasites and their hosts. In: Fraser R.S.S. (ed.), Mechanisms of Resistance to Plant Diseases, Dr W. Junk Publishers, Dordrecht, The Netherlands, pp. 80-142.

Czapla, T.H. and Lang, B.A., 1990, Effects of plant lectins on the larval development of European corn borer (Lepidoptera: Pyralidae) and southern corn rootworm (Coleoptera: Chrysomelidae). J. Econ. Entomol., **83**, 2480-2485.

Daniell, H., Datta, R., Varma, S., Gray, S. and Lee, S.B., 1998, Containment of herbicide resistance through genetic engineering of the chloroplast genome. Nature Biotechnol., **4**, 345-348.

Datta, S.K., 2000, Transgenic rice: Development and products for environmentally friendly sustainable agricultu al Sciences to the Crisis of Biosphere on the Earth in the 21st Century, pp. 237-246.

De Block, M., Botterman, T., Vandewiele, M., Dockx, J., Thoen, C., Gossele, V., Movva, N.R., Thompson, C., vanMontagu, M. and Leemans, J., 1987, Engineering herbicide resistance in plants by expression of detoxifying enzyme. EMBO J., **6**, 2513-2518.

De Wit, P.J.G.M., 1997, Pathogen avirulence and plant resistance: a key role for recognition. Trends Plant Sci., 2, 452-458.

Della-Cioppa, G., Bauer, S.C., Taylore, M.L., Rochester, D.E. and Klien, B.K., 1987, Targeting a herbicide resistant enzyme from *Escherichia coli* to chloroplasts of higher plants. BioTechnol., **5**, 579-586.

Ding, X., Gopalakrishnan, B., Johnson, L.B., White, F.F., Wang, X., Morgan, T.D., Kramer, K.J. and Muthukrishnan, S., 1998, Insect resistance of transgenic tobacco expressing an insect chitinase gene. Transgenic Research, **7**, 77-84.

Duan, X., Li, X., Xue, Q., Abo el Saad, M., Xu, D. and Wu, R., 1996, Transgenic rice plants harbouring an introduced potato proteinase inhibitor II gene are insect resistant. Nature Biotechnol., **14**, 494-498.

Dulmage, H.T., 1981, Insecticidal activity of isolates of *Bacillus thuringiensis* and their potential for pest control. In: Burges H.D. (ed.), Microbial Control of Pests and Plant Diseases, Academic Press, London, pp. 129-141.

Dunwell, J.M., 2000, Transgenic approaches to crop improvement. J. Expt. Bot., **51**, 487-496.

Edmonds, H.S., Gatehouse, L.N., Hilder, V.A. and Gatehouse, J.A., 1996, The inhibitory effects of the cysteine protease inhibitor, oryzacystatin, on digestive proteases and on larval survival and development of the southern corn rootworm (*Diabrotica undecimpunctata* Howard). Entomol. Exp. Appl., **78**, 83-94.

Erickson, F.L., Dinesh-Kumar, S.P., Holzberg, S., Ustach, C.V., Dutton, M., Handley, V., Corr, C. and Baker, B.J., 1999, Interactions between tobacco mosaic virus and the tobacco N gene. Phil.Trans. R. Soc. Lond. B Biol. Sci., 354, 653-658.

Estruch, J.J., Kadwell, S., Merlin, E. and Crossland, L., 1994, Cloning and characterization of a maize pollen-specific calcium-dependent calmodulin-independent protein kinase. Proc. Natl. Acad. Sci. (USA), **91**, 8837-8841.

Estruch, J.J., Warren, G.W., Mullins, M.A., Nye, G.J., Craig, J.A. and Koziel, M.G., 1996, Vip3A, a novel *Bacillus thuringiensis* vegetative insecticidal protein with a wide spectrum of activities against lepidopteran insects. Proc. Natl. Acad. Sci. (USA), **93**, 5389-5394.

Estruch, J.J., Carozz, N.B., Desai, N., Duck, N.B., Warren, G.W. and Koziel, M.G., 1997, Transgenic plants: an emerging approach to pest control. Nature Biotechnol., **15**, 137-141.

Finer, J.J., Finer, K.R. and Ponnapa, T., 1999, Particle bombardment mediated transformation. Curr. Topics Microbiol. Immunol., 240, 59-80.

Fitches, E., Gatehouse, A.M.R. and Gatehouse, J.A., 1997, Effects of snowdrop lectin (*GNA*) delivered via artificial diet and tansgenic plants on the development of tomato moth (*Lacanobia oleracea*) larvae in laboratory and glasshouse trials. J. Insect Physiol., **43**, 727-739.

Flor, H.H., 1971, Current status of the gene-for-gene concept. Annu. Rev. Phytopathol., **9**, 275-296.

Florack D.E.A., Allefs, S., Bollen, R., Bosch, D., Visser, B., Stiekema, W., 1995, Expression of giant silkmoth cecropin B genes in tobacco. Transgenic Res., **4**, 132-141.

Fujimoto, H., Itoh, K., Yamamoto, M., Kayozuka, J. and Shimamoto, K., 1993, Insect resistant rice generated by a modified delta-endotoxin genes of *Bacillus thuringiensis*. BioTechnol., **11**, 1151-1155.

Gallun, R.L., 1978, Genetics of biotypes B and C of the Hessian fly. Ann. Entomol. Soc. Am., **71**, 481-486.

Gatehouse, A.M.R., Boulter, D. and Hilder, V.A., 1991, Potential of plant-derived genes in the genetic manipulation of crops for insect resistance. In: Gatehouse A.M.R., Hilder V.A. and Boulter D. (eds), Plant Genetic Manipulation for Crop Protection, Vol. 1, Redwood Press Ltd., Melksham, UK, pp. 155-160.

Gatehouse, L.N., 1995, Novel genes for insect resistance in transgenic plants. Ph.D. Thesis, Durham University, UK.

Gerlach, W.L., Llewellyn, D., Haseloff, J.P., 1987, Construction of a plant disease resistance gene from the satellite RNA of tobacco ringspot virus. Nature, **328**, 802-805.

Gill, S.S., Cowles, E.A. and Pietrantonio, F.V., 1992, The mode of action of *Bacillus thuringiensis* endotoxins. Annu. Rev. Entomol., **37**, 615-636.

Golemboski, D.B., Lomonossoff, G.P. and Zatlin, M., 1990, Plants transformed with a tobacco mosaic virus nonstructural gene sequence are resistant to the virus. Proc. Natl. Acad. Sci. (USA), **87**, 6311-6315.

Habibi, J., Backus, E.A. and Czapla, T.H., 1993, Plant lectins affect survival of the potato leafhopper (Homoptera: Cicadellidae). J. Econ. Entomol., **86**, 945-951.

Hadi, M.Z., McMullen, M.D. and Finer, J.J., 1996, transformation of 12 different plasmids into soybean via particle bombardment. Plant Cell Reports, **15**, 500-505.

Hamilton, C.M., 1997, A binarry-BAC system for plant transformation with high-molecular-weight DNA. Gene, **200**, 107-116.

Hammond, J. and Kamo, K.K., 1995, Effective resistance to potyvirus infection confered by expression of antisense RNA in transgenic plants. Mol. Plant Microbe Interact., **5**, 674-682.

Hannay, C.L., 1953, Crystalline inclusions in aerobic spore-forming bacteria. Nature, **172**, 1004.

Harrison, B.D., Mayo, M.A. and Baulcombe, D.C., 1987, Virus resistance in transgenic plants that express cucumber mosaic virus satellite RNA. Nature, **328**, 799-802.

Hatchett, J.H. and Gallun, R.L., 1970, Genetics of the ability of the Hessian fly, *Mayetiola destructor* to survive on wheats having different genes for resistance. Ann. Entomol. Soc. Am., **63**, 1400-1407.

Hattori, J., Brown, D., Mourad, G., Labbe, H., Ouellet, T., Sunohara, G., Rutledge, R., King, J. and Miki, B., 1995, An acetohydroxy acid synthase mutant reveals a single site involved in multiple herbicide resistance. Mol. Gen. Genet., **246**, 419-425.

Hayakawa, T., Zhu, Y., Itoh, K., Kimura, Y., Izawa, T., Shimamoto, K. and Toriyama, S., 1992, Genetically-engineered rice resistant to rice stripe virus, an insect-transmitted virus. Proc. Natl. Acad. Sci. *(USA)*, **89**, 9865-9869.

Heifetz, P.B., 2000, Genetic engineering of the chloroplast. Biochimie, **82**, 655-666.

Heinrichs, E.A. and Mochida, O., 1984, From secondary to major pests' status: The case of insecticide-induced rice brown planthopper, *Nilaparvata lugens* resurgence. Prot. Ecol., **7**, 201-218.

Hemenway, C., Fang, R.X., Kaniewski, J.J., Chua, N.H. and Tumer, N.E., 1988, Analysis of the mechanism of protection in transgenic plants expressing the potato virus X coat protein or its antisense RNA. EMBO J., **7**, 1273-1280.

Hilder, V.A. and Boulter, D., 1999, Genetic engineering of crop plants for insect resistance—a critical review. Crop Protection, **18**, 177-191.

Hilder, V.A. and Gatehouse, A.M.R., 1991, Phenotypic costs to plants of an extra gene. Transgenic Res., **1**, 54-60.

Hilder, V.A., Gatehouse, A.M.R., Sheerman, S.E., Baker, R.F. and Boulter, D., 1987, A novel mechanism of insect resistance engineered into tobacco. Nature, **330**, 160-163.

Hoekema, A.B.R., Huisman, M.J., Molendijk, L., van den Elzen, P.J.M. and Cornelissen, B.J.C., 1989, The genetic engineering of two commercial potato cultivars for resistance to potato virus X. BioTechnol., **7**, 273-278.

Hoffmann, M.P., Zalom, F.G., Wilson, L.T., Smilanick, J.M., Malyj, L.D., Kiser, J., Hilder, V.A. and Barnes, W.M., 1992, Field evaluation of transgenic tobacco containing genes encoding *Bacillus thuringiensis* δ-endotoxin or cowpea trypsin inhibitor: Efficacy against *Helicoverpa zea* (Lepidoptera: Noctuidae*)*. J. Econ. Entomol., **85**, 2516-2522.

Hoffmann, C., Vanderbruggen, H., Hofte, H., Van Rie, J., Jansens, S. and Van Mellaert, H., 1988, Specificity of *Bacillus thuringiensis* δ-endotoxins is correlated with the presence of high affinity binding sites in the brush border membrane of target insect midguts. Proc. Natl. Acad. Sci. (USA), **85**, 7844-7848.

Horsch, R. B., Fry, J. E., Hoffmann, N.I., Eichholtz, D. Rogers, S. G. and Fraely, R.T., 1985, A simple and general method for transferring genes into plants. Science, **227**, 1229-1231.

Howe, G.A., Lightner, J., Browse, J. and Ryan, C.A., 1996, An octadecanoid pathway mutant (*JL5*) of tomato is compromised in signaling for defense against insect attack. Plant Cell, **8**, 2067-2077.

Hudspeth, R.L. and Grula, J.W., 1989, Structure and expression of maize gene encoding the phosphoenolpyruvate carboxylase isozyme involved in C4 photosynthesis. Plant Mol. Biol., **12**, 579-589.

Ishimoto, M., Yamada, T. and Kaga, K., 1999, Insecticidal activity of an a-amylase inhibitor-like protein resembling a putative precursor of α-amylase inhibitor in the common bean, *Phaseolus vulgaris* L. Biochem. Biophys. Acta, **1432**, 104-112.

Jansens, S., Cornelissen, M., Clercq R de, Reynaerts, A. and Peferoen, M., 1995, *Phthorimaea opercullella* (Lepiddoptera: Gelechiidae) resistance in potato by expression of *Bacillus thuringiensis CryIA(b)* insecticidal crystal protein. J. Econ. Entomol., **88**, 1469-1476.

Kaniewski, W., Lawson, C., Sammons, B., Haley, L., Hart, J., Delannay, X. and Tumer, N.E., 1990, Field resistance of transgenic Russet Burbank potato to effects of infection by potato virus X and potato virus Y. BioTechnol., **8**, 750-754.

Komari, T., Hiei, Y., Ishida, Y., Kumashiro, T. and Kubo, T., 1998, Advances in cereal gene transfer. Curr. Opin. Plant Biol., **1**, 161-165.

Kota, M., Daniell, H., Varma, S., Garczynski, S.F., Gould, F. and Moar, W.J., 1999, Overexpression of the *Bt Cry2Aa2* protein in chloroplasts confers resistance to plants against susceptible and Bt resistant insects. Proc. Natl. Acad. Sci. (USA), **96**, 1840-1845.

Koziel, M.G., Beland, G.L., Bowman, C., Carozzi, N.B., Crenshaw, R., Crossland, L., Dawson, J., Desai, N., Hill, M., Kadwell, S., Launis, K., Maddox, D., McPherson, K, Meghji, M.R., Merlin, E., Rhodes, R., Warren, G.W., Wright, M. and Evola, S.V., 1993, Field performance of elite transgenic maize plants expressing an insecticidal protein derived from *Bacillus thuringiensis*. BioTechnol., **11**, 194-200.

Khush, G.S., 1999, Green revolution: preparing for the 21st century. Genome, **42**, 646-655.

Lapidot, M., Gafny, R., Ding, B., Wolf, S., Lucas, W.J. and Beachy, R.N., 1993, A dysfunctional movement protein of tobacco mosaic virus that partially modifies the plasmodesmata and limits virus spread in transgenic plants. Plant J., **2**, 959-970.

Lawson, C., Kaniewski, W., Haley, L., Rozman, R., Newell, C., Sanders, P. and Tumer, N.E., 1990, Engineering resistance to multiple virus infection in a commercial potato cultivar: resistance to potato virus X and potato virus Y in transgenic Russet Burbank Potato. BioTechnol., **8**, 127-134.

Lee, S.I., Lee, S.H., Koo, J.C., Chun, H.J., Lim, C.O., Mun, J.H., Song, Y.H. and Cho, M.J., 1999, Soybean Kunitz trypsin inhibitor (*SKTI*) confers resistance to the brown planthopper (*Nilaparvata lugens* Stal.) in transgenic rice. Mol. Breeding, **5**, 1-9.

Lermontova, I. and Grimm, B., 2000, Overexpression of plastidic protoporphyrinogen IX oxidase leads to resistance to the diphenylether herbicide acifluorfen. Plant Physiol., **122**, 75-84.

Li, W.B., Zarka, K.A., Douches, D.S., Coombs, J.J., Pett, W.L., Grafius, E.J., Li, W.B., 1999, Coexpression of potato PVY coat protein and *CryV-Bt* genes in potato. *J. Am. Soc. Hortic. Sci.*, **124**, 218-223.

Lilley, C.J., Urwin, P.E. and Atkinson, H.J., 1999, Characterization of plant nematode genes: identifying targets for a transgenic defence. Parasitology, **118**, 63-72.

MacIntosh, S.C., Kishore, G.M., Perlak, F.J., Marrone, P.G., Stone, T.B., Sims, S.R. and Fuchs R.L., 1990, Potentiation of *Bacillus thuringiensis* insecticidal activity by serine protease inhibitors. J. Agric. Food Chem., **38**, 1145-1152.

Maenpaa, P., Gonzalez, E.B., Ahlandsberg, S. and Jansson, C., 1999, Transformation of nuclear and plastomic plant genomes by biolistic particle bombardment. Mol. Biotechnol., **13**, 67-72.

Malone, L.A., Giacon, H.A., Burgess, E.P.J., Maxwell, J.Z., Christeller, J.T. and Liang, W.A., 1995, Toxicity of trypsin endopeptidase inhibitor to honey bees (Hymenoptera: Apidae). J. Econ. Entomol., **88**, 46-50.

Martin, G.B., 1999, Functional analysis of plant disease resistance genes and their downstream effectors. Curr. Opin. Plant Biol., **2**, 273-279.

Matthews, R.E., 1981, Portraits of viruses: turnip yellow mosaic virus. Intervirology, **15**, 121-144.

Michelmore, R., 2000, Genomic approaches to plant disease resistance. Curr. Opin. Plant Biol., **3**, 125-131.

Michelmore, R.W. and Meyers, B.C., 1998, Clusters of resistance genes in plants evolve by divergent selection and a birth-and-death process. Genome Res., **8**, 1113-1130.

Miller, E.D. and Hemenway, C., 1998, History of coat protein-mediated protection. Methods Mol. Biol., **8**, 25-38.

Milne, R. and Kaplan, H., 1993, Purification and characterization of a trypsin like digestive enzyme from spruce budworm (*Christoneura fumiferana*) responsible for the activation of δ-endotoxin from *Bacillus thuringiensis*. Insect Biochem. Mol. Biol., **23**, 663-673.

Mitra, A. and Zhang, Z., 1994, Expression of a human lactoferrin cDNA in tobacco cells produces antibacterial protein(s). Plant Physiol., **106**, 977-981.

Mohammed, A., Douches, D.S., Pett, W., Grafius, E., Coombs, J.L., Li, W. and Madkour, M.A., 2000, Evaluation of potato tuber moth (Lepidoptera: Gelechiidae) resistance in tubers of *Bt-cry5* transgenic potato lines. J. Econ. Entomol., **93**, 472-476.

Mohan, M., Nair, S., Bentur, J.S., Prasad Rao, U. and Bennett, J., 1994, RFLP and RAPD mapping of the rice *Gm2* gene that confers resistance to biotype 1 of gall midge (*Orseolia oryzae*). Theor. Appl. Genet., **87**, 782-788.

Mohan, M., Nair, S., Bhagwat, A., Krishna, T.G., Yano, M., Bhatia, C.R. and Sasaki, T., 1997a, Genome mapping, molecular markers and marker-assisted selection in crop plants. Mol. Breed., **3**, 87-103.

Mohan, M., Satyanarayanan, P.V., Kumar, A., Srivastava, M.N. and Nair, S., 1997b, Molecular mapping of a resistance specific PCR based marker linked to a gall midge resistance gene (*Gm4t*), in rice. Theor. Appl. Genet., **95**, 777-782.

Morton, R.L., Schroeder, H.E., Bateman, K.S., Chrispeels, M.J., Armstrong, E. and Higgins, T.J., 2000, Bean a-amylase inhibitor 1 in transgenic peas (*Pisum sativum*) provides complete protection from pea weevil (*Bruchus pisorum*) under field conditions. Proc. Natl. Acad. Sci. (USA), **97**, 3820-3825.

Mourgues, F., Brisset, M.N. and Chevreau, E., 1998, Strategies to improve plant resistance to bacterial diseases through genetic engineering. Tibtech, **16**, 203-210.

Nair, S., Bentur, J.S., Prasada Rao, U. and Mohan, M., 1995, DNA markers tightly linked to gall midge resistance gene (*Gm2*) are potentially useful for marker-aided selection in rice breeding. Theor. Appl. Genet., **91**, 68-73.

Nair, S., Kumar, A., Srivastava, M.N. and Mohan, M., 1996, PCR-based DNA markers for *Gm4t* gene that confers resistance against gall midge and their potential in marker-aided selection in rice. Theor. Appl. Genet., **92**, 660-665.

Nayak, P., Basu, D., Das, S., Basu, A., Ghosh, D., Ramakrishnan, N.A., Ghosh, M. and Sen, S.K., 1997, Transgenic elite indica rice plants expressing *CryIAc* δ-endotoxin of *Bacillus thuringiensis* are resistant against yellow stem borer (*Scirpophaga incertulas*). Proc. Natl. Acad. Sci. (USA), **94**, 2111-2116.

Nejidat, A. and Beachy, R.N., 1990, Transgenic tobacco plants expressing a coat protein gene of tobacco mosaic virus are resistant to some other tobamovirus. Mol. Plant Microbe Interact., **3**, 247-351.

Norelli, J.L., Aldwinckle, H.S., Destephano-Beltran, L. and Jaynes, J.M., 1994, Transgenic "Malling 26" apple expressing the attacin E gene has increased resistance to *Erwinia amylovora*. Euphytica, **77**, 123-128.

Nutt, K.A., Allsopp, P.G., McGhie, T.K., Shepherd, K.M., Joyce, P.A., Taylo, G.O., McQualter, R.B., Smith, G.R. and Ogarth, D.M., 1999, Transgenic sugarcane with increased resistance to canegrubs. Proc. Conf. Australian Society of Sugar Cane Technologists, Townsville, Brisbane, Australia, pp. 171-176.

Oerke, E.C., Dehne, H.W., Schonbeck, F. and Weber, A., 1994, Crop Production and Crop Protection: Estimated Losses in Major Food and Cash Crops. Elsvier, Amsterdam, Netherlands, 808 pp.

Parniske, M., Hammond-Dosack, K.E., Golstein, C., Thomas, C.M., Jones, D.A., Harrison, K., Wulff, B.B. and Jones, J.D., 1997, Novel disease resistance specificities result from sequence exchange between tandemly repeated genes at the *Cf-4/9* locus of tomato. Cell, **91**, 821-832.

Perlak, F.J., Deaton, R.W., Armstrong, T.A., Fuchs, R.L., Sims, S.R., Greenplate, J.T. and Fischhoff, D.A., 1990, Insect resistant cotton plants. Biotech., **8**, 939-943.

Perlak, F.J., Fuchs, R.L., Dean, D.A., McPherson, S.L. and Fischhoff, D.A., 1991, Modification of coding sequence enhances plant expression of insect control protein genes, Proc. Natl. Acad. Sci. (USA), **88**, 3324-3328.

Perlak, F.J., Stone, T.B., Muskopf, Y.N., Petersen, L.J., Parker, G.B., McPherson, S.A., Wyman, J., Love, S., Biever, D. and Fischhoff, D.A., 1993, Genetically-improved potatoes: protection from damage by Colorado potato beetle. Plant Mol. Biol., **22**, 313-321.

Pinto, Y.M., Kok, R.A. and Baulcombe, D.C., 1999, Resistance to rice yellow mottle virus (RYMV) in cultivated African rice varieties containing RYMV transgenes. Nature Biotechnol., **7**, 702-707.

Powell, P.A., Sanders, P.R., Tumer, N., Farley, R.T. and Beachy, R.N., 1990, Protection against tobacco mosaic virus infection in transgenic plants requires accumulation of coat protein rather than coat protein RNA sequences. Virology, **175**, 124-130.

Powell, K.S., Gatehouse, A.M.R., Hilder, V.A. and Gatehouse, J.A., 1993, Antimetabolic effects of plant lectins and plant and fungal enzymes on the nymphal stages of two important rice pests, *Nilaparvata lugens* and *Nephotettix ciniteps*. Entomol. Exp. Appl., **66**, 119-126.

Powell, K.S., Gatehouse, A.M.R., Hilder, V.A. and Gatehouse, J.A. (1995) Antifeedant effects of plant lectins and an enzyme on the adult stage of the rice brown planthopper, *Nilaparvata lugens*. Entomol. Exp. Appl., **75**, 51-59.

Prins, M., Kikkert, M., Ismayadi, C., de Graauw, W., de Haan, P. and Goldbach, R., 1997, Characterization of RNA-mediated resistance to tomato spotted wilt virus in transgenic tobacco plants expressing *NSm* gene sequences. Plant Mol. Biol., **33**, 235-243.

Purcell, J.P., Greenplate, J.T., Jennings, M.G., Ryers, J.S., Pershing, J.C., Sims, S.R., Prinsen, M.J., Corbin, D.R., Tran, M., Sammons, R.D. and Stonard, R.J., 1993, Cholesterol oxidase: A potent insecticidal protein active against bollweevil larvae. Biochem. Biophy. Res. Comm., **196**, 1406-1413.

Purrington, C.B. and Bergelson, J., 1997, Fitness consequences of genetically engineered herbicide and antibiotic resistance in *Arabidopsis thaliana*. Genetics, **145**, 807-814.

Rao, K.V., Rathore, K.S., Hodges, T.K., Fu, X., Stoger, E., Sudhakar, D., Williams, S., Christou, P., Bharathi, M., Brown, D.P., Powell, K.S., Spence, J., Gatehouse, A.M.R. and Gatehouse, J.A., 1998, Expression of snowdrop lectin (*GNA*) in transgenic rice plants confers resistance to rice brown planthopper. Plant J., **15**, 469-477.

Riggs, R.D. and Wrather, J.A., 1992, Biology and Management of the Soybean Cyst Nematode, ASP Press, St. Paul, MN.

Rola, A.C. and Pingali, P.L., 1993, Pesticides, Rice productivity and Farmer's Health, Report IRRI, Phillipines.

Rommens, C.M. and Kishore, G.M., 2000, exploiting the full potential of disease resistance genes for agricultural use. Curr. Opin. Biotech., **11**, 120-125.

Rossi, M., Goggin, F., Milligan, S., Kaloshian, I., Ultman, D. and Williamson, V., 1998, The nematode resistance gene *Mi* of tomato confers resistance against the potato aphid. Proc. Natl. Acad. Sci. (USA), **95**, 9750-9754.

Ross, H., 1986, Potato Breeding—Problems and Perspectives. J. Plant Breeding Supplement 13, Advances in Plant Breeding, Verlag Paul Parey, Berlin, 132 pp.

Sardesai, N., Rajyashri, K.R., Behura, S.K., Nair, S. and Mohan, M., 2001, Genetic, physiological and molecular interactions of rice with its major dipteran pest, gall midge (*Orseolia oryza*)—a review. Plant Cell Tissue & Organ Culture (in press).

Satyendra, R., Stewart, J.M. and Wilkins, T., 1998, Assessment of resistance of cotton transformed with lectin genes to tobacco budworm. Special Report, Arkansas Agricultural Experimental Station, No. 188, Ark. Agric. Expt. Stn., University of Arkansas, Fayetteville, pp. 95-98.

Saxena, D., Flores, S. and Stotzky, G., 1999, Insecticidal toxin in root exudates from Bt corn. Nature, **402**, 480.

Schroeder, H.E., Gollasch, S., Moore, A., Tabe, L.M., Craig, S., Hardie, D.C., Chrispeels, M.J., Spencer, D. and Higgins, T.J.V., 1995, Bean α-amylase inhibitor confers resistance to pea weevil (*Bruchus pisorum*) in transgenic peas (*Pisum sativum* L.). Plant Physiol., **107**, 1233-1239.

Shade, R.E., Schroeder, H.E., Pueyo, J.J., Tabe, L.M., Murdock, L.L., Higgins, T.J.V., Chrispeels, M.J., 1994, Transgenic pea seeds expressing α-amylase inhibitor of the common bean are resistant to bruchid beetles. BioTechnol., **12**, 793-796.

Shah, D.M., Horsch, R.B., Klee, H.J., Kishore, G.M., Winter, J.A., Tumer, N.E., Hironaka, C.M., Sanders, P.R., Gasser, C.S., Aykent, S., Siegel, N.R., Rogers, S.R. and Fraley, R.T., 1986, Engineering herbicide tolerance in transgenic plants. Science, **233**, 478-481.

Sharma, H.C. and Agarwal, R.A., 1983, Role of some chemical components and leaf hairs in varietal resistance in cotton to jassid, *Amrasca biguttula buguttula* Ishida. J. Entomol. Res., **7**, 145-149.

Sharma, H.C. and Norris, D.M., 1991, Chemical basis of resistance in soybean to cabbage looper, *Trichoplusia ni*, J. Sci. Food Agric., **55**, 353-365.

Sharma, H.C., Sharma, K.K., Seetharama, N. and Ortiz, R., 2000, Prospects for using transgenic resistance to insects in crop improvement. Elect. J. Biotechnol., **3**, ISSN: 0717-3458.

Shukle, R.H. and Murdock, L.L., 1983, Lipoxygenase, trypsin inhibitor, and lectin from soybean: effect on larval growth of *Manduca sexta* (Lepidoptera: Sphingidae). Environ. Entomol., **12**, 787-791.

Singsit, C., Adang, M.J., Lynch, R.E., Anderson, W.F., Wang, A., Cardineau, G. and Ozias-Akins, P., 1997, Expression of a *Bacillus thuringiensis CryIA(c)* gene in transgenic peanut plants and its efficacy against lesser cornstalk borer. Transgen. Res., **6**, 169-176.

Song, W.Y., Wang, G.L., Chen, L.L., Kim, H.S., Pi, L.Y., Holsten, T., Gardner, J., Wang, B., Zhai, W.X., Zhu, L.H., Fauquet, C. and Ronald, P.C., 1995, A receptor kinase-like protein encoded by the rice disease resistance gene, Xa21. Science, **270**, 1804-1806.

Staskawicz, B.J., Ausubel, F.M., Baker, B.J., Ellis, J.G. and Jones, J.D., 1995, Molecular genetics of plant disease resistance. Science, **268**, 661-667.

Tacke, E., Salamini, F. and Rohde, W., 1996, Genetic engineering of potato for broad-spectrum protection against virus infection. Nature Biotechnol., **14**, 1597-1601.

Tai, T.H., Dahlbeck, D., Clark, E.T., Gajiwala, P., Pasion, R., Whalen, M.C., Stall, R.E. and Staskawicz, B.J., 1999, Expression of the *Bs2* pepper gene confers resistance to bacterial spot disease in tomato. Proc. Natl. Acad. Sci. USA, **96**, 14153-14158.

Tan, Y., Pan, Y, Zhang, Y., Lixia, Z. and Xu, Y., 1993, Resistance to gall midge *Orseolia oryzae* in chinese rice varieties compared with varieties from other countries. Int. Rice Res. Newslett., **18**, 13-14.

Tang, X., xie, M., Kim, Y.J., Zhou, J., Klessig, D.F. and Martin, G.B., 1999, Overexpression of *Pto* activates defense responses and confers broad resistance. Plant Cell, **11**, 15-29.

Terras, F.R.G., Eggermont, K., Kovaleva, V., Raikhel, N.V., Osborn, R.W., Kester, A., Rees, S.B., Torrekens, S., van Leuven, F., Vanderleyden, J., Cammue, B.P.A.and Broekaert, W.F., 1995, Small cysteine-rich antifungal proteins from radish: their role in host defence. Plant Cell, **7**, 573-588.

Thomas, J.C., Adams, D.G., Keppenne, V.D., Wasmann, C.C., Brown, J.K., Kanosh, M.R. and Bohnert, H.J., 1995, Proteinase inhibitors of *Manduca sexta* expressed in transgenic cotton. Plant Cell Reports, **14**, 758-762.

Tojo, A. and Aizawa, K., 1983, Dissolution and degradation of δ-endotoxin by gut juice protease of silkworm, *Bombyx mori*. Appl. Environ. Microbiol., **45**, 576-580.

Tuli, R., Bhatia, C.R., Singh, R.K. and Chaturvedi, R., 2000, Release of insecticidal transgenic crops and gap areas in developing approaches for more durable resistance. Curr. Sci., **79**, 163-165.

Tumer, N.E., O'Connell, K.M., Nelson, R.S., Sanders, P.R., Beachy, R.N., Fraley, R.T. and Shah, D.M., 1987, Expression of alfalfa mosaic virus coat protein gene confers cross protection in transgenic tobacco and tomato plants, EMBO J., **6**, 1181-1188.

Vaeck, M., Reynaerts, A., Hoftey, H., Jansens, S., DeBeuckleeer, M., Dean, C., Zabeau, M., Van Montagu, M. and Leemans, J., 1987, Transgenic plants protected from insect attack. Nature, **327**, 33-37.

Vain, P., Worland, B., Clarke, M.C., Richard, G., Beavis, M., Liu, H., Kohli, A., Leech, M., Snape, J. and Christou, P., 1998, Expression of an engineered cysteine proteinase inhibitor (oryzacystatin-I δ-D86) for nematode resistance in transgenic rice plants. Theor. Appl. Genet., **96**, 266-271.

Van der Salm, T., Bosch, D., Honee, G., Feng, I., Munsterman, E., Bakker, P., Stiekema, W.J. and Visser, B., 1994, Insect resistance of transgenic plants that express modified *cryIA(b)* and *cryIC* genes: A resistance management strategy. Plant Mol. Biol., **26**, 51-59.

Van Dun C.M.P. and Bol, J.F., 1988, Transgenic tobacco plants accumulating tobacco rattle virus coat protein resist infection with tobacco rattle virus and pea early browning virus. Virology, **167**, 649-652.

Van Dun, C.M.P., Overduin, B., van Vloten-Doting, L. and Bol, J.F., 1988, Transgenic tobacco expressing tobacco streak virus or mutated alfalfa virus coat protein does not cross-protect against alfalfa mosaic virus infection. Virology, **164**, 383-389.

Van Rie, J., Jansens, S., Hoftey, H., Degheele, D. and van Mellaert, H., 1989, Specificity of *Bacillus thuringiensis* δ-endotoxin. Importance of specific receptors on the brush border membrane of the mid gut of target insects. Eur. J. Biochem., **186**, 239-247.

Verberne, M.C., Verpoorte, R., Boll, J.F., Mercado-Blanco, J., Linthorst, H.J., 2000, Overproduction of salicylic acid in plants by bacterial transgenes enhances pathogen resistance. Nature Biotechnol., **18**, 779-783.

Vrieling, K., van Wijk C.A.M. and Swa, Y.J., 1991, Costs assessment of the production of pyrrolizidine alkaloids in ragwort (*Senecio jacobaca* L.) Oecologia, **97**, 541-546.

Warren, R.A., 1996, Microbial hydrolysis of polysaccharides, Annu. Rev. Microbiol., **50**, 183-212.

Wasmann, C.C., Echt, C., Dunn, R.L., Bohneert, H.J. and McCoy, T.J., 1994, Introduction and expression of an insect proteinase inhibitor in alfalfa (*Medicago sativa* L.). Plant Cell reports, **14**, 31-36.

Whitham, S., McCormick, S. and Baker, B., 1996, The N gene of tobacco confers resistance to tobacco mosaic virus in transgenic tomato. Proc. Natl. Acad. Sci. (USA), **93**, 8776-8781.

Wilson, W.D., Flint, H.M., Deaton, R.W., Fischhoff, D.A., Perlak, F.J., Armstrong, T.A., Fuchs, R.L., Berberich, S.A., Parks, N.J. and Stapp, B.R., 1992, Resistance of cotton lines containing a *Bacillus thuringiensis* toxin to pink bollworm (Lepidoptera: Gelechiidae) and other insects. J. Econ. Entomol., **85**, 1516-1521.

Wong, E.Y., Hironaka, C.M.and Fischhoff, D.A., 1992, *Arabidopsis thaliana* small subunit leader and transit peptide enhance expression of *Bacillus thuringiensis* proteins in transgenic plants. Plant Mol. Biol., **20**, 81-93.

Wraight, C.L., Zangerl, A.R., Carrol, M.J. and Berenbaum, M.R., 2000, Absence of toxicity of *Bacillus thuringiensis* pollen to black swallowtails under field conditions. Proc. Natl. Acad. Sci. (USA), **97**, 7700-7703.

Wu, G., Short, B.J., Lawrence, E.B., Levine, E.B., Fitzimmons, K.C. and Shah, D.M., 1995, Disease resistance conferred by expression of a gene encoding H2O2-generating glucose oxidase in transgenic potato plants. Plant Cell, **7**, 1357-1368.

Xu, D.P., Xue, Q.Z., McElroy, D., Mawal, Y., Hilder, V.A. and Wu, R., 1996, Constitutive expression of a cowpea trypsin inhibitor gene *CpTi*, in transgenic rice plants confers resistance to two major rice insect pests. Mol. Breeding, **2**, 167-173.

Zaitlin, M., Anderson, J.M., Perry, K.L., Zhang, L. and Palukaitis, P., 1994, Specificity of replicase-mediated resistance to cucumber mosaic virus. Virology, **201**, 200-205.

Zhao, J.Z., Zhao, K.J., Lu, M.G., Fan, X.L. and Guo, S.D., 1998, interactions between *Helicoverpa armigera* and transgenic Bt cotton in North China. Scientia Agric. Sinica, **31**, 1-6.

8

CURRENT RESISTANCE MANAGEMENT STRATEGIES FOR Bt CORN IN THE UNITED STATES

SHARLENE M. MATTEN, RICHARD L. HELLMICH* AND ALAN REYNOLDS

Unites States Environmental Protection Agency, Office of Pesticide Programs, Biopesticides and Pollution Prevention Division (7511C) 1200 Pennsylvania Ave. NW, Washington, D.C. 20460, USA
USDA–ARS, Corn Insects and Crop Genetics Research Unit, and Department of Entomology, Genetics Laboratory c/o Insectary, Iowa State University, Ames, IA 50011, USA

INTRODUCTION

Agriculture has been plagued by pests since the dawn of time. Pesticides control pests—insects, fungi, and weeds—until the pests begin to resist the pesticides. As resistance develops, more pesticide is needed to control the pest until that particular pesticide fails. Integrated pest management (IPM) was designed to prevent insects from developing resistance to insecticides, and resistance management remains a cornerstone of many IPM programs today. A critical component of IPM is the monitoring of increases in a pest's tolerance to a particular pesticide. Such monitoring, however, is rarely conducted in a proactive manner.

Insect resistance management (IRM) describes the practices aimed at reducing the potential for insect pests to become resistant to a pesticide. IRM is important for transgenic crops expressing *Bacillus thuringiensis* (Bt) insecticidal proteins (commonly called Bt crops) because insect resistance poses a threat to the future use of microbial Bt formulations in primarily organic farming and Bt technology as a whole. Academic and government scientists, public interest groups, and organic and other farmers have expressed concern that the widespread planting of these genetically-modified plants will hasten insect resistance to Bt endotoxins.

Under the Federal Insecticide, Fungicide, and Rodenticide Act (FIFRA), the United States Environmental Protection Agency (EPA) ensures there will be no unreasonable and adverse effects from the use of a pesticide when all the economic factors are fully considered. With Bt technology, EPA has stated that it is working to prevent what would happen if Bt could no longer control insect pests and more toxic insecticides had to be used in its place. In 2000 EPA imposed new IRM requirements on registered Bt crops in order to combat insect resistance to Bt endotoxins. Sound IRM will prolong the life of Bt insecticides, and adherence to IRM plans benefits growers, producers, researchers, and consumers. EPA's strategy to address insect resistance to Bt is twofold: (i) they mitigate any significant potential for pest resistance by instituting IRM plans; and (ii) Bt toxins better understand how pest resistance happens and how it can best be stopped.

The development of scientifically sound and sustainable IRM strategies for Bt field corn has required a number of stakeholders (growers, seed suppliers, scientists, and regulators) to be engaged in the process. The EPA has held multiple fora that have focused on insect resistance management for Bt Crops: three Federal Insecticide Fungicide and Rodenticide Act Scientific Advisory Panel Meetings (1995, 1998, 2000), six public workshops (1999, 2001), two public hearings (1997), two Office of Pesticide Program Dialogue Committee Meetings (1996, 1999), and one technical briefing (2001). In addition, a regional research committee that has been in existence since 1954 has provided a forum for these stakeholders. Currently, the NC205 committee formally addresses research on the ecology and management of the European corn borer and other stalk-boring Lepidoptera. It includes scientists from 20 US states, Mexico, and Canada and is supported by land grant universities, and the United States Department of Agriculture, Cooperative State Research, Education, and Extension Service (USDA-CSREES) and Agricultural Research Service (USDA-ARS). NC205 has sponsored annual resistance management meetings with industry, academics, and the EPA and several symposia and conferences in order to discuss IRM issues. The meetings have provided opportunities for sharing information, establishing research priorities, and building trust among the participants. The overall goal of all of these stakeholders' fora has been to identify the most viable science-based practical resistance management strategies. Members of the NC205 committee emphasize the importance of 'practical' IRM, because they recognize that the ultimate stewards of the Bt technology are the growers. So, whenever possible and without compromising the scientific integrity of resistance management, grower realities are considered. Members of the National Corn Growers Association (NCGA) and the industry group Agricultural Biotechnology Stewardship Technical Committee (ABSTC), currently representing Dow AgroSciences, Monsanto, Pioneer and Syngenta who have been regular participants in the meetings, have welcomed this approach. Grower participation in IRM meetings has been particularly valuable because their interest in Bt corn (*Zea mays*) is high (Fig. 8.1; USEPA, 2001, p. I 23) and they often bring common-sense approaches to the forefront.

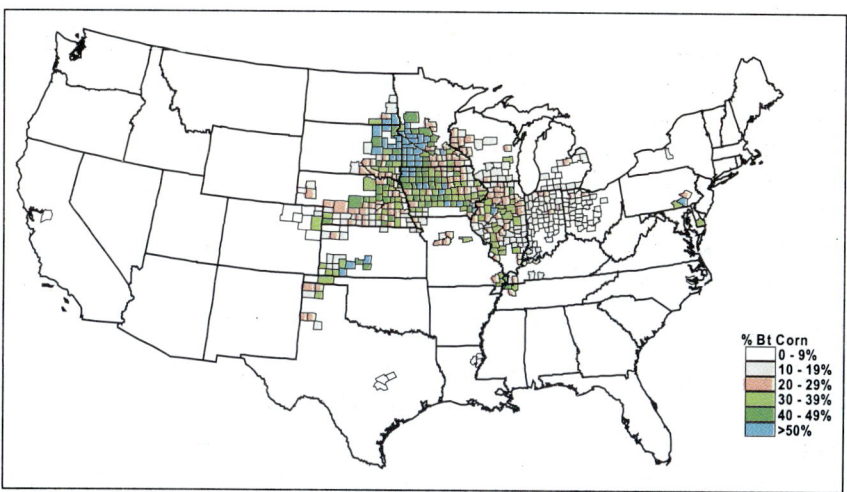

Fig. 8.1 Percentage of total corn acreage planted to Bt corn hybrids in countries in which greater than 50,000 total acres of corn was planted. (*Source:* Bt corn industry 1999 sales data)

Resistance management for Bt crops has focused on the use of the high-dose/structured refuge strategy to mitigate insect resistance to specific Bt endotoxins. Scientists believe that a high dose and the planting of a refuge (a portion of the total acreage using non-Bt plants) delays the development of insect resistance to Bt crops by maintaining insect susceptibility.

This chapter considers IRM for field corn with an emphasis on: (i) appropriate dose expression and refuge; (ii) resistance monitoring; (iii) remedial action plan; and (iv) grower participation. This IRM approach assumes that the pest biology and ecology are well known and that the Bt crops are used in a way that complements an overall IPM plan.

HIGH DOSE AND STRUCTURED REFUGE STRATEGY

Several resistance management strategies that minimize selection pressures for resistant insects when feeding on Bt crops have been proposed. The strategy that has received the most attention, however, involves two components: high dosage and refuge (Roush, 1997a,b; Gould, 1998a,b). The high-dose/refuge strategy assumes that resistance to Bt plants is recessive and is conferred by a single locus with two alleles, resulting in three insect genotypes: susceptible homozygotes (SS); heterozygotes (RS); and resistant homozygotes (RR) (Roush, 1994; Gould, 1998b; USEPA, 1998; Bourguet et al., 2000). It also assumes that there will be a low initial frequency of resistance alleles and that there will be random mating between resistant and susceptible adults. Under ideal circumstances, only rare RR individuals will survive a high dose produced by the Bt crop, and both SS and RS individuals will be susceptible to the Bt toxin. Susceptible insects SS from refuge will mate with rare resistant RR insects surviving the Bt crop so as to produce susceptible RS heterozygotes that will be killed by the Bt crop. This strategy should dilute resistant (R) alleles from the insect populations and delay the evolution of resistance.

High Dose

An EPA Science Advisory Panel (SAP, 1998) noted that a Bt plant-incorporated protectant could be considered to provide a high dose if verified by at least two of the following five approaches:

- Serial dilutions bioassay with artificial diet containing lyophilized tissues of Bt plants with tissues from non-Bt plants as controls.
- Bioassays using plant lines with expression levels approximately 25-fold lower than the commercial cultivar determined by quantitative enzyme linked immunosorbent assay (ELISA) or some more reliable technique.
- Survey large numbers of commercial plants in the field to make sure that the cultivar is at the lethal dose $(LD)_{99.9}$ or higher to ensure that 95% of heterozygotes would be killed (Andow and Hutchison, 1998).
- Similar to approach 3 but would use controlled infestation with a laboratory strain of the pest that had an LD_{50} value similar to the field strains.
- Determine whether a later instar of the targeted pest could be found with an LD_{50} that was approximately 25-fold higher than that

of the neonate larvae. If so, the later stage could be tested on the Bt crop plants to determine whether ≥95% of the larvae were killed.

EPA has adopted the 25 × definition of 'high dose,' but agrees with a 2000 SAP Subpanel that this definition is 'imprecise, provisional, and may require modification as more knowledge becomes available about the inheritance of resistance' (SAP, 2001).

The high-dose/refuge strategy assumes that resistance alleles are initially rare. That is, it is assumed that Bt resistance alleles are $<10^{-3}$ for the high-dose/refuge strategies currently used for Bt crops. Studies using the F_2 screen by Andow et al. (1998, 2000) and Andow and Alstad (1998, 1999) indicate that resistance alleles may be present at frequencies $<9 \times 10^{-3}$ in southern Minnesota and $<3.9 \times 10^{-3}$ in central Iowa. These data support the assumption that the frequency of Bt resistance alleles in natural populations of *Ostrinia nubilalis* (Hubner) is $<10^{-3}$, validating one of the key assumptions of the high-dose/refuge strategy.

Most of the currently-registered Bt corn products have been evaluated to determine high dose (via the 1998 SAP verification techniques) for *O. nubilalis*, the primary target pest. It is likely that BT11, MON810, and TC1507 corn types have a high dose for *O. nubilalis* (Walker et al., 2000). None of the currently-registered Bt corn products have been known to express a high dose for corn earworm, *Helicoverpa zea* (Boddie), which is known to be less susceptible to Bt proteins than other targeted lepidopteran pests. High-dose evaluations for other secondary pests like southwestern corn borer, *Diatraea grandiosella* Dyar and fall armyworm, *Spodoptera frugiperda* (J. E. Smith) have been sporadic.

The lack of a high dose could allow partially resistant (i.e. heterozygous RS insects) to survive, thus increasing the frequency of resistance genes in an insect population. For this reason, numerous IRM researchers and expert groups have concurred that non-high dose Bt expression presents a substantial resistance risk relative to high-dose expression (Roush, 1994; ILSI, 1998; Gould, 1998b; Onstad and Gould, 1998; SAP, 1998, 2001). Although the high-dose/refuge strategy is the preferred strategy for IRM in Bt crops, effective IRM may still be possible under non-high dose conditions, for example, by increasing the refuge size, or limiting the total acres.

Refuge

The size, placement, and management of refuge are critical to the success of strategies relying on high-dose/structured refuge to mitigate insect resistance to the Bt proteins. Structured refuges include all suitable non-Bt host plants for targeted pests that are planted and managed by people

(SAP, 1998). A 500:1 ratio of susceptible-to-resistant insects has been suggested as a suitable goal, assuming a resistance allele frequency of 5 × 10^{-2} (SAP, 1998, 2001). Planting date, size, and placement of the structured refuge should be based on the current pest biology and should also maximize the overlap of susceptible insects with possible resistant insects.

Non-Bt field corn provides the best refuge to increase the probability that susceptible insects will mate with potentially-resistant *O. nubilalis* from Bt corn. Current refuge requirements for corn in the United States are shown in Table 8.1. Non-Bt corn hybrids used as refuges should be selected for growth, maturity, fertility, irrigation, weed management, planting date, and yield traits similar to the Bt corn hybrid. Hybrids that are not agronomically similar may result in different developmental times in corn pests that could lead to assortative (nonrandom) mating between insects from refuge and Bt plants. Recent research has shown that temporal and alternate host, noncorn refuges (e.g. weeds, oats, alfalfa, and soybean) are inadequate strategies (Losey et al., 2001; USEPA, 2001).

Proximity

Refuge proximity is a critical variable for resistance management. Refuges must be located so that the potential for random mating between susceptible moths from the refuge and possible resistant survivors from the Bt cornfield is maximized. Therefore, insect flight and ovipositional behavior are critical variables to consider for refuge proximity. Refuges planted as external blocks should either be adjacent or in proximity to the Bt cornfield (Onstad and Gould, 1998). The current requirement for corn is that refuges should be within ~800 m (0.5 mi) of the Bt field, although within ~400 m (0.25 mi) is the preferred distance (Table 8.1).

There are trade-offs when considering refuge placement that involve movement of insect larvae and adults. In general, proximity of Bt and refuge plants benefits the random mating of adults, but it increases the chances that the larvae may move from Bt plants to refuge plants or from refuge plants to Bt plants. Distant and temporal plantings of Bt and refuge plants eliminate problems with larval movement, but potentially compromise. The random mating of susceptible and resistant adults. Refuge configurations that have been considered include external blocks, in-field strips, seed mixes, temporal refuge, and non-corn hosts. Several research projects have been conducted to identify the benefits and limitations for each configuration.

Many microhabitat factors are positively correlated with the population density of *O. nubilalis* adults, including relative humidity and

Table 8.1 Summary of Bt field corn refuge requirements

Protein/Event	Refuge size Corn belt	Refuge size Cotton areas	Proximity
Cry1Ab BT11	20% sprayed* or unsprayed	50% sprayed* or unsprayed	External blocks: 0.5 mi (0.25 mi or closer preferred)
MON810			
Cry1F			In-field strips: Must be at least four-rows in width
TC1507			(>6 rows preferred)

*Spraying is based on whether economic thresholds are reached for one or more target pests.
Source: USEPA (2001)

plant density (DeRozari et al., 1977; Hellmich et al., 1998). Mark-release-recapture *O. nubilalis* studies indicated a tendency for more adults to disperse away from the irrigated cornfields as compared with non-irrigated cornfields, presumably because of higher moisture levels in the irrigated fields (Hunt et al., 2001). These results suggest that non-random mating, which may compromise IRM plans, may occur more often in irrigated areas if the refuge is not placed in proximity or within the fields. For male *O. nubilalis* dispersal, another mark-recapture study by Showers et al. (2001) showed that males dispersing in search of mates may move considerable distances (>800 m). The authors suggested that, for male movement, the current refuge proximity guidelines of ~800 m (0.5 mi) should be adequate to ensure mating between susceptible individuals and resistant survivors from the Bt cornfield.

Seed mixes versus in-field strips versus external blocks

Refuge for Bt corn can be planted as blocks adjacent to the fields (edges or headlands), blocks within the fields, or even strips within the fields (Ostlie et al., 1997). Research has shown that *O. nubilalis* larvae are capable of moving upto six corn plants within or between the rows, with the majority of movement occurring within a single row. Older larvae (4th and 5th instars) *O. nubilalis* are more likely to move within rows than between rows (Ross and Ostlie, 1990). Movement between rows is a cause for concern because heterozygous (partially resistant) *O. nubilalis* larvae may begin feeding on Bt plants then move to nearby non-Bt plants in order to complete their development, thus possibly defeating the high-dose strategy and increasing the risk of resistance. Thus, seed mixes have been eliminated as possible *O. nubilalis* refuges (Mallet and Porter, 1992, Davis and Onstad, 2000).

Putting non-Bt seed in adjacent outside planter boxes and Bt seed in the remaining boxes can produce in-field strips. These resulting strips should extend the full length of the field and include a minimum of six rows planted with non-Bt corn alternating with a Bt corn hybrid. Due to the concerns with larval movement, wider refuge strips (≥6 rows) are preferred to narrower strips. Growers with planters eight rows or less may use a minimum of four rows (SAP, 2001). (Refuge seed in two adjacent outside rows of an eight-row planter produces four-row refuge/ 12-row Bt strips, i.e. 25% refuge.) In-field strips may offer the greatest potential to ensure random mating between susceptible and resistant adults because they can maximize random mating of adults. Modeling indicates that strips at least six rows in width are as effective for *O. nubilalis* IRM as adjacent blocks when a 20% refuge is used (Onstad and Guse, 1999).

Given the trade-offs with larval movement and adult random mating, either external blocks or in-field strips (across the entire field, at least six rows in width) may be the optimal refuge configurations. Seed mixing is the least preferred configuration and is not allowed under the current IRM requirements mandated by EPA (USEPA, 2001).

Temporal and spatial refuge

The use of temporal or spatial mosaics has received some attention as alternate strategies to structured refuge to delay resistance. A temporal refuge, in theory, would manipulate the life cycle of *O. nubilalis* by having the Bt portion of the crop planted at a time in which it would be most attractive to *O. nubilalis*. For example, transgenic cornfields would be planted before conventional corn. Because *O. nubilalis* are thought to preferentially oviposit on taller corn plants, the hypothesis is that the Bt corn will be infested instead of the shorter, less attractive conventional corn. However, there are indications that temporal refuges are an inferior alternative to structured refuges (SAP, 1998). Research has shown that planting date cannot be used to accurately predict and manipulate *O. nubilalis* oviposition rates (Pilcher and Rice, 2001; USEPA, 2001). Local climatic effects on corn phenology render the planting date a difficult variable to manipulate for managing *O. nubilalis*. Additional studies will have to be conducted under a broad range of conditions to fully answer this question. In addition, a temporal mosaic may lead to assortative mating in which resistant moths from the Bt crop mate with each other because their developmental time differs from susceptible moths emerging from the refuge (Gould, 1994).

Spatial mosaics involve the planting of two separate Bt corn events, with different modes of action. The idea is that insect populations will be exposed to multiple proteins, reducing the likelihood of resistance to either protein. The primary pests of corn (*O. nubilalis*, *H. zea*, and *D. grandiosella*) generally remain on the same plant throughout the larval feeding stages, individual insects will be exposed to only one of the proteins. In the absence of structured refuges producing susceptible insects, resistance may still have the potential to develop in such a system as it would in a single protein monoculture. Future products may have multiple Bt or other proteins (pyramiding the genes) expressed in the same corn hybrids that may decrease the probability of resistance evolving by a single (or even) multiple target insect species. When resistance mechanisms are independent, pyramiding of toxin genes may be the most effective way to manage resistance to Bt and other insecticidal transgenic toxins. According to Roush (1998), pyramids have the potential to greatly reduce the refuge size requirements for successful resistance management.

RESISTANCE MONITORING

Bt corn registrants are required to monitor for insect resistance (shifts in the frequency of resistance-conferring alleles) to the Bt toxins as an important early warning sign to resistance development in the field and also to determine whether IRM strategies are working. An additional value of resistance monitoring is that it may provide validation of parameters used in IRM models. Effective monitoring programs should have well-established baseline susceptibility data, sensitive detection methods, and a reliable collection network. Chances of finding resistant larvae in Bt corn depend on level of pest pressure, frequency of resistant individuals, number of samples, and sensitivity of the detection technique. Therefore, as the frequency of resistant individuals or the number of collected samples increases, so does the likelihood of sampling a resistant individual (Roush and Miller, 1986). The goal is to detect resistance in an insect population before the occurrence of widespread crop failures, and if possible, in time, so that mitigation practices can delay the development of resistance.

Monitoring for resistance should be undertaken in areas where the pests are known to regularly overwinter. Other secondary pests also may need to be monitored (on an individual basis), because these pests may be of local or regional significance. Previous experience with conventional insecticides has shown than once the resistant phenotypes are detected at a frequency >10%, control or crop failures are common (Roush and Miller, 1986). Due to sampling and sensitivity limitations, resistance could develop to Bt toxins before it is easily detected in the field. Sampling locations should be selected to reflect all crop production practices and should be separated by a sufficient distance to reflect distinct populations but should focus on intensively-planted Bt crop areas in which selection pressure is expected to be high.

Resistance detection and monitoring can be difficult and imprecise because rare genes are hard to detect. For example, if the phenotypic frequency of resistance is 1 in 1,000 then >3,000 individuals must be sampled to have a 95% probability of one resistant individual (Roush and Miller, 1986). Several methods have been proposed: (i) grower reports of unexpected damage; (ii) systematic field surveying of Bt corn; (iii) diagnostic concentration assay; (iv) F_2 screen; (v) screening against resistant colonies; and (vi) sentinel Bt crop. field plots.

Grower Reports of Unexpected Damage

Grower participation by reporting unexpected damage in their Bt corn is an important first step for monitoring. Growers are encouraged to report any unsuspected control problems to a local technical expert. Registrants provide growers with toll-free telephone numbers or a Web site to report any unusual control problems. A confirmed grower report of unexpected pest damage in a Bt crop may be a way to document a control failure and may be a useful monitoring system for determining the success or failure of the existing resistance management strategies.

Systematic Field Surveillance

Registrant-sponsored surveys of grower Bt fields for damaged plants could be used to monitor resistance phenotypes and gauge the geographic area where resistant populations exist. In-field detection systems (for quick determination of the presence or absence of Cry proteins in corn plants) have been developed for Cry1Ab and Cry1F proteins.

Diagnostic Dose

The EPA currently requires the diagnostic dose bioassay. The diagnostic dose does not completely discriminate between resistant and susceptible individuals, but it is used as an indication of resistance. A discriminating dose, on the other hand, discriminates individuals 100% of the time but will not be available until a resistant population can be characterized (Blair Siegfried, personal communication). Diagnostic dose bioassays are most efficient when resistance is common or conferred by a dominant allele (resistance allele frequency >1%) (Andow and Alstad, 1998). If resistance is inherited as a recessive trait, the frequency of individuals in a population that demonstrate resistance will equal the square of the allele frequency. For example, if the initial resistance allele frequency is

1 in 1,000, then only 1 in a million larvae would be expected to be homozygous resistant. Typically, diagnostic dose assays are based on 100-300 larvae to detect resistance at a frequency of 1-3% (Roush and Miller, 1986). It should be considered as one of the central components of any monitoring plan, but other monitoring methods may prove valuable in conjunction with the diagnostic concentration assay.

F_2 Screen

The F_2 screen may be a useful monitoring technique for Bt corn, especially for the detection of resistance alleles that are rare and recessive. The technique also allows fewer samples to be collected in order to detect potential susceptibility shifts rather than the discriminating dose assay. The F_2 screen is conducted by sampling mated females from natural populations, rearing the progeny of each female as an isofemale line, and sib-mating F_1 progeny. The F_2 larvae resulting from sib matings are assayed using an appropriate screening procedure such as a discriminating concentration assay or Bt crop, followed by statistical analyses of the family data. This method may be the most useful to analyze populations that are expected to be at high risk for resistance development. Each isofemale line allows for characterization of four genomes, thus improving the sensitivity over the discriminating dose assay by 10-fold (Andow and Alstad, 1998). The F_2 screen could be an effective method for detecting changes in the allele frequency of a recessive or partially-recessive allele and can be used to verify some of the assumptions underlying high-dose/refuge resistance management (Andow and Alstad, 1998; Andow et al., 1998). If resistance alleles are found, they can be characterized to estimate the fitness of the genotypes, determine whether there is a cost of resistance, and enable predictions of the evolution of resistance. The F_2 screen has been used to estimate the frequency of resistance to *Cry* toxins from *B. thuringiensis* in *O. nubilalis* (Andow et al., 1998, 2000); rice stem borer, *Scirpophaga incertulas* (Walker) (Bentur et al., 2000); and diamondback moth, *Plutella xylostella* (Linnaeus) (Zhao et al., 2002).

Zhao et al. (2002) tested the sensitivity of the F_2 screen to detect the frequency of rare resistance alleles using resistant populations of the diamondback moth. The moths were previously selected in the laboratory for resistance to Cry1Ac and Cry1C toxins of Bt. On using Bt broccoli (*Brassica oleracea* var. *italica*) as the diagnostic substrate, only one F_2 family was detected for Cry1Ac resistance and no family was detected for Cry1C resistance. Six families were detected for either Cry1Ac or Cry1C resistance by using the diagnostic diet bioassay. Four F_2 families were confirmed to contain one copy of an allele resistant to Cry1Ac in the original single-pair matings and four other F_2 families contained an allele resistant to Cry1C. These results suggest that transgenic plants expressing a high level of a Bt toxin in an F_2 screen may underestimate the frequency of resistance alleles with false negatives, or fail to detect true resistance alleles. Zhao et al. (2002) concluded that the diagnostic diet assay was a better F_2 screen method to detect resistance alleles, especially for the Cry1Ac resistance in diamondback moth. They also concluded that further validation of the F_2 screen method for each insect-crop system should be conducted before the procedures used in the F_2 screen can be used routinely to detect rare Bt resistance alleles in field populations.

Screening Against Test Stocks

Gould et al. (1997) used a series of genetic crosses of field-collected males mated to laboratory-selected females from a highly-resistant line of tobacco budworm, *Heliothis virescens* (Fabricius), to estimate the frequency of a Cry1Ac resistance allele in a natural population. The method can identify recessive or incompletely dominant resistance alleles from field-collected males. Using a colony of *H. virescens* that can survive on transgenic Bt cotton (*Gossypium hirsutum*) producing the Cry1Ac delta endotoxin, they crossed field-collected males with virgin-resistant females. Individuals homozygous for resistance were produced when field-collected males had one of more copies of the resistance allele. By using an assay that discriminates between heterozygous- and homozygous-resistant individuals, they could determine which wild males carried a resistance allele. Using this allelic recovery method, Gould et al. (1997) estimated the resistance allele frequency to be 1.5×10^{-3}. This method is only useful when there are previously isolated resistance alleles.

Sentinel Bt Crop Field Plots

Venette et al. (2000) proposed the use of an in-field screen to examine resistance allele frequency. This method uses Bt sweet corn to screen for *O. nubilalis* and *H. zea* that are resistant to the Bt protein. That is, the Bt crop is used to screen for resistant individuals. By sampling large numbers of Bt-expressing plants for live corn borer larvae, the frequency of resistance can be estimated and resistant individuals can be collected for documentation of resistance. For example, Venette et al. (2000) suggest that sampling ears of Bt sweet corn (18-21 days postsilking stage) for European corn borer can increase the sampling efficiency by two orders of magnitude over splitting stalks. Late-planted sentinel Bt sweet corn would provide a highly attractive oviposition site for females and reduce the number of plants required to attain an acceptable sample size. If the Bt sweet corn is planted at the appropriate time, larval attack will cause extensive damage, and large areas of Bt sweet corn can be sampled rapidly by examining this damage. If potentially resistant individuals or populations are identified in the field, they must be brought to the laboratory so that the resistance can be verified.

There are potential problems with this method that must be addressed before its widespread adoption. A high number of false positives, e.g. larvae from off-type corn, would reduce the efficiency and accuracy of resistance allele measurement. Also, false positives could occur if the larvae moved between Bt and non-Bt off-types or weeds. The method is also limited if Bt genes do not occur in both sweet corn and field corn. Currently, there is only BT11 Cry1Ab field corn and sweet corn.

Current Monitoring Practices

The EPA has imposed specific monitoring requirements on all registrants of Bt field corn products (USEPA, 2001). This body has mandated that registrants will monitor for resistance and/or trends in increased tolerance for European corn borer, southwestern corn borer, or corn earworm. Sampling should be focused in those areas in which there is the highest risk of resistance development.

The ABSTC plan concentrates resistance monitoring in areas where Bt corn market penetration is highest as well as areas with the highest insecticide use. The ABSTC plan applies to both Cry1Ab (MON810 and BT11) and Cry1F (TC1507) Bt field corn hybrids. The plan includes the identification of counties growing >50,000 acres of field corn (Bt and non-Bt) to focus monitoring efforts. The ABSTC's proposed plan is designed to detect resistance when it reaches 1-5% of the population. Four corn-growing regions were identified and monitoring for each pest is ongoing in the regions in which the pests are prevalent (Fig. 8.2). The ABSTC proposed a sampling goal of four to six locations in areas predominantly infested with *O. nubilalis* (regions I and II) and two to three locations in areas coinfested with *O. nubilalis* and *D. grandiosella* (region III) or in areas predominantly infested with *D. grandiosella* (region IV). When possible, at least 200 first or second flight adults (100 females), 100 second flight egg masses, or 100 diapausing larvae per site are collected in each region, although insect population levels may limit the number collected. The ABSTC monitoring plan proposes to use diagnostic dose or discriminating dose bioassays in order to detect the resistance alleles once they reach a frequency of 1 in 100. An annual resistance monitoring report to the EPA is required every year.

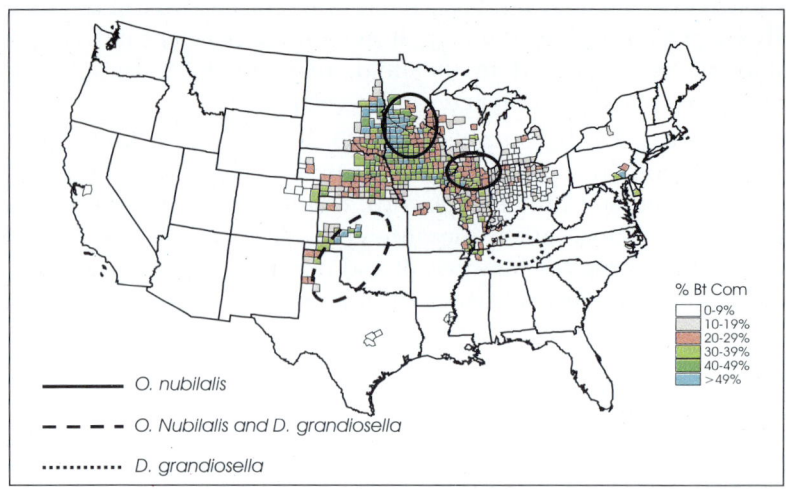

Fig. 8.2 Resistance monitoring regions for *O. nubilalis* and *D. grandiosella*

Baseline Susceptibility Data and Monitoring Results

The EPA currently mandates that both baseline susceptibility and diagnostic concentration are developed for certain primary target pests, including *O. nubilalis* and *H. zea*. This information is essential to managing resistance in pest populations, especially in assessing whether a field control failure was due to actual resistance or other factors affecting the expression of the Bt protein.

European corn borer and corn earworm

Blair Siegfried (University of Nebraska) has coordinated a standardized monitoring program for two corn pests involving lethal concentration $(LC)_{50}$ susceptibility determinations and diagnostic concentration (LC_{99}) bioassays to determine the susceptibility levels to Bt corn. For baseline susceptibility (LC_{50}), bioassays have been conducted for *O. nubilalis* (Siegfried et al., 1995; unpublished studies submitted to EPA, Master Record Identification Document (MRID) # 450369-02 and 453205-02, see discussion in USEPA, 2001) and *H. zea* (Siegfried et al., 2000). In 1999, *O. nubilalis* were collected from 14 separate sites and F_1, F_2 generations or both were bioassayed to determine LC_{50} values. Bioassays used dilutions of purified *Cry*1Ab obtained from *B. thuringiensis kurstaki* strain HD1-9 (Syngenta Biotechnology, Inc., Research Triangle Park, NC) spread on an artificial diet. In 2000, 13 *O. nubilalis* populations were sampled using similar procedures with formulated *Cry*1Ab protein (CellCap, Dow

AgroSciences, Indianapolis, IN). *O. nubilalis* are more sensitive to the CellCap *Cry*1Ab formulation; therefore, susceptibility results from 2000 are not directly comparable with those from 1995 to 1999. The results for *O. nubilalis* are displayed in Table 8.2 and show no significant change in *O. nubilalis* susceptibility (LC_{50} and effective concentration ($EC)_{50}$) to *Cry*1Ab over the first 5 years (1995-1999) of testing. Baseline susceptibility studies conducted by Marçon et al., (2000) were used to determine the diagnostic concentration (LC_{99}) for *O. nubilalis*. These tests with the diagnostic concentrations were conducted in a similar manner to the bioassays to determine LC_{50} values. For 2000 analyses, a new diagnostic dose (10 ng/cm^2) was established for the CellCap *Cry*1Ab formulation. The results for populations in both years showed nearly 100% mortality for *O. nubilalis* at the diagnostic dose (LC_{99}) (MRID # 450369-02 and 453205-02). For *H. zea*, baseline susceptibility (LC_{50}) values ranged from 70.3 (laboratory colony) to 221.3 ng/cm^2 (field colony). A separate diagnostic concentration analysis (using similar methods to those used for *O. nubilalis*) was conducted for *H. zea* (dose of 6600 ng/cm^2), which showed nearly 100% mortality (Siegfried et al., 2000). None of the *O. nubilalis* and *H. zea* populations show <99% mortality at a diagnostic concentration, and the LC_{50} for *O. nubilalis* has not significantly changed in 5 years. Thus, *Cry*1Ab susceptibility of *O. nubilalis* and *H. zea* populations has not changed.

Southwestern corn borer

Additional monitoring work has been done with *D. grandiosella*. Based on collections from 1998 and 1999, studies were conducted by Andi Trisyono and Michael Chippendale in order to determine *D. grandiosella* susceptibility to *Cry*1Ab and to establish a diagnostic concentration

Table 8.2 Mean susceptibility of *O. nubilalis* to Cry1Ab from 1995 to 2000 (unpublished study submitted to EPA, MRID # 450369-02)

Year	LC_{50} (ng Cry1Ab/cm^2) ± SEM	EC_{50} (ng Cry1Ab/cm^2) ± SEM
1995	4.34 ± 0.68	0.37 ± 0.007
1996	6.25 ± 1.25	1.25 ± 0.14
1997	2.12 ± 0.53	0.42 ± 0.007
1998	2.57 ± 0.28	0.43 ± 0.05
1999	4.01 ± 0.49	0.62 ± 0.11
2000*	0.12 - 0.49 **	Not Reported

* Data for 2000 from MRID # 453205-02.

** Data collected for 2000 were obtained using a different *Cry*1Ab formulation (CellCap) that is more toxic to *O. nubilalis*. As such, the results from 2000 are not directly comparable with results from previous years (1995-1999). LC_{50} values are given as a range (without SEM).

(unpublished study submitted to EPA, MRID # 450369-02). A bioassay was conducted that established a diagnostic concentration for *D. grandiosella* of 110 ng of Cry1Ab protein/g diet. Susceptibility data (LC_{50} values and EC_{50} values), determined after 7 and 14 days of exposure to Cry1Ab, are summarized in Table 8.3. *D. grandiosella* monitoring was also conducted for the 2000 growing season by Qisheng Song, using similar methodology (unpublished study submitted to EPA, MRID # 453205-02) to obtain susceptibility data (LC_{50} values and EC_{50} values). The susceptibility data are summarized in Table 8.3. A diagnostic concentration assay was performed (7-d test dose = 0.35 µg Cry1Ab/g diet, 14-d test dose = 5 µg Cry1Ab/g diet), which resulted in 100% mortality for all of the tested populations.

Table 8.3 *D. grandiosella* susceptibility to Cry1Ab from 1998 to 2000 (MRID # 450369-02 and 453205-02)

Year	LC_{50} (µg Cry1Ab/g diet)		EC_{50} (ng Cry1Ab/g diet)	
	Field populations	Laboratory colony	Field populations	Laboratory colony
1998	7-d: 0.22-1.09	7-d: 1.01	7-d: 2.2-6.6	7-d: 7.6
	14-d: 0.04-0.09	14-d: 0.28	14-d: 2.4-5.4	14-d: 6.2
1999	7-d: 0.07-0.17	7-d: 1.06-1.12	7-d: 2.6-3.7	7-d: 4.2-6.3
	14-d: 0.02-0.05	14-d: 0.26-0.34	14-d: 1.9-3.3	14-d: 4.9-5.1
2000*	7-d: 0.08-0.15	7-d: 0.98	14-d: 2.51-4.88	14-d: 4.97
	14-d: 0.04-0.09	14-d: 0.27		

* Units for the 2000 data are µg Cry1Ab/ml diet for LC_{50} values and ng Cry1Ab/ml diet for EC_{50} values.

Taken together, the *D. grandiosella* monitoring results show that, to date, no appreciable increase in susceptibility has resulted from exposure to Cry1Ab corn. Although the susceptibility data were variable and require further refinement, results indicated that the laboratory colonies evaluated were not as susceptible to Cry1Ab at the field-collected populations. Furthermore, the results from 1998 and 1999 indicated that a bioassay using growth inhibition is more sensitive than one based on larval mortality. A. Trisyono and M. Chippendale suggested that bioassays based on growth inhibition rather than larval mortality may have greater benefits because such bioassays require smaller amounts of Bt protein, sublethal effects can be observed, the time of observation is flexible (weight gain is being compared with a control), and variation is minimized (MRID # 450369-02).

Cross Resistance

Cross resistance occurs when a pest becomes resistant to one Bt protein that then allows the pest to resist other, separate Bt proteins. Efforts are underway to assess whether corn insects, especially *O. nubilalis*, have cross resistance to various Bt proteins. Future monitoring methods may incorporate such information because cross resistance is an area of major concern for resistance management and poses risks to both transgenic Bt crops and microbial Bt insecticides. Cross resistance also poses a risk to pyramid strategies, in which multiple proteins are deployed simultaneously in the same hybrid. To date, the development of cross resistance has not been shown in insect pests exposed in the field to Bt crops producing different Bt proteins. In general, it is possible for resistance to Bt proteins to occur through several different mechanisms, some of which may result in cross-resistance to other proteins. A well-documented mechanism of resistance is reduced (midgut) binding affinity to Bt proteins. Different Cry proteins may bind to distinct receptors in an insect gut. Modifications to these insect crystalline protein receptors have been implicated in resistance to Cry proteins. Other mechanisms that may lead to resistance (and ultimately cross-resistance) include protease inhibition, metabolic adaptations, gut recovery, and behavioral adaptations (Heckel, 1994; Tabashnik, 1994).

Cross resistance may result for binding sites, if two proteins share the same binding site (receptor) in the insect midgut. Therefore, if exposure to one Bt protein results in a modification of the receptor, other proteins sharing this site will be affected as well. An example of a possible shared binding site resulting in cross resistance was observed with tobacco budworm. *H. virescens* selected for resistance to Cry1Ac also were found to be resistant to the Cry1Aa, Cry1Ab, and Cry1F proteins (Gould et al., 1995).

Cross-resistance patterns in *O. nubilalis* have proven to be complicated. The binding of three Bt insecticidal crystal proteins to the midgut epithelium of *O. nubilalis* larvae was characterized by performing binding experiments with both isolated brush-border membrane vesicles and gut tissue sections (Denolf et al., 1993). Results demonstrated that two independent insecticidal crystal protein receptors are present in the brush border of *O. nubilalis* gut epithelium. From competition-binding experiments, it was concluded that Cry1Ab and Cry1Ac are recognized by the same receptor. Also, the Cry1B protein did not compete for the binding site of Cry1Ab and Cry1Ac and was determined to have a different receptor. Cry1D and Cry1E, two proteins that are not toxic to *O. nubilalis*, do not bind to gut epithelial cells. Other experiments using laboratory-selected resistant strains to predict survival and cross resistance in the field on Bt corn with *O. nubilalis* have provided different results. A *Cry*1Ac-resistant *O. nubilalis* strain (produced by Bill Hutchinson, University of Minnesota) and a *Cry*1Ab-resistant *O. nubilalis* strain (produced by Cliff Keil, University of Delaware) had a moderate level of resistance, approximately 30 to 60´. None of the resistant larvae survived on Bt corn beyond the second instar. Interestingly, the Cry1Ac-resistant *O. nubilalis* were not cross-resistant to Cry1Ab and Cry1Ab-resistant *O. nubilalis* were not cross-resistant to Cry1Ac (B. Hutchison, personal communication; reviewed by USEPA, 1998). Based on receptor-binding studies, both resistant strains would have been expected to survive on Bt corn. Thus, although the two proteins are closely related, there may be different binding mechanisms or binding affinity in *O. nubilalis* relative to other pests, such as *P. xylostella* or *H. virescens*.

Based upon the binding properties of Cry1A and Cry2A proteins in *H. zea*, *H. virescens*, and *O. nubilalis* larvae, there seems to be a much lower probability of cross resistance developing to Cry2A delta endotoxins from resistance to Cry1Ab or Cry1Ac. Since the Cry1A and Cry2A proteins exhibit different binding characteristics and very low amino acid homology, they probably possess different modes of action. However, there is some evidence for the development of broad cross-resistance to Cry1 and Cry2A in at least two laboratory-selected strains: beet armyworm, *Spodoptera exigua* (Hubner) (Moar et al., 1995) and *H. virescens* (Gould et al., 1992).

Overall, cross-resistance patterns and their underlying physiological mechanisms are very complex and somewhat unpredictable, even within a closely-related group of proteins and susceptible insects. To mitigate the risks of cross resistance to Bt corn, additional research will be needed to assess the potential for cross resistance with each Bt protein and the targeted pest. To date, research has been focused primarily on shared binding site studies with a limited subset of Bt protein and corn pests (notably *O. nubilalis*). Further mitigation measures could include the restrictions of certain hybrids determined to be at risk for cross resistance. Such measures have been undertaken in southern cotton-growing regions where *H. zea*, a pest of corn and cotton, may be exposed to multiple Bt toxins in both Bt corn and Bt cotton. Given the unpredictability of cross resistance among pest species, it would be useful to generate additional cross-resistance data for other species to gain a more complete understanding of the implications for Bt field corn.

REMEDIAL ACTION

Remedial action plans consist of response measures that are intended to contain Bt-resistant insects and perhaps eliminate them before they become widespread. This assumes that resistance will develop in localized populations. Bt corn registrants (applicable to MON810, BT11, and TC1507) are required to develop strategies for suspected resistance and confirmed resistance (USEPA, 2001, see Appendix 3).

For suspected resistance, the registrant must instruct growers to: (i) use alternate measures to control the pest suspected of resistance to Bt corn in the affected region; and (ii) destroy crop residues in the affected region immediately after harvest (i.e. within 1 mo) with a technique appropriate for local production practices.

Confirmed resistance must be reported to EPA within 30 days. The registrant must immediately stop the sale and distribution of Bt corn in the remedial action zone where resistance occurs until an effective local mitigation plan approved by the EPA has been implemented. A resistance event becomes confirmed if the progeny of the sampled *O. nubilalis*, *H. zea*, or *D. grandiosella* population exhibit all of the following characteristics in neonate bioassays:

- If there is >30% survival and >25% leaf area damaged in a 5-day bioassay using Cry1Ab-positive or Cry1F-positive leaf tissue under controlled laboratory conditions.
- If standardized laboratory bioassays using diagnostic doses for *O. nubilalis* (Marçon et al., 2000), *D. grandiosella* (MRID # 450369-02), or *H. zea* (USDA–ARS, Southern Insect Management Research Unit, unpublished data) demonstrate that the resistance has a genetic basis and survivorship in excess of 1% (gene frequency of population >0.1).
- If an LC_{50} in a standard Cry1Ab or Cry1F diet bioassay exceeds the upper limit of the 95% confidence interval of the standard unselected laboratory population LC_{50} for susceptible *O. nubilalis*, *D. grandiosella*, or *H. zea* populations, as established by the ongoing baseline monitoring program.

Eradication of a resistance gene (as part of a remedial action plan) may prove to be difficult. Rather, a plan based on slowing the spread of resistance genes (and possibly causing their decline) may prove to be more practical. As part of a plan to slow resistance genes, the following elements should be considered:

(i) education of growers and crop consultants to look for unexpected pest damage;

(ii) monitoring for plant damage, pest susceptibility, and resistance, allele frequency (with rapid verification and alternate control strategies for verified resistance);

(iii) sales suspensions of the affected product in that particular region until it can be shown that product benefits outweigh its risks;
(iv) continual monitoring to determine the effectiveness of the remedial action plan; and
(v) an assessment of how the resistance problem occurred (SAP, 2001).

GROWER EDUCATION AND COMPLIANCE

Growers are perhaps the most essential component of a successful IRM program because, ultimately, they are responsible for planting refuges and carrying out the details of an IRM plan. Thus, a program that educates growers about the importance of IRM and follows with compliance monitoring is an integral part of any resistance management strategy.

Education

For growers to plant refuges according to guidelines, they must be presented with consistent and up-to-date messages. When Bt corn was first commercialized in the United States, there was some confusion among growers because the various seed suppliers and academic groups recommended different refuge amounts. Subsequently, there was a concerted effort by members of the NC205 committee and biotech companies to coordinate messages. The Agricultural Biotechnology Stewardship Technical Committee was formed to coordinate industry IRM recommendations. Members of this committee worked closely with the members of NC205 and NCGA to communicate a consistent IRM message. Consequently, several efforts, notably those by NC205, NCGA, individual seed and technology companies, and ABSTC have developed grower education materials regarding IRM that have been useful to Bt corn users.

Specific examples of educational tools for growers include grower guides, technical bulletins, sales materials, training sessions, Web sites, toll-free numbers for questions or further information, and educational publications (Ostlie et al., 1997; Anderson and Hellmich, 2001). Most of these IRM education materials warn that the misuse of Bt technology will result in losing a valuable tool, hence product stewardship is always promoted. There is hope that growers will prove to be stewards of the Bt technology is much the same way they are the stewards of their land.

The NC205 committee has recognized that IRM strategies should be scientifically sound, but they must also consider the practical, logistical, and economic needs of the growers (a fine balance). There is no question that IRM plans should be scientifically valid. But, there is a question about whether growers will implement IRM plans that are not practical. High grower adoption of IRM strategies is ultimately the most important aspect of IRM. The best of plans will fail, if most growers cannot or will not use them. Generally, field specialists have taken a common-sense approach to IRM and try to work within the growers' equipment and field limitations. One approach is to offer growers a 'toolbox' of options that all fulfill the IRM requirements so that they can choose IRM plans that best fit their region of the country or individual operation. Growers should be allowed some flexibility with their plans, but any flagrant consistent misuse of the Bt technology must be dealt with accordingly.

Compliance

In addition to carrying out effective IRM education for growers, Bt corn registrants are required to establish a broad compliance program as part of the IRM requirements (USEPA, 2001, see V. Bt Corn Confirmatory Data and Terms and Conditions of the Amendment). Ideally, a compliance program should: (i) establish an enforcement structure that will maximize compliance; (ii) monitor the level of compliance; and (iii) investigate effects of noncompliance on IRM.

Enforcement structure

The first element of a system to ensure a high level of compliance is a mechanism to create a legally-enforceable obligation on Bt corn growers to comply with the refuge program. This is accomplished through grower agreements (USEPA, 2001, see V11-12). Registrants have flexibility to design grower agreement programs that fit their own business practices. As part of the compliance assurance plan, each registrant must establish and publicize a phased compliance approach. In other words, a guidance document will indicate how the registrant will address noncompliance within the terms of the IRM program and general criteria for choosing among options for responding to noncompliant growers (USEPA, 2001, see V12-16). Although recognizing that for reasons of difference in business practices, there are needs for flexibility between different companies, all Bt corn registrants must use a consistent set of standards for responding to noncompliance. The options will include withdrawal of the right to purchase Bt corn for an individual grower or for all growers in a specific region. An individual grower found to be significantly out of compliance 2 years in a row will be denied sales of the product the following year. Similarly, seed dealers who are not fulfilling their obligations to inform or educate growers of their IRM obligations will lose their opportunity to sell Bt corn.

Compliance monitoring

The IRM compliance assurance program includes an annual survey of a statistically-representative sample of Bt corn growers conducted by an independent third party. The survey should measure the degree of compliance with the IRM program by growers in different regions of the country and consider the potential impact of non-response. Each registrant is required to provide the EPA with compliance monitoring report and plans for updating the compliance assurance plan on an annual basis. Registrants are also required to follow up on tips and complaints concerning noncompliance with IRM requirements.

There have been several surveys and estimates of the level of grower compliance for Bt corn IRM. Marlin Rice (Iowa State University) and colleagues have conducted regular grower surveys to measure the grower attitudes toward various aspects of Bt corn, including compliance with IRM guidelines (Pilcher et al., 2002). These surveys have shown that the majority of growers understand and are receptive to the need for refuge and resistance management. However, they also demonstrate that some level of noncompliance is expected. The results from the 1996 grower survey conducted by Clint Pilcher and M. Rice showed that 23.5% of the sampled growers would follow a prescribed IRM strategy, 57.1% would do so if compatible with their growing practices, 7.2% would not follow IRM, and 12.2% "didn't know" (unpublished data submitted to EPA, MRID # 444754-01). Results from the 1998 grower survey showed that 25.5% of growers would implement recommended IRM, 58.9% would do so if it proved compatible with their growing practices, 2.6% would not follow IRM recommendations, and 12.9% 'didn't know' (unpublished data submitted to EPA, MRID # 450568-01).

For compliance information submitted by industry, ABSTC conducted a compliance survey for the 2000-growing season (USEPA 2001). The ABSTC compliance plan consists of grower agreements and contracts intensified education for regions showing low compliance and restrictions on future use of Bt corn for individual growers repeatedly out of compliance. The compliance survey was conducted by an independent marketing research firm and included anonymous telephone surveys of 501 total growers, each farming at least 200 acres. This survey did not involve visits to individual farms (i.e. grower audits). Compliance was assessed for two Bt corn IRM requirements: percentage of refuge (required ≥20%) and refuge proximity (required within 800 m of the Bt field). Survey respondents indicated that 87% planted an appropriate amount of refuge (at least 20%), whereas 13% had less than the required amount or no refuge. For proximity, 82% of growers reported refuges planted within 800 m of the Bt field (18% reported refuges planted greater than 1.6 km from the Bt field). When both refuge percentage and proximity are considered together, 71% of growers were in total compliance.

Collectively, these surveys indicate that 100% compliance is unlikely and that some level of noncompliance must be expected. An expectation of 30% (or greater) noncompliance may be reasonable, given these survey results. Without confirmatory visits to individual farms (i.e. audits), it may be impossible to verify the accuracy of the anonymous telephone survey on a regional or local basis. The end result could be increased false-positives, which may artificially inflate the estimates of grower compliance.

Effects of noncompliance

A significant noncompliance with IRM among growers may increase the risk of resistance for Bt corn. However, compliance is a complex issue for Bt crops and IRM. Currently, there is disagreement as to the appropriate refuge size or deployment, level of adoption, and level of compliance necessary to achieve risk reduction. Currently, the financial burden of implementing refuge requirements is borne primarily by the growers. Increasing the refuge size or limiting refuge deployment to better mitigate the risk of resistance is likely to increase the costs to growers and could result in a higher rate of grower noncompliance. Ultimately, predictive IRM models will need to be updated in order to reflect some degree of noncompliance, so that the potential impact can be more thoroughly understood.

OTHER IRM RESEARCH TOPICS

Research to improve the understanding of corn insect biology and ecology is underway at several institutions. In particular, research on adult mating biology and larval movement is emphasized. Two other important areas of research include IRM in high-sprayed areas and implication of *H. zea* north-south movement for IRM.

High-Spray IRM

In some parts of the western corn belt, particularly in areas with high *D. grandiosella* populations, there are concerns that regular spraying of refuges will result in too few susceptible adults. This is a complex problem that involves considering insecticide timing (i.e. which insects are targeted), and whether Bt corn is sprayed when the refuge corn is sprayed. Growers in some of these areas have realized that secondary pests, such as spider mites, *Tetranychus urticae* Koch and *Oligonychus pratensis* (Banks), may require insecticide treatments. Rather than pay the premium for Bt corn, some growers have opted to use non-Bt corn and time their insecticide treatments so as to control *O. nubilalis* or *D. grandiosella*. Also, there is speculation about whether areawide suppression of *O. nubilalis* and *D. grandiosella* due to high percentages of Bt corn could reduce the number of required insecticide treatments.

Corn Earworm Movement

During the summer, *H. zea* adults move northward from southern overwintering sites to corn-growing regions in the corn belt. There is a possibility that a large number of *H. zea* move from corn-growing regions in the north to cotton-growing regions in the south. Research is needed to assess whether this phenomenon occurs, and if it does in large enough numbers to compromise resistance management of *H. zea*, i.e. additional exposure to Bt crops and increased selection pressure for *H. zea* resistance. This effect is compounded by the fact that neither Bt cotton nor any registered Bt corn event contains a high dose for *H. zea*.

In considering this issue, the 2000 SAP indicated that *H. zea* refuge is best considered on a regional scale (instead of structured refuge on an individual farm basis), due to the long distance movements typical of this pest (i.e. refuge proximity is not as important for *H. zea*). According to the SAP, a 20% refuge (per farm) would be adequate for *H. zea*, provided the amount of Bt corn in the region does not exceed 50% of the total corn crop. If the regional Bt corn crop exceeds 50%, however, additional structured refuge may be necessary (SAP, 2001). However, the SAP did not define what a region should be (i.e. county, state, or any other division). Additional research will likely be needed to fully determine the risk of *H. zea* north-south movement and appropriate mitigation measures.

CONCLUSIONS

The EPA requires an unprecedented IRM program for Bt crops. The specific IRM strategies and requirements for Bt corn in the United States have been developed by a coalition of stakeholders including EPA, USDA, academic researchers, industry, seed companies, public interest groups, and growers. Many of these stakeholders recognize that IRM strategies need to be scientifically sound, practical, flexible, implementable, and sustainable. IRM requirements for Bt corn include a 20% mandatory non-Bt corn refuge in the corn belt and a 50% mandatory non-Bt corn refuge in cotton-growing areas to be planted within 800 m (400 m or closer preferred) in order to mitigate insect resistance. There are also requirements for annual resistance monitoring, remedial action plan, grower education, grower compliance, research, and annual reporting. The Cry1Ab registrations for Bt11 and MON810 field corn hybrids and the Cry1F registrations for TC1507 field corn hybrids will automatically expire on midnight 15 October 2008. Additional IRM research on the effect of north-south movement by corn earworm and high use of insecticide sprays will allow current IRM strategies to be further improved for greater long-term sustainability.

ACKNOWLEDGEMENTS

The authors wish to thank the following individuals for their time and assistance in the development and review of this manuscript: Janet Andersen, Director, Michael Glikes, and Robyn Rose, USEPA/Office of Pesticide Programs/Biopesticides and Pollution Prevention Division; and Douglas Sumerford USDA–ARS, Corn Insects and Crop Genetics Research Unit.

REFERENCES

Anderson, P. L. and Hellmich, R. L., 2001, Bt corn and insect resistance management: What are they? Site-Specific Management Guidelines, Potash and Phosphate Institute, South Dakota State University.

Andow, D. A. and Alstad, D. N., 1998, The F_2 screen for rare resistance alleles. J. Econ. Entomol., **91**, 572-578.

Andow, D. A. and Alstad, D. N., 1999, Credibility interval for rare resistance allele frequencies. J. Econ. Entomol., **92**, 755-758.

Andow, D. A., Alstad, D. N., Pang, Y. -H., Bolin, P. C. and Hutchison, W. D., 1998, Using a F_2 screen to search for resistance alleles to *Bacillus thuringiensis* toxin in European corn borer (Lepidoptera: Crambidae). J. Econ. Entomol., **91**, 579-584.

Andow, D. A. and Hutchison, W. D., 1998, Bt corn resistance management. In: Mellon M. and Rissler J. (eds), Now or never: Serious new plans to save a natural pest control, Union of Concerned Scientists, Washington, DC, pp. 19-66.

Andow, D. A., Olson, D. M., Hellmich, R. L., Alstad, D. N. and Hutchison, W. D., 2000, Frequency of resistance to *Bacillus thuringiensis* toxin Cry1Ab in the Iowa population of European corn borer (Lepidoptera: Crambidae). J. Econ. Entomol., **93**, 26-30.

Bentur, J. S., Andow, D. A., Cohen, M. B., Romena, A. M. and Gould, F., 2000, Frequency of alleles conferring resistance to a *Bacillus thuringiensis* toxin in a Philippine population of *Scripophaga incertulas* (Lepidoptera: Pyralidae). J. Econ. Entomol., **93**, 1515-1521.

Bourguet, D., Genissel, A. and Raymond, M., 2000, Insecticide resistance and dominance levels. *J. Econ. Entomol.*, **93**, 1588-1595.

Davis, P. M. and Onstad, D. W., 2000 Seed mixtures as a resistance management strategy for European corn borers (Lepidoptera: Crambidae) infesting transgenic corn expressing Cry1Ab protein. J. Econ. Entomol., **93**, 937-948.

Denholf, P., Jansens, S., Peferoen, M., Degheele, D. and Van Rie, J., 1993, Two different *Bacillus thuringiensis* delta-endotoxin receptors in the midgut brush border membrane of the European corn borer, *Ostrinia nubilalis* (Hübner) (Lepidoptera: Pyralidae). Appl. Environ. Microbiol., **59**, 1828-1837.

DeRozari, M. B., Showers, W. B. and Shaw, R. H., 1977, Environment and the sexual activity of the European corn borer. Environ. Entomol., 6, 657-665.

Gould, F., 1994, Potential and problems with high-dose strategies for pesticidal crops. Biocontr. Sci. Technol., 4, 535-548.

Gould, F., 1998a, Evolutionary biology and genetically engineered crops. BioScience, 38, 26-33.

Gould, F., 1998b, Sustainability of transgenic insecticidal cultivars: Integrating pest genetics and ecology. Annu. Rev. Entomol., 43, 701-726.

Gould, F., Anderson, A., Jones, A., Sumerford, D., Heckel, D., Lopez, J., Micinski, S., Leonard, R. and Laster, M., 1997, Initial frequency of alleles for resistance to *Bacillus thuringiensis* toxins in field populations of *Heliothis virescens*. Proc. Natl. Acad. Sci. USA, 94, 3519-3523.

Gould, F., Anderson, A., Reynolds, A., Bumgarner, L. and Moar, W., 1995, Selection and genetic analysis of a *Heliothis virescens* (Lepidoptera: Noctuidae) strain with high levels of resistance to *Bacillus thuringiensis* toxins. J. Econ. Entomol., 88, 1545-1559.

Gould, F., Martinez-Ramirez, A., Anderson, A., Ferré, J., Silva, F. and Moar, W., 1992, Broad spectrum Bt resistance. Proc. Natl. Acad. Sci. USA, 89, 1545-1559.

Heckel, D. G., 1994, The complex genetic basis of resistance to *Bacillus thuringiensis* toxin in insects. Biocontr. Sci. Technol., 4, 405-417.

Hellmich, R. L., Pingel, R. L. and Hansen, W. R., 1998, Influencing European corn borer (Lepidoptera: Crambidae) aggregation sites in small grain crops. Environ. Entomol., 27, 253-259.

Hunt, T. E., Higley, L. G., Witkowski, J. F., Young, L. J. and Hellmich, R. L., 2001, Dispersal of adult European corn borer (Lepidoptera: Crambidae) within and proximal to irrigated and non-irrigated corn. J. Econ. Entomol., 94, 1369-1377.

International Life Science Institute (ILSI), 1998, An Evaluation of Insect Resistance Management in Bt Field Corn: A Science-based Framework for Risk Assessment and Risk Management. ILSI Press, Washington, D.C., 78 pp.

Losey, J. E., Calvin, D. D., Carter, M. E. and Mason, C. E., 2001, Evaluation of non-corn host plants as a refuge in a resistance management program for European corn borer (Lepidoptera: Crambidae) on Bt-corn. Environ. Entomol., 30, 728-735.

Mallet, J. and Porter, P., 1992, Preventing insect adaptation to insect-resistant crops: are seed mixtures or refugia the best strategy? Proc. Roy. Soc. Lond. B, 255, 165-169.

Marçon, P., Siegfried, B., Spencer, T. and Hutchinson, W., 2000, Development of diagnostic concentrations for monitoring *Bacillus thuringiensis* resistance in European corn borer (Lepidoptera: Crambidae). J. Econ. Entomol., 93, 925-930.

Moar, W. J., Pusztai-Carey, M., van Faasen, H., Bosch, D., Frutos, R., Rang, C., Luo, K. and Adang, M. J., 1995, Development of *Bacillus thuringiensis* CryIC resistance by *Spodoptera exigua* (Hübner) (Lepidoptera: Noctuidae). Appl. Environ. Microbiol., 61, 2086-2092.

Onstad, D. W. and Gould, F., 1998, Modeling the dynamics of adaptation to transgenic maize by European corn borer (Lepidoptera: Pyralidae). J. Econ. Entomol., 91, 585-593.

Onstad, D. W. and Guse, C. A., 1999, Economic analysis of transgenic maize and nontrangenic refuges for managing European corn borer (Lepidoptera: Pyralidae). J. Econ. Entomol., 92, 1256-1265.

Ostlie, K. R., Hutchinson, W. D. and Hellmich, R. L., 1997, Bt Corn and European Corn Borer: Long-term Success Through Resistance Management, North Central Regional Extension Publication, NCR 602, USA.

Pilcher, C. D. and Rice, M. E., 2001, Effect of planting dates and *Bacillus thuringiensis* corn on the population dynamics of European corn borer (Lepidoptera: Crambidae). J. Econ. Entomol., **93**, 730-742.

Pilcher, C. D., Rice, M. E., Higgins, R. A., Hellmich, R. L., Witkowski, J. F., Calvin, D. D., Ostlie, K. R. and Steffey, K. L., 2002, Biotechnology and the European corn borer: Measuring farmer perceptions and the adoption of transgenic Bt corn as a pest management strategy. J. Econ. Entomol., **95**, 878-892.

Ross, S. E. and Ostlie, K. R., 1990, Dispersal and survival of early instars of European corn borer (Lepidoptera: Pyralidae) in field corn. J. Econ. Entomol., **83**, 831-836.

Roush, R. T., 1994, Managing pests and their resistance to *Bacillus thuringiensis*: Can transgenic crops be better than sprays? Biocontr. Sci.Technol., **4**, 501-516.

Roush, R. T., 1997a, Managing resistance to transgenic crops. In: Carozzi N. and M. Koziel (eds), Advances in insect control: the role of transgenic plants, Taylor & Francis, London, pp. 271-294.

Roush, R. T., 1997b, Bt-transgenic crops: Just another pretty insecticide or a chance for a new start in resistance management? Pestic. Sci., **61**, 328-334.

Roush R. T., 1998, Two-toxin strategies for management of insecticidal transgenic crops: Can pyramiding succeed where pesticide mixture have not? Proc. Roy. Soc. Lond. B, **353**, 1777-1786.

Roush, R. T. and Miller, G. L., 1986, Considerations for design of insecticide resistance monitoring programs. J. Econ. Entomol., **79**, 293-298.

Scientific Advisory Panel (SAP), 1998, Subpanel on *Bacillus thuringiensis* (Bt) Plant-Pesticides (February 9-10, 1998). Transmittal of the final report of the FIFRA Scientific Advisory Panel Subpanel on *Bacillus thuringiensis* (Bt) Plant-Pesticides and Resistance Management, Report dated, April 28, 1998. (Docket Number: OPPTS-00231).

Scientific Advisory Panel (SAP), 2001, Subpanel on Insect Resistance Management (October 18-20, 2000). Report: Sets of scientific issues being considered by the Environmental Protection Agency regarding: Bt plant-pesticides risk and benefit assessments. Report dated, March 12, 2001. (pp. 5-33).

Showers, W., Hellmich, R. L., Derrick-Robinson, M. and Hendrix III, W., 2001,Aggregation and dispersal behavior of marked and released European corn borer (Lepidoptera: Crambidae) adults. Environ. Entomol., **30**, 700-710.

Siegfried, B. D., Marcon, P. C. R. G., Witkowski, J. R., Wright, R. J., and Warren, G. W., 1995, Susceptibility of field populations of European corn borer, *Ostrinia nubilalis* (Hübner) (Lepidoptera: Pyralidae), to *Bacillus thuringiensis* (Berliner). J. Agric. Entomol., **12**, 267-273.

Siegfried, B. D., Spencer, T. and Nearman, J., 2000, Baseline susceptibility of the corn earworm (Lepidoptera: Noctuidae) to the Cry1Ab toxin from *Bacillus thuringiensis*. J. Econ. Entomol., **93**, 1265-1268.

Tabashnik, B. E., 1994, Evolution of resistance to *Bacillus thuringiensis*. Annu. Rev. Entomol., **39**, 47-79.

U.S. Environmental Protection Agency (USEPA), 1998,The Environmental Protection Agency White paper on Bt Plant-Pesticide Resistance Management. U.S. EPA, Biopesticides and Pollution Prevention Division (7511W) 14 January 1998. [EPA Publication 739-S-98-001]

U.S. Environmental Protection Agency (USEPA), 2001, Biopesticides Registration Action Document: *Bacillus thuringiensis* Plant-Incorporated Protectants (10/16/01), posted at http://www.epa.gov/pesticides/biopesticides/Pips/bt_brad.htm.

Venette, R. C., Hutchison, W. D. and Andow, D. A., 2000, An in-field screen for early detection and monitoring of insect resistance to *Bacillus thuringiensis* in transgenic crops. J. Econ. Entomol., **94**, 1055-1064.

Walker, K. A., Hellmich, R. L. and Lewis. L. C., 2000, Late-instar European corn borer (Lepidoptera: Crambidae) tunneling and survival in transgenic corn hybrids. J. Econ. Entomol., **93**, 1276-1285.

Zhao, J. -Z., Li, Y. -X., Collins, H. L. and Shelton, A. M., 2002, Examination of the F_2 screen for rare resistant alleles to *Bacillus thuringiensis* toxins in the diamondback moth (Lepidoptera: Plutellidae). J. Econ. Entomol. **95**, 14-21.

9

ROLE OF TRANSGENIC MICROBES AND ENDOPHYTES IN CROP PROTECTION

SARVJEET KAUR, RHITU RAI AND AQBAL SINGH
National Research Centre on Plant Biotechnology
Indian Agricultural Research Institute,
New Delhi-110 012, India

INTRODUCTION

The vocal concerns of the public at the extensive employment of chemical pesticides for global crop protection has intensified efforts to evolve strategies that reduce the damage to crop plants and also ensure protection to human health. *Bacillus thuringiensis* (Bt) gene products are an important element of such an alternative approach. Bt is a gram positive, spore-forming bacterium that produces parasporal insecticidal crystal proteins (ICP), encoded by *cry* genes, that have specific toxicity towards several insect orders while being harmless to non-target species (Schnepf et al., 1998). Delivery of Bt gene products to plants basically involves three broad strategies: (a) the Bt gene product ICP is administered to the plant in the form of foliar sprays (Gaertner et al., 1993; Chet et al., 1993); (b) the Bt gene is cloned into the plant genome to impart genetic resistance against insect damage (Estruch et al., 1997); (c) micro-organisms are involved in the delivery of Bt gene products to crop plants (Dimock et al., 1989; Kim, 1993; Turner et al., 1993).

The experience with these strategies has been a mixed one. The foliar sprays of ICP have their limitations because of their decreased persistence due to inactivation by ultraviolet light, heat, reaction with leaf exudates and surface pH changes. The development of Bt transgenic plants have also not found the expected popularity among the farmers due to environmental concerns arising from the persistence of such genes in the environment and concerns pertaining to the development of resistance in insects to Bt gene products (Shelton et al., 2002). Microorganisms are, therefore, being considered as an alternative for delivery of Bt gene products as they compensate for the limitations of the other approaches of delivering Bt genes to the plants. Enhancement of on-plant persistence as well as insecticidal efficacy of Bt-based biopesticides is possible through several molecular approaches (Kaur, 2000). Insecticidal *cry* genes have been transferred to other Bt strains in order to create recombinant Bt strains having increased insecticidal activity and spectrum of targets insect pests. The longevity of insect control has been increased by transferring the insecticidal *cry* genes from Bt to better-persisting microbes such as other *Bacillus* sps. residing in the phylloplanes of crops, non-pathogenic strains of P*seudomonas* sps., root-nodulating bacteria and endophytes. Bt genes have been cloned in microorganisms capable of withstanding vagaries of radiation and other abiotic factors and thus foliar sprays with improved foliar persistence are possible (Glick and Bashan, 1997).

Endophytic microorganisms represent a niche that they occupy in plants (Hallman et al., 1997; James and Olivares, 1997), that can be exploited for the expression of Bt gene in the plants, with features more acceptable than Bt transgenic crop plants (Turner et al., 1993; Saxena and Stotzky, 2000). The endophyte-based Bt delivery systems provide a safeguard against degradation of microbial pesticides; pollution of the environment from plants residues; localized delivery of the Bt gene product and, more importantly, this technology ensures repeated sales for the retailer since the seeds are to be treated prior to the sowing season with Bt gene-cloned endophyte. These products with their intrinsic advantage are, therefore, likely to have greater acceptability with the users. Recent advances in the development of Bt transgenic microbes and endophytes for crop production have been discussed below.

Bt TRANSGENIC MICROBES

Recombinant Bt Strains

Recombinant Bt strains having a broader spectrum and greater potency have been created through site-specific recombination. This elegant strategy makes use of site-specific recombination vectors, which are specially designed to selectively eliminate the antibiotic resistance marker genes after introduction into the host Bt strain (Baum et al., 1996). Such recombinant strains are desirable from the standpoint of ecological safety as they do not carry any 'foreign' non-Bt DNA (Sanchis et al., 1997). Furthermore, since such vectors do not carry any origin of replication with conjugative function, the risk of their transfer to other strains does not exist. A coleopteran-active recombinant Bt strain EG7653, producing three to four times more Cry3A protein, was constructed through site-specific recombination and was approved as the active ingredient of 'Raven OF, bioinsecticide by the US Environmental Protection Agency (EPA) [Baum et al., 1996]. The *cry1C* gene toxic to Egyptian cotton leaf worm, *Spodoptera littoralis* (Boisduval) was transferred to Bt Kto strain, having *cry1Ac* gene which has proved to be toxic to European corn borer, *Ostrinia nubilalis* (Hubner) through site-specific recombination. The resulting strain had a broader activity spectrum than the parental strain (Sanchis et al., 1997).

In a further improvement of this strategy, a chimeric *cry1C/Ab* gene carrying C-terminal domain of *cry1Ab* and N-terminal domains of *cry1C* gene, which was more toxic to *S. littoralis* than the unmodified *cry1C* gene, was used. The recipient strain was an asporogenic Bt mutant, which did not produce viable spores, an environmentally-advantageous feature (Sanchis et al., 1999). The recombinant Bt strain protected cabbage and broccoli from their pest complex under natural infestation conditions. The hybrid *cry* genes consisting of insecticidal specificity-conferring domains from different *cry* genes, can thus be used to broaden the insecticidal spectrum of the recombinant Bt strains. Fusion of toxic domain of *cry1Ab* from Bt subsp. *aizawai* and *cry1Ac* from Bt subsp. *entomocidus*, yielded a hybrid with the toxicity of both the genes (Honee et al., 1991). A chimeric gene, comprising Cry3Aa domain toxicity determining the region towards coleopteran pest *Leptinotarsa texana* and Cry1Ac C-terminal region, was expressed as a 140 kDa Cry protein in *E.coli* (Carmona and Ibarra, 1999).

Some Cry proteins can synergize the activity of other Cry proteins (Sayyed et al., 2001). Synergistic insecticidal activity of different Bt genes can also be brought about in recombinant Bt strains. High level of resistance to Cry3Aa toxin in the cotton wood leaf beetle, *Chrysomela scripta* Fabricius was suppressed by recombinant bacteria expressing the dipteran toxic *cyt1Aa* gene together with the coleopteran-toxic *cry3Aa* gene, suggesting synergistic effect of *cyt* 1Aa toxin (Federici and Bauer, 1998). Transformation of *cyt1A* gene from a Bt strain into a recombinant *B. sphaericus* strain restored its toxicity towards a resistant population of *Culex quinquefasciatus* Say (Wirth et al., 2000). However, instead of synergism, antagonistic interaction occurred between Cry1Ac and Cyt1A toxins, expressed together in a recombinant strain tested against *Trichoplusia ni* (Hubner) (Del Rancon-Castro et al., 1999). Thus, the net insecticidal activity of a recombinant strain would depend on the combinatorial interactions between individual genes.

Another approach for augmenting the insecticidal potential of Bt strains is by way of genetic manipulation of the regulatory elements of the *cry* gene expression (Kaur, 2002). Combining different genetic elements involved in *cry* gene transcription and translation can increase the *cry* gene expression in the recombinants. Use of sporulation-independent vegetative promoter such as that of *cry3A* has resulted in increased yield of Cry proteins in the recombinant strains (Malvar et al., 1994). The cry1A promotor, along with the mRNA stabilizing sequence STAB-SD, resulted in fourfold enhancement of toxicity of Bt subsp. *kurstaki* HD-1 to *Spodoptera exigua* (Hubner) (Park et al., 2000). Such strategies are promising in developing Bt strains of greater potency.

Bt Transgenic *Bacillus* spp.

The on-plant persistence of Bt sprays can be increased by transferring *cry* genes to other longer persisting *Bacillus* sps. The *cry1Ac* gene was transferred by conjugation in a *Bacillus megaterium* strain, which persisted in the cotton phyllosphere (Bora et al., 1994). The transipient strain persisted in the cotton phyllosphere for more than 28 days, as compared with the donor Bt subsp. *kurstaki* HD-1, which persisted for only upto 7 days. Protection from *Helicoverpa armigera* first instar larvae was observed in leaf bioassays upto 21 days post spray. The *cry1Ac* gene of Bt subsp. *kurstaki* was transferred to *Bacillus polymyxa* strain colonizing the rice phyllosphere (Sudha et al., 1999). The gene was efficiently expressed in the recombinant *B. polymyxa*, which was toxic to yellow stem borer of rice, *Scirpophaga incertulas* (Walker). Inoculation of rice plants with *B. polymyxa* increased the plant shoot and root growth, suggesting a dual biofertilizer and biopesticidal benefit of the application of these transgenic microbes. Transgenic *B. subtilis* and *B. licheniformis*, that naturally colonize the phylloplane of tomato plants, were developed using *cry1Ab* gene from a native Bt strain for protection from a South American tomato moth (*Tuta absoluta*) (Theoduloz et al., 2003) Bt isolates naturally occurring on the phylloplane of crop plants also hold the promise of development as biopesticides (Kaur and Singh, 2000).

The dipteran-active *cry* genes from Bt subsp. *israelensis* have been transferred to the longer persisting *B. sphaericus* in order to increase the persistence in the aquatic feeding zones for mosquito-control (Bar et al., 1998; Panbangred et al., 2000). The recombinant *B. sphaericus* produced a high level of Cry protein and was toxic to *Aedes*, *Anopheles* and *Culex* larvae (Poncet et al., 1997).

Bt Transgenic *Pseudomonas* spp.

To improve the foliar persistence of Bt biopesticide, scientists at the Mycogen Corporation (USA) have introduced the *cry* genes into a non-pathogenic strain of *Pseudomonas flourescens*. The recombinant *P. flourescens* strain produced crystal proteins encapsulated within the cell wall. By a proprietary chemical treatment, the cell wall was made more rigid through cross-linking for better persistence. This biopesticide was considered safe, being devoid of spores and live cells and was the first recombinant product approved for field tests and marketing by the USEPA. There was an approximate twofold increase in the foliar persistence of this biopesticide as compared to traditional Bt sprays for the control of lepidopteran and coleopteran pests of cabbage and potato (Gaertner et al., 1993). Leaf and root colonizing strains of *Pseudomonas* sps. also have the potential for development as transgenic bacteria-carrying *cry* genes. The *cry4Ba* gene from *B. thuringiensis* var. *morrisoni*, PG-14 was transferred into a grassroot colonizing *P. flourescens* strain PI, by using suicide vectors such that the integration was dependent upon homologous recombination (Waalwijk et al., 1991). The sequences flanking the *cry4Ba* gene are relatively specific to PI strain and the probability of homologous recombination upon encountering these sequences on other plasmids in other bacteria may be low. Nevertheless, the deployment of a live, recombinant strain necessitates appropriate environmental risk assessment. In another instance, the *cry* gene was transferred into the leaf colonizing *P. cepacia* but the recombinant strain had a slower rate of growth (Stock et al., 1990).

Bt Transgenic Nodulating Bacteria

Root-nodulating bacteria have been used as the alternate delivery system for *cry* genes to protect the nodules from soil-dwelling pests. Root nodule infestation of pigeonpea by the dipteran pest *Rivella angulata* was decreased by upto 40% by a recombinant *Bradyrhizobium*-carrying *cry*11Aa gene (Nambiar et al., 1990). Transgenic *Bradyrhizobium* cells expressing the *cry* gene have been constructed (da Costa Lima et al., 2000). The *cry1C* gene from Bt subsp. *aizwai* was introduced into nitrogen-fixing *Azospirillum lipoferum* by transformation (Gounder and Rajendran, 2001). The *cry3Aa* gene from Bt subsp. *tenebrionis* was transferred to *Rhizobium meliloti* and *R. leguminosarum* to protect the alfalfa and pea from coleopteran pests clover root cuculio, *Sitona hispidulus* (Fabricius) and the pea leaf weevil, *Sitona lineatus* (Linnaeus), respectively (Bezdicek et al., 1994). The recombinant *R. leguminosarum* strains were found to be equally competitive with the wild type strain for nodule occupancy. Interestingly, a better nodule occupancy by the recombinant *R. leguminosarum* strain carrying *cry3Aa* gene has been reported (Giddings et al., 1997).

Apart from insect control, the potential of transgenic *Rhizobium* spp. is also being exploited for the control of pathogenic fungi. *Rhizobium meliloti* cells, transgenic for chitinase gene on lysis, were able to inhibit the growth of *Rhizoctoria solani* (Sufnit et al., 1993). *Rhizobium*, besides inducing root nodules in legumes, can attach to the roots of non-host plant in the same way as they attach to the host plant (Chen and Phillips, 1976; Terouchi and Syno, 1990). This phenomenon, plus the ability to vastly colonize the rhizosphere of a wide variety of plants, make *Rhizobium* spp. the obvious choice for attempts to generate new biocontrol agents.

In a similar experiment, Chet and Inbar (1994) transferred the *chiA* gene from *Serratia marcescens* to *Trichoderma harzianim* and *R. meliloti* cells. In both the cases the transformed microorganisms expressed the chitinase gene and subsequently displayed increased antifungal activity in the soil. In a related experiment, Koby et al. (1994) found that the *chi A* gene endows *P. fluorescens* with the capacity to control pathogen *R. solani* in the soil.

Other Bt Transgenic Bacteria

The development of resistance in target insects to Bt genes can be delayed by the deployment of multiple insecticidal genes that have different receptors or mode of action in the insect gut (Frutos et al., 1999). The *cry* genes, in combination, can have a synergestic effect on the insect pests (Poncet et al., 1995). In addition, the genes *cyt1Aa* and *p20* encoding cytolytic and accessory proteins, respectively, of Bt subsp. *israelensis*, also impart synergism to *cry4Aa* and *cry 11Aa* genes expressed in transgenic *E. coli* against *Aedes agyptii* (Linnaeus) larvae (Khasdan et al., 2001).

Cry proteins from Bt subsp. *israelensis* have been expressed in the nitrogen-fixing cyanobacterium *Anabaena* PCC 7120 for mosquito control (Buossiba et al., 2000). However, apart from the difficulty of achieving an adequate level of *cry* gene expression in the recombinant cyanobacteria, the release of such transgenic bacteria in the environment is a major obstacle in the exploitation of these strains. Baculoviruses have also been tried as an expression system for *cry* genes (Je et al., 1997). However, baculoviruses are not an efficient vehicle for delivery of Cry proteins.

Environmental Impact of Bt Transgenic Microbes

Potential environmental consequences need to be ascertained prior to commercialization of Bt transgenic microbes. Safe options include recombinant Bt strains created through the conjugation or site-specific recombination (Sanchis et al., 1999) and asporogenous Bt strains, which have limited persistence at the delivery site and beyond (Vilas Boas et al., 2000). The recombinant Bt strains, resulting upon conjugative transfer of plasmids between Bt strains, represent only the natural process of genetic recombination and are not deemed genetically engineered *per se*. Furthermore, dissemination of Bt spores into the environment is curtailed with the use of asporogenous Bt strains as biopesticides. Much potential exists for the development of novel Bt strains into biopesticides (Kaur, 2002). At the same time, identification of novel insecticidal genes from other bacteria and plants is required to tackle the development of resistance in pests. Physical movement of Bt, away from the site of delivery, can be largely restricted with the use of transgenic endophytes, as discussed in the following section.

ENDOPHYTES OF PLANTS

The term 'endophyte' is derived from the Greek word 'endon' (within) and 'phyte' (plant). Endophytes have recently been variously defined; the bacteria that reside within the living plant tissue without doing substantive harm or giving benefits other than securing residency (Kado, 1992). Hallmann et al. (1997) have referred to them as bacteria isolated from the surface of disinfected plant tissue; Petrini (1998) has referred to the endophytes as those organisms that at some time of their life cycle live symptomlessly within the plant tissue.

Endophytes, nevertheless, having been variously defined, are extensively spread throughout the plant kingdom. They have been reported from a large number of plant taxa, including crop plants. Since 1940, there have been numerous reports on indigenous endophyte bacteria in various plant tissues (Table 9.1).

Endophytes as Gene Delivery Agents

With the advent of r-DNA technology, the potential use of endophytes as gene-delivering agents in plants have been examined (Fahey et al., 1991). Their systemic distribution via metabolic translocation of the endophytes makes them potential candidates for delivery of the desired genes to the plants (Misaghi and Donndelinger, 1990).

Endophytes can be variously exploited for delivering biopesticides, fungicides, plant growth factors, etc., within the plant tissue with several major advantages: (i) interspecific gene movement that bypasses the limitations of traditional plant breeding; (ii) applying the technology of engineering prokaryotes, which is still more rapid, cheap and advanced than that for the eukaryotes to produce transgenic plants; (iii) the approach has potentially broad practical use since some endophyte bacteria can be gainfully applied to different crop cultivars and species (Dimock et al., 1993); (iv) mobility of the gene is limited and minimizes the potential horizontal spread of the introduced gene (Obukowicz et al.,

Table 9.1 Most common indigenous endophytic bacterial genera of plant parts of different crops

Plant part	Bacterial taxa	Plant species	Reference(s)
Seed	*Bacillus, Erwinia, Flavobacterium, Pseudomonas*	Species (27), including cereals, vegetables and woody plants	Mundt and Hinkle (1976)
Root	*Pseudomonas, Erwinia*	Alfalfa (*Medicago sativa* L.)	Gagne et al. (1987)
	Bacillus, Corynebacterium	Corn (*Zea mays* L.)	Lalende et al. (1989)
Root, radicle, stem, unopened flowers, boll	*Erwinia, Bacillus, Clavibacter, Xanthomonas*	Cotton (*Gossypium hirsutum* L.)	Misaghi and Donndelinger (1990)
Root	*Agrobacterium, Burkholderia, Serratia*	Cotton (*Gossypium hirsutum* L.)	McInroy and Kloepper (1995a)
Root	*Burkholderia, Enterobacter*	Corn (*Zea mays* L.)	McInroy and Kloepper (1995a)
Root	*Pseudomonas, Bacillus, Enterobacter Agrobacterium, Chrysobacterium, Burkholderia, Arthrobacter, Stenotrophomonas*	Cucumber (*Cucumis sativis* L.)	Mahaffee et al. (1997)
Root	*Pseudomonas, Enterobacter, Bacillus, Corynebacterium* and other Gram-positive bacteria (e.g. *Serratia*)	Rough lemon (*Citrus jambhiri* Lush.)	Gardner et al. (1982)
Root	*Bacillus, Erwinia, Pseudomonas Corynebacterium, Lactobacillus, Xanthomonas*	Sugar-beet (*Beta vulgaris* L.)	Jacobs et al. (1985)
Tuber	*Bacillus*	Potato	Hollis (1951)

Table 9.1 contd.

Table 9.1 contd.

Plant part	Bacterial taxa	Plant species	Reference(s)
Tuber	Micococcus, Pseudomonas Bacillus, Flavobacterium, Xanthomonas, Agrobacterium, Coryneformis	Potato (Solanum tuberosum L.)	de Boer and Copeman (1974)
Stem	Enterobacter, Klebsiella, Pseudomonas	Corn (Zea mays L.)	Fisher et al. (1992)
Stem	Bacillus	Corn (Zea mays L.)	McInroy and Kloepper (1995)
Stem	Bacillus	Cotton (Gossypium hirsutum L.)	McInroy and Kloepper (1995)
Stem	Enterobacter, Pseudomonas, Pantoea, Rhodococcus	Grapevine	Bell et al. (1995)
Fruit	Pseudomonadaceae Enterobacteriaceae Achromobacteriaceae Micrococcaceae	Tomato Cucumber	Samish et al. (1961)
Root	Azospirillum spp., Herbaspirillum	Rice, wheat	Christianen-Weniger (1998) Elbeltagy et al. (2001)
Stem	Azorhizobium caulinodans, Rhizobium spp. Enterobacter cloaceae.	Maize	Hinton and Bacon (1995)
Root	Bacillus sp.	Pine	Chanway (1998)
Stem	Agrobacterium tumefaciens	Rose	Marti et al. (1999)

1986a,b); and (v) it has a promising application in those plants, specially grasses, where transferring foreign DNA is difficult due to recalcitrance of these plants to tissue culture (Obukowicz et al., 1987).

Transgenic Endophytes for Crops

Maize

The first endophyte based product 'Incide' was developed for maize by the Crop Genetic International (CGI). The gram-positive bacteria *Clavibacter xyli* subsp. *cynodontis* (*Cxc*) was the endophyte exploited for delivering of Bt gene to the plant. This endophyte, besides maize, also colonizes a large number of crops like rice, sorghum, white millet, oats, etc. In addition, it is a fairly localized endophyte, being confined to stem xylem and, consequently, absent in the progeny (Dimock et al., 1993).

The Incide product has been developed by inserting the gene coding for a 130 kD Cry1Ac protein from Bt subsp. *kurstaki* strain HD-73, into the chromosome of wild type *Cxc* isolate by the plasmid integration vector containing the toxin coding region of an antibiotic-resistant marker, a *Cxc* sequence for homologous recombination with chromosome and an *E.coli* plasmid *ori* gene of replication. The resulting recombinants produce the 130 kD Cry1Ac protein and are toxic to the European corn borer (ECB) larvae, besides a number of other larval lepidoptera. This product has the potential of imparting 80% resistance to plants against *O. nubilalis* (ECB). Efforts are in progress to increase the toxin expression levels above the current estimate of approximately 1% of the total plant protein (Tomasino et al., 1995).

From our laboratory (Rhitu et al., 2002), *B. subtilis* has been found to be a prominent maize endophyte in maize cultivar PEHM-1. The Cry1Ab crystal protein-coding gene under control of a strong *cry1C* promoter was cloned in a high copy number (70 ± 20 per cell) in the shuttle vector pHT370. This chimeric plasmid (pRA-1), upon introduction into *B. subtilis*, was found to produce 3% of the total soluble proteins of *B. subtilis* recombinant strain (RA-1). This was effective against

Table 9.2 Mortality of *chilo partellus* larvae on artificial diet

Source of crystal protein	Mortality (%) at different days				
	4	5	6	7	8
E. coli (DH5α)	–	–	–	–	–
Bacillus subtilis	–	–	–	–	–
Clone RA-1	6.66	46.6	73.26	86.58	100
E. coli (pSB033b)	19.98	59.96	93.24	100	–

Chilo partellus (Swinhoe) larvae in an insect bioassay experiment using *B. subtilis* recombinant strain RA-1 protein coated on an artificial diet (Table 9.2).

The transgenic endophyte (RA-1) was introduced into the plant by inoculum seed coating. The level of *cry1Ab* expression in the plant, estimated by ELISA, was found to be low (0.01% of the total soluble proteins). The RA-1-inoculated plants nevertheless showed high tolerance to *C. partellus* as compared to the control plants, which were free of endophyte. There was lesser foliar damage, smaller tunnel length and reduction in weight of larvae in the treated plants (Table 9.3, Fig. 9.1). The low *in planta* expression of *cry1Ab* gene was due to the

Table 9.3 *In planta* evaluation of *cry* positive clone against corn stem borer

Inoculation	Tunnel length (cm)	Larval weight (g)	Reduction (%) in weight
Control	12.4 ± 2. 61	0.38 ± 0.004	–
Transgenic	9.7 ± 1.38	0.252 ± 0.0005	34

Fig. 9.1 Tunneling damage due to borer in control and where transgenic bacteria RA-1 have been introduced

increasing incidence of phenotype segregants of the RA-1 population. Forty days after emergence (DAE), it was as much as 64%.

Sugarcane

The *cry1C* gene has been cloned from the native Bt strain with activity against the sugarcane borer, *Eldana saccharina* Walker (Herrera et al., 1994). The gene was introduced into an isolate of *Pseudomonas fluorescens*, capable of colonizing sugarcane in two broad host range plasmids pDER405 and pKT240. By using the Omegon Km vector, the *cry* gene was introduced into the chromosome of *P. fluorescens*. Trials in greenhouses have indicated that sugarcane treated with *P. fluorescens*: Omegon Km *cry* were more resistant to the *Eldana* damage than the untreated sugarcane (Herrera et al., 1994).

In another experiment, Downing et al., (2000) have cloned the *cry*1Ac gene under the control of *tac* promoter. The fusion product was introduced into the integration vector pJFF350. The plasmid was introduced into a *P. fluorescens* strain isolated from the phylloplane and the endophyte bacterium *Herbaspirillum seropedicae* found in sugarcane. The ptac-*cry1Ac7* construct was introduced into the chromosome of *P. fluorescens* using the integration vector pJFF350 carrying the artificial interposon Omegon-Km. It was observed that expression levels of the integrated *cry1Ac7* gene were much higher under the control of *tac* promoter than under the control of its endogenous promoter. In *H. seropedicae*, more Cry1Ac toxin was produced when the gene was cloned under the control of the *Nmr* promoter on pML122 and bioassays showed that the former resulted in higher mortality of *E. saccharina* larvae. An increased toxic effect was observed when *P. fluorescens* 14: ptac-tox, combined with *P. fluorescens* carrying the *Serratia marcescens* chitinase gene *chiA* under the control of *tac* promoter, was integrated into the chromosome.

Phaseolus

An endophyte strain of *P. fluorescens* was isolated from micropropagated apple plantlets and introduced into beans (*P. vulgaris*) via their root tips. The gene coding for the major chitinase of *Serratia marcescens chiA* was cloned under the control of *tac* promoter in the broad host range plasmid pKT240 and integration vector pJFF357. Bioassays revealed that *P. fluorescens* carrying the *tac-chiA*—either in the plasmid or integrated into the chromosome—is an effective biocontrol agent of the phytopathogenic fungus *R. solani* in bean seedlings under plant growth chamber conditions when it is present as endophyte in the bean (Downing and Thomson, 2000).

Tobacco

Insertion sequence of IS50L of transposon Tn5 was used as a non-self transposable vector to integrate the endotoxin gene from Bt subsp. *kurstaki* HD-1 into the chromosome of two corn root-colonizing strains of *P. fluorescens*. A DNA fragment with Km^R gene from Tn5 and tox was inserted into an IS50L element (IS50L-tox) contained in a suicidal plasmid. Transposition of IS50L tox into the chromosome of *P. fluorescens* occurred by selecting for Km^R transconjugants and supplying the transposase in *cis* from a linked IS50R element. Bioassays for toxicity were done against larvae of tobacco horn worm, *Manduca sexta* (Johannsen) and a significantly increased mortality of larvae was observed (Obukowicz et al., 1987).

Grasses

A cassette of *cry4B* gene from Bt subsp. *morrisoni* that placed the gene under the control of *tac* promoter was constructed. The kanamycin resistance gene npt-II and the *cry4B* gene cassette were cloned with the *Pseudomonas* sequences. These constructs were introduced into the root colonizing strain of *P. fluorescens*. Bioassays revealed that this transgenic bacterial strain was toxic towards the leather jacket larvae of *Tipula oleraceae*. These larvae cause severe yield reduction in grasses (Waalwijk et al., 1991). A spodoptera-active protein expressed by *Cxc*/Bt recombinant was found to be a potent inoculant to protect grasses from armyworm damage. Other lepidoptera pests of turf such as sodwel worms are also potential agents (Dimock et al., 1993).

Other crops

Cxc Bt has the potential to control stem borer (*Chilo* spp.) in rice. Crop Genetics International has conducted preliminary field trials with promising results.

CONCLUSIONS

Microorganisms, as delivery system for biopesticidal genes, are still in their formative stages, but with their inherent advantages over transgenic plants, they will have a wider acceptability and role in crop protection. Bacteria such as Cxc, *P. fluorescens*; *Bacillus* sps., etc., may be engineered in the near future to produce other molecules in addition to biopesticides. These may include antibiotics, biofungicides; bacteriocin and other anti-microbial agents conferring disesase resistance to the host plant (Glick and Bashan, 1997). This may be followed by production of plant growth regulators by engineered microorganisms (Yue et al., 2001) that would enhance yield and quality, increase harvest efficiency, and even improve host plant adaptation to environmental stress. The endophyte bacterial delivery system would theoretically be applicable to any crop for which an acceptable endophyte can be identified (Dimock et al., 1993) and thus have a wider applicability.

REFERENCES

Bar, D. E., Sandler, V., Makayoto, M. and Kenyanb, A., 1998, Expression of chromosomally inserted *Bacillus thuringiensis israelensis* toxin genes in *Bacillus sphericus*. J. Invertebr. Pathol., **72**, 206-213.

Baum, J. A., Kakefuda, M. and Gawron-Burke, C., 1996, Engineering *Bacillus thuringiensis* bioinsecticide with an indigenous site specific recombination system. Appl. Environ. Microbiol., **62**, 4367-4373.

Bell, C. R., Dickie, G. A., Harvey, W. L. G. and Chan, J. W. Y. F., 1995, Endophyte bacteria in grapevine. Can. J. Microbiol., **41**, 46-53.

Bezdicek, D. F., Quin, M. A., Forse, L., Heron, D. and Kahn, M. L., 1994, Insecticidal activity and competitiveness of *Rhizobium* spp. containing the *Bacillus thuringiensis* subsp. *tenebrionis* endotoxin gene (*cry III*) in legume nodules. Soil. Biol. Biochem., **26**, 1637-1646.

Bora, R. S., Murty, M. G., Shenbagarthi, R. and Sekar, V., 1994, Introduction of a lepidopteran specific insecticidal protein gene of *Bacillus thuringiensis* subsp. *kurstaki* by conjugal transfer into *Bacillus megaterium* strain that persists in cotton phyllosphere. Appl. Environ. Microbiol., **60**, 214-222.

Boussiba, S., Wu, X. Q., Ben-Dov, E., Zarka, A. and Zaritsky, A., 2000, Nitrogen-fixing cyanobacteria as gene delivery system for expressing mosquitocidal toxins of *Bacillus thurigiensis* sp. *israelensis*. J. Appl. Physiol., **12**, 461-467.

Carmona, A. A. and Ibarra, J. E., 1999, Expression and crystallization of *cry*3Aa-*cry*1Ac chimerical protein of *Bacillus thuringiensis*. World J. Microbiol. Biotech., **15**, 455-463.

Chanway, C. P., 1998, Endophytes: they are not just fungi. Can. J. Microbiol., **74**, 321-322.

Chen, A. P. T. and Phillips, D. A., 1976, Attachment of *Rhizobium* to legume roots as the basis for specific interaction. Physiol. Plant, **38**, 83-88.

Chet, I. and Inbar, J., 1994, Biological control of fungal pathogens. Appl. Biochem. Biotechnol., **48**, 37-43.

Chet, I., Barak, Z. and Oppenheimmer, A. B., 1993, Genetic engineering of microorganisms for improved biocontrol activity. In: Chek I. (ed.), Biotechnology in Plant Disease Control. Wiley-Less, New York, pp. 211-235.

Christianen-Weniger, C., 1998, Endophytic establishment of diazotrophic bacteria in auxin induceed tumors of cereal crops. Crit. Rev. Plant Sci., **17**, 55-76.

da Costa Lima, Lemos, M. V. F., Lemos, E. G. M. and Alves, L. M. C., 2000, Transference of a crystal protein gene from *Bacillus thuringiensis* and its expression in *Bradyrhizobium* sp. cell. World J. Microbiol. Biotech., **16**, 361-365.

deBoer, S. H. and Copeman, R. J., 1974, Endophyteic bacterial flora in *Solanum tuberosum* and its significance in bacterial ring rot diagnosis. Can. J. Plant Sci., **54**, 115-122.

Del Rancon-Casteo, M. C., Barajas-Huerta, J. and Ibarra, J. E., 1999, Antagonism between *cry* 1Ac1 and *cyt* 1A1 toxins of *Bacillus thruingiensis*. Appl. Environ. Microbiol., **65**, 2049-2053.

Dimock, M., Turner, J. and Lanpel, J., 1993, Endophyte micro-organism for delivery of genetically engineered microbial pesticides in plants. *In*. L. Kim (ed.), Advanced Engineered Biopesticides, Marcel Dekker Inc., New York. pp. 85-97.

Dimock, M. B., Beach, R. M. and Charlson, P. C., 1989, Endophyte bacteria for the delivery of crop protection agents. In: Roberts D.N. and Grandis R.R. (eds), Biotechnology, Biological Plant Pesticide and Novel Plant: Pest Resistance For Insect Pest Management. Boyce Thompson Institute for Plant Research, Ithaca, NewYork., pp. 89-92.

Downing, K. J. and Thomson, J. A., 2000, Introduction of *Serratia marcescens* chiA gene into an endophyte *Psuedomonas fluoroescens* for the biocontrol of phytopathogenic fungi. Can. J. Microbiol., **46**, 363-369.

Downing, K. J., Leslie, G. and Thomson, J. A., 2000, Biocontrol of the sugarcane borer *Eldana saccharina* by expression of the *Bacillus thuringiensis* cry IAC7 and *Serratia marcescens*. ChiA genes in sugarcane associated bacteria. Appl. Environ. Microbiol., **66**, 2804-2810.

Elbeltagy A., Nishioka K., Salo T., Ye B, Hamada, T., Isawa, T., Mitsam, H. and Minomusawa, K., 2001, Endophytic colonization and *in planta* N_2 fixation by *Herbaspirillium sp.* isolated from wild rice sp. Appl. Environ. Microbiol., **67**, 5285-5293.

Estruch, J. J., Carozzi, N. B., Desai, N., Duck, N. B., Warren, G. W. and Koziel, M. G., 1997, Transgenic plants: an emerging approach to pest control. Nature Biotechnol., **15**, 137-141.

Fahey, J. W., Dimock, M. B., Tomasino, S. F., Taylor, J. M. and Carlson, P. S. 1991. Genetically-engineered endophytes as biocontrol agents: a case study from industry. In: Andrews J. H. and Hiramd S. S. (eds), Microbial Ecology of Leaves, Springer Verlag, UK, pp. 401-411.

Federici, B. A. and Bauer, L. S., 1998, Cyt 1Aa protein of *Bacillus thuringiensis* is toxic to the cotton wood leaf beetle, *Chrysomela scripta*, and suppresses high levels of resistance to Cry3Aa. Appl. Environ. Microbiol., **64**, 4368-4371.

Fisher, P. J., Petrini, O. and Scott, H. M. L., 1992, The distribution of some fungal and bacterial endophytes in maize (*Zea mays* L.). New Phytol., **122**, 299-305.

Frutos, R., Rang, C. and Royer, M., 1999, Managing insect resistance to plants producing *Bacillus thuringiensis* toxins. Crit. Rev. Biotechnol., **19**, 227-276.

Gaertner, F. H., Quick, T. C. and Thompson, M. A., 1993, Cell cap: an encapsulation system for insecticidal biotoxin proteins. In: Kim L. (ed.), Advanced Engineered Pesticides, Marcel Dekker Inc., New York, pp. 73-83.

Gagne S., Richard C., Lousseau, H. and Btoun, H., 1987, Xylem residing bacteria in alfalfa roots. Can. J. Microbiol., **33**, 996-1000.

Gardner, T. M., Fiedman, A. W. and Zablatowic, R. M., 1982, Identification and behaviour of xylem residing bacteria in rough lemon roots of Florida citrus stress. Appl. Environ. Microbiol., **43**, 1335-1342.

Giddings, G., Mytton, L., Griffiths, M., McCarthy, A., Morgan, C. and Skot, LK., 1997, A secondary effect of transformation in *Rhizobium leguminosarum* transgenic for *Bacillus thuringiensis* subsp. *tenebrionis* delta endotoxin (*cry IIA*) genes. Theor. Appl. Genet., **45**, 1062-1068.

Glick, B. R. and Bashan, Y., 1997, Genetic manipulation of plant growth promoting bacteria to enhance biocontrol of plant pathogen. Biotechnol. Adv., **15**, 353-378.

Gounder, R. and Rajendran, N., 2001, Transformation studies of *Bacillus thuringiensis* cry1C gene into nitrogen-fixing *Azospirillum lipoferum*. Z. Naturforsch. [C], **56**, 245-248.

Hallmann, J., Quadt-Hallmann, A., Mahaffee, W. F. Q. and Kloepper, J. W., 1997, Bacterial endophytes in agricultural crops. Can. J. Microbiol., **43**, 895-914.

Herrera, G., Synman, S.J., Thomson, J.A., 1994, Construction of a bioinsecticidal strain of *Pseudomonas fluorescens* active against the sugarcane borer, *Eldena saccharina*. Appl. Environ. Microbiol., **60**, 682-690.

Hinton, D. M. and Bacon, C. W., 1995, *Enterobacter cloaceae* is an endophyte symbiont of corn. Mycopathologia, **129**, 117-125.

Hollis, J. P., 1951, Bacteria in healthy potato tissue. Phytopathology, **41**, 350-366.

Honee, G., Convents, D., van Rie, J., Jansens, S., Peferoen, M. and Visser, B., 1991, The carboxyl terminal domain of the toxic fragment of a *Bacillus thuringiensis* crystal protein determines receptor binding. Mol. Microbiol., **5**, 2799-2806.

Jacobs, M. J., Bugbee, W. M. and Gabrielson, D. A., 1985, Enumeration, location and characterization of endophyte bacteria within sugar beet roots. Can. J. Bot., **63**, 1262-1265.

James, E. K. and Olivares, F. L., 1997, Infection, colonization of sugarcane and other gramminaceous plants by endophyte diazotrophs. Crit. Rev. Plant Sci., **17**, 77-110.

Je, Y. H., Jin, B. R., Park, H. W., Roh, J. Y., Chang, J. H., Woo, S. D. and Kang, S. K., 1997, Expression of fusion protein with *Autographa californica* nuclear polyhedrosis virus polyhedrin and *Bacillus thuringiensis* Cry 1Ac crystal protein in insect cells. Korean J. Appl. Entomol., **36**, 3341-3350.

Kado, C. I., 1992, Plant pathogenic bacteria. In: Balows A., Truper H. G., Dowrkim M. and Schleifer K. H. (eds), Springer Verlag, New York, pp. 660-662.

Kaur, S., 2000, Molecular approaches towards development of novel *Bacillus thuringiensis* biopesticides. World J. Microbiol. Biotechnol., **16**, 781-793.

Kaur, S., 2002, Potential for developing novel *Bacillus thuringinensis* strains and transgenic crops and their implications for Indian agriculture. *Agbiotech Net*. 4th May, ABN 088, pp. 1-10. http://www.agbiotechnet.com CAB International.

Kaur, S. and Singh, A., 2000, Natural occurrence of *Bacillus thuringiensis* leguminous phylloplanes in the New Delhi region of India. World J. Microbiol. Biotechnol., **16**, 679-682.

Khasdan, V., Ben-Dov, E., Manasherob, R., Boussiba, S. and Zaritsky, A., 2001, Toxicity and synergism in transgenic *Escherichia coli* expressing four genes from *Bacillus thuringiensis* subsp. *israelensis*. Environ. Microbiol., **3**, 798-806.

Kim, L., 1993, Advanced Engineered Pesticides, Marcel Dekker Inc., New York, 430pp.

Koby, S., Schickler, M., Chet, I. and Oppenhesmer, A. B., 1994, The chitinase encoding T7 based chiA gene endows *P. fluorescens* with capacity to control the pathogen in the soil. Gene, **147**, 81-83.

Lalende, R., Bissonnette, N., Couttee, D. and Antoun, H., 1989, Identifiication of rhizobacteria from maize and determination of their plant growth promoting potential. Plant Soil, **115**, 7-11.

Mahaffee, W. F., Bauskce, E. M., Van Vuurde, J. W. L., Vander Wolf, J. M. and Kloepper, J. W., 1997, Comparative analysis of antibiotic resistance, immunoflorescent colony staining and a transgenic marker (bioluminescence) for monitoring the environmental fate of rhizobacterium. Appl. Environ. Microbiol., **63**, 1617-1622.

Malvar, T., Gawron-Burke, C. and Baum, J. A., 1994, Over-expression of *Bacillus thuringiensis* HKn A, a histidine protein kinase homolog bypassing early S_{po} mutations that result in *cryIIIA* overproduction. J. Bacteriol., **176**, 4742-4749.

Marti, R., Cubero, J., Daza, A., Piques, J., Sakedo, C. I., Morente C. and Lopez, M. M., 1999, Evidence of migration and endophytic presence of *Agrobacterium tumefaciens* in rice plants. European J. Plant Pathol., **105**, 39-50.

McInroy, J. A. and Kloepper, J. W., 1995, Survey of indigenous bacterial endophytes from cotton and sweet corn. Plant & Soil, **173**, 337-342.

Misaghi, I. and Donndelinger, C. R., 1990, Endophyte bacteria in symptom free cotton plant. Phytopathology, **80**, 808-811.

Mundt, J. P. and Hinkle, N. F. 1976 Bacteria within ovules and seed. Appl. Environ. Microbiol,. **32**, 694-698.

Nambiar, P. T. C., Ma, S. W. and Iyer, V. N., 1990, Limiting an insect infestation of N_2 fixing root nodules of the pigeonpea (*Cajanus cajan*) by engineering the expression of an entomocidal gene in its root nodules. Appl. Environ. Microbiol., **56**, 2866-2869.

Obukowicz, M. G., Perlak, F. J., Bolten, C. L., Kusan-Keetzneis, K., Meyer E. J. and Watrud, W., 1987, IS50L as a non-self transposable vector used to integrate the *B. thuringiensis* delta endotoxin gene into the chromosome of root colonizing *Pseudomonas*. Gene, **51**: 95-96.

Obukowicz, M. G., Perlak, F. J., Kusamo-Kertzoner, K., Mayer, E. J. and Watrud, L. S., 1986a, Integration of delta endotoxin, gene of *B. thuringinensis* into the chromosome of root colonizing strains of *Psuedomonas* using Tn5. Gene, **45**, 327-331.

Obukowicz, M. G., Perlak, F. J., Kusamo-Kertzoner, K., Mayer, E. J. and Watrud, L. S., 1986b. The Tn5 mediated integration of the delta8 endo-toxin gene for *B. thuringiensis* into the chromosome of root colonizing *Psuedomonads*. J. Bacteriol., **168**, 982-989.

Panbangred, W., Panjaisee, S. and Tantimavanich, S., 2000, Expression of the mosquitocidal *cry IVB* gene under the control of different promoters in *Bacillus thuringiensis* 2362 and acrystalliferous *Bacillus thuringiensis* subsp. *israelelnsis* C4Q2-72. World J. Microbiol. Biotech., **16**, 163-169.

Park, H. W., Bideshi, D. K. and Federici, B. A., 2000, Molecular genetic manipulation of truncated *cry*1Ac protein synthesis in *Bacillus thuringiensis* to improve stability and yield. Appl. Environ. Microbiol., **66**, 4449-4455.

Park, H. W., Delecluse, A. and Federici, B. A., 2001, Construction and characterization of a recombinant *Bacillus thuringiensis* subsp. *israelensis* strain that produces *cry IIB*. J. Invertebrate Pathol., **78**, 37-44.

Petrini, O., 1998, What are endophytes anyway ? In, Proceedings of the 7th ICCP, Edinburgh, U.K. 2.9.15

Poncet, S., Delecluse, A., Klier, A. and Rapoport, G., 1995, Evaluation of synergistic interactions between the *cry* IV A, *cry* IV B and *cry* IV D toxic components of *Bacillus thuringiensis* subsp. *israelensis* crystals. J. Invertebrate Pathol., **66**, 131-135.

Poncet, S., Bernhard, C., Dervyn, E., Cayley, J., Klier, A. and Rapoport, G., 1997, Improvement of *Bacillus thuringiensis* toxicity against dipteran larvae by integration via homologous recombination of the *cry II A* toxin gene from *Bacillus thuringiensis* subsp. *israelensis*. Appl. Environ. Microbiol., **63**, 4413-4420.

Rhitu R., Prasanta, D. and Aqbal, S., 2002, Cloning of Bt gene in maize endophyte for control of shoot borer, *Chilo partellus* (Swinhoe). In: Koul O., Dhaliwal G. S., Marwaha S. S. and Arora J. K. (eds), Biopesticides and Pest Management Vol. 2, Campus Books International, New Delhi, pp.116-121.

Samish, Z., Etinger, T. R. and Bick, M., 1961, Microflora within healthy tomatoes. Appl. Environ. Microbiol., **9**, 20-25.

Sanchis, V., Aggaisse, H., Chaufaux, J. and Lereclus, D., 1997, A recombinase mediated system for Elimination of antibiotic resistance gene markers from genetically engineered *Bacillus thuringiensis* strains. Appl. Environ. Microbiol., **63**, 779-784.

Sanchis, V., Gohar, M., Chaufaux, J., Arantes, O., Meier, A., Agaisse, H., Caylay, J. and Lereclus, D., 1999, Development and field performance of broad spectrum nonviable asporogenic recombinant strain of *Bacillus thuringiensis* with greater potency and UV resistance. Appl. Environ. Microbiol., **65**, 4032-4039.

Saxena, D. and Stotzky G., 2000, Insecticidal toxin from *B. thuringiensis* is released from root of transgenic Bt corn in vitro and in situ., FEMS Microbiol. Ecol., **33**, 35-39.

Sayyed, A. H., Crickmore, N., Wright, D. J. (2001) Cyt1Ac from *Bacillus thringiensis* subsp. *israelensis* is toxic to the diamondback moth, *Plutella xylostella*, and synergizes the activity of *cry* 1Ac towards a resistant strain. Appl. Environ. Microbiol., **67**, 5859-5861.

Schnepf, E., Crickmore, N., Van Lie, J., Lereclus, D., Baum, J., Feitelson, J., Zeigler, D. R. and Dean, D. H., 1998, *Bacillus thuringiensis* and its pesticidal crystal protein. Microbiol. Mol. Biol. Rev., **62**, 775-806.

Shelton, A. M., Zhao, J. Z. and Roush, R. J., 2002, Economic, ecological, food safety and social consequences of the deployment of Bt transgenic plants. Ann. Rev. Entomol., **47**, 845-881.

Stock, C. A., McLoughlin, T. J., Klein, J. A. and Adang, M. J., 1990, Expression of a *Bacillus thuringiensis* crystal protein gene in *Pseudomonas cepacia*. Can. J. Microbiol., **36**, 879-884.

Sturz, A. V., Christe, R. R. and Nowak, J., 2000, Bacterial Endophytes: Potential role in developing sustainable systems of crop production. Crit. Rev. Plant Sci., **19**, 1-30.

Sudha, S. N., Jayakumar, R. and Sekar, V., 1999, Introduction and expression of the *cry 1Ac* gene of *Bacillus thuringiensis* in a cereal associated bacterium *Bacillus polymyxa*. Curr. Microbiol., **38**,163-167.

Sufnit, Y., Barah, Z., Kapuluk, Y., Oppenheim, A. B. and Chet, I., 1993, Expression of *Serratia marcescens* chitnase genes in *R. meliloti* during symbiosis on alfalfa roots. MPMI , **6**, 293-298.

Terouchi, N. and Syno, K., 1990, *Rhizobium* attachment and curling in asparagus, rice and oat plants. Plant Cell Physiol., **31**, 119-127.

Theoduloz, C., Vega, A., Salazar, M., Gonzalez, E. and Meza-Basso, L., 2003, Expression of a *Bacillus thuringiensis* delta-endotoxin *cry*1Ab gene in *Bacillus subtilis* and *Bacillus licheniformis* strains that naturally colonize the phylloplane of tomato plants (*Lycopersicon esculentum* Mills). J. Appl. Microbiol., **94**, 375-381.

Tomasino, S. F., Leister, R. T., Dimock, M. B., Beach, R. M. and Kelly, J. L., 1995, Field performance of *Clavibacter xyli* subsp. *cynodontis* expressing the insecticidal crystal protein genes *cry*1Ac of *Bacillus thuringiensis* against European corn borer in field corn. Biol. Cont., **5**, 442-448.

Turner, J. T., Kelly, J. I. and Carlson, P. S., 1993, Endophytes an alternative genome for crop improvement. International Crop Science Congress, Ames, Iowa, July 14-22.

Vilas-Boas, L. A., Vilas Boas, G. F. L. T., Saridakis, H. O., Lemos, M. V. F., Lereclus, O., Arantes, O. M. N., 2000, Survival and conjugation of *Bacillus thuringiensis* in a soil microcosm. FEMS Microbiol. Ecol., **31**, 255-258.

Waalwijk, C., Dullemans, A. and Maa, C., 1991, Construction of a bioinsecticidal rhizosphere isolate of *Pseudomonas fluorescens*. FEMS Microbiol. Lett., **77**, 257-264.

Wirth, M. C., Walton, W. E. and Federici, B. A., 2000, Cyt 1A from *Bacillus thruingiensis* restricts toxicity of *Bacillus sphaericus* against resistant *Culex quignefasciatus* (Diptera: Culicidae). J. Med. Entomol, **37**, 401-407.

Yue, Q., Wang, C., Giantagna, J. J. and Meyer, W. A., 2001, Volatile compounds of endophyte free and infected tall fescue. *Phytochemistry*, **58**, 935-941.

10

IMPACTS OF TRANSGENIC Bt CROPS ON NON-TARGET ANIMAL SPECIES

GRAHAM HEAD* AND GALEN DIVELY**
*Monsanto LLC, St. Louis,MO, USA
**Department of Entomology,University of Maryland
College Park, MD, USA

INTRODUCTION

Recent advances in molecular biology and genetics have led to the creation of exciting new opportunities in the field of agriculture. Some of the first and most widely-used genetically-engineered crops have been modified to express insecticidal crystalline (Cry) proteins derived from the bacterium *Bacillus thuringiensis* (Bt). These so-called Bt crops are protected from the feeding of various groups of herbivorous pest insects. In 1995 and 1996, varieties of potato, cotton and corn expressing various Bt proteins were approved for commercial use in the United States. In Bt potato, the Cry3Aa protein provides protection against damage from the Colorado potato beetle. In Bt cotton and Bt corn, the proteins expressed (Cry1Ac in cotton and Cry1Ab or Cry1F in corn) confer protection against a number of lepidopteran herbivores. The Bt cotton and Bt corn products, in particular, are widely used (James, 2001). Bt cotton has since been registered for commercial use in Argentina, Australia, China, Columbia, India, Indonesia, Mexico, and South Africa. Bt corn has a similar potential; in 2001, about 20 million acres of Bt corn were planted in the United States, almost 40% of corn acres in Argentina are occupied with Bt corn, and smaller amounts were planted commercially in Canada, Spain, and South Africa. A critical part of the introduction of such products is to ensure their environmental safety. In this chapter, we shall describe the safety assessment process used to evaluate the potential impact of Bt crops on non-target species, and we review the laboratory and field research that has been conducted in this area (see also Shelton et al., 2002 for a broader review of the impacts of Bt crops).

MODE OF ACTION AND SPECTRUM OF ACTIVITY OF Bt PROTEINS

Bt Cry (crystalline) proteins are produced by strains of the common bacterium *Bacillus thuringiensis*, and are entomopathogenic in nature. Different sets of proteins are produced by different strains of the bacterium. These proteins are structurally diverse but have a common, well-understood mode of action that consists of several steps: ingestion of the protein, proteolysis, binding to receptor sites in the gut, and pore formation (Schnepf et al., 1998). Since the steps in the insecticidal activity are relatively specific, any particular Bt protein is active against a very limited and different set of animal species. The Cry proteins have been classified into a hierarchical scheme of classes and subclasses based upon their structure and pattern of insecticidal activity.

A large amount of in vitro work has defined the activity spectrum of many Cry proteins; particularly those used in commercial pesticidal products. The most heavily-studied examples are Classes 1 to 4: Cry1 proteins have been demonstrated to be active against a subset of lepidopteran insects, Class 2 proteins display activity against lepidopteran and dipteran insects, Class 3 proteins are active against certain coleopteran insects, and Class 4 proteins are active against some dipteran insects. Currently, commercialized transgenic Bt crops contain either Class 1 or Class 3 proteins, and are targeted against lepidopteran and coleopteran herbivores, respectively (MacIntosh et al., 1990; Schnepf et al., 1998, and references cited therein). Some isolated cases have been reported of these Cry proteins having in vitro activity against species from other orders than those described, but all of these cases appear to involve confounding factors that have impacted the development of the test organism within the study, such as preliminary processing of the proteins in ways that do not naturally occur (see specific discussions below).

When the gene encoding a Bt protein is engineered into a crop plant, the structure of the expressed protein and its spectrum of activity have been demonstrated to be comparable to the proteins found in *Bacillus thuringienisis* strains in nature (Perlak et al., 1991; Betz et al., 2000). Some scientists have suggested that the modified Bt proteins expressed in transgenic plants may possess broader insecticidal activity than the wild type proteins (for example, Hilbeck et al., 1998b), but there is no evidence to support this contention and the broad range of studies described below suggest otherwise. Obviously the 'method and duration of presentation' differ between these engineered Bt proteins and the wild types, with plant-expressed proteins being present in certain plant cells only, higher in concentration in some plant tissues than others, and expressed throughout the entire crop cycle (Jepson et al., 1994; EPA, 2001). For an insect to be affected by the plant-expressed protein, it must be susceptible to the relevant Bt Cry protein and must somehow ingest the appropriate plant tissues. Thus, though the spectrum of activity of plant-expressed Bt Cry proteins is similar or identical to bacterially-expressed proteins, the routes of non-target exposure are limited and must involve direct or indirect consumption of particular plant tissues. However, if an animal does ingest the appropriate plant tissues, the potential level of exposure may be relatively high because of the extended period of protein availability and the protein expression levels achieved in transgenic crops.

SCOPE OF NON-TARGET STUDIES WITH Bt CROPS

As part of the regulatory packages submitted by biotechnology companies for each Bt crop product prior to commercialization, a comprehensive set of studies are performed in order to assess potential non-target effects. An initial set of laboratory tests (known as Tier 1 tests) are carried out with a variety of non-target species primarily using exposure to pure protein, with testing concentrations based on the maximum possible environmental exposure plus a safety factor. Where appropriate, testing occurs with relevant plant tissues. Test species are chosen based on considerations of the product and region, and typically include insect predators, parasitoids and pollinators, as well as soil-dwelling and aquatic invertebrates. These species are selected to be representative of different taxa and ecological guilds, and often constitute economically-important species. The results of these tests can be compared to predictions from the known properties of the protein (mode of action information, as described above), and have been found to be very consistent with these predictions. Different routes of exposure to the insecticidal protein are assessed, including direct consumption of leaf tissue by herbivores, deliberate or incidental feeding on pollen, and ingestion of plant material that has become incorporated into the soil. Subsequent tests carried out when sufficient transgenic seed is available may be more field based in nature. In these studies, the product is compared with reasonable agronomic alternatives for control of the target insects. Impacts on the population dynamics of important guilds are examined. After commercialization, where the need is indicated by the initial tests, follow-up work can take place in commercial-sized fields managed with standard grower practices. At that point, larger scale effects and emergent properties (system functioning) can be studied.

The organisms most commonly chosen for laboratory testing have been those previously used in the testing of microbial insecticides. For any given product, testing typically includes a representative avian species (usually quail or chickens because they are seed feeders), a sensitive aquatic invertebrate (*Daphnia*), representative soil invertebrates important in decomposition processes (one earthworm species and one species of Collembola), an insect pollinator (generally an adult and a larval honeybee test), an insect parasitoid species, and two or more species of insect predators [usually a neuropteran like *Chrysoperla carnea* (Stephens) and a coccinellid]. These species are chosen to offer a broad taxonomic and ecological representation, while primarily targeting insects because of what is known about the spectrum of activity of Bt Cry proteins. At the same time, the list of species chosen for these laboratory tests can never be truly comprehensive, which is why field studies of these ecological communities and processes generally are carried out to confirm and extend the results of the laboratory tests.

Most of the non-target field studies that have been carried out thus far have involved Bt corn, and most of these have used transformation events expressing Cry1Ab, particularly E-176 corn. It should be noted that E-176 has much higher expression of Cry1Ab in pollen than the other commercialized Cry1Ab events MON 810 and Bt-11. In this sense, E-176 could be viewed as a conservative surrogate for the other more widely-used Bt corn events from the point of view of non-target testing for potential pollen impacts. Similar work has been carried out with the Bt corn event TC1507 expressing the Cry1F protein. Most of the corn studies have taken place in the United States with the remainder being conducted in Europe. Earlier studies sometimes involved relatively small plot sizes and limited replication, but some of the later studies used plots or fields of one hectare or more. Comparable field studies also have been performed with Cry1Ac-expressing cotton in the United States, Australia and China, and with transgenic potato expressing the Cry3Aa protein in the United States. The results of all of these studies have been very consistent across all crops and Bt Cry proteins, both in terms of general trends in overall arthropod abundance and diversity, and the observed effects (or lack of effects) on those particular species that are regarded as economically or ecologically important.

GENERAL TRENDS IN THE RESULTS OF NON-TARGET STUDIES

The Cry1 proteins in currently-available Bt corn and cotton products are active against Lepidoptera, while the Cry3 protein present in Bt potato is active primarily (and maybe solely) against chrysomelid Coleoptera. These proteins would be predicted to display negligible activity outside these orders (see above), and this has been the case in all of the laboratory tests using pure protein. No adverse impact of Cry1Ab, Cry1Ac or Cry1F has been observed on any non-lepidopteran species in these tests at concentrations at or near the maximum predicted field exposure. Similarly, no adverse impact of Cry3Aa has been observed on any non-coleopteran species, or even with Coleoptera from families other than the Chrysomelidae.

Furthermore, the toxins are expressed within the plant tissues, minimizing the exposure of non-phytophagous insects to these proteins. Therefore, the obvious prediction is that there should be minimal impact on the fauna in transgenic Bt crop fields compared with similarly-treated non-transgenic fields. In fact, across the large number of field studies that have been conducted, few or no differences have been seen with respect to community structure or individual species abundance where fields of Bt crops have been compared to conventional crops that have not been treated with insecticides. Where they have been calculated, indices of species diversity and community structure have not differed significantly for Bt cornfields compared to conventional cornfields (e.g. Lozzia, 1999; Dively and Rose, 2003), or for Bt cotton fields compared to conventional cotton fields (Xia et al., 1999; Fitt and Wilson, 2002; Naranjo and Ellsworth, 2003).

The only species that have been observed to be significantly and consistently less abundant in fields of Bt crops relative to fields of conventional crops are the target pest Lepidoptera or Coleoptera. All varieties of Bt field corn and sweet corn, and Bt cotton, provide control or suppression of certain lepidopteran pests, including larvae that bore into corn stalks, i.e. *Ostrinia nubilalis* (Hubner) and certain *Diatraea* spp. or cotton bolls, *Pectinophora gossypiella* (Saunders). Certain other species that feed on corn or cotton foliage and reproductive structures, *i.e.* specific *Spodoptera* spp. and *Helicoverpa zea* (Boddie) in the case of corn and *Heliothis virescens* (Fabricius) and *Helicoverpa* spp. in the case of cotton. Transgenic Bt potato controls the chrysomelid beetle, *Leptinotarsa deciemlineata* (Say) (Reed et al., 2001). These impacts on the primary pest populations are reflected in reduced in-field populations of these pest species. In addition, several field studies have shown lower population densities of certain saprophytic beetles and flies in Bt corn compared to conventional corn, and this effect was linked to the absence of plant injury by the target species in the Bt corn (Dively et al., 2002; Dively and Rose, 2003).

In studies where the conventional crop fields have been sprayed for the target pest species of the Bt crop, many non-target species have been observed to have an adverse impact, leading to significantly lower non-target populations in sprayed conventional fields as compared to Bt crop fields. With cornfields, this is particularly obvious for foliage-dwelling species because of the method of application of these insecticides, but ground-dwelling species like carabids and cursorial spiders are also often affected, directly or indirectly, by the insecticidal sprays and are apparently not affected by Bt corn (Dively and Rose, 2003; Candolfi et al., 2004). Similarly, a variety of studies of Bt cotton in the United States, Australia and China have all demonstrated that populations of many non-target species are higher in Bt cotton fields than in sprayed conventional cotton fields (Xia et al., 1999; Head et al., 2001b; Fitt and Wilson, 2002; Naranjo and Ellsworth, 2003). With conventional cotton, lepidopteran pests generally require one or more insecticide applications, so these results are likely to be more typical than the results of studies where the conventional cotton fields are left unsprayed. Likewise, work on potato fields in the northeastern US has found larger populations of many generalist predators in Bt potato fields than in conventional potato fields treated with appropriate broad-spectrum insecticides (Reed et al., 2001).

IMPACTS OF Bt CROPS ON PARTICULAR GUILDS

Due to their economic and/or ecological importance, many studies have concentrated on a number of specified species or a particular guild. Below, we consider the potential hazard to these taxa and possible routes of exposure in each case, and then review the results of the relevant tests.

Arthropod Predators

Natural enemies, and particularly generalist predators, have been the focus of many studies because of their role in the biological control of various agricultural pests (Symondson et al., 2002). Based on what is known about the limited spectrum of activity of the Bt Cry proteins expressed in currently commercialized Bt crops, no direct toxic effects from Bt crops would be expected for any of these species. Furthermore, because the larvae of these groups feed primarily on other arthopods, direct exposure will be limited to whatever opportunistic feeding occurs on the nectar or pollen of Bt crops. For a few insect predators like coccinellids, substantial feeding on corn pollen may take place with larvae or adults. However, secondary exposure to Bt proteins may occur if the natural enemies feed upon other animals that have fed upon a Bt crop plant. In addition, indirect effects may occur at the population level if the prey or host species of the predators are a target of the Bt crop and are depressed in numbers.

The so-called Tier 1 laboratory studies that have been conducted by companies as a part of the regulatory packages for Bt crops have not found any direct toxic effects of Cry1 or Cry3 proteins against insect predators for Bt protein concentrations at or much greater than maximum possible exposure under natural conditions (see reviews in Betz et al., 2000; EPA, 2001). Obviously, these tests are not meant to mimic natural exposure, nor do they test all possible species that could be exposed but they do represent highly conservative tests of possible hazard using carefully-chosen surrogate species.

Researchers interested in the fate of particular predatory species have carried out additional laboratory and semi-field tests of potential non-target impacts. These tests have used a variety of designs, with differing degrees of realism in terms of the route and level of Bt exposure. Given that many predators feed on some amount of pollen at some point in their life cycle, many of these studies have involved feeding predatory species pollen from Bt crops and comparable control lines. None of these studies have found any adverse impacts of Bt pollen on the survival or development of various insect predators (for example, Pilcher et al., 1997; Riddick and Barbosa, 1998). Comparable studies using Bt corn silks with a hemipteran predator also found no such effect (Al-Deeb et al., 2001).

Obviously, the above studies involved direct exposure and—under field conditions—exposure also can occur through secondary pathways with predators feeding upon herbivores that had fed on a Bt crop plant. However, any secondary exposure of this sort should have relatively little impact on arthropod predators for several reasons. First, the level of Bt protein present in herbivores that have fed on Bt plants is far lower than the level of Bt protein present in the plant tissues, presumably because of dilution effects (Head et al., 2001a; Dutton et al., 2002). Some insects, particularly phloem feeders like aphids, do not appear to pick up any Bt protein at all probably because no Bt protein is present in the parts of the plant where they are feeding (Head et al., 2001a; Raps et al., 2001). Thus, predators feeding on these different prey species will be exposed to very little Bt protein. Second, arthropod predators usually prey upon a variety of species, some or all of which may not be feeding on the Bt crop at all. These arguments indicate that predators should not be adversely affected through secondary exposure pathways. However, one set of studies has been presented as a possible example of adverse impacts through secondary exposure. Hilbeck et al. (1998a,b, 1999) performed a number of laboratory studies with the predatory lacewing *C. carnea*, feeding the larvae of this species on various herbivore species that were fed an on artificial diet containing Bt protein. They found higher mortality and slower development of lacewings exposed to Bt-intoxicated insects than for lacewings fed on comparable controls. Subsequent studies by other researchers indicate that these results actually reflect feeding on nutritionally-poorer prey than any toxic effect of the Bt protein. Dutton et al. (2002) provided lacewings with a variety of prey species, some of which were susceptible to the Bt protein and others of which were not. The prey species not susceptible to the Bt protein fed and developed normally on an artificial diet containing the Bt protein, and accumulated significant amounts of Bt protein within them through this feeding. When they were used as a food source for lacewings, the lacewings showed no adverse effects. In contrast, susceptible herbivores fed significantly less, developed more slowly and suffered higher mortality on a diet containing Bt protein than on a control diet, and surviving individuals had relatively little Bt protein within them. Nevertheless, when lacewings fed upon these prey insects, their development was adversely affected. Thus, the effects observed by Hilbeck et al., appear to reflect the poor nutritional quality of Bt-susceptible prey rather than any toxic effect of the Bt protein on lacewings. Such a situation should have little relevance to the field because other prey sources that are not affected by Bt crops will be more widely available and probably preferred under natural conditions. Field observations support this conclusion (see below). Furthermore, comparable tri-trophic studies by Al-Deeb et al., (2001) with the hemipteran predator *Orius insidiosus* (Say) saw no effect when feeding on Bt-intoxicated European corn borers. In this case, the results were confirmed with direct feeding studies on Bt corn silks and observations of populations in non-Bt and Bt cornfields.

Numerous field studies also have focused on generalist predators, particularly *Coleomegilla maculata* (DeGeer) (Coleoptera: Coccinellidae), *C. carnea* (Neuroptera: Chrysopidae), *O. insidiosus* (Heteroptera: Anthocoridae), and guilds of carabids, because of their abundance in cornfields and also their perceived importance. No adverse effects have been seen for any of these species in these studies or in the broader, community-level studies of Bt corn (Pilcher et al., 1997; Lozzia, 1999; Manachini et al., 1999; Manachini, 2000; Candolfi et al., 2004; Castanera and Ortega, 2002; Dively et al., 2002; Dively and Rose, 2002) and Bt cotton (Head et al., 2001b; Hagerty et al., 2001; Fitt and Wilson, 2002; Naranjo and Ellsworth, 2003). The absence of even indirect trophic effects of Bt corn and Bt cotton in these studies is not surprising, because most of these predatory species feed on many different prey species, the vast majority of which are not directly impacted by Bt corn or Bt cotton, e.g. sucking insects like aphids and whiteflies. In contrast, the insecticidal sprays used in conventional corn had clear adverse impacts—at least transiently—on almost all common predators, and particularly those species foraging above ground (Dively et al., 2002; Dively and Rose, 2003; Candolfi et al., 2004). Similarly, the insecticidal sprays used in conventional cotton also had clear adverse impacts on almost all of the important arthropod predators (Xia et al., 1999; Head et al., 2001b; Fitt and Wilson, 2002). In addition, work on potato fields in the northeastern US has found larger populations of many generalist predators in Bt potato fields than in conventional potato fields treated with appropriate broad-spectrum insecticides (Reed et al., 2001).

Insect Parasitoids

As with arthropod predators, no direct toxic effects from Bt crops would be expected for any parasitoid species, given what is known about the spectrum of activity of the Bt proteins expressed in currently-commercialized Bt crops. Furthermore, because the larvae of these groups feed solely on other arthropods, larval parasitoids will not face any direct exposure. Adult exposure also will be very limited because of their occasional feeding on pollen or nectar. However, secondary exposure to Bt proteins may occur if the parasitoids feed on herbivore larvae that have fed upon a Bt crop plant. In addition, indirect effects may occur at the population level if the host species of the natural enemies are a target of the Bt crop and are depressed in numbers.

As with predatory species, the Tier 1 laboratory studies have not found any direct toxic effects of Cry1 or Cry3 proteins against parasitoids for Bt protein concentrations at or much greater than maximum possible exposure under natural conditions (see reviews in Betz et al., 2000; EPA, 2001). On the other hand, secondary exposure studies indicate that parasitoids that develop on hosts exposed to Bt may be adversely impacted. When reared on Bt-susceptible insects that had fed on Bt corn, the larval development and mortality of the parasitoid *Parallorhogas pyralophagus* was adversely affected, but the fitness of the emerging adults was not impacted (Bernal et al., 2002). However, parasitoids appear to locate, parasitize and develop normally on Bt-resistant insects, whether or not these insects have been feeding on Bt plants (Schuler et al., 1999).

It should also be remembered that fundamental differences in the manner in which Bt plants act relative to conventional insecticides would be a major determinant of the relative impact that these products have on non-target species. With Bt plants, having expression of the insecticidal protein only within the plant and preferentially within certain tissues means that many parasitoids will never be exposed to any Bt protein. For example, egg parasitoids will not be exposed to the Bt proteins in Bt crops because herbivores that have fed on these crops do not incorporate any of the consumed Bt into their eggs. Furthermore, host density-mediated effects will not impact the populations of egg parasitoids because the adults of the target pest species do not distinguish between transgenic and non-transgenic fields when they are laying eggs. In contrast, Consoli et al. (2001) demonstrated that many commonly-used insecticides are toxic to the egg parasitoid *Trichogramma galloi*.

A number of field studies have looked at the impacts on parasitoids or the level of parasitism in Bt cornfields. Due to their specificity, species that parasitize the larval stages of target pests of Bt crops would be expected to be rarer in fields of Bt crops than in comparable fields of conventional crops. As expected, the few specialist parasitoids that parasitize *O. nubilalis* and certain other stalk boring Lepidoptera in corn have been found to be rarer in Bt corn than in conventional corn, e.g. *Macrocentrus cingulum* (Hymenoptera: Braconidae) (Candolfi et al., 2004; Venditti and Steffey, 2003), though even this result is not consistent across studies or even within studies (Orr and Landis, 1997; Venditti and Steffey, 2003). Similarly, the few specialist parasitoids that parasitize foliage-feeding Lepidoptera like *H. armigera* in cotton have been found to be rarer in Bt cotton than in non- Bt cotton (Xia et al., 1999; Fitt and Wilson, 2002). Of course, it is important to consider these results in the context of alternative practices. As mentioned earlier, the insecticidal sprays used in conventional corn (Candolfi et al., 2004; Dively et al., 2002; Dively and Rose, 2003) and cotton (Xia et al., 1999; Fitt and Wilson, 2002) had clear adverse impacts, at least transiently, on these same parasitoid species. Furthermore, any effective pest control practice that decreases the abundance of the host species will have comparable effects.

Secondary Arthropod Pests

Secondary non-target pests, by definition, will not be controlled directly by Bt crops. However, because natural enemies can be important for the control of these species, indirect effects may occur because natural enemy populations will be better conserved in Bt crops than in sprayed conventional crops. Indeed, Symondson et al. (2002) argued that generalist predators (one of the guilds that appears to benefit the most from removal of broad spectrum insecticides—see above) are particularly important for natural control of a variety of agricultural pests. If this is the case, then we should expect to see fewer secondary pest outbreaks in systems where broad-spectrum insecticide use is decreased. Those studies that have included the sampling of secondary pests of corn have not found any consistent significant differences between the abundance observed in Bt corn and untreated conventional corn. Dowd (2000) observed slightly higher numbers of nitidulids (dusky sap beetles) on Bt sweet corn than on conventional sweet corn, but this appears to be related to varietal differences in ear morphology rather than anything to do with the Bt protein. In cotton, those studies that have included sampling of secondary pests of cotton have either found no significant differences between the abundances observed in Bt cotton and conventional cotton, or have found some suppression of secondary pests like aphids and mites by the natural enemy populations on Bt cotton (Fitt and Wilson, 2002). Work on potato fields in the northeastern US has found larger populations of many generalist predators in Bt potato fields than in conventional potato fields treated with appropriate broad-spectrum insecticides (Reed et al., 2001). In this system, the generalist predators are important for control of green peach aphids, which in turn, transmit certain viral diseases. Due to their impact on aphid natural enemies, broad-spectrum insecticide use leads to significantly higher populations of this important pest, relative to transgenic potato fields.

Of course, where Bt crops are replacing broad-spectrum insecticides, some secondary pests that were directly controlled by the insecticides may no longer be controlled on the Bt crops, necessitating some insecticide use on the Bt crop. For example, plant bugs have been noted as a significant pest on Bt cotton, and the pest status of European corn borers and potato leafhoppers has been elevated with the use of Bt potato. In conventional cropping systems, the broad-spectrum insecticides used to control the primary pests also sometimes control these secondary pests.

Arthropod Decomposers and Other Soil Organisms

Any exposure of species important in decomposition and other soil processes to the Bt protein in Bt crops will necessarily be limited because of their feeding habits and microhabitat preferences. Despite evidence suggesting that some amount of Bt protein may be exuded from the roots of Bt corn plants (Saxena et al., 1999), this amount seems to be small, as does the amount moving into the soil through tillage of plant material and other comparable mechanisms. Furthermore, all of the Bt proteins currently found in commercial Bt products break down rapidly in soil (see reviews in EPA, 2001), though some small amount of Bt protein may bind to clay particles in the soil (Tapp and Stotzky, 1995). Consequently, the levels of Bt protein found in the soil of Bt crop fields will be low even after several consecutive years of growing Bt crops (see Head et al., 2002).

Even if substantial amounts of Bt protein were to persist and accumulate in soil (by mechanisms not previously observed), given what is known about the spectrum of activity of Cry1Ab, Cry1Ac, Cry1F and Cry3Aa, no activity is expected against the invertebrate species that are important to soil processes. Various laboratory and greenhouse studies confirm this to be the case. For example, Sims and Martin (1997) demonstrated that two species of Collembola are not susceptible to a variety of Cry1, Cry2 and Cry3 Bt proteins. Similarly, Saxena and Stotzky (2001) found that representative species of earthworms, nematodes, protozoa, bacteria, and fungi were not impacted by the Cry1Ab protein found in the most commonly-used Bt corn products. In contrast, many of the insecticides commonly used in corn and cotton have substantial negative effects on influential taxa like earthworms (Mostert et al., 2000).

Several of the non-target field studies on Bt corn have included sampling of either ground dwelling insects and/or pitfall trapping (Candolfi et al., 2004; Dively et al., 2002; Dively and Rose, 2002). As predicted, no adverse impacts on soil-dwelling taxa have been detected. Similarly, several of the non-target field studies on Bt cotton have included sampling of either ground-dwelling insects and/or pitfall trapping (Fitt and Wilson, 2002; Naranjo and Ellsworth, 2003), and no adverse impacts on ground-dwelling or soil-dwelling taxa have been detected in these cases.

Non-target Relatives of the Target Pests

Until now, we have focused upon guilds of invertebrates that play important roles in agricultural systems. For these guilds, we do not expect Bt crops to pose a hazard, but the impact of Bt crops on these guilds must be assessed because of their ecological and economic importance. However, within any agricultural system, there will usually be some number of non-target species that are related to the target pest species and which may be susceptible to the Bt protein expressed in the Bt crops. For example, in the case of Bt corn or Bt corn expressing Cry1 proteins for control of lepidopteran pests, it is important to understand the potential risks to non-target Lepidoptera. Similarly, for products like Bt potato that express coleopteran-active proteins, an assessment of potential impacts on non-target Coleoptera is necessary.

Because all of the lepidopteran species commonly observed feeding as larvae in cotton fields are viewed as pests, and insecticide use on cotton fields is high relative to other crops, the issue of the impact of Bt cotton on non-target non-pest Lepidoptera does not arise. No non-target non-pest lepidopterans are observed in significant numbers feeding in or around cotton fields. In field corn, where insecticide use is relatively low (much lower than with conventional cotton fields), non-target Lepidoptera theoretically could exist around cornfields and might be impacted by a Bt corn product. Because no non-target Lepidoptera feed on corn leaves, stalks, ears or roots, the only reasonable route of exposure would be through incidental feeding on corn pollen. The monarch butterfly represents a worst-case scenario with respect to its susceptibility to the Cry1Ab protein and its occurrence in and around agricultural fields. A substantial amount of work on monarch butterflies has demonstrated that even larval monarch butterflies developing on host plants within cornfields will be minimally impacted (Hellmich et al., 2001; Sears et al., 2001). This is primarily because the exposure to pollen of non-target Lepidoptera is very limited in time and space and, for Cry1Ab corn pollen, the expression level of Bt in the pollen of MON810 and Bt11 corn is very low. Monarchs are relatively insensitive to Cry1F protein and thus are unlikely to be impacted by consuming pollen from Cry1F corn. This indicates that Bt corn will have little or no impact on non-target Lepidoptera as a group because the monarch butterfly represents a likely worst-case scenario with respect to its susceptibility to the Cry1Ab protein and its occurrence in and around agricultural fields (Vaituzis and Tomimatsu, 2001). Comparable studies of the impact of Bt corn pollen on other Lepidoptera have also found no adverse impacts for these species, e.g. the milkweed tiger moth (Jesse and Obrycki, 2002) and swallowtails (Wraight et al., 2000)), supporting the notion that monarchs are a highly sensitive surrogate for other such potentially-exposed Lepidoptera.

In the case of the Cry3Aa protein expressed in potato, the spectrum of activity appears to relatively narrow (compared to the Cry1 proteins and their lepidopteran activity) and no susceptible non-target beetle species appear to exist in the potato agro-ecosystems. The only observed insecticidal activity with Cry3Aa has been against beetles in the family Chrysomelidae, and no non-pest chrysomelids are found in these environments. Tier 1 laboratory studies have not found any direct toxic effects of Cry3 proteins against other non-target beetle families (see reviews in Betz et al., 2000; EPA, 2001). Furthermore, field studies in potato agro-ecosystems have not observed any adverse impacts of Bt potato on any non-target beetle species (Riddick et al., 2000; Reed et al., 2001).

Potential Broader Impacts on Wildlife, Particularly on Higher Trophic Levels

The broad differences in the invertebrate communities that are seen between Bt crop systems and conventional crops treated with broad-spectrum insecticides will have obvious implications for higher trophic levels. Many small vertebrates depend on invertebrates as food sources for at least a part of each year and/or at particular stages in their development. Agricultural practices like the use of broad-spectrum insecticide decrease the abundance of these invertebrate prey species and will have detrimental effects on the vertebrate populations. For example, Brickle et al. (2000) demonstrated that the breeding success of corn buntings in the United Kingdom was affected by such practices. Similarly, Tremblay et al. (2001) showed that a number of bird species found in agricultural systems were strongly dependent on the availability of invertebrate species and that these bird species were also capable of depressing the populations of certain pest species. Thus the loss of suitable food sources for these insectivorous vertebrate species can affect their population size, which may in turn, reduce the positive impacts that these species have in curbing arthropod pest outbreaks. This will apply not to bird species but also to various small mammals found in agro-ecosystems and even to some amphibians. Many of the broad-spectrum insecticides also have deterrent effects on vertebrate species. This too will serve to decrease the abundance of these insectivorous species in agricultural fields. Removing the adverse impacts of insecticides on these species should prove to be beneficial to these systems as a whole.

CONCLUSIONS

Collectively, the non-target studies performed to date demonstrate that Bt crops do not have any unexpected toxic effects on non-target species; only the targeted pest species are directly impacted by Bt crops, as would be predicted from knowledge of the mode of action and specificity of Bt proteins. Due to this specificity, Bt crops effectively preserve the local populations of various economically-important biological control organisms that can be adversely impacted, at least transiently, by broad-spectrum chemical insecticides. The only indirect adverse effects on non-target organisms that have been observed with Bt crops are local reductions in the numbers of certain specialist parasitoids whose hosts are the primary targets of Bt crops. Such trophic effects will be associated with any effective pest control technology, be it transgenic, chemical, or cultural, as well as with natural fluctuations in lepidopteran host populations. Moreover, specialist parasitoids represent a very small portion of the non-target fauna, both in terms of the numbers of individuals and the number of taxa that have been recorded in agricultural ecosystems.

Because of these characteristics, Bt crops often will fit well into integrated pest management programs, complementing their agenda of natural and supplemented biological control. The insecticides that Bt crops replace routinely have adverse effects on many important beneficial species (Amano and Haseeb, 2001). Already, the introduction of certain Bt crops has led to unprecedented reductions in insecticide inputs with clear benefits in terms of limiting secondary pest outbreaks (see, for example, Turnipseed et al., 2001; Naranjo and Ellsworth, 2002; Shelton et al., 2002).

REFERENCES

Al-Deeb, M. A., Wilde, G. E. and Higgins, R. A., 2001, No effect of *Bacillus thuringiensis* corn and *Bacillus thuringiensis* on the predator *Orius insidiosus* (Hemiptera: Anthocoridae). Environ. Entomol., **30**, 625-629.

Amano, H. and Haseeb, M., 2001, Recently-proposed methods and concepts of testing the effects of pesticides on the beneficial mite and insect species: study limitations and implications in IPM. Appl. Entomol. Zool., **36**, 1-11.

Bernal, J. S., Griset, J. G. and Gillogly, P. O., 2002, Impacts of developing on Bt maize-intoxicated hosts on fitness parameters of a stem borer parasitoid. J. Entomol. Sci., **37**, 27-40.

Betz, F. S., Hammond, B. G. and Fuchs, R. L., 2000, Safety and advantages of *Bacillus thuringiensis*-protected plants to control insect pests. Reg. Toxicol. Pharm., **32**, 156-173.

Brickle, N. W., Harper, D. G. C., Aebischer, N. J. and Cockayne, S. H., 2000, Effects of agricultural intensification on the breeding success of corn buntings *Miliaria calandra*. J. Appl. Ecol., **37**, 742-755.

Candolfi, M., Brown, K., Reber, B. and Schmidli, H., 2004, A faunistic approach to assess potential side-effects of genetically modified Bt-corn on non-target arthropods under field conditions. Biocontr. Sci. Technol. (In Press)

Castanera, P. and Ortega, F., 2002, Environmental implications of Bt-maize in Spain: monitoring corn borers resistance and nontarget impacts. http://www.rki.de/GENTEC/GENENG/FORUM/FORUM.HTML

Consoli, F. L., Botelho, P. S .M. and Parra, J. R. P., 2001, Selectivity of insecticides to the egg parasitoid *Trichogramma galloi* Zucchi, 1988, (Hym.,Trichogrammatidae). J. Appl. Ecol., **125**, 37-43.

Dively, G. P., Patton, T. W., Miller, A., Nelson, J. and Embrey, M., 2002, Effects of transgenic field corn expressing two lepidopteran-specific insecticidal proteins on the invertebrate community. In: Proceedings of the Society of Environmental Toxicology and Chemistry, European 12th Annual Meeting, Vienna, Austria.

Dively, G. P. and Rose, R., 2003, Effects of Bt transgenic and conventional insecticide control on the non-target invertebrate community in sweet corn. In: Proceedings of the First International Symposium of Biological Control of Arthropods, U.S. Forest Service, Amherst, MA.

Dowd, P. F., 2000, Dusky sap beetles (Coleoptera: Nitidulidae) and other kernel damaging insects in Bt and non- Bt sweet corn in Illinois. J. Econ. Entomol., **93**, 1714-1720.

Dutton, A., Klein, H., Romeis, J. and Bigler, F., 2002, Uptake of Bt-toxin by herbivores feeding on transgenic maize and consequences for the predator *Chrysoperla carnea*. Ecol. Entomol. (In press)

EPA, 2001, Registration Action Document for *Bacillus thuringiensis* Plant-Incorporated Protectants. U.S. EPA, October 16, 2001. http://www.epa.gov/pesticides/biopesticides/reds/brad_bt_pip2.htm

Fitt, G. P. and Wilson, L. J., 2002, Non-target effects of Bt-cotton: a case study from Australia. In: Akhurst R. J., Beard C. E. and Hughes P. (eds), Proceedings of the 4th Pacific Conference on Biotechnology of *Bacillus thuringiensis* and its Environmental Impact, CSIRO, Canberra, Australia, pp. 175-182.

Hagerty, A., Turnipseed, S. G. and Sullivan, M. J., 2001, Impact of predaceous arthropods in cotton IPM. In Proceedings of the Beltwide Cotton Conferences, National Cotton Council, Memphis, TN, pp. 812-815.

Head, G., Brown, C. R., Groth, M. E. and Duan, J. J., 2001a, Cry1Ab protein levels in phytophagous insects feeding on transgenic corn: implications for secondary exposure risk assessment. Entomol. Exp. Appl., **99**, 37-45.

Head, G., Freeman, B., Moar, W., Ruberson, J. and Turnipseed, S., 2001b, Natural enemy abundance in commercial Bollgard and conventional cotton fields. In Proceedings of the Beltwide Cotton Conferences, National Cotton Council, Memphis, TN, pp. 796-798.

Head, G., Surber, J. B., Watson, J. A., Martin, J. W. and Duan, J. J., 2002, No detection of Cry1Ac protein in soil after multiple years of transgenic cotton (Bollgard®) use. Environ. Entomol., **31**, 30-36.

Hellmich, R. L., Siegfried, B. D., Sears, M. K., Stanley-Horn, D. E., Daniels, M. J., Mattila, H. R., Spencer, T., Bidne, K. G. and Lewis, L. C., 2001, Monarch larvae sensitivity to *Bacillus thuringiensis*-purified proteins and pollen. Proc. Natl. Acad. Sci. USA (published online Sept. 14, 2001) 10.1073/pnas.211297698.

Hilbeck, A., Moar, W. J., Pusztai-Carey, M., Filippini, A. and Bigler, F., 1998a, Toxicity of *Bacillus thuringiensis* Cry1Ab toxin to the predator *Chrysoperla carnea* (Neuroptera: Chrysopidae). Environ. Entomol., **27**, 1255-1263.

Hilbeck, A., Baumgartner, M., Fried, P.M. and Bigler, F., 1998b, Effects of transgenic *Bacillus thuringiensis* corn-fed prey on mortality and development time of immature *Chrysoperla carnea* (Neuroptera: Chrysopidae). Environ. Entomol., **27**, 480-487.

Hilbeck, A., Moar, W.J., Pusztai-Carey, M., Filippini, A. and Bigler, F., 1999, Prey-mediated effects of *Cry*1Ab toxin and protoxin and *Cry*2A protoxin on the predator *Chrysoperla carnea*. Entomol Exp. Appl., **91**, 305-316.

James, C., 2001, Global Review Of Commercialized Transgenic Crops: 2001. ISAAA Brief No. 24, ISAAA, Ithaca, NY.

Jepson, P. C., Croft, B. A. and Pratt, G. E., 1994, Test systems to determine the ecological risks posed by toxin release from *Bacillus thuringiensis* genes in crop plants. Molec. Ecol. **3**, 81-89.

Jesse, L. C. H. and Obrycki, J. J., 2002, Assessment of the non-target effects of transgenic Bt corn pollen and anthers on the milkweed tiger moth, *Euchatias egle* Drury (Lepidoptera: Arctiidae). J. Kansas Entomol. Soc., **75**, 55-58.

Lozzia, G. C. (1999) Biodiversity and structure of ground beetle assemblages (Coleoptera: Carabidae) in *Bt* corn and its effects on non-target insects. Bollettino di Zoologia Agraria e di Bachicoltura, **31**, 37-58.

MacIntosh, S. C., Stone, T. B., Sims, S. R., Hunst, P. L., Greenplate, J. T., Marrone, P. G., Perlak, F. J., Fischhoff, D. A. and Fuchs, R. L., 1990, Specificity and efficacy of purified *Bacillus thuringiensis* proteins against agronomically important insects. J. Invert. Pathol., **56**, 258-266.

Manachini, B., 2000, Ground beetle assemblages (Coleoptera, Carabidae) and plant dwelling non-target arthropods in isogenic and transgenic corn crops. Bollettino di Zoologia Agraria e di Bachicoltura, **32**, 181-198.

Manachini, B., Agosti, M. and Rigamonti, I., 1999, Environmental impact of Bt-corn on non target entomofauna: Synthesis of field and laboratory studies. In: Brown C., Capri E., Errera, G., Evans, S. P. and Trevisan, M. (eds), Human and Environmental Exposure to Xenobiotics. Proceedings of the XI Symposium Pesticide Chemistry, La Goliardica Pavese, Cremona, Italy, pp. 873-882.

Mostert, M. A., Schoeman, A. S. and van der Merwe, M., 2000, The toxicity of five insecticides to earthworms of the Pheretima group, using an artificial soil test. Pest Manage. Sci., **56**, 1093-1097.

Naranjo, S. E. and Ellsworth, P. C., 2003, Arthropod communities and transgenic cotton in the western United States: implications for biological control. In Proceedings of the First International Symposium of Biological Control of Arthropods, U.S. Forest Service, Amherst, MA.

Orr, D. B. and Landis, D. A., 1997, Oviposition of European corn borer (Lepidoptera: Pyralidae) and impact of natural enemy populations in transgenic versus isogenic corn. J. Econ. Entomol., **90**, 905-909.

Perlak, F. J., Fuchs, R. L., Dean, D. A., McPherson, S. L. and Fischhoff, D. A., 1991, Modification of the coding sequence enhances plant expression of insect cotton protein genes. Proc. Natl. Acad. Sci. USA, **88**, 3324-3328.

Pilcher, C. D., Obrycki, J. J., Rice, M. E. and Lewis, L. C., 1997, Preimaginal development, survival and field abundance of insect predators on transgenic *Bacillus thuringiensis* corn. Environ. Entomol., **26**, 446-454.

Raps, A., Kehr, J., Gugerli, P., Moar, W. J., Bigler, F. and Hilbeck, A., 2001, Immunological analysis of phloem sap of *Bacillus thuringiensis* corn and of the nontarget herbivore *Rhopalosiphum padi* (Homoptera: Aphididae) for the presence of Cry1Ab. Molec. Ecol., **10**, 525-533.

Reed, G. L., Jensen, A. S., Riebe, J., Head, G. and Duan, J. J., 2001, Transgenic Bt potato and conventional insecticides for Colorado potato beetle (Coleoptera: Chrysomelidae) management: Comparative efficacy and non-target impacts. Entomol. Exp. Appl., **100**, 89-100.

Riddick, E. W. and Barbosa, P., 1998, Impact of Cry3A-intoxicated *Leptinotarsa decemlineata* (Coleoptera: Chrysomelidae) and pollen on consumption, development, and fecundity of *Coleomegilla maculata* (Coleoptera: Coccinellidae). Ann. Entomol. Soc. Amer., **91**, 303-307.

Riddick, E. W., Dively, G. and Barbosa, P., 2000, Season-long abundance of generalist predators in transgenic versus nontransgenic potato fields. J. Entomol. Sci., **35**, 349-359.

Saxena, D., Flores, S. and Stotzky, G., 1999, Transgenic plants: insecticidal toxin in root exudates from Bt corn. Nature, **402**, 480.

Saxena, D. and Stotzky, G., 2001, *Bacillus thuringiensis* (Bt) toxin released from root exudates and biomass of Bt corn has no apparent effect on earthworms, nematodes, protozoa, bacteria, and fungi in soil. Soil Biol. Biochem., **33**, 1225-1230.

Schnepf, E., Crickmore, N., van Rie, J., Lereclus, D., Baum, J., Feitelson, J., Zeigler, D. R. and Dean, D. H., 1998, *Bacillus thuringiensis* and its pesticidal crystal proteins. Microbiol. Molec. Biol. Rev., **62**, 775-806.

Schuler, T. H., Potting, R. P., Denholm, J. I. and Poppy, G. M., 1999, Parasitoid behaviour and Bt plants. Nature, **400**, 825.

Sears, M. K., Hellmich, R. L., Stanley-Horn, D. E., Oberhauser, K. S., Pleasants, J. M., Mattila, H. R., Siegfried, B. D. and Dively, G. P., 2001, Impact of Bt corn pollen on monarch butterfly populations: A risk assessment. Proc. Natl. Acad. Sci. USA, (published online Sept. 14, 2001) 10.1073/pnas.211329998.

Shelton, A. M., Zhao, J. Z. and Roush, R. T., 2002, Economic, ecological, food safety, and social consequences of the deployment of Bt transgenic plants. Annu. Rev. Entomol., **47**, 845-881.

Sims, S. R. and Martin, J. W., 1997, Effect of the *Bacillus thuringiensis* insecticidal proteins *Cry*1Ab, *Cry*1Ac, *Cry*2A, and *Cry*3A on *Folsomia candida* and *Xenylla grisea* (Insecta: Collembola). Pedobiol., **41**, 412-416.

Symondson, W. O. C., Sunderland, K. D. and Greenstone, M. H. (2002) Can generalist predators be effective biocontrol agents? Annu. Rev. Entomol., **47**, 561-594.

Tapp, H. and Stotzky, G., 1995, Insecticidal activity of the toxins from *Bacillus thuringiensis* subspecies *kurstaki* and *tenebrionis* adsorbed and bound on pure and soil clays. Appl. Environ. Microbiol., **61**, 1786-1790.

Tremblay, A., Mineau, P. and Stewart, R. K., 2001, Effects of bird predation on some pest insect populations in corn. Agric. Ecosyst. Environ., **83**, 143-152.

Turnipseed, S., Sullivan, M. J., Hagerty, A. and Ridge, R., 2001, Cotton as a model IPM system in the southeast: A dream or potential reality? In Proceedings of the Beltwide Cotton Conferences, National Cotton Council, Memphis, TN, pp. 1009-1010.

Vaituzis, Z. and Tomimatsu, G., 2001, Risk Analysis to the Karner Blue Butterfly and Other Endangered *Lepidoptera* from Bt *Cry*1Ab event MON810 and Bt11, and Bt *Cry*1F corn TC1507, Biopesticides and Pollution Prevention Division, U.S. Environmental Protection Agency. Washington, D.C.

Venditti, M. E. and Steffey, K. L., 2003, Field effects of Bt corn on the impact of parasitoids and pathogens on European corn borer in Illinois. In Proceedings of the First International Symposium of Biological Control of Arthropods, U.S. Forest Service, Amherst, MA.

Wraight, C. L., Zangerl, A. R., Carroll, M. J. and Berenbaum, M. R., 2000, Absence of toxicity of *Bacillus thuringiensis* pollen to black swallowtails under field conditions. Proc. Natl. Acad. Sci. USA, **97**, 7700-7703.

Xia. J. Y., Cui, J. J., Ma, L. H., Dong, S. X. and Cu, X. F., 1999, The role of transgenic *Bt* cotton in integrated insect pest management. Acta Gossypii Sinica, **11**, 57-64.

REGULATION OF GENETICALLY-MODIFIED CROPS: A SCIENTIFIC PERSPECTIVE

KEES HULSMAN
Australian School of Environmental Studies
Faculty of Environmental Sciences, Griffith University
Nathan. Q. 4111, Australia

INTRODUCTION

There has been a noticeable increase in the environmental release of transgenic or genetically-modified organisms (GMOs) over the past decade. Biotechnology is a multi-billion global industry and governments perceive it as an income earner for the future in the new globalized economy. Many governments are encouraging biotechnology businesses to set up headquarters in their jurisdiction. For example in Australia, Queensland Premier Peter Beattie and his government are marketing Queensland as "The Smart State" and are investing in the development of biotechnology. Research Institutions and Universities are directing research funds towards biotechnology because of its financial returns via research grants and market returns resulting from the development of new products and processes.

One of the major problems associated with GMOs is the uncertainty of their impacts on human and environmental health. These two concerns are brought together in the application of gene technology to agriculture. There is an on going conflict between the proponents and critics of the application of gene technology in agriculture.

Proponents of GMOs have argued that biotechnology is necessary to feed the world. They claim that it is more environmentally friendly than chemical sprays are to control weeds and animal pests and will also reduce agricultural reliance on toxic chemicals. In addition, gene technology is a natural extension of conventional breeding techniques but it has the advantage of making precise genetic changes and, therefore, may be used to solve problems caused by intensive agriculture in a more environmentally-friendly manner (Levidow and Carr, 1997). On the other hand, critics have argued that we currently produce sufficient food to feed the world, the problem is the distribution of food and GE will not fix that (e.g. Ho, 1999). GM crops will introduce new ecological risks and decrease the diversity of plant cultivars (Ho, 1999). Research and development in this area is driven more by economics than by environmental health criteria. This will not reduce our reliance on technology, agrochemicals or genetic solutions (Levidow and Carr, 1997). In fact, gene technology will generate additional problems that will require added genetic changes to deal with them. This has been described as a genetic treadmill (Hindmarsh et al., 1991a; Levidow and Carr, 1997). For example, GM crops may accelerate the resistance amongst plant and insect pests to pesticides. Wide-spectrum herbicides, in combination with herbicide-resistant crops, will create a loss of indigenous agricultural and natural biodiversity (Ho, 1999). This threatens the basis of food security not only in developing countries but also globally. Moreover, seed royalties, restrictive practices of seed certification and competition with subsidized produce from developed nations threaten the livelihood of small farmers (Ho, 1999), who form the basis of agricultural production in these countries. Indeed, GE agriculture will simply foster the poverty cycle in developing countries. It seems that GE agriculture is a form of neo-colonialism with developed countries exploiting the natural resources of developing countries. Thus, it is not in the interests of developing countries to adopt GE agriculture at this stage of their development.

Some critics have argued that GM crops deal with the symptoms and do not treat the cause of the problem. Sustainable agriculture requires a reorganization of agricultural systems so that we can avoid monocultures (Levidow and Carr, 1997). After all it is the monoculture that creates suitable conditions for pest populations to grow large. Finally, GE is not a natural extension of conventional breeding; molecular biologists have oversimplified the genetic system to the point that their predictions about phenotypic outcomes are grossly inadequate. These points will be developed further in the section on What Is The Degree of Uncertainty?

There needs to be an effective means of evaluating what impacts may result from releasing GMOs into the environment. Growing GM crops have the potential to have adverse effects on the:

- environment by increasing the weediness and, hence, the invasion of natural ecosystems by exotic plants;
- agricultural system by increasing weediness as well as the speed of development of resistance of pests to pesticides and, thus, decrease their effectiveness to control pests elsewhere; and
- effects of the food on human health, such as increasing the exposure of people to allergens (see Table 11.1).

A lack of information about the likelihood of impacts, especially ecological ones, exists because these have been given a lower priority as compared to direct effects on human health. However, it needs to be recognized that human health is indirectly affected by the health of our environment. Another problem is that once released, GMOs can reproduce and spread but not be recalled (Regal, 1996). Therefore it is important to make the correct decision at the very outset.

These differences between the proponents and critics of gene technology also emerge in the debate as to the nature and degree of regulation of biotechnology required. Proponents argue that the regulation of GMOs should be based on sound scientific evidence, whereas many critics argue that it should also be based on the Precautionary Principle (see Levidow and Carr, 1997). The proponents of biotechnology maintain that the precautionary principle discriminates against GMOs because it imposes the burden of proof for their safety on the applicant and ignores the lower perceived risks associated with GMOs compared with the risks associated with the use of chemicals on unmodified crops (Levidow and Carr, 1997). Clearly, the conflict pertaining to regulation must be resolved in the near future.

The basis of the conflict is embedded in disagreement as to what constitutes sound science. These disagreements include the appropriateness of experimental design for monitoring effects, but they even extend to something more fundamental. That is, the different paradigms and theories used by various stakeholders in the debate to interpret and predict outcomes resulting from the application of gene technology. In this chapter, I explore the two opposing paradigms (Genetic Determinism and Fluid Genome) that give rise to conflict between the proponents and some of the critics of the application of gene technology.

The chapter is structured in the following manner: First, I shall deal with some of the problems associated with gene technology—specifically the degree of uncertainty about the effects of GMOs released into the environment, and the controversy about what to regulate: the process, the product or the impacts. Then I shall address how scientific

Table 11.1 Summary of perceived hazards resulting from the application of biotechnology to agriculture according to Fluid Genome paradigm

Type of impact	Perceived hazards according to Fluid Genome
Human and animal health	Allergenic and toxic effects of GM products caused by interactions between introduced gene(s) and host genome
	Increase in pesticide-related illness amongst farm workers, contamination of food and drinking water resulting from the increased use of herbicides on herbicide-resistant crops
	Spread of antibiotic resistant marker genes by horizontal gene transfer to other bacteria such as pathogens
	Potential for horizontal gene transfer and recombination producing novel pathogenic bacteria and viruses
	Potential for transgenic DNA to infect cells after ingesting transgenic food; regenerate disease virus inserts itself into the cell's genome and cause harmful or even lethal effects such as cancer
Agricultural and natural biodiversity	Spread of transgenes etc., to related weed species and thereby create superweeds
	Increase in the indiscriminate use of broad-spectrum herbicides in association with herbicide-resistant plants causing the large scale extinction of indigenous and natural species
	Increased use of other herbicides to control herbicide-resistant 'volunteers', decreasing the indigenous biodiversity
	Destroy soil fertility and thus decrease yield per unit area
	Accelerate the evolution of biopesticide-resistance in major pest species and thus affect organic farming industry through the loss of the effectiveness of biopesticides
	Harm beneficial insect species
	Increased use of natural biopesticides in GMOs and an increase in the development of resistance. This would deprive the ecosystem of its natural pest controls and thus its ability to rehabilitate itself after being disturbed
	Horizontal transfer of transgenes and marker genes to unrelated species via bacteria and viruses, which may create new weed species
	Increased potential to create new virulent strains of virus Transgenic DNA can be perpetuated and amplified under the right environmental conditions. As a result, it could unleash cross-species epidemics of infectious plant and animal diseases that would be impossible to control or recall

Source: Based on Ho (1999)

controversies can be resolved before delving into how to regulate GMOs. The emphasis in this part is on the role of science in regulation; why it is the key to regulation and quality control. The debate is global and the issues raised in this chapter are applicable across the world, especially to those countries in which the approach is mainly science-based.

DEGREE OF UNCERTAINTY

The issue of uncertainty has been subject to intense debate between the proponents and critics of the application of gene technology. What is the cause of this controversy between the scientists? This is a complex issue and reflects the history of the disciplines from which the two main protagonists appear. GE molecular biologists have emerged from the laboratory, whereas their opponents—the ecologists—have emerged from the field. These different origins are reflected in the approach used by each group in its discipline. The molecular biologists tend to be reductionist, whereas the ecologists tend to have a more holistic approach in dealing with the organism in its environment (Kasanmoentalib, 1996).

GE molecular biologists have broken from the confines of their controlled environment laboratories to apply their technology in the broader uncontrollable outside world. Whilst they were in their laboratories where they could control the environmental factors, they were confident about the outcomes. However, when they could not extend the laboratory conditions to the scale of their applications outside the laboratory, uncertainty increased because they could not control the numerous environmental factors (Latour, 1999). This thesis is at odds with the views of the more enthusiastic proponents of the application of gene technology (Davis 1987, 1989; van Dommelen, 1999).

The critics of genetic engineering are not only ecologists. There are also molecular biologists who are stridently critical of the theoretical basis of genetic engineering (e.g. Ho, 1997, 1999; Strohman, 2001; Commoner, 2002). There has been an accumulation of observations since the 1940s that are not consistent with the Central Dogmas of Genetic Determinism, e.g. transposable genetic elements, jumping genes (Whitehouse, 1969; Wheale and McNally, 1988) and inheritance of acquired characteristics (Steele, 1979). As a result of this, the criticisms by this group are particularly damaging to the future of genetic engineering's application. Their criticisms will be examined in detail in this chapter because they affect the viability of the current type of application of gene technology.

GE molecular biologists pride themselves on being able to place a specific gene at a specific locus on a chromosome (Persley and Peacock, 1990). That is precision! They claim with certainty that any gene from any organism can be placed into the genome of another organism from completely different taxa and then produce a specific desired result (Davis 1987, 1989; Persley and Peacock, 1990; van Dommelen, 1999). This certainty arises from the perceived precision of the theory that guides their thinking. The Central Dogma of Genetic Determinism is that one gene produces a protein (Commoner, 2002). On this basis, most plants and animals were expected to have about 100,000 genes (Persley and Peacock, 1990). However, the results of the Human Genome Project have shown that there are insufficient genes to code for all of the proteins that exist within a cell (Commoner, 2002). For example, humans have 30,000 genes (see Commoner, 2002) and in the cell membrane, there are 10 billion proteins consisting of 10,000 different types (Alberts et al., 1994). Therefore, according to Genetic Determinism, 10,000 genes would be required to provide the protein of the cell membrane. That would leave 20,000 genes to code for everything else. That is not enough.

An alternate explanation is that one gene codes for more than one protein. This is achieved by a variety of means—first for example, the reading part of the gene, and/or the use of fragments of different genes to code for different proteins or variants thereof (Ho, 1999). Second, where 'spliceosome' proteins cut the messenger RNA into fragments and then rearrange them into a number of different combinations, thereby changing the nucleotide base sequence from that of the original DNA (Commoner, 2002). Thus, the types of conclusions that one draws when using Genetic Determinism and its associated theories are very different from those drawn when using the Fluid Genome Paradigm and Alternate Gene Splicing Theory.

Let us consider the implications of the use of the two above-mentioned paradigms, because criticism of the science-based approach is really directed at the conclusions drawn from Genetic Determinism. For example, there is a great deal of certainty about the outcome of the insertion of a specific gene into another organism on the basis of the one gene-one protein scenario of Genetic Determinism. This is because the gene codes for a specific protein and does not interact with its neighboring genes to produce other proteins. In contrast, it follows from the Alternate Gene Splicing Theory that a simple insertion of a specific gene from a different taxon can lead to the production of a different set of proteins than it did in its parent cell (Commoner, 2002). Exactly which proteins it will produce is uncertain because it depends on which fragments of DNA from the other genes it interacts with in producing proteins or which spliceosome proteins reconfigure the messenger RNA. This may not be possible to predict *a priori*. In this context, the perceived certainty of the outcome of even the insertion of a single gene decreases markedly. This outcome has implications for what should be regulated. There are several approaches to deal with this problem and these will be dealt with in the following pages.

The Fluid Genome Paradigm and Alternate Gene Splicing Theory also have implications for the Principle of Substantial Equivalence, i.e. when a GM product and its conventional counterpart are considered to be essentially the same, then the GM product need not be labeled as being genetically modified. The Principle of Substantial Equivalence can help fast track products through the regulatory system. However, there is a problem in determining which criteria should be used to establish substantial equivalence. In the case of food, it means that its molecular, composition and nutritional information are compared with those of its conventional counterpart in order to establish substantial equivalence (ARMCANZ, 1997; van Dommelen, 1999). According to Genetic Determinism, a GMO and a conventionally-bred organism are essentially the same (Persley and Peacock, 1990), whereas according to Alternate Gene Splicing Theory, there is a potentially large difference between them. Given the large numbers of proteins that a single gene can be responsible for, what is actually measured and what should be measured to determine equivalence? Is it a question of don't look, don't find? (Mellon and Rissler, 1995). If it is, then biosafety tests are a disaster waiting to happen.

Miller and Gunary (1993) argued that risk is primarily a function of the product's characteristics and not the method of genetic modification. Thus, some GE molecular biologists think that environmental release of GMOs is over-regulated because its justification is based on assumptions that contradict evolutionary principles (Davis, 1989). However, according to the Fluid Genome Paradigm, the product's characteristics are a function of the method of genetic modification. For example, the insertion of a single gene into a genome by recombinant DNA techniques leads to increased instability of the genome and, thus, unpredictable outcomes (Ho, 1999).

There are at least three flaws in the argument that the application of gene technology is over-regulated. The first flaw is that this over-regulation argument is based on Genetic Determinism, a flawed paradigm which predicts that a GMO with a single gene inserted into it is essentially the same as its unmodified relative (Scott, 1987; Wheale and McNally; 1988, Ho, 1997, 1999; Strohman, 2001; Commoner, 2002). If one uses the Fluid Genome paradigm, then one comes to a different conclusion that justifies the regulation of the release of GMOs into the environment. For example, the insertion of a single gene from another taxa creates a set of potential hazards not normally encountered with conventionally-bred organisms. In conventional breeding, each parent contributes 50% of its genome to its progeny. Therefore, any gene will be accompanied by 50% of its parental neighborhood. Consequently, it will also have a whole suite of genes with which it usually interacts to produce proteins. Given that its genes have evolved together in a gene pool, they are more likely to be compatible with genes from the other parent than with any gene that comes from another taxon. The introduction of a foreign gene into the genome disrupts the stability of the genome that has taken 100s and possibly 1000s of years to stabilize under specific selection pressures (Ho, 1999). Thus, GE may increase the number of novel combinations realized and may produce ecologically-harmful organisms (Ninio, 1983).

The second flaw is that the over-regulation viewpoint ignores the characteristics of the environment into which an organism is introduced. These characteristics are important because they affect the selection pressures to which the introduced organism would be subjected. I will use a non-GE example to illustrate the point. Sickle Cell Anaemia (SCA) is a fatal disease controlled by a recessive gene. Individuals who are heterozygous for SCA have some sickle-shaped red blood cells amongst their normal ones and although they survive, they are at a disadvantage with respect to individuals who have normal red blood cells only. However, in areas prone to malaria, individuals who are heterozygous for SCA have a selective advantage over normal because they are more resistant to malaria than individuals who have normal red blood cells only (Campbell, 1990). The point is that an organism that is at a selective disadvantage in one set of circumstances may have a selective advantage in a different set of circumstances (Tiedje et al., 1989). Therefore, the risks that an organism poses to human and/or environmental health cannot be realistically considered without reference to the environment into which it is placed and lives.

The third flaw in the argument is the misunderstanding of evolutionary principles as used by Davis (1989). A number of proponents (Brill, 1985; De Silva and Bailey, 1986; Davis 1987, 1989) have specifically argued that genetic manipulation would decrease a GMO's competitiveness. Tiedje et al. (1989) gave references of examples of genetic manipulation increasing the competitiveness of GMOs, which shows that the proponents' argument is incorrect. Van Dommelen (1999) pointed out that Davis' (1989) argument was based on the mistaken view that organisms have maximum fitness and ignored the environment in which the organisms live. Davis (1989) also ignored the fact that the fitness landscape can change and, thus, lead to changes in the organism's fitness. These concepts about how fitness changes in response to environmental change provide the theoretical bases of evolution.

THE REGULATION

Regulation can focus on one or more of three aspects of GMOs. These aspects are the process, the product and the impact of the product. The arguments for and against focusing on these aspects of GMOs will be developed in the following paragraphs:

Miller and Gunary (1993) and van Dommelen (1999) argued that if there is to be regulation, then the product should be regulated because it is the characteristics of the product that pose the risk and not the means of genetic manipulation. They claim that the means of genetic manipulation does not affect the characteristics of the product. However, there is evidence that the means of genetic manipulation does affect the characteristics of the product. For example, GM yeast, which was engineered with multiple copies of one of its own genes in order to increase the rate of fermentation, accumulated the metabolite methylglyoxal at toxic, mutagenic concentrations (Inose and Murata, 1995). Multiple copies of the same genes are created and inserted by recombinant DNA techniques. In addition, there is the problem of GMOs and their potential to be toxic or allergenic (Nordlee et al., 1996). Moreover, there is evidence starting to emerge that allergenicity of plants is connected to proteins that are involved of the defence against diseases and pests (Ho, 1999). Thus, plants that have been genetically modified to combat diseases and/or pests may possess a greater potential for allergenicity than unmodified plants (Ho, 1999).

Regulation on the basis of the process alone seems to be very difficult because of the complexity of protein production within a cell. Before the advent of the Fluid Genome Paradigm, regulation on the basis of the process seemed to be a viable option. Products were produced by recombinant DNA technology, cell fusion, somaclonal variation and other techniques and could well have very different results because such techniques did not have the precision of recombinant DNA techniques. These techniques place numerous other genes into a cell with little control over which genes were inserted into a chromosome. Further processing was required to determine which cells had the desired traits before they were replicated en masse (Wheale and McNally, 1988; Persley and Peacock, 1990).

Focusing on the process as the means of regulation is feasible if it used to classify products into specific groups for a specific vetting process (Teidje et al., 1989). However, irrespective of the process, a product can be monitored to determine whether its properties have changed in more than the desired outcome. These changes in the product's properties could be tested to determine what effect they had on human and environmental health before any decision was made about environmental release of the GMO. However, the problem will be: what should the testers be looking for? The outcomes of even insertions of a single gene can be unpredictable. For example, a gene that occurs in the inner ear of chicks as well as humans is known to code for 576 variant proteins (Black, 1998). The largest number of proteins that a gene can code for by alternate gene splicing to date is in a fruitfly, *Drosophila*. One gene is responsible for over 38 000 variant proteins (Schumucker et al., 2000). Do testers check every protein produced (as a result of the change in the organism's genome) or merely a subset of them? This is a complex problem still awaiting a satisfactory solution.

Focusing on the impacts of a product on human and environmental health in order to vet products for environmental release may take a long time to determine its long-term impacts. One of the main problems here is what to look for when monitoring the environment, given the complex and unpredictable nature of the outcomes of some of the applications. Properly and well-designed experiments are needed to obtain the necessary information so as to facilitate informed decision-making. In these experiments, specific variables are measured, and if the appropriate ones are not measured, then potential problems will not be detected.

So, we are left with the question of what to regulate—the product, the process, the impact or a combination of two or more of the three aspects. Obviously, the impact of the product on health (human and or environmental) is important but that sort of information is not usually available in the short term. Therefore, this has to come into play later, during the long-term evaluative process. So, what is the role of the product and/or process in the evaluative process? On the basis of Genetic Determinism, the process is irrelevant and the product should be regulated on the basis of its phenotypic characteristics. On the basis of the Fluid Genome Paradigm, the process should be used to classify products into various categories and then subject them to an evaluation of their phenotypic characteristics. GMOs ought to be subjected to greater scrutiny than the products produced by traditional techniques (Teidje et al., 1989), because the technique used to produce them does affect their phenotypic characteristics. How can decision makers such as politicians make an informed decision as to which approach to use? Even scientists disagree on the correct approach. To answer that question, we should consider how non-scientists can resolve scientific controversies.

HOW CAN SCIENTIFIC CONTROVERSIES BE RESOLVED?

van Dommelen (1999) argued that established scientific principles should be used to identify significant gaps in our understanding and address them with properly-designed experiments. This is an extremely logical way to proceed. However, proponents such as Davis and Miller would see Genetic Determinism as being an established scientific principle, whereas critics such as Commoner, Ho and Strohman would not accept this ideology as an established scientific principle because it is fatally flawed. It seems that to resolve this impasse, researchers who use Genetic Determinism to guide their thinking should conduct experiments designed to detect unexpected proteins and show that there are none, thus disproving the Fluid Genome paradigm. Whereas researchers who use the Fluid Genome Paradigm to guide their thinking may conduct similar experiments to demonstrate the presence of unexpected proteins, as is predicted by the paradigm.

van Dommelen (1999) has proposed a mechanism for resolving scientific controversies by comparing the relevant research questions generated by various points of view. For example, the controversy over adequate definition of the problem as the basis for biosafety testing can be evaluated in terms of the possible specific relevant questions (SRQs). van Dommelen (1999) argued that the SRQs can be evaluated by discussing the individual relevant questions in terms of the larger SRQ that they make up (see Table 11.2).

> All scientific research is centered around raising and answering questions. The quality...of the answers...depends on the quality of our process of questioning. Science can be seen as the art of questioning. Research questions can be methodologically inadequate in a given context because they are: *badly chosen in view of the research purpose, inadequately phrased* and/or *improperly addressed experimentally.*

(van Dommelen, 1999).

This strategy requires an analysis of the methodological basis of particular claims. If the relevant research questions that arise from a specific claim cannot be justified by a biological theory, then the burden of proof for the claim should fall on the person making the claim and not on those questioning its validity and reliability. Thus, van Dommelen's approach overcomes the immediate need to generate evidence to support a specific viewpoint. The decision-making process does not have to grind to a halt while scientists conduct experiments to obtain the necessary evidence to show which view is the better approximation of reality. Nor

Table 11.2 Possible relevant questions about agent(s) of GMO hazard and the reasons for their relevance

Question	Agent for GMO hazard	Argument for relevance
1	Are the vector, the donor, the recipient, or the genetic insert known to be a cause for concern?	Previous experiences may be indications of future hazards
2	What are the gene products of the transgene in the GMO?	New gene products may constitute new hazardous agents
3	Can the transgene survive outside the GMO?	The transgene may become a hazardous agent all by itself
4	Will the copy number of an integrated transgene affect the GMOs characteristics?	The number and place of integration of the transgenes into the genome may affect the traits of the GMO
5	Can the transgene be transferred to non-target organisms?	HGT may provide a non-target organism with characteristics that confer a set advantage and enable it to become a pest species
6	Will the insertion of a transgene increase the instability of the GMO's genome?	The transgene may increase genome instability, leading to a rearrangement of genes and hence affect their expression. The new products may be unexpected hazardous agents.
	...other relevant questions...	...include in SRQ...?

Source: van Dommelen (1999)

does decision making have to occur in an information vacuum on a specific issue. At this point, I must emphasize that van Dommelen's approach to resolve scientific controversies in the short term does not eliminate the need to collect evidence to support a specific viewpoint. The approach is simply an interim means of dealing with a serious problem that often impedes decision making.

Theories guide the questions that should be asked. This is the main reason for the central role that science should play in the regulation of biotechnology. The theories needed will come from genetics, physiology, ecology and medicine. Thus, an interdisciplinary and multidisciplinary approach is needed to integrate the relevant theories in a meaningful way with respect to the problem. If the combination of theories does not ask all of the relevant research questions, then a revised theory and in some cases a new theory is required. Scientists and decision makers can distinguish between the appropriateness of specific theories by utilizing the identification of the SRQs and, thus, not have to enter a potentially unending debate about the suitability of a specific paradigm or theory (van Dommelen, 1999).

The paradigm that accounts for the results explained by its rival and also clarifies additional outcomes will be preferred. Therefore, let us consider the possible relevant questions about the agent of GMO hazard, based on the two competing paradigms: Genetic Determinism and Fluid Genome. The first four questions in Table 11.2 are relevant to both paradigms. However, the predictions from the respective paradigms are different. For example, Genetic Determinism predicts that the gene products are predictable because the genetic changes are precise and controlled and the expression of a gene is independent of other genes. In contrast, Fluid Genome predicts that the gene products are unpredictable because the position where the transgene is placed on the chromosome, or where it ends up, affects its expression since it interacts with its neighbouring genes. The fifth question concerning horizontal gene transfer (HGT) is more likely to occur under Fluid Genome than Genetic Determinism because of the mobility of genes in the genome and, thus, has a greater likelihood of passing a gene to another organism. The sixth question regarding the stability of the genome is really only relevant to the Fluid Genome paradigm because it predicts that the insertion of a foreign gene may destabilize the genome (Ho, 1999). According to Genetic Determinism, the insertion of a foreign gene should have no effect on the stability of the genome. Thus, on the basis of van Dommelen's approach, the Fluid Genome Paradigm would be the preferred option because it poses a greater number of relevant questions than its alternate.

Another means of choosing between competing paradigms is to check the assumptions. Table 11.3 presents the assumptions underlying Genetic Determinism. On considering the evidence against each assumption, it is clear that Genetic Determinism is not viable. The assumption that one gene produces one protein is clearly untenable, given the results of studies by Black (1998) and Schmucker et al. and (2000), the Human Genome Project (Commoner, 2002) the details of which are presented in the Section 'What to Regulate'. The Fluid Genome Paradigm accounts for all the same criteria as does Genetic Determinism, as well as explains some of the unexpected results such as jumping genes, instability of the transgenic lines, and the results of the Human Genome Project as to shortfall in the expected number of genes in

Table 11.3 Assumptions underlying Genetic Determinism and the evidence against them

Assumption	Genetic determinism	Evidence against
1	Each gene is an independent unit of information. Each gene codes for one protein	A gene may code for multiple proteins (Black, 1998; Schumucker et al., 2000)
2	Each gene is expressed without any kind of interaction	Expression of a gene is affected by its interaction with other genes (Riedle, 1996; Ho, 1997, 1999)
3	Genes are stable. They only change if mutation occurs	❖ Genes may change in response if mutation occurs to conditions of the organism and environment. (Cullis, 1983; Ho, 1987 Foster, 1992; Symonds, 1994).
4	Genes or sets of genes do not change in response to the environment	❖ These adaptive changes can be passed onto the next generation (Campbell et al., 1973; Hall and Hartl, 1974; Steele, 1979; Ho et al., 1983; Ho, 1987; Cairns et al., 1988; Rothenfluh et al., 1995; Riede, 1996)
5	Genes stay where they are placed	❖ Genes may jump within a genome (Temin, 1980; McClintock, 1984). ❖ They may also be transferred to another taxa (Horizontal Gene Transfer) (Hoffmann et al., 1994; Lorenz and Wackernagel, 1994; Tschape, 1994; Whatmore and Kehoe, 1994; Bik et al., 1995; Kapur et al., 1995; Barinaga, 1996)

Source: Based on Ho (1999)

humans. Thus, on the basis of both approaches, the Fluid Genome Paradigm is preferred to Genetic Determinism.

It is interesting to note that genetics is one of the last disciplines in Biology to opt for a dynamic model as its paradigm. For example, the cell membrane was thought of as being like a sandwich with two layers of phosolipids, held between two further layers of protein. This static model was proposed in 1930s and was accepted until about the early 1970s (Campbell, 1990), when, the Fluid Mosaic model was proposed as a better approximation of the structure of the cell membrane (Campbell, 1990). Even in ecology, populations are considered to be governed by stochastic processes. Static models are not good approximations of biological reality.

How to Regulate?

Approaches to dealing with the uncertainty associated with the environmental release of transgenics tend to fall along a continuum, ranging from a narrow science-based one to the broader socio-economic and ethical one. The science-based approach focuses mainly on the technical aspects of safety for humans and the environment, whereas the broader socio-economic and ethical one deals with not only the technical information but also the social, economic, ethical, political and environmental perspectives. In addition, the socio-economic and ethical approach is a more transparent process than its science-based counterpart (Levidow and Carr, 2000). It has to be because it would not work if the various stakeholders could not talk openly and honestly with one another (De Marchi and Ravetz, 1999; Ravetz, 1999). The USA regulatory framework for GMOs is usually called risk-based or science-based regulations in which decisions are based on scientific evidence. In contrast, the European Union (EU) regulatory framework differs from the USA framework in that although it is based on 'sound' science, it also coexists with the Precautionary Principle (Levidow and Carr, 2000).

The Precautionary Principle is an approach designed to deal with the uncertainty about the outcomes from the application of this technology. This principle comes in a variety of shades of green, ranging from a light green through to a deep green version. The light green version is really a cautionary principle, i.e. if in doubt proceed with caution, whereas the deep green one is emphatically precautionary, if in doubt do not proceed until the relevant safety information is available (see Hindmarsh and Hulsman, 2004).

The version of the Precautionary Principle which is used by the EU delays the approval of many GM crops for release because of the concerns about risks and/or its requiring more evidence about their safety (Levidow and Carr, 2000). This is a darker green than the one used by Australia. In the Australian regulatory system, the Precautionary Principle is used in a cautionary manner (i.e. light green). This means that if there are reasonable concerns about risks and uncertainty, then one progresses with caution and does not stop until better information upon which to base a decision is available (Hindmarsh and Hulsman, 2004). The Precautionary Principle helps identify the research necessary (Levidow and Carr, 1997).

Regulation of GMOs seems to be going through a very slow transition phase from the science-based system in which the experts made the decisions, to a broader framework that includes social, ethical, political, and economic perspectives in the decision-making process. The inadequacies of the science-based approach (Levidow and Carr, 1997) have led to the development of Post-Normal Science, which involves a broader network of peer review and, more importantly, is not encumbered by the unrealistic view that science is certain, objective and value-free (Ravetz, 1999). It is clear that the alleged sound-science approach has been unsuccessful because of the criteria used and the mindset of the people applying the criteria to determine what is sound science. There seems to be a double standard in operation when evaluating some experimental trials (Levidow and Carr, 1997), e.g. studies to determine the effects of Bt on non-target predators such as lacewings. Few regulators questioned the reliability of using microbial Bt to test for harm to non-target species. After this method was challenged, regulators failed to question the adequacy of the other methods such as the eggshell-coat or the seven-day exposure period (Levidow and Carr, 2000).

A general example to illustrate the double standard in operation in evaluating GMOs is that of the regulators in the USA and France using the concept of sound science 'to assign a weak burden of evidence for safety and a strong burden for risk, thus facilitating commercial approval (Levidow and Carr, 2000). This indicates that commercial interests were given higher priority than health ones. If human and environmental health were given top priority, then the precautionary principle would have been applied and the evidential onus would fall on the proposers to provide evidence of the safety of their product and its production.

The Australian regulatory framework is an example of a system in transition, albeit very slow. We have moved from a solely science-based approach to a broader one, in name at least. In name because although ethical, environmental and societal perspectives are being recognized, they do not seem to be given as much weight in the decision-making process as the technical perspective. It is a question of who controls the Regulatory Agenda. During the 1980s and the early 1990s, the answer was clearly the molecular biologists (Hindmarsh et al., 1991b). Now, in the early 2000s with a restructuring of the regulatory system, ethics and impacts on society may have attained a higher weighting in the decision-making process. Just how much weighting they are given will depend very heavily on the individuals who have been appointed to key positions such as the Gene Technology Regulator and membership of the technical committee that is responsible for advising the decision makers whether a GMO should be approved for environmental release.

ROLE OF SCIENCE IN REGULATION

At present, scientists often disagree about the nature and degree of threats that GMOs pose to human and environmental health. So, while the experts disagree, the politicians are expected to find practical ways of dealing with the lack of scientific certainty (van Dommelen, 1999).

The paradigm and theories used to interpret the data guide the posing and selecting of the relevant questions. Thus, if relevant questions are missed by one of the parties using a specific paradigm and theories, then those used by the other party to generate the relevant questions are likely to be better. In this way, decision makers can determine which scientific approach to use to improve the decision-making process.

Why is Science the Key to the Regulatory Agenda?

What specific theories (genetic and ecological) are guiding the thinking of the scientists assessing the risks? If they are using inadequate theories, then they will not be asking all of the relevant questions. The reason for this is that science guides the overall approach used to risk assessment.

Regulatory systems, in both Australia and the UK for example, are focused on hazard identification and characterization. This is a direct outcome of the belief of molecular biologists that the 'behaviour of the phenotype can be inferred from precise knowledge of the genotype, since we are dealing with minor genetic changes' (Kasanmoentalib, 1996). This viewpoint is incorrect for at least two reasons. First, Alternate Gene Splicing Theory would predict that small genetic changes may lead to large changes in the phenotype. It is difficult to predict the behavior of the phenotype from the genotype, if one does not know precisely how a gene interacts with others in its neighborhood. Second, it ignores the selection pressures operating in the habitats into which the GMOs are introduced and or will disperse. Even on the basis of ecological theory, it is unlikely (van Dommelen, 1999).

Current regulatory systems need to be expanded to include questions about horizontal gene transfer, about the unexpected outcomes from the interaction of the 'novel' gene with its neighbors. The questions that can be posed are:

- under what conditions would worst-case scenarios occur?
- what are the assessment endpoints, which would serve as early warning sentinels? and
- what are the stopping protocols? (Barnes and Hulsman 1995).

Answering these queries will require lateral thinking and the inclusion of all of the stakeholders in the decision-making process. In the case of monitoring, it is a matter of determining what exactly needs to be measured in order to detect the highly unlikely or the unexpected.

Kasanmoentalib (1996) concluded that the existing regulatory system in Europe was 'inherently biased in favour of one view, that of the genetic engineering community.' This conclusion is also likely to apply to those of Australia, the USA and any other country with a regulatory system. The reason for this is simple. Molecular biologists have controlled the science that has been used to evaluate gene technology and advise their governments about the correct policy for its application (Hindmarsh et al., 1991b). Indeed, the alliance between the business world and molecular biologists formed a powerful economic and political block that opposed effective regulation of gene technology's application (Krimsky, 1991; Wright, 1994). Australia could serve as a case study of what has occurred elsewhere in the world.

Perhaps science should be regarded as one aspect of the input into the decision-making process? There is a growing number of scientists who believe that science cannot provide all of the answers and, indeed, science's success has created the circumstances in which it cannot provide the answers. The approach known as Post-Normal Science advocates the input from all of the stakeholders, in which every group operates within its limitations and forms an extended community or network of peer review (De Marchi and Ravetz, 1999; Ravetz, 1999). This approach works on a consensual approach rather than an adversarial one. It seems to be a means of trying to arrive at the best decision and avoiding the bias that results from the adversarial approach in which stakeholders who have the 'winning' argument dominate the outcome. In the adversarial approach, many problems and issues may be glossed over because information, which is damaging to one's case, may not be presented and spurious arguments are put forward instead (Dale, 1993; van Dommelen, 1999), regarding the safety of the GMOs. The Post-Normal approach is based more on trust between the various stakeholders playing within the limitations of their roles and decisions are made through conversation and dialogue, (De Marchi and Ravetz, 1999).

The adoption of the Post-Normal approach to decision making may well be opposed by the economic interests which provide the major driving force behind the development of gene technology. A major cause for concern is the strong partnership between molecular biologists and the industry in developing and aggressively marketing GM products (Kasamoentalib, 1996; De Marchi and Ravetz, 1999). It is extremely important that big business interests do not dictate the decisions of governments and regulatory bodies, as was attempted in Europe over the GM maize during the second half of the 1990s (De Marchi and Ravetz, 1999). The priorities of big business do not necessarily coincide with those of the general community or environment, especially in the short term. However, in the long term, priorities of big business and the community should coincide, particularly if we accept Ecologically-Sustainable Development as a first priority. After all, the long-term survival of humans on earth depends on our utilizing our resources in an ecologically sustainable way.

Controlling Science and the Regulatory Agenda

In Australia, during the late 1980s and early 1990s, the Genetic Manipulation Advisory Committee (GMAC), which was the committee that made recommendations to government for the approval for field trials and commercial release of GMOs, was dominated by molecular biologists, a group of scientists who were deeply embedded in the Positivist paradigm (Persely and Peacock, 1990). Indeed, during this period, any criticism of gene technology was regarded as being anti-science, because its practitioners perceived molecular biology, upon which gene technology was based as a certain, objective and value-free science. The weakest link in the gene technology edifice is the paradigm and the genetic theory upon which it is based.

As we have seen, the Central Dogma of Crick's Gene Theory upon which the thinking of many molecular biologists has been based, is incorrect (Commoner, 2002). This theory has now been replaced by the Alternate Gene Splicing Theory, which clearly demonstrates the certainty about the outcomes that many molecular biologists had about the changes that were being made to genomes was a mirage because it was based on a grossly inadequate theory.

Progress in science occurs in two different ways. Normally, scientists identify what needs to be known and then identify the order of priority for what needs to be known. Effectively, scientists build upon existing knowledge. The second way is the manner of scientific revolutions whereupon discoveries—sometimes in an unrelated discipline—cause a change in the frame-of-reference and, hence, theories used to explain specific phenomena (Gould, 1977; Loehle, 1988). Once scientists have adopted a specific frame-of-reference; we return to accumulating small amounts of information to test the currently-accepted theory.

The theories used affect the approach to regulation. At present, the regulations of Australia and the UK, for example, are trait-oriented and reflect the conclusions drawn from theories used by laboratory-based scientists. Biotechnological scientists appear to believe that recombinant DNA techniques are refined and precise and transfer well-defined genes into target organisms. Therefore, the risks are predictable and can be controlled (Kasanmoentalib, 1996). It follows from this that risks are primarily a function of the characteristics of the product of the genetic manipulation (Dale, 1993; van Dommelen, 1999). Indeed, the better that one understands the function of the gene being inserted, the better one can understand and predict the risks and, thus, control them better.

In contrast, many ecologists take a different view about the predictability of the behavior of the GMOs in the environment. This is exemplified in van Dommelen (1999) compilation that we were never likely to be able to predict the consequences of small genetic changes on behavior at a population level.

These opposing viewpoints are locked in a stalemate that has resulted from different disciplinary paradigms (Krimsky, 1991) and different theories operating at different scales or levels of resolution. The stalemate reveals the chasm between the reductionist approach used by laboratory–based scientists (molecular biologists) and the holistic approach used by field-based ones (ecologists) (Kasanmoentalib, 1996).

Laboratory-based scientists have controlled what theories may be used to interpret the facts and, therefore, they have controlled the approach to regulation since the Asilomar Conference in 1972 (Hindmarsh et al., 1991b). As a consequence of this, the regulations are trait-based (Kasanmoentalib, 1996). However, regulation is continuing in a transition or evolutionary phase. Initially, biotechnology was self regulated and contained in the laboratory. When biotechnology spread beyond the confines of the laboratory and into the field, some ecologists were amongst those who were critical of self-regulation by the biotechnology industry (Alexander, 1985; Regal, 1986). It seems as though there is some movement on the part of regulators to include community input and consider the ethical issues arising from the application of biotechnology, e.g. in agriculture. However, the current regulatory system still gives more weighting to the laboratory-based scientists than to the field-based one and, therefore, the regulatory system is not precautionary (Kasanmoentalib, 1996). This statement is likely to be true on a global basis.

Theoretical models 'are a prerequisite for the rational design of' empirical studies (van Dommelen, 1999). The models that we use affect which questions are asked (van Dommelen, 1999). This is why science is so important in guiding what to look for and where to look for it in each phase of Ecological Risk Assessment.

QUALITY CONTROL

The key to the regulation of GMOs is quality control. The way in which to achieve the degree of quality control required is use a Post-Normal Science Approach. Such an approach recognizes the complexity of the problem being dealt with and, thus, extends the peer review process to include a great variety of expertise in the decision-making process. The role of those engaged in the peer review process is to utilize the extended facts and take an active part in solving the problem that they face (Ravetz, 1999). Decision-making should not be restricted to the technical experts only. The application of gene technology has far broader implications than simply for the science of molecular biology and molecular genetics. It has the potential to impact upon society, upon the natural environment as well as human health. Therefore, the various stakeholders should have an input in the decision making. By taking this broad-based approach to the regulation of biotechnology, we are more likely to optimize the benefits to society and the environment while minimizing the risks. Where it is deemed necessary to take risks, the extended peer review process makes it more likely that the risks taken are acceptable from a broader group of society than if it were left to the technical experts to decide what is an acceptable risk and what is not.

It is truly a challenge to design a set of experiments and associated monitoring schemes to provide the necessary information to make an informed decision about the environmental release of GMOs and or their products. The application of biotechnology must be based on good science if humanity is to benefit both directly and indirectly from it. At present, poor science or even plain bad science is sometimes used to justify a specific position, be it in support or in opposition to the application of gene technology. The use of poor or bad science will undermine everyone's confidence in the regulatory system and the regulators. Therefore, it needs to be avoided at all costs.

Part of the good science approach is that we could utilize the progression from containment to small-scale field trials, to large-scale field trials and, finally, to commercial release in order to collect data necessary to inform decisions about the suitability of progressing to the next stage. This progression provides a means of obtaining specific data to test the suitability of the theories guiding the process. It provides the basis for learning and developing a more effective and efficient model to vet GMOs for environmental release. This is why it is so important that the experiments and trials be designed in such a way as to provide valid and reliable data. The data required would depend on the research questions asked which are shaped by the paradigm and the theories that one uses to predict and interpret the outcomes. Thus, who controls the science—which guides the regulation of gene technology and its applications—controls the agenda.

CONCLUSIONS

In conclusion, the controversy surrounding the use of biotechnology in agriculture has raged between GE molecular biologists on one hand and some molecular biologists and ecologists on the other. The conflict has arisen because of the differences in the paradigm and the theories used to interpret what is happening. The science-based approach is only as strong as the theories used to predict and interpret the data associated with the phenomenon of interest. The precautionary approach may be used to prioritize research that needs to be carried out to reduce the uncertainty regarding the safety of GMOs. The situation in biology today is analogous to Physics 100 years ago. The clockwork universe was known and all physicists had to do was to tidy up a few of the loose ends. Many physicists did not expect any major new discoveries to be made. Then Relativity and Quantum Mechanics arrived and changed the mindset of physicists. The reductionist approach championed by molecular biologists was brought into Biology by atomic physicists, who after WW II, left physics and began doing research in molecular biology (Regal, 1996). The Human Genome Project was supposed to help clear up some of the loose ends in a very well understood system. Instead, this project identified the shortcomings of the theoretical basis of genetics.

We live in exciting times and the challenge that faces biologists, molecular and ecologists alike, is to work together to solve a particularly fascinating and extremely important problem. How can gene technology be used so that we receive the optimal benefits while minimizing the costs to human and environmental health? Even in a Post-Normal Science approach, science has a very important role to play in guiding decision makers to essential questions and to provide answers to as many of them as possible.

ACKNOWLEDGEMENTS

I thank Richard Hindmarsh, Sue Quinnell and Jacinta Zalucki for providing constructive feedback on the manuscript. I thank Dr Pat Dale, Head of Australian School of Environmental Studies, for making the facilities and resources of the school available to me for this work.

REFERENCES

Alberts, B., Bray, D., Lewis, J., Raff, M., Roberts, K. & Watson, J.D., 1994, Molecular Biology of the Cell, 3rd edition. Garland Publications, NY.

Alexander, M., 1985, Ecological consequences reducing uncertainties. Issues Sci. Technol., **1**, 57-68.

ARMCANZ., 1997, Regulation of Gene Technology. What to regulate. URL http://www.affa.gov.au/docs/operating_environment/armcanz/gene/attributes2.html 15 December 1997, 2 pages.

Barinaga, M., 1996 A shared strategy for virulence. Science, **272**, 1261-1263.

Barnes, P. and Hulsman, K., 1995, Ecological risk assessment of transgenic plants. Paper delivered at ANZAAS 1995 Congress, Newcastle 25 September 1995. Abridged version. Search, **26**, 277-280.

Bik, E.M., Bunschoten, A.E., Gouw, R.D. and Mooi, F.R., 1995, Gensis of novel epidemic vibrio-cholerae-O139 strain: evidence for horizontal transfer of genes involved in polysaccaride synthesis. EMBO, **14**, 209-216.

Black, D.L., 1998, Splicing in the inner ear : a familiar tune but what are the instruments? Neuron, **20**, 165-168.

Brill, W.J., 1985, Safety concerns and genetic engineering in agriculture. Science, **227**, 381-384.

Cairns, J. Overbaugh, J. and Miller, S., 1988, The origin of mutants. Nature, **335**, 142-145.

Campbell, J.H., Lengyd, J.A.and Langridge, J., 1973, Evolution of a second gene for b-galactosidase in *E. coli*. Proc. Nat. Acad. Sci. USA, **70**, 1841-1845.

Campbell, N., 1990, Biology, Second Edition, Benjamin Cummings, Redwood City.

Commoner, B., 2002, Unravelling the DNA Myth. The spurious foundation of genetic engineering. Australian Financial Review. 1 March 2002. Or Wysiwyg://4/http://www.commondreams.org/views02/0209-01.htm

Cullis, C.A., 1983, Environmentally induced DNA changes in plants. Crit. Rev. Pant Sci., **1**, 117-131.

Dale, P.J., 1993, The release of transgenic plants into agriculture. J. Agric. Sci., **120**, 1-5.

Davis, B.D., 1987, Bacterial domestication underlying assumptions. Science, **235**, 1329-1335

Davis, B.D., 1989, Evolutionary principles and regulation of engineered bacteria. Genome, **31**, 864-869.

Dommelen, van A., 1999, Hazard Identification of Agricultural Biotechnology. Finding Relevant Questions. International Books, Utrecht.

De Marchi, B. and Ravetz. J., 1999, Risk management and governance: a post-normal science approach. Futures, 31, 743-757.

De Silva, N. A. and Bailey, J. E., 1986, Theoritical growth yield estimates for recombinant cells. Biotechnol. Bioengineering, 28, 741-746.

Foster, P.L., 1992, Directed mutation: between unicorns and goats. J. Bacteriol., 174, 1711-1716.

Gould, S.J., 1977, Ever Since Darwin, Penguin, Suffolk.

Hall, B.G. and Hartl, D.L., 1974, Regulation of newly evolved enzymes, I: Selection of a novel lactase regulated by lactose in *Escherichia coli*. Genetics, 76, 391-400.

Hindmarsh, R. and Hulsman, K. 2004. Beyond the Lab! Precautionary weakness for GMO releases. In: R. Hindmarsh and G. Lawrence (eds), *Recoding Nature: Critical Perspectives of Genetic Engineering*. University of NSW Press, Sydney.

Hindmarsh, R. Burch, D.F. and Hulsman, K., 1991a, Agrobiotechnology in Australia: Issues of control, collaboration and sustainability. Prometheus, 9, 221-248.

Hindmarsh, R., Hulsman, K., Burch, D. and Brownlea, A., 1991b, Setting the regulatory agenda on genetic engineering in Australia 1974-1981. A case of scientific control. Proceedings of the Australian Political Studies Association Conference, Griffith University, July 1991.

Ho, M.-W., 1987, Evolution by process not by consequence : implications of the new molecular genetics for development and evolution. Int. J. Comp. Psychol., 1, 3-27.

Ho, M.-W., 1997, DNA and the new oganicism. In: Wirz J. and van Bueren E. Lammets (eds), The Future of DNA, Kluwer Academic Publishers, Dordrecht, pp. 71-93.

Ho, M,-W., 1999, Genetic Engineering. Dream or Nightmare? Turning the Tide on the Brave New World of Bad Science and Big Business, 2nd Edition, Continuum, New York.

Ho, M.-W, Tucker, C., Keeley, D. and Saunders, P.T., 1983, Effects of successive generations of ether treatment on penetrance and expression of bithorax phenocopy in *Drosophila melanogaster*. J. Expt. Zool., 225, 357-368.

Hoffmann, T., Golz, C. and Scheider, O., 1994, Foreign DNA sequences are received by a wild-type strain of *Aspergillus niger* after co-culture with transgenic higher plants. Curr. Genet., 27, 70-76.

Inose, T., and Murata, K., 1995, Enhanced accumulation of toxic compounds in yeast cells having high glycotic activity: a case study on the safety of genetically-engineered yeast. Int. J. Food Sci. Technol., 30, 141-146.

Kapur, V., Kanjilal, S., Hamrick, M.R., Li, L.L, Whittam, T.A., Sawyer, S.A. and Musser J.M., 1995, Molecular population genetic analysis of the streptokinase gene of *Streptococcus pyogenes*: mosaic alleles generated by recombination. Mol. Microbiol., 16, 509-519.

Kasanmoentalib, S., 1996, Deliberature release of genetically modified organisms. Applying the Precautionary Principle. In A. van Dommelen (ed.), Coping with Deliberate Release - The Limits of Risk Assessment, International Centre for Human and Public Affairs, Tilburg, pp. 137-146.

Krimsky, S., 1991, Biotechnics and Society: The Rise of Industrial Genetics, Praeger, New York.

Latour, B., 1999, Give me a laboratory and I will raise the world. In Mario Biagioli (ed.), The Science Studies Reader, Routledge, New York, pp. 258-275.

Levidow, L. and Carr, S., 1997, How biotechnology regulation sets a risk/ethics boundary. Agric. Human Values, **14**, 29-43.

Levidow, L and Carr, S., 2000, Sound Science or Ideology? Forum for Applied Research and Public Policy, Fall 2000, 44-50.

Loehle, C., 1988, Philosophical tools: potential contributions to ecology. Oikos, **51**, 97-104.

Lorenz, M.G. and Wackernagel, W., 1994, Bacterial gene transfer by natural genetic transformation in the environment. Microbiol. Rev., **58**, 563-602.

McClintock, B., 1984, The significance of responses of the genome to challenge. Science, **226**, 792-801.

Mellon, M. and Rissler, J., 1995, Transgenic crops : USDA Data on small scale tests contribute little to commercial risk assessment. Bio/Technology, **13**, 96.

Miller, H.I. and Gunary, D., 1993, Serious flaws in the horizontal approach to biotechnology risk. Science, **262**, 1500-1501.

Ninio, J., 1983,. MolecularApproaches to Evolution. Princeton University Press, New Jersey.

Nordlee, J.A., Taylor, S.L., Townsend, J.A., Thomas, L.A. and Bush R.K., 1996, Identification of a brazil nut allergen in transgenic soybeans. New Eng. J. Med., **14**, 688-728.

Persley, G.J. and Peacock, W.J., 1990, Biotechnology for bankers. In: Persley G.J. (ed.), Agricultural Biotechnology: Opportunities for International Development, CAB International, Wallingford, pp. 3-24.

Ravetz, J.R., 1999, What is Post-Normal Science. Futures, **31**, 647-653.

Regal, P.J., 1986, Models of genetically engineered organisms and their ecological impact. In: Mooney H.A. and Drake J.A. (eds), Ecology of Biological Invasions of North America and Hawaii, Springer-Verlag, New York, pp. 111-129.

Regal, P.J., 1996, Metaphysics in genetic engineering : cryptic philosophy and ideology in the 'Science' of Risk Assessment. In: van Dommelen A. (ed.), Coping with Deliberate Release – The Limits of Risk Assessment, International Centre for Human and Public Affairs, Tilburg, pp. 15-32.

Riedle, I., 1996, Three mutant genes co-operatively induce brain tumor formation in *Drosophila* malignant brain tumor. Cancer Genet. Cytogenet., **90**,135-141.

Rothenfluh, H.S., Blanden, R.V. and Steele, E.J., 1995, Hypothesis: a memory lymphocyte-specific soma-to-germline genetic feedback loop. Immunol. Cell Biol., **71**, 227-232.

Scott, A., 1987, Pirates of the Cell—The Story of Viruses from Molecule to Microbe, 2nd Edition, Basil Blackwell, Oxford.

Schmucker, D., Clemens, J.C., Shu, H., Worby, C.A., Xiao, J., Muda, M., Dixon, J.E. and Zipursky, S.L., 2000, *Drosophila* dscam is an axon guidance receptor exhibiting extraordinary molecular diversity. Cell, **101**, 671-684.

Steele, E.J., 1979, Somatic Selection and Adaptive Evolution- On the Inheritance of Acquired Characters, Williams -Wallace, Toronto.

Strohman, R., 2001, Towards a new paradigm for life. Beyond Genetic Determinism. 22 March 2001. http://www.psrast.org/strohmnewgen.htm

Symonds, N., 1994, Directed mutation: a current perspective. J. Theor. Biol., **169**, 317-322.

Tiedje, J.M., Colwell, R.K., Grossman, Y.L., Hodson, R.E., Lenski, R.E., Mack, R.N. and Regal, P.J., 1989, The planned introduction of genetically engineered organisms : ecological consideration and recommendation. Ecology, **70**, 298-315.

Temin, H.M., 1980, Origin of retroviruses from cellular moveable genetic elements. Cell, **21**, 599-600.

Tschape, H., 1994, The spread of plasmids as a function of bacterial adapability. FEMS Microb. Ecol., **15**, 23-32.

Whatmore, A.M. and Kehoe, M.A., 1994, Horizontal gene transfer in the evolution of group A Streptococcal emm-like genes: gene mosaics and variation in vir regulons. Mol. Microbiol., **15**, 1039-1048.

Wheale, P. and McNally, R., 1988, Genetic Engineering : Catastrophe or Utopia? Harvester Wheatsheaf, Hemel Hampstead.

Whitehouse, H.L.K., 1969, Towards and Understanding of the Mechanism of Heredity, 2nd Edition, Edward Arnold, London.

Wright, S., 1994, Molecular Politics: Developing American and British Regulatory Policy for Genetic Engineering, 1972-1982, Chicago University Press, Chicago.

12

TRANSGENIC CROPS FOR SMALL FARMERS: A DREAM OR A NIGHTMARE?

YOLANDA MASSIEU TRIGO
*Fuenters 18A, Col. Toriello Guerra, C.P. 14050, México
D.F., Mexico*

INTRODUCTION

Two thirds of the so-called Third World depends on agriculture to survive with 75% of all the farmers in these countries comprising small-scale farmers (Broerse and Bunders,1991). With some differences—as in Mexico, 25% of the population lives from agriculture and in Ghana, this sector is of about 57%—small farmers are still an important part of the world population that survives (sometimes in acute poverty conditions) on auto-consumption agriculture.

Small-scale agriculture has often been seen as merely low productive, and frequently as indigenous and peasant, environmental and traditional knowledge whose genetic resources have been ignored or stolen. Industrial agriculture, as it has developed since the beginning of the Green Revolution, implies to seek the highest yields of a single crop, despite environmental and socio-economic costs. Although small-scale agriculture has low yields considering just one crop, it produces a lot of food if other peasant products are taken into account, in a more diverse and sustainable production model. Just to give some examples: in Java, small farmers cultivate 607 plant species in their home gardens; in Subsaharian Africa, women cultivate 230 species and African gardens have more than 60 tree species. Rural families in Congo eat leaves from more than 50 different tree species from their lands. In Chiapas, Mexico, Mayan peasants are characterized as low productive because they have corn yields of two tonnes per acre, but they produce 20 tonnes of different foods considering their combined crops, including beans and squash. A study in western Nigeria showed that family orchards occupied only 2% of the land but they produced half the total agricultural production. In Indonesia, 20 percent of the family income and 40% of the domestic food supply came from home orchards which were managed by women (Shiva, 2000). In small-scale agriculture, food production is linked with other surviving activities, including the use of medical plants or manufacturing some kind of clothing or kitchen needs.

Meanwhile, production conditions for small farmers are steadily worsening: with few exceptions, small-scale farmers in poor countries live in areas with low and irregular rainfall, short rainy seasons, midsummer droughts, high intensity of rainfall and/or high temperature and consequent, a high rate of evaporation. Most small farmers cultivate very poor soils and are confronted not only with soil of diminished quality but also with the reduction of soil quantity through erosion. Concerning to rainforest, in the Selva Lacandona, Mexico—one of the Northern hemisphere's last rainforests—small-scale peasants have faced both the decrease of their corn harvests and the lost of the soil, so they cannot survive from this kind of agriculture, while the rich biodiversity of the place is disappearing. With the Zapatista rebellion in 1994 and the violent conditions in the region, to seek a sustainable use of this rainforest becomes more complicated. Between 1978 and 1993, rainforests and tropical forests have decreased by 41.8%. If this destruction rhythm continues, the original rainforest and tropical forest surface will disappear in 25 to 30 years. In 1978, the Montes Azules Biological Reserve was created, with 331,200 ha. (Hernández and Sadé, 1998).

While some indigenous small-scale farmers' knowledge does pertain to traditional biotechnology, like fermentations, genetic engineering is a more complex agricultural biotechnology owned almost completely by big private corporations, due to the investments that are needed. There is a strong debate about its possibilities to benefit small-scale farmers in the Third World countries.

To give an idea about the controversial issue, we could talk about the *Terminator* case and the recently-discovered genetic engineering technology for drought resistance. Another recent and outstanding example concerns the recent transgenic pollution among the small-scale corn producers in Central and South Mexico, which has been mentioned in the corn case study.

TERMINATOR TECHNOLOGY

Terminator technology has been created to reinforce the firms' control over seeds. It has been considered as a threat both for biodiversity and small farmers. Agro-biotechnology companies argue that this technology could be applied for biosafety reasons as well. Terminator technology consists of seed sterility and it is not clear yet if there is possibility of transgenes to flow and transmit the sterility to the other crops. This technology has been strongly criticized because it means that farmers who buy these new seeds will have to buy it again each cycle. Also, it can have repercussions for those farmers who cannot buy it and whose crops can acquire the sterility because of transgenes flux. This is a serious threat to the traditional practice of obtaining seed from harvest, very common for small-scale farmers in underdeveloped countries. This practice becomes illegal with increasing intellectual property rights over commercial crops. Besides, the damage to the surrounding plants can be serious too if this sterility spreads.

Nevertheless, in the United States and Europe patents have been authorized for this technology since 1994. These patants belong to Monsanto, DuPont-Pioneer, BASF, Delta&Pine, Astra Zeneca, ExSeed Genetics and Novartis companies, to the United States Department of Agriculture (USDA) and to 8 research centres and institutes. Patents have been given in the United States, France, Switzerland, the United Kingdom, The Netherlands and Germany. Such patents can be used in all plants.

Until now, *Terminator* has not been commercialized and there are no commercial crops being sold at this moment. There has been an interesting social movement against it, in which there can be north-south alliances, mainly of NGO (Non-government Organizations) from the north and peasants organizations from the south. In 1998, the Canadian Rural Advancement Foundation International (RAFI-today ETC Group) made a call by internet to many countries, in order to send letters to the USDA and stop the *Terminator* technology. They gathered approximately 4000 firms and Monsanto declared in November of that year that they would not bring to the market any *Terminator* crop. In spite of this, seven new patents were given to *Terminator* in 1999 and the company that owns more of these patents is Novartis, not Monsanto (RAFI, 1999).

The use of sterile and privately-owned seeds, a practise which has been ethically questioned by authors like Mooney (1978) since the 1970s and Shiva (2000) in more recent times, is part of the agrobiotechnology global firms' strategy to recover their investments. Considering that for each transgenic crop that succeeds, ten thousands do not, the companies' urgency becomes clear. So, the strategy adopted by global firms consists in 'leasing' the genetically-modified seed, with all its inserted characteristics, instead of selling it. This can be done with seed sterility and intellectual property rights. This way, seeds can be used a single time and global agrobiotech firms control food production and the possibility of reproducing useful plants (Pollan, 1998).

The *Terminator* debate has become so tough that all transgenic crops are criticized and are called by RAFI "Traitor", because of their new traits. They are considered as traitors to those farmers who row them. For RAFI, this extreme seed control, implying that to obtain all the benefits, farmers need some of the chemicals sold by the owner company, leads the farmer to 'bioserving'. In this sense, the Traitor technology offers the possibility to insert some commercial characters to the seed and the company can then choose to activate or desactivate them after selling it. This transforms the Traitor in the platform to bring the characters the firm has already patented. Farmers can buy the seed in the same way they buy machinery, with or without accessories. Depending of the added characters to which the farmer can have access—or the ones the company wishes to discover in a certain moment—certain chemicals can be used, like fumigation or immersing seeds, in order to activate the desirable characters in the seed. This way, the biotech firms 'marry' Traitor seed with their own chemicals; one is useless without the other. Commercial and economic realities clearly show how the technology's evolution works (RAFI, 1999).

For Sally Mile Hayes, from the USDA, the *Terminator* new method was developed to study plant genes' expression, but many critical researchers think that its primary future use will be the development of the 'technology protection systems', directly related with intellectual property, that act against free access to technology and genetic resources (Lehman, 1998). In this case, besides technology property protection, the aim is to possess a genetic richness, which belongs to the countries where genetic resources are placed.

Other recent approaches consider this technology as a clear example of how the new technologies do not take into account traditional knowledge and techniques (Viniegra, 1999). In case *Terminator* technology could spread in the world's food production, it will mean a waste of traditional knowledge. It also poses a threat to biodiversity and small-scale agriculture.

Drought Resistance Technology

Another example about how possible biotechnology benefits will aid small-scale farmers dependent on the economic interests of big corporations and political decisions is the finding of a drought-resistance technology in Toronto University, Canada, in 1998. This discovery has a positively great potential to benefit small–scale farmers who farm dry lands or have been affected by drought. It could also contribute to biodiversity conservation in arid and desert areas. Until now, the technology is being tested in Toronto in *Arabidopsis taliana*, an experimental plant with no commercial value. The researcher who identified the genetic modification works for a private company (Performance Plants Inc.), founded in 1995 by members of the Plant Biotechnology Group of Queen's University. The discovery was announced in June 1998 and three months later, the firm Dow AgroSciences announced that it had started an alliance of production-commercialization with Performance Plants Inc. to introduce the new technology in commercial crops. The alliance means $1.2 million dollars financing for Performance and it seek to sell the new varieties. Dow calculates that farmers who plant canola will have a 10% yield increase on using this technology (www.innovationplace.com, www.library.utoronto.ca). This example shows clearly how private corporate interests decide whether the possible benefits of agricultural genetic engineering will benefit small farmers. In this case, only the farmers who can acquire the new varieties sold by Dow will be able to plant arid lands and/or conserve water.

For many developing countries who have scarce resources and scientific-technological structure, the fact that agricultural biotechnology is being primarily developed by private corporations means serious risks: the capacity to asses if a new transgenic crop is risky either for the environment or the small farmers is limited; there are scarce germplasm banks and lack of knowledge about local biodiversity. In some of these countries—as Mexico—the recent economic policies concerning free trade for agriculture, have led to the destruction of many small farms and the migration of many small farmers.

The recently-approved Cartagena Biosafety Protocol considers that a country in such a situation can ask for resources to make the risk assessments of a transgenic crop and that a country can also deny imports of transgenic crops if there is doubt that it can harm its environment or the small-scale farmers' socioeoconomic conditions (www.biodiversidadla.org).

CASE STUDIES IN MEXICO

In Mexico, as in most of the underdeveloped countries, transgenic crops have not grown as swiftly as in the USA, Canada and China (see Table 12.1). However, the detailed global scenario of transgenic crops has already been discussed in Chapters 1 and 2 of this volume. Mexico has

Table 12.1 Global area of transgenic crops 1996-2001 (million ha)

COUNTRY	1996 Ha	1996 %	1997 Ha	1997 %	1998 Ha	1998 %	1999 Ha	1999 %	2001 Ha	2001 %
USA	1.5	52	8.1	64	20.5	74	28.5	72	35.7	68
China	1.1	39	1.8	14	<0.1	<1	0.3	1	1.5	3
Argentina	0.1	4	1.3	10	4.3	15	6.7	17	11.8	22
Canada	0.1	4	1.3	10	2.8	10	4	10	3.2	6
Australia	<0.1	1	<0.1	<1	<0.1	<1	<0.1	<1		
Mexico	<0.1	<1	<0.1	<1	<0.1	<1	<0.1	<1		
South Africa					<0.1	<1	<0.1	<1		
Spain							<0.1	<1		
France							<0.1	<1		
Portugal							0	0		
Romania							0	0		
Ucrain							0	0		
TOTAL	2.8	100	12.8	100	27.8	100	39.9	100	52.2	100

Source: James (1998, 1999); www.isaa.org, 'Global Gm crop area continues to grow and exceeds 51 million ha for the first time in 2001'

an Agriculture Biosafety Committee established since 1988 and, in 2000, the President of the Republic ordered an Interministries Biosafety Commission (IBC), which is working as per the advice of an expert Consultative Council. In this country, the only crop that was completely liberated was FlvrSvr tomato from Calgene Company in 1995. It was an export product to the USA and did not have the commercial success as expected, leading to its termination. The other crop occupying about 76,000 ha. in 2000, as a pre-commercial trial, is Bt cotton from Monsanto. It is grown in the North of the country and not in the south, where there are still wild relatives of the crop and biodiversity is higher.

As socioeconomic impacts of biotechnology are quite specific, it depends on the ecosystem, whether a product is meant for export or for local market; if it is produced by small farmers or by big ones; if it is a food crop or not, etc. We shall study some examples of agricultural products in Mexico. These cases already have biotechnology applications and small farmers have been affected.

Potato

Potato production in Mexico can be distinguished in two regions:
   (a) The centre of the country, where the majority of potatoes are produced in rainfed lands, in mountain zones above 2000 metres altitude. Producers are mainly small-scale farmers who use

predominately coloured potato varieties, with low yields (3 to 10 tonnes per ha) but they have plague resistance and can be grown in heights such as 3,500 metres over sea level.

(b) Some states of the north and of the Bajío region (west-centre of the country). These areas represent about 59% of the acreage cultivated with potatoes, under irrigation conditions and high yields (the highest ones: 40 tonnes per hectare, average 20 tonnes), twice that of the central small-farmer region. Producers can be classified as agricultural entrepreneurs. They cultivate white varieties, mainly Alpha, that are bigger and have higher yields.

Only 25% of potato seeds are certified, non-certified varieties and local varieties are grown in most areas. Certified seeds come from 10 tissue culture laboratories and 17 propagation greenhouses.

The total potato area has decreased; in 1989 it was 73,500 ha which has become 63,000 ha in 1999. This loss in surface has been compensated with higher yields (average 14.5 tonnes per ha). The annual Mexican potato production is about 1.4 million tonnes with a value of $300 million US dollars. Approximately, 17% of the potato harvest is used as seed and 73% is for direct consumption. Potato production is highly speculative and NAFTA has granted Mexican potato producers a protection period for fresh potato imports, until 2004 (Massieu et al., 2000).

Potato is the only crop in which transgenic virus resistance varieties have been produced within Mexico, by Mexican researchers from Centro de Investigación y Estudios Avanzados (CINVESTAV), in Irapuato, Guanajuato State, which is part of a network of public research institutes. The project began as a collaboration with Monsanto, who donated the resistance gene and the training of Mexican researchers. This project has been mentioned as an example of a new way of technology transfer: private-public and north-south (Commandeur, 1996). Small farmers have been specifically mentioned as a target group.

Until now, the new varieties have been tested in field trials and as yet, are not commercialized. Its possible benefit could be a yield increase, as there is the presence of the resistance viruses (PVX, PVY and PLRV) in Mexico. Nevertheless, these viruses are not the primary concern of potato production in Mexico. The most important problems, which cause a lot of losses are late blight disease and golden nematode. Concerning the benefits for small farmers, it is not clear how they are going to avail of the seed, although they are interested, as can be confirmed in interviews of small farmers in the main production areas. If corporate companies will propagate the seed and sell it, the small-scale farmers are unable to afford it. Also, there are no public agencies who are in a position to do so. So, in spite of the good will of CINVESTAV researchers about their varieties benefits reaching small farmers and some recent optimistic analysis concerning this case (Qaim, 1998), these producers have been absent of the project since the beginning and there are no technology transfer paths to bring them the seed, as they are none in the formal potato seed market (Massieu et al., 2000).

Flower Production

Mexico's flower production appears divided in two producer sectors:
  (a) Intensive greenhouse production, which uses cloned materials from the Netherlands, France and the USA and has bigger yields and exports (85% to the USA) (Bancomext, 1999); and
  (b) Traditional producers, some of whom have been producing flowers since pre-colonial times. They use native varieties and their market for seed and propagation materials is a more informal one. Mexico has a dynamic national market for flowers, specially in cities and turist areas, where these producers sell their flowers (Massieu, 1997)

Mexico has 10,000 hectares of both greenhouse and traditional production, the main exporter being Mexico State, with 93% of the exports. The total production amounts to 15.1 thousand tonnes. The main export varieties are roses, carnations, margarets, statice, gladiolas, lilium, iris and gypsophyla. Exporters are increasing their growth, although they make a very small portion of the total Mexican agricultural exports. Rose is the most important exported variety in terms of both volume and value. Exports were worth 10.4 million US dollars, that touched 23.1 million in 1996. Mexico is the fourth target flower supplier to the USA (with 4.5% of the market), after Colombia (62.8%), The Netherlands (12%) and Ecuador (9.9%) (Bancomext, 1999). Mexican participation in this international market has been growing constantly: it was 0.1% in 1981, and 2.5% in 1994 (Lara, 1998). The flower export is a highly competitive market, catering to luxury and fashion. To compete, multinational companies producing genetic materials have a capacity to frequently bring new products into the market. This is an expensive endeavour and local companies do not have such resources, so the Mexican flower production invests a lot of money the acquiring genetic materials. This is a paradox in a country with a very antique tradition of flower production and a vast flower biodiversity (Massieu, 1997).

A flower plantation creates a lot of employment, compared with some mechanized crops like wheat or sorghum (only 10 laboned days in all the cycle). Intensive flower production requires 3,992 laboned days per-year and a traditional one, 1,248 (Massieu, 1997). This high labour use changed in 1989, when the flower sector suffered a crisis, production concentrated in only a few companies and nearly half of the workers were fired. Although it is difficult to estimate the number of small farmers, in 1988, 25,000 were calculated, while the most important entrepreneur groups dedicated to intensive flower production (Visaflor, Megaflor, Monrog and Villa Guerrero Group) planted a 30 ha. greenhouse surface, offering employment to 3000 workers and 250 technicians (Lara, 1998).

Another paradox is that intensive export flower production needs to import expensive technology, including greenhouses, and this is possible because of low salaries in Mexican agriculture. Salaries are 88% of the costs in traditional agriculture and 24% in intensive one (Massieu, 1997).

In Mexico intensive flower production using biotechnology through cloned materials sold by the companies from the USA, France and the Netherlands, started to develop in the 1980s. It grew separately from the traditional flower production. The traditional farmers produce for the national market and do not have access to biotechnology and greenhouses. The advantage of intensive export flower production from the use of cloned materials is outstanding because it allows all the plants to blossom at the same time. Considering that in flower production, in some special days, like 14 February, the price of a bloom can reach five times the normal price, it is understandable why these flower producers make big investments to get the cloned materials and build expensive greenhouses. Producers are now beginning to make field trials with transgenic flowers. Meanwhile, small flower farmers cannot compete or export and they continue producing with traditional technology for the internal market.

Henequen

Henequen (*Agave fourcroydes*) (Eastmond and Robert, 2000) is a fibre-producing cactus plant used for the production of rope, cloth, sacks and carpets. Its centre of origin is in Yucatan, Mexico. The Mayans domesticated it in the prehispanic period. It is a slow-growing plant with a long life cycle of upto twenty years, flowering only once towards the end of its life. Most seeds are infertile and propagation occurs mainly through subterranean shoots. It is a special plant because it can be highly productive under harsh ecological conditions of water and soil scarcity such as is found in the henequen-growing area in Yucatan.

Henequen is also a socially-relevant plantation, as it was initiated in the Yucatan in the nineteenth century in haciendas (big properties), with very hard exploitation conditions for Mayan workers. In the Agrarian Reform times, primarily 1930s, all these big haciendas were divided and distributed among the Mayan peasants. Its 'golden times' were in the beginning of the twentieth century; in 1921, Mexico provided 88% of the world supply of this fibre. In 1988, the country only provided 12.5% of this offer and nowadays, Mexico does not export anymore. Local production is inadequate to provide the fiber industry, which has recently imported Brazilian fibre. From 1990 to 1993, henequen production decreased from 35 thousand tonnes per year to 27 thousand tonnes. At present, 14,000 small farmers depend on henequen production, most of whom live in poverty conditions. It seems that in the coming years, there can be a slight increase in the fibre's demand, as there is an ongoing concern about using non-polluting materials (Noffil, 2000).

In the henequen industry, technological innovation is slow. First, because traditional ideas about how henequen should be cultivated are very deeply rooted, specially among small farmers. Second, there is no effective technical assistance system to transfer new ideas to the producers. To implement a typical breeding programme is hard, because of the long life cycle (12 to 20 years to blossom) and its extremely limited sexual reproduction.

Micropropagation, a technique for multiplying plants vegetatively through the in vitro culture of their cells, tissues or organs, offers the possibility of speeding up multiplication of outstanding individuals by reducing the time taken to produce new generations from years to mere months. This technology makes the large-scale production of elite clonal lines possible. It produces pathogen-free plants that are more vigorous and faster growing than the field-produced plants.

Micropropagation is more expensive than traditional adventitious shoot production. Clones multiplied in nurseries cost only 20% of the cost of laboratory produced plants. This makes the in vitro production of millions of plants economically unfeasible. However, in vitro plants grow stronger and generate about three times more shoots in the nursery than the field-propagated plants.

By combining both these methods, micropropagation can be used to produce vigorous elite lines as a starting material for a cheap multiplication in nurseries, in order to provide sufficient, high quality planting material to re-establish henequen plantations.

There is an in vitro culture of agaves programme in the Centro de Investigación Científica de Yucatán (Centre of Scientific Research of Yucatan—CICY), a public research institute, which has produced technologies for its efficient propagation. The project tests the benefits of combining tissue culture (micropropagation) with nursery propagation by developing experimental nurseries and plantations together with the growers.

In 1994, field-testing of 150 clonal lines began through a collaborative agreement with a group of henequen growers. After selecting the most outstanding plants from the growers' plantations, CICY cloned and propagated 300,000 micropropagated plants, which were then propagated in nurseries. These plants have been monthly evaluated in eight nurseries and eight experimental fields, in an area of 65 hectares.

CICY is implementing a system to distribute elite micropropagated plants to the nurseries of the four associations of small- and large-scale farmers, so they can propagate the plants themselves. As CICY is a public-funded research centre, the plants have not been protected in anyway and they will be available to the growers for nursery propagation and distribution.

The results obtained so far in the nursery and the plantation seem promising. The main effect is the greater propagation capacity of the clonal lines with a yield average of 5.6 shoots per plant per year, compared to only 1.5 produced by the field propagation ones.

The clonal lines also exhibit faster growth in the plantation. After three years, they have almost doubled their height and produced an average of 81 leaves per plant, while the field plants have only increased their height by 48% and produced 60 leaves. The total foliar area of the clonal plants is more than double that of the field-produced ones.

The main impacts derived from the clonal lines at this moment are:
- The 300% increase in the adventitious shoots allows growers to replant their plantation with elite material more rapidly than expected.
- The elite materials are disease-free and will help to prevent the spread of diseases.
- The overall faster growth of the clonal lines has enabled these plants to reach harvesting earlier than normal. Instead of five years of a normal harvest time, the clonal lines take only three years.
- In vitro multiplication produces plants that are more homogeneous morphologically as also physiologically. This synchronizes growth so that the plants reach the various production stages at the same time, making all the agricultural procedures more simple and efficient.
- The most important measure of productivity in henequen plants is fibre yield. The first harvest promises an increase.
- As elite individuals grow faster, their first leaves can also be harvested, adding 8% to the total production.

Although the initial costs of planting elite henequen are 25% above the price of field propagated adventitious shoots, economic studies show that the additional costs will be compensated for by the increased production.

Most small henequen growers will not be able to stay in the business if they cannot improve their productivity. This CICY project can contribute to this effort, as "the increased income from using the elite materials can have a much-needed multiplication effect, because in a situation of lacking employment opportunities and low purchasing power, any extra money that stays in the communities will benefit the local economy" (Eastmond and Robert, 2000).

Corn

Mexico is the centre of origin of corn as well as the main food crop in the country. Its participation is the most important in surface, value and employment. Corn is produced in different geographic, ecologic and social contexts. It is the base of the national diet, which supplies a majority of proteins and calories consumed by the Mexican population. Corn is basically a small-farmers crop, produced with family work. In 1995, 92% of corn farmers produced in farms of less than 5 ha. (FIDA-IICA, 1995). It is calculated that these small farmers consume approximately 35% of the production in these areas (Fristcher, 1999). Corn production in Mexico is in a difficult situation since the 1960s, which was the last decade when the country produced enough supply and even exported this produce. Since then, corn production has been neglected by economic policies and has faced varying competences. In the 1960s and 1970s, Mexican agriculture was strongly influenced by a increase in the livestock production. This meant that many corn areas were substituted with forage crops, mainly sorghum and that the country lost its capacity to supply enough corn for the internal market since 1970. Meanwhile, state-established prices (called guarantee prices) did not increase in 15 years (Contreras et al., 1999). In the 1980s, commercial opening had destructive effects in corn production. Although it is still protected in NAFTA (North America Free Trade Agreement) and the frontier will be completely opened by 2008, imports are increasing and now amounts to between 30% and 50 per cent of the national consume. This is a serious situation, because all Mexicans eat corn (mainly in the processed *tortillas* form) and for the poor people (over 40 millions, of the total 90 millions population), it is sometimes the only food. For the United States of America (USA), corn is one of the most important agricultural products and Mexican imports come from this country. Transgenic corn is non-regulated in the USA and in 1999, it meant 30% of the corn surface in the country. Corn enters in Mexico with the imports, as it is no separated from the non-transgenic corn. This caused a recent transgenic pollution problem.

Transgenic corn has generated a strong debate in the country. Mexico's Agricultural Biosafety Committee and the Inter-Minister Commission have been assessing the applications to make transgenic crops field trials in the country and in 2000, this Commission and its Consultative Committee decided not to allow more field trials of transgenic corn. The reason of this decision was that corn is a highly sensible crop, because of the presence of local varieties and wild relatives and the many subsistence peasants depending on the crop. If Mexico liberalizes transgenic corn cultivation, genetic flux can affect both biodiversity and small farmers. So, the Commission has decided to make a wide national research about corn biodiversity and socioeconomic conditions before taking decisions of more trials or transgenic corn liberalization. In 2003 there is permission again for field trials for research purposes.

In spite of the prohibition of any transgenic corn seed imports and any planting of transgenic corn varieties even for field trials since 1999 due to serious considerations for risks to biodiversity, there are transgenic grains in corn's imports from the USA. In October 2001, the presence of transgenic corn in small farmers' landrace cultivars in Oaxaca was reported for the first time. This was predictable, as food imports are sold to these peasants as grain and for them it is a common practice to plant new seeds and make some kind of improvement and experiments. This transgenic presence was first reported by Chapela and Quist (2001) and confirmed by Mexican public research institutions for Puebla region also. The degree of transgenic pollution spread is as yet undocumented.

Transgenic pollution's implications are very important because this is the main crop and food in the country and it is not only a commodity, but also a cultural source of many kinds. Modern agro-biotechnology, which means the use of genetic engineering to design new crops, has one of its main products in corn. There exist two types of transgenic corn in the international market: Bt resistant to insects (7.5 million ha in the world in 1999) and Bt with herbicide resistance included (21.6 million ha) (James, 1999).

Biotechnology offers controversial alternatives to Mexican corn's structural, agro-ecological and cultural characteristics. The risk aspect concerns the adverse effects in the crop's biodiversity, biosafety aspects, intellectual property rights, technological development, etc., mainly in the USA. In Canada and China, there exist the two mentioned varieties of transgenic corn.

In the USA, the main transgenic corn variety cultivated is insect-resistant Bt Corn, from Monsanto Company. There is already a controversy about the convenience of planting this corn, as studies show the possibility that the crop can affect beneficial insects, like Monarch butterflies, which migrate from Canada to Mexico in winters and are an environmental symbol (Losey et al., 1999). Recently, a transgenic variety named StarLink, which is forbidden to human consumption and only allowed for animals in the USA, was detected in some processed corn foods in the USA. This food had been manufactured in Mexico. Starlink variety has also been found in Japan's corn imports from the USA. From the 16.5 millions of tonnes that Japan imports form the USA, 1,500 tonnes were Starlink. Japan is demanding that the USA separate transgenic corn form non-transgenic ones (www.biodiversidadla.org., 2000). These two examples show that control is not quite possible when transgenic varieties are completely liberalized.

The companies that produce transgenic corn argue that these varieties increase productivity as they diminish plague losses and reduce production costs because they need less pesticide. In addition, they also protect the environment. Their yield increases would make opening more lands to agriculture unnecessary, therefore, contributing to preservation (Serratos, 1998).

On the other hand, environmental, peasant, indigenous and some academic groups argue that genetically-modified plants are dangerous to ecosystems' sustainability, produce genetic erosion and are an obstacle to seed's free access, as this is controlled by big multinational agro-biotechnology corporations. Besides, genetic flux from transgenic corn could have unpredictable consequences for the crop's diversity.

Due to this concern, Mexico has banned the planting of transgenic corn seed. Besides, it seems that the plagues to which transgenic corn commercial varieties are resistant are not present in Mexican territory and herbicide tolerance would not be accessible to most producers. It is necessary to make rigorous assessments before liberating transgenic corn to the environment and it is urgent to develop technical-scientific capacities for this, as well a biosafety legislation, that has not existed until now.

This type of biotechnological pollution is a serious problem because now transgenic genes are spreading with no control nor biosafety actions in small peasants' farms, where corn survives a high biodiversity.

The other controversial aspects concern the consumers, as Mexican consumers are eating transgenic imported corn unknowingly. Although it has not been demonstrated that this consumption can damage human health, there is an ethical question about the right of the consumers to know what they eat. In 2000, Greenpeace and some tortilla manufacturers from Mexico City implemented a tortilla manufacturing net to make tortillas form non-transgenic, national and white corn. This is an important alternative for those consumers who do not want to eat transgenic corn and because Mexican food culture prefers white corn to the yellow imported one.

Biotechnology applied to corn can be seen in the following ways in Mexico:

- The research made about the crops, mainly in public research centres, that from 1999 to 2002 could not make field trials and can do it again since 2003.
- The possibility of importing transgenic seeds to cultivate in the country.
- Mexican imports for consumption of transgenic Bt corn from the USA.

The first aspect considers what is done by the CIMMYT (International Corn and Wheat Improvement Centre, of the CGIAR system) and other public institutions, like CINVESTAV. In the CIMMYT, there is the biggest germplasm corn bank in the world, where many local varieties are stored. It is a public centre and access to the varieties is free. This organization made field trials to find a drought-resistant transgenic variety until 1999, now the institution has plans to continue this project due to field trials ban termination in 2003. In CINVESTAV, a promising research is being conducted about an acidic soil tolerant transgenic variety, which can mean cultivating corn in many deforested and useless tropical areas (Herrera, 1999).

BIODIVERSITY, BIOPIRACY AND SMALL FARMERS' RIGHTS

Biodiversity concerns all the living organisms in this world and their genes, including people of different races and cultures who live in a determined territory. Since the beginning of new biotechnology and genetic engineering, its importance has grown, as it is the main source of genetic material for new biotechnology products, like food and medicines.

Mexico is one of the greater biodiverse centres in the world, but this richness is being rapidly lost: if destruction is not stopped, in the next ten years 96 birds, mammals, reptiles, fish and amphibian species, as well as 66 plants and fungi will disappear from the Mexican territory, as was warned by environmental groups in 1998 (La Jornada, 1998). The majority and most important biodiversity centres are in tropical and subtropical regions, where original crops have been developed and agriculture commenced. Mexico is one of the ten Vavilov Biodiversity

Centres, called so because the botanic Vavilov was the first to identify them (Greenpeace, 1999). These centres and the crops originated in them are:

- Central America: Corn, common pea, sweet potato
- Andes: Potato, lima pea, peanuts
- South Brazil, Paraguay: Mandioca
- Southwest Asia: Rye, barley, wheat, green pea
- Abyssinia: Barley, sorghum, white wheat
- Central Asia: Wheat
- Indo-Myanmar: Rice, dwarf wheat
- Southeast Asia: Banana, sugar cane, yam, rice
- China: Fox tail white wheat, soybean, rice (Vélez and Rojas,1998)

Megadiverse countries placed in these Vavilov regions are: Mexico, Colombia, Brazil, Zaire, Madagascar, India and Indonesia. For example, the greatest biodiversity is placed in tropical forests; although they represent only 7% of the planet surface, they maintain 90% of the biological diversity. A group of eighteen countries has 98% of the tropical forests, all of them underdeveloped and nine of them Latin American. In the Mexican territory there are all the main types of echosystems. Diversity of fish and amphibious reaches 1000 species, the highest in the world; there are 439 mammal species, the highest in the American continent; vascular plants are more or less 25,000, higher than Europe or the USA and Canada together. Only referring to flowers, Tuxtlas reserve in Veracruz, has 1,300 species, more than all British Islands (Massieu et al., 1993).

Modern agriculture has not paid enough attention to biological diversity, which led to more homogeneity and genetic erosion, because its aim has been to increased yields of a few useful crops. This has happened since the development of improved varieties with the so-called Green Revolution. Now it seems to be more acute with the new transgenic crops, which apparently solve plague and yield problems, but are genetically poor. In Mexico, the agricultural ecological situation is serious: of the existing 24 million of agriculture ha., 72% presents some erosion degree. Of the irrigated area, 12% is polluted with salts and there is over exploitation of the underwater sources. Extensive livestock exploitation has destroyed 90% of the tropical forests and currently occupies between 90 and 100 million ha. (Massieu et al., 1993).

Biotechnology could contribute to solve this situation if research is aimed to store germplasm, reforest the degraded areas or find transgenic crops with drought and polluted soil resistance. The problem is that in Mexico, almost all agrobiotechnology research is made in public centres with scarce funds and agrobiotechnology multinational firms are the ones which increasingly finance the research.

Talking about biodiversity, biotechnology and small farmers' rights mean analysing the access to genetic resources. In agriculture, genetic engineering has generated new economic values of biological diversity, being the main source of genes to research and obtaining new genetically-modified varieties. This means that underdeveloped countries possessing such a diversity (like Mexico) should have adequate laws and policies to make a rational use of these resources. Pharmaceutical companies have taken plants and animals to make new medicines for many decades, but now biodiversity is a straw material for modern biotechnology. Large agro-biotechnology companies are increasingly interested in having access to biodiversity. This happens in a world context that favours private owning of living creatures and their genes, through intellectual property rights. Private interests touch local and indigenous communities' rights in this sense, as frequently, biological diversity is located in these group territories.

The international concept about biodiversity property rights has changed in recent years: in the 1960s, they were considered mankind's patrimony which meant that access to them was free. Since the 1970s and 80s, many countries changed their laws in order to make it possible patenting living organisms. As soon as the first genetically-modified organisms began to appear, the first patents were given. Meanwhile, some countries developed another kind of property right, aiming to protect the farmers' rights upon the plant varieties they developed. In this case, farm breeders' rights are working in certain regions of the planet, yet they are not available to small and poor peasants, who have to be well informed and prepared to register their varieties. In contrast, big agro-biotech companies are eager to protect their varieties somehow. In Mexico, a Plant Breeders Right Law exists since 1996 and most of the applications for protecting plants (279) have been made by multinational companies, mainly pertaining to flowers (90) (SAGAR, 1999). The country still does not have Biosafety and Genetic Resource Laws. As Mexico is one of the signatories of the Cartagena Biosafety Protocol, it is a future task of the Parliament to approve this law. There are already some initiatives from political parties and civil organizations.

In the described context, biopiracy as a crime has appeared concerning the illegal use of patented or protected plant varieties. In Mexico, piracy is rather common, specially in video and music industry, although it exists in other products such as beauty creams, parfums and liquors as well. There is an ongoing concern for many industrial sectors and companies, like Microsoft Mexico, Disney Mexico, Parfumeire

Versalles (Chanel), Deportes Martí, Levi Strauss Mexico, Fila Mexico, Warner Brothers and Tycoon Enterprises, who have begun a campaign against illegal copying, production and reproduction of their patented products.

In agriculture, it is not so common, as many farmers in Mexico (nearly 85%) do not buy seed, they keep it from their harvest. This practice, common in the underdeveloped world where many poor peasants ensure at least part of their feeding from their crops, can be seen as a crime if the harvest from which they obtain seed is from patented material. Monsanto has recently demanded two farmers from Arkansas who dared to plant soybean from Monsanto seed they had bought earliler. Agro-biodiversity means food future for humankind and its increasing private appropriation is ethically controversial.

Some modern productions in Mexico export flowers, causing worry for the people reproducing patented materials and the companies selling it. Until now, as most of their applications to protect their plants are in process, they can't take legal actions against the offenders. If they discover that their materials are being planted without permission, the only thing they can do is to stop selling their varieties to the people discovered doing so (Massieu, 1997).

There is another face of the problem—large pharmaceutical and agro-biotechnology companies take microorganisms, plants and animals from Mexican territory without information, permission or compensation for the owners. In Mexico, as there are no specific laws to regulate the access to biodiversity, this access has quite often been very easy to private interests and the local communities haven't been even informed. This represents a disadvantageous impact of biotechnology in small farmers' resources. It is scientifically complex to assess a bioprospection application, as strong inequalities exist between multinational firms and the poverty-stricken conditions of the small farmers and local communities involved. In Mexico there are increasing cases of biopiracy. One of them is directly related to indigenous medical knowledge and small farmers' use of medicinal plants; the case of ICGB project in Chiapas.

Chiapas Medical Plants

Since 1998, the English biotech companies Molecular Nature Limited, the University of Georgia and EcoSur (Colegio de la Frontera Sur—a Mexican public research centre) have been subtracting medical herbs and millenary knowledge from Chiapas communities. This was denounced by OMIECh (Organización de Médicos Indígenas del Estado de Chiapas), a traditional Medical Association in February 2000. Three engaged actors invited them to participate in the project. They conditioned their acceptance to the existence of a legal framework, which could protect their knowledge and natural resources, and this hasn't been done as yet. So, the OMIECh rejected participating in the project and presented the case to the UN President of the Working Group about Indian People, Erika Irene Daes, asking her for help. So the Mexican government stopped the project, considering the act as stealing their plants and knowledge. RAFI informed in March 2000 that the project was in its second year and was going to receive a US $2.5 million donation from the International Cooperative Biodiversity Group (CIGB) that includes the US National Health Institutes, National Foundation for Science and US Department of Agriculture. The Altos of Chiapas (the region studied) is one of the richest biodiverse zones of the planet. Through centuries, Mayan culture has accumulated knowledge about their environment and they have developed a complex and wide rage of medication. There are approximately 6,000 plants in the area, many of whom have medicinal uses. They can be the cure for many diseases. The project could identify 2,000 unique components of the plants, which would be chemically identified by Molecular Nature Limited and a copy of all the collected samples would be stored in the University of Georgia-Athens. There are already some samples of living plants in Georgia by now.

While the project was running, OMIECh struggled to bring it to a halt. They argued that they did not give their consent and they doubted whether the project's benefits will ever reach the starving indigenous population of Chiapas. Besides, there is an ethical issue: traditionally, medical herbs and knowledge about them has been collectively used by Mayans in Chiapas. This project could mean private property over Mayan resources, which Mayan people, through OMIECh, have openly rejected (RAFI, 2000). At the moment, it seems that indigenous people and RAFI's struggle has borne some fruits: The Environment and Natural Resources Minister has denied permission to continue the collections since October 2000. In 2000, there was also an initiative of RAFI, CECCAM (Centre of Studies for the Change in Rural Mexico), COMPITCh (Chiapas Traditional Medicine Council) and other civil organizations of a moratory to any bioprospection contract in the absence of any laws and information of the local communities involved.

CONCLUSIONS

Concerning biotechnology's impact over small farmers, it is important for underdeveloped and biodiverse countries to invest more funds in their own biotechnological research. This is necessary to assess the risks of transgenic crops and face the future challenges.

The recently-signed Cartagena Biosafety Protocol represents an international consensus about how to deal with transgenic crops. In this document, it is established that poor countries can ask for resources if they cannot conduct risk assessments. This means an opportunity for these countries to develop and reinforce their own biotechnology research capacity. It is also important that in Mexico, most of the scarce agr-obiotechnology research is made in public research centres and institutes. This means that the concept about genetic resources is not their privatization, but their use for collective benefit and as a sovereign issue. The problem is that these institutes increasingly depend on the funds of multinational agro-biotechnology companies.

Certain reflections exist about the exposed case of biopiracy concerning the ethical objections about nature and food privatization. It means a danger not only for nature itself, but for future mankind's feeding and local people's rights. Nevertheless, it is clear that if biological megadiverse countries will take care of biodiversity and preserve it for the future, they need funds for research. The question is: Is making bioprospection contracts with industrialized countries companies and institutions the only way?

In order to achieve benefits of biotechnology for small farmers, specially from less developed countries, it is necessary that governments have a clear policy towards stimulating small scale sustainable and diverse agriculture production. Unfortunately, during last decades the dominating policy in many of these countries, like Mexico, has been focusing in free trade and competitiveness, which means in many cases small farmers production's extinction.

REFERENCES

Bancomext (Banco Mexicano de Comercio Exterior), 1999, Importaciones de Flores Frescas Mexicanas por País de Destino, Secretaría de Comercio y Fomento Industrial, Mexico.

Broerse, J. and Bounders, J., 1991, The potential of biotechnology for small scale agriculture: How to reorient research and development. In: Bunders J. and Broerse J. (eds), Appropriate Biotechnology in Small-scale Agriculture, CAB International, UK, pp. 1-22.

Commandeur, P., 1996, Private-public cooperation in transgenic virus resistant potatoes. Biotech. Develop. Monitor, **28**, 25-29.

Contreras, E., Camacho, D. and Jarquín, M. E., 1999, Entre la explotación y la exclusión: la producción de alimentos básicos en Chiapas y Oaxaca. In: Espinosa L. M. (ed.), Sector agropecuario y alternativas comunitarias de seguridad alimentaria y nutrición en México, Ed. UAM, INNSZ, CECIPROC, Plaza y Valdés, México, pp. 105-140.

Eastmond, A. and Robert, M., 2000, Henequen and the challenge of sustainable development in Yucatan, Mexico. Biotech. Develop. Monitor, **41**, 11-15.

FIDA-IICA (Instituto Iberoamericano de Capacitación para la Agricultura), 1995, Reformas del Sector Agrícola y el Campesinado en México, ED. IICA, Costa Rica

Fristcher, M., 1999, El maíz en México: auge y crisis en los noventa. Cuadernos Agrarios, **17-18**, 142-163.

Greenpeace, 1999, Centros de Diversidad, Greenpeace, Mexico.

Hernández, V.H. and Sadé, E., 1998, *Selva Lacandona.* Un paraíso en extinción, Ed. Pulsar, México.

Herrera, L., 1999, Transgenic plants for tropical regions: some considerations about their development and their transfer to small farmers. Proc. Natl. Acad. Sc. USA, **96**, 59-81.

James, C., 1999, Global status of Commercialized Transgenic Crops: ISAAA Briefs, **12**, Preview. ISAAA, Ithaca, NY, p.VI.

La Jornada, 1998, En menos de diez años desaparecerán 96 especies de animales, La Jornada newspaper, 25 June, 1998.

Lara, S., 1998. Nuevas Experiencias Productivas y Nuevas Formas de Flexibilización del Trabajo en la Agricultura Mexicana, Ed. Procuraduría Agraria-Juan Pablos, Mexico.

Lehman, V., 1998, Patent on seed sterility threatens seed saving. Biotech. Develop. Monitor, **35**, 6-8

Losey, J. E., Rayos, J. S. and Cartes, M. E., 1999, Transgenic pollen harms Monarch larvae. Nature, **399**, 214

Massieu, Y., Castañeda, Y. and Barajas, R. E., 1993, Biodiversidad: pócima mágica o económica. In: González C. (ed.), La agricultura 500 años después, Ed. IIEc-UNAM, Mexico, pp. 425-442.

Massieu, Y., 1997, Biotecnología y Empleo en la Floricultura Mexicana, Ed. Universidad Autónoma Metropolitana-Azcapotzalco, México.

Massieu, Y., González, R. L., Chauvet, M., Castañeda, Y. and Barajas, R. E., 2000, Transgenic potatoes for small scale farmers: a case study in Mexico. Biotech. Develop. Monitor, **41**, 6-9.

Mooney, P., 1978, Seeds of the Earth: a Private or a Public Resource? International Coalition for Development, Canada.

Noffil, L., 2000, Henequén: símbolo de pobreza para 14 mil campesinos yucatecos, La Jornada newspaper, 19 September 2000, p.34

Pollan, M., 1998, 'Playing God in the garden', New York Times Magazine, 25 October 1998.

Qaim, M., 1998, Transgenic Virus Resistant Potatoes in Mexico: Potential Socioeconomic Implications of North-South Biotechnology *Transfer*, *ISAAA Briefs*, **7**, 48p.

Quist, D. and Chapela, I., 2001, Transgenic DNA introgressed into traditional maize landraces in Oaxaca, Mexico. Nature, **414**, 541-543

RAFI (Rural Advancement Foundation International), 1999, Tecnologías Traitor. Nuevas Implicaciones de Terminator (www.rafi.ca).

RAFI, 2000, Stop biopiracy in Mexico, Indigenous organizations claim immediate moratory. Mexican authorities denied bioprospection permission, Genotypes (www.rafi.org).

SAGAR (Minister of Agriculture, Livestock and Rural Development), 1999, Gaceta Oficial de los Derechos de Obtentor de Variedades Vegetales, SNICS, Mexico

Serratos, J. A., 1998, El maíz transgénico en México, Los vegetales transgénicos, el ambiente y la salud, La Jornada newspaper, La Jornada Ecológica magazine, **70**, p.8.

Shiva, V., 2000, La globalización del hambre. Una guerra contra la naturaleza y los pobres, La Jornada newspaper, Masiosare magazine, **143**, México, 3-6.

Vélez, G. and Rojas, M., 1998, Definiciones y Conceptos Básicos sobre Biodiversidad, Biodiversidad, Sustento y Culturas, **Cuadernillo 1**, Programa Semillas, Bogotá, Colombia.

Viniegra, G., 1999, Monsanto, el gen Terminator y las plagas de la agricultura, La Jornada Newspaper, Lunes en la Ciencia magazine, **67**, México, 1.

www.innovationplace.com/General/Newsletter/un.1998/b2.html, www.library.utoronto.ca.

www.biodiversidadla.org/documentos36.htm, Biological Diversity Convention, 2000, Cartagena Protocol about Biotechnology Safety. United Nations.

www.biodiversidadla.org.prensa/prensa1125.html, 2000, Prensa Rural Net Noticias, 10 October 2000.

13

ECOLOGICAL, ECONOMIC AND SOCIAL PERSPECTIVES ON TRANSGENIC CROP PROTECTION: PATH FOR THE DEVELOPING WORLD

SARVJEET KAUR
National Research Centre on Plant Biotechnology
Indian Agricultural Research Institute
New Delhi-110 012, India

INTRODUCTION

Global population, having exceeded 6 billion in 2000, is expected to reach 9 billion in the next fifty years (James, 2000). This will aggravate the food insecurity in the developing countries. Despite technological advances in food production, a staggering 840 million people, comprising 13% of the world population, including 200 million children, are poor, food-insecure and mainly concentrated in the developing countries of Asia, Africa and Latin America. Disparity across the population in terms of economic access makes even the available food unreachable to the poor. The challenge before the developing world is to provide affordable low-cost nutrition. This has to be achieved both by an increase in the food production on the arable land available and by making the food production systems sustainable, i.e. maximizing production without bringing about environmental degradation. The elusive goal of achieving food security depends upon the complex interplay of several political, social and technical factors.

The 'Gene Revolution', ushering in the transgenic or genetically-modified (GM) plants, has tremendous potential in augmenting both the quantity and quality of food produced, particularly in the area of crop protection. Since 1985, more than 50 plant species have been genetically modified worldwide. The most frequent transgenic traits are herbicide tolerance (40%), insect resistance (24%), product quality (21%), viral resistance (10%) and fungal resistance (4%). The global area under transgenic crops in 2000 was 44 million hectares in 13 countries, including 5 developing countries (James, 2000). In 2001, the global area under transgenic crops increased to 52.6 million hectares (Table 13.1). The global area under transgenic crops increased by 12% in 2002 and with a further increase of 15% in 2003, the global area under transgenic crops is estimated to be 67.7 million hectares in 2003 (James, 2003). Four countries—USA and Canada, among the developed countries, and China and Argentina among the developing countries—grew 99% of the area under transgenic cultivation in 2000 (James, 2001). In addition, there was an increase in area under GM crops in 2001 in nine other countries—South Africa, Australia, Mexico, Bulgaria, Uruguay, Romania, Spain, Indonesia and Germany. In 2003, eighteen countries were growing GM

Table 13.1 Global area under transgenic crops (million ha)

Crop	1996	1997	1998	1999	2000	2001
Soybean	0.5	5.1	14.5	21.6*	25.8*	33.3*
Maize	0.3	3.2	8.3	11.1	10.3	9.8
Cotton	0.8	1.4	2.5	3.7	5.3	6.8
Rapeseed	0.1	1.2	2.4	3.4	2.8	2.7
Others**	1.1	1.9	0.1	0.1	~0	~0
Total	2.8	12.8	27.8	39.9	44.2	52.6

* Without transgenic soybean grown in Brazil on more than 1 million ha.
**Figures for 1998 and 1999 don't include possible acreage of transgenic tobacco in China, which amounted to more than 1 million ha in 1996 and 1997
Sources: James (2001) and http://www.worldseed.org/statistics.html

crops. During 1999 to 2001, herbicide resistance has remained the dominant transgenic trait, followed by insect-resistance and stacked herbicide and insect resistance traits in commercially-grown transgenic crops.

Transgenic plants harbouring pest-resistance genes offer effective pest control and reduce the need of harmful chemical pesticides. Many types of pest resistance genes are being used for the development of transgenic crops such as the insecticidal crystal protein from the bacterium *Bacillus thuringiensis* (Bt), protease inhibitors, α-amylase inhibitors, lectins, fungal chitinase and bacterial secondary metabolites (Gatehouse and Gatehouse, 1999). Novel designer proteins and plant-derived pesticidal genes are also being considered (Krattiger, 1997). Only the insect-tolerant transgenic crops carrying the Bt genes have been commercially released so far. Bt transgenic corn, cotton, tomato, potato and soybean have been developed (Letourneau et al., 2001). The most widely-grown Bt crop is corn, accounting for over 70% of the global area devoted to Bt crops, followed by cotton (Table 13.2). The acreage of Bt-potatoes is relatively less. Currently, corn, cotton and potato are the only commercial Bt crops and they are under a review by the Environmental Protection Agency (EPA) of the USA for the removal of their deregulated status (Letourneau et al., 2001).

Table 13.2 Global area of commercialized Bt crops (million ha)

Year	Potato	Corn	Cotton	Total Bt crop
1996	0.005	0.29	0.68	0.97
1997	0.14	2.79	0.80	3.61
1998	0.02	6.71	2.42	9.16
1999	0.02	9.59	2.48	11.80
2000	0.02	8.20	3.23	11.49
2001	0.02	5.90	5.90	12.00

Source: Letourneau et al. (2001), James (2001)

Although three quarters of the global transgenic acreage is in the developed countries, there has been a steady increase in the area under transgenic crops in the developing countries. Between 2000 and 2001, the percentage growth of the area under transgenic crops has been higher in the developing countries as compared to the developed countries, although the absolute growth in the transgenic area was twice as high in the developed countries (5.6 million ha), compared to the developing countries (2.8 million ha) (James, 2001). The increase in transgenic crops area between 2002 and 2003 was almost the same in developing countries (4.4 million ha) and in developed countries (4.6 million ha) with % growth more than twice as high (28%) in the developing countries as compared with that in developed countries (11%) (James, 2003). Almost one third of the global transgenic area in 2003 was in the developing countries. Nevertheless, the issue of large-scale commercial cultivation of transgenic plants is fraught with seemingly complex ecological, economic and social ramifications, especially in the agricultural scenario of the developing world. This chapter is an attempt to discuss the various perspectives on the significant issues relating to transgenic crop cultivation in the developing countries.

THE ECOLOGICAL PERSPECTIVE

For many people, the first introduction to the disastrous consequences of the use of chemical pesticides came through Rachel Carson's exceedingly popular book 'The Silent Spring', which highlighted the toxic effects of pesticides on beneficial insects and man (Carson, 1960). Over reliance on chemical pesticides has led to environmental degradation, the development of resistance in pests and elimination of natural predators of pests. Pesticides—though effective for short-term pest control—have long-term detrimental effects, which make them unsustainable. Insect resistant-transgenic crops can give a positive environmental benefit by reduction in the pesticide usage but certain concerns over their large-scale cultivation still exist.

Perceived Risks to the Environment

There is a great concern over the environmental safety of the 'directed' genetic modification of plants. Some perceived ecological risks of transgenic crops warrant a thorough assessment.

Direct impact of transgenic crops in cultivated fields and natural ecosystems

The effects of introduced gene on the ecologically-significant phenotypic traits of the transgenic crop, e.g. the safety of transgene towards non-target organisms, the environmental fitness of the crop, potential for invasiveness or weediness and any unpredicted effects of genetic engineering, such as changes in the levels of secondary metabolites need to be thoroughly ascertained. A concern is expressed that the transgenic product released from a transgenic plant by secretion or upon cell death may be toxic to the broad range of non-target organisms, e.g. the fauna that consume the transgenic crop parts and the beneficial natural predators of the pests that ingest toxin-containing plant parts. To avoid any residual buildup of Bt toxin in the food chain, possible precautions include tissue-specific expression of the transgene in only the plant parts targeted by the insect pests such as the leaves and stems and not in the floral parts. It is expected that the toxin would get rapidly biodegraded in the beneficial predators who have fed on the insects that had ingested the toxin. In this regard, the specificity of Bt toxicity is advantageous from the point of view of being harmless to the beneficial insects. The deployment of insect-resistant transgenic plants needs to be compatible with the biological control in order to ensure an ecological benefit over the use of chemical pesticides. Bt transgenic plants had no adverse effect on the ladybirds which feed on the aphids, which were raised on transgenic plants (Dogan et al., 1996). No detrimental effects on the ability of the hymenopteran parasitoid, *Diaeretiella rapae* to control its aphid host, *Myzus persicae* (Sulza) was observed in two *Brassica napus* transgenic lines expressing lepidopteran-toxic *cry1Ac* gene or coleopteran-toxic proteinase inhibitor oryzastatin I (OZ-I) gene from rice (Schuler et al., 2001). However, the biological control of the host pest *Helicoverpa armigera* (Hubner) by its parasitoid was said to be not fully compatible with the use of Bt toxin due to early and rapid mortality of the pest (Blumberg et al., 1997). In another instance, Bt toxin was found to have some toxicity towards a beneficial predator insect (Hilbeck et al., 1998). Such instances emphasize the need for detailed data on the toxicity of Bt towards beneficial predators. The impact of pollen from commercial cultivation of Bt-transgenic corn hybrids on Monarch butterfly, *Danaus plexippus* population was found to be negligible in a two-year study in Canada (Sears et al., 2001).

The protease inhibitor transgenic plants, expressing the cowpea trypsin inhibitor (CpTI) gene, were said to be unlikely to have any deleterious effect on the ladybirds. The ladybirds have cysteine rather than serine proteases in their guts, and will not suffer from protease inhibition (Walker et al., 1998). In addition, CpTI by itself is not toxic enough to cause acute detriment to either the pest or the predator. Protease inhibitors were not only less effective in protecting *Brassica* transgenic plants from diamondback moths, they may actually lead to increased plant damage by the moths (Winterer and Bergelson, 2001). Nevertheless, the potential non-target effects of the transgenic crop cultivation appear to be manageable (Wraight et al., 2000).

A concern is also voiced about the possible release of the transgenic product, e.g. the Bt toxin into the soil after the harvest. This may cause soil and groundwater contamination. The Bt toxin can persist in the soil after release from root exudates of the growing transgenic crops, by binding to the surface-active particles in the soil and may later contribute to the selection of resistant soil insects (Saxena and Stotzky, 2000). A thorough evaluation of the persistence of such toxins and their effects on soil inhabitants is warranted (Stotzky, 2001; Zwahlen et al., 2003). Such effects can be exceedingly difficult to predict and some phenomena may also have a cascading effect on the ecological dynamics of communities of an area. However, since biological molecules are generally biodegradable, they may not pose a severe threat of environmental contamination.

A concern is expressed that the transgene in the new genetic background may endow the plant with extra combativeness, making it more persistent so that the plant may itself become an invasive weed. A comforting thought, though, is that the weediness characteristics are several, each being controlled by at least one gene. The crop plants contain only some of these genes. The introduction of a single gene or even two genes into a crop plant is unlikely to convert it into a significantly greater weed than the parent it was derived from. To create an aggressive weed, all the required factors need to be present. Genetically, modified plants behaved substantially the same as non-transgenic parental lines in over 400 field trials worldwide (Miller and Powell, 1994). Most genetic changes tend to reduce rather than increase the fitness. The wild type genotypes are generally more weedy and fitter. Nevertheless, fitness will ultimately depend on the environmental context in which the plant is grown.

Interspecific and intraspecific gene flow

Unintended lateral transfer of a transgene between related and unrelated species is a potentially worrisome aspect of transgenic technology. It is feared that the escape of a transgene to its related species or weeds growing near the transgenic crop may occur by pollen dispersal, thereby creating 'superweeds', endowed with, for instance insect resistance, which may eventually invade new habitats. While introgression of transgenes resulting in enhanced weediness is unlikely to happen in many cases—especially the gene flow between different species—it is theoretically possible. Concern over weediness depends on the type of transgene and the plant. Plants differ in their out-crosssing potential, either intraspecific or interspecific, to their few or many relatives. Movement of transgenes from the crop *Brassica napus* to the wild relative *Brassica rapa* has been reported (Mikkelson et al., 1996). US Department of Agriculture (USDA) has become increasingly concerned that virus genes in the virus-resistant transgenics may get hijacked or swapped with the natural viruses, thus creating new recombinant virus strains.

Considering the vast diversity of crop plants and the types of genes that can be introduced, the environmental impact of the transgenics needs to be assessed on a case-by-case basis. Genes conferring the advantage of reproduction or survival such as the genes for insect resistance, drought or salt resistances are of concern. Theoretically—ecological balance could be influenced by the unintended transfer of, say, an insecticidal gene from the transgenic crop to its weedy relative, in the rather unlikely scenario of insect damage being the major limiting factor preventing the weed's spread. A certain amount of genetic fluidity is tolerated in nature. Breeding for insect resistance, drawn from the gene pool of wild relatives, has been historically, a fairly common activity without any significant ecological consequences. Interspecific hybrids are generally sterile and less adapted to survival. Still, the ecological risks need to be considered in the context of large-scale cultivation of transgenics in the proximity of sexually-compatible relatives (Bergelson et al., 1999). All the relevant data regarding important crops targeted for transgenic development is required in order to successfully predict whether the transgene could enable the plant to extend beyond its geographical range. There is very little possibility of the horizontal gene transfer from transgenic plants to the terrestrial bacteria as its frequency is extremely low (Nielson et al., 1998).

Development of resistant pests

An important ecological issue pertaining to the deployment of Bt transgenics is the development of resistant pests. Development of resistance in insects exposed to transgenic plants expressing a single insecticidal gene season-long is a disconcerting possibility. It is a serious matter of concern as it can lead to new virulent forms. Such an eventuality will render ineffective, the Bt biopesticides also, which have been in use for more than three decades. Consortia of regulators, industry, academia and extension workers connected with the Bt projects have come together to evolve management practices for the appropriate deployment of Bt transgenics for delaying the onset of resistance development. The recommendations, emerging from the ecological simulation studies carried out to evolve strategies for deployment of Bt transgenic crops advocate as-high-as-possible toxin dose to minimize the chances of survival of the resistant heterozygotes, planting non-transgenic refuges to sustain the homozygous susceptible insect population and deployment of multiple toxin genes acting on different receptor sites in the insect midgut (Roush, 1997). Transgenic plants must only be deployed in an integrated pest management (IPM) strategy. Furthermore, each insect-Bt crop system may have a unique management requirement because of the biology of the insects, although refuges that produce an adequate number of susceptible alleles are a must (Shelton et al., 2000; Zhao et al., 2002).

The sustainability of these strategies requires an intense gene flow between the insect population feeding on transgenic and the one feeding on non-transgenic plants (Bourguet et al., 2000). Development of resistant insects usually occurs in the form of localized outbreaks. Some insects are highly mobile and can affect large areas. Studies on the long-range movement of resistant insects to other crops are required. The USDA has issued guidelines, wherein Bt cotton must be grown along with non-Bt cotton and sprayed with only non-foliar Bt insecticides. This puts a limit on the use of foliar Bt insecticides. These strategies can, at best, delay but not prevent the onset of resistance, which is a natural evolutionary phenomenon. Employing a multi-tactic program of IPM and effective economic practices along with the appropriate pest control measures to maintain a healthy crop, though, can decelerate its pace. This will require monitoring for pest densities and evaluation of economic injury levels so that the pesticides are applied only when necessary. Deployment and conservation of biological control agents, host plant resistance, cultural, bio-rational and genetic methods of pest control are also envisaged (Hoy et al., 1998). Guidelines for deployment of insect tolerant transgenic plants must include resistance-management tactics. In the scenario of developing countries, the feasibility of follow up on on-farm monitoring of all the agricultural practices required for delaying resistance needs to be seriously ascertained before implementing large-scale cultivation of the Bt transgenic crops. Enforcement of the practice to include non-transgenic refugia in transgenic plantation could be a problem, particularly in the high output-demanding scenario of the developing countries. At the same time, it will be important to keep refining the resistance strategies. In a large-scale resistance management plan in Arizona, USA, a remedial action plan for monitoring of compliance with the proposed guidelines, involved direct participation of cotton growers to provide them with incentives for financially supporting this program (Carriere et al., 2001a).

Insecticidal gene pyramiding, using more than one type of Bt transgenes having different binding site specificity to provide multiple levels of protection, is a useful strategy for delaying the onset of resistance in the target insects. The mortality, larval weight and pupation percentages were significantly higher on the transgenic tobacco having both Bt and CpTI genes (Fan et al., 1999). Development of resistance in *Plutella xylostella* (Linnaeus), diamondback moth, was significantly delayed on transgenic broccoli expressing both *cry1AC* and *cry1C* genes (Zhao et al., 2003). The inclusion of novel fusion protein genes or synthetic modification of hybrid Bt genes for broader range of insecticidal activity has been attempted (De Cosa et al., 2001; Tu et al., 2000; Naimov et al., 2003). High expression of Bt transgene is required for functionally-recessive inheritance of resistance (Liu et al., 2001). Overexpression of Bt in chloroplast can also delay the development of broad-spectrum Bt resistance in the field (Kota et al., 1999). Options of tissue-specific expression of Bt gene or expression triggered through spray with a chemical such as salicylic acid have also been considered (Christov et al., 1999). Differential expression of Bt toxin gene in various plant parts in commercial transgenic cotton varieties was found to affect larval survival and development which, in turn, may impact the population dynamics of these pests and thus become a critical factor in resistance management (Adamczyk et al., 2001). Adoption of cultural practices e.g. early harvest varieties, which are harvested from the fields before the larvae can complete development, can also be of help in delaying resistance.

Furthermore, separate refuges delayed resistance better than mixed refuges because they conserved relatively more susceptible alleles than the resistance alleles and did not increase the effective dominance of resistance (Tang et al., 2001). The usable life of Bt transgenic crops would be prolonged in multiple-cropping situations, where crops with different transgenes can serve as refuges for each other and maintain susceptibility in the system as a whole (Caprio and Suckling, 2000). Fortunately, the fitness costs associated with the evolution of resistance to Bt are substantial (Carriere et al., 2001b). Homologuos resistance loci are predicted in *P. xylostella* and *Heliothis virescens* (Fabricius) based on the linkage of Cry 1A resistance to mannose phosphate isoenzymes (Herrero et al., 2001). Large refuges, low initial resistance allele frequency, incomplete resistance and density-dependent population growth in refuges are the factors favouring the reversal of inheritance of resistance (Carriere and Tabashnik, 2001). A detailed understanding of these resistance mechanisms, increased knowledge of pest biology and experimental evaluation of resistance management strategies will enable improving upon the existing strategies (Ferre and Van Rie, 2002).

Yield compensation and adequate expression of the transgene

Would there be any yield compensation in the transgenic plants having multiple copies of transgenes? Would there be enough toxin produced in plant at the maturity level? These are also matters of concern. To be effective, Bt transgenics need to express toxins at a high level. The possible situations of the transgenes interacting with the other transgenes or resident genes *in planta* due to sequence homologies and thereby, inactivating expression, also need to be considered. Transgenics can also become unstable, down regulated or altogether silenced. Such transgenics would then be similar to the non-transgenic variety and hence of negligible environmental impact, except in the case of transgenics developed for down regulation of an allergen or toxic substance. Greater knowledge of the organization of plant genomes will be helpful in assessing the positional effects of the introduced transgene. Nevertheless, complex interactions also occur naturally in the resident genes. By subjecting transgenics to proper assessment before commercial release, the transgenic variety with the unpredictable gene expression can be rejected.

Monoculture and eventual loss of biodiversity

Another eco-concern over the large-scale cultivation of transgenic crops is that it may lead to monoculture and eventual erosion of the genetic diversity of crops. This is especially relevant in the context of developing countries, which are often the centres of crop diversity. In many of these places, the cultivars are grown in the proximity of the wild relatives. Gene flow in such situations is well documented (Bhat and Chopra, 1999). Far fewer wild relatives exist in the developed countries. In the USA and Australia, the release of transgenic plants has been kept out of the area where wild relatives exist. Because of lesser crop diversity, the risk ratio to transgenics for the developed world is smaller than that for the developing countries. Hence, the developing countries need to make a more pragmatic risk assessment for addressing the biosafety issues in their own particular environment.

Concerns Related to Food Safety and Human Health

The main concern over transgenic plants with regard to food safety and human health is whether the transgenic plant is likely to pose a greater risk than the non-transgenic plant variety it is derived from. Food safety risks associated with transgenic plants include the spread of antibiotic resistance and the production of toxic proteins and allergens. Allergic reactions may differ among different people. In nature, transgenomic combinations have occurred over thousands of years. Some of the cereals that we eat today are an outcome of genetic recombination. Crops modified to resist pests or with allergens removed or even with enhanced nutritional quality will be beneficial to human health. Nevertheless, there is a concern that the antibiotic resistance gene, if used as a selection marker for transgenic development, may get transferred to bacterial pathogens in the human gut, thereby spreading drug resistance. This phenomenon needs to be carefully assessed to rule out any eventual negative health impact. Novartis Seeds, a unit of Novartis AG (now Syngenta Seeds), has developed an alternative positive selection process called Positech, based on a naturally-derived enzyme, which is being tested on maize, wheat, barley, sugar beet and vegetables (Nair-Ghaswalla, 2000). It has been licensed to more than a hundred advanced academic and industry research laboratories around the world. Positive selection strategies use cytokinins, xylose isomerase gene and phosphomannose isomerase gene for selection of transformed plants (Veluthambi et al., 2003). For the development of 'golden rice' engineered to synthesize provitamin A, mannose has been used as a selective agent (Lucca et al., 2001).

The assessment of food safety has been based largely on experience and a long history of use rather than actual testing. A clear basis for food safety assessment is generally lacking at both the national and international level. In the USA, the acronym GRAS—generally recognized as safe—is used to describe the safety of a product. The US Environmental Protection Agency (EPA) and the Food and Drug Administration (FDA) have contended that Bt potatoes and maize are safe to consume. CpTI and GNA (*Galanthus nivalis*) are both plant-derived genes and did not exhibit any growth depressive effects in rat feeding trials (Pusztai et al., 1996). Toxicological tests, which are sensitive and comprehensive enough to detect any unintended effects of the GM foods, need to be developed. Many innovations in the evaluation of unintended secondary metabolic changes in transgenic food crops have been achieved, such as multi-component analysis of low molecular weight compounds and chemical fingerprinting of transgenic crops to compare any possible compositional alterations with respect to non-transgenic isogenic parental and closely-related species bred at identical as well as multiple sites (Noteborn et al., 2000). An international regulation on safe genetic modification of plants for food use is required. Furthermore, food safety criteria should be relevant for conditions in the developing countries, where under-nutrition and malnutrition are widespread.

Fail-safe procedures for regulating the transgenic food are needed so that the public is assured of no health risks or suspected allergies. Public awareness on the merits of transgenic food is necessary in order to clear the rhetorical confusion created by anti-GM food lobbies. Garnering consumer acceptance by food labelling will depend on generating public awareness by adequately explaining the benefits of transgenic technology.

Assessment of Benefits-to-Risks Ratio

Ensuring the complete biosafety of transgenic crops is a formidable task. Many uncertainities merit continued caution. Success of transgenic technology will require a steady and conscientious assessment and management of risks. This risk assessment will dictate the regulations that need to be adopted and guide strategies for development and deployment of future transgenics. Safety parameters include analysis of risk-benefit ratio, efficacy of transgenics, spread or dispersal, human health and the environmental fate (Cohen and Chambers, 1991). Weighing of potential risks and benefits is an important aspect of biosafety assessment of the transgenic crops. Risk assessment should focus on high probability risk rather than hypothetical or unrecognizable risk.

Approaches towards addressing the professed ecological risks

Ecological risk assessment requires a thorough understanding of the factors governing the abundance and distribution of the species affected directly by transgenic crop cultivation as well as the indirect influence of these species on the communities they inhabit. Risk assessment experiments are methodically challenging, due to several sources of variation affecting the potential risk. The experiments need to be performed over a period of time for consistent monitoring. Given the higher crop genetic diversity of the developing countries, the local genetic background into which the transgene might introgress, needs to be assessed for the effect of this introgression. The performance of the transgenic crop can vary in different environments due to differences in pollination or cross-pollination between transgenic crop and related species, which may vary in different environments. Also, associated pests of each area may be different, necessitating specific IPM strategies. Regionally-based modifications in some agricultural practices may be required to beget optimum gain from transgenic cultivation. Hence, multi-location trials are necessary. Such field trials should be conducted to determine the impact of transgenic plant on the wild relatives and land races as well as on the agronomic and natural environment. Specific alterations, if any, in the environment associated with the transgenic plants, can be detected through such trials. While multi-location trials would include location-specific bioassays, a global monitoring and bioassays conducted in different countries will help in relating the assessment data properly.

A marker gene such as the green fluorescent protein can be introduced, at no fitness cost, into the host plant along with the agronomically-important gene for the infield monitoring of the expression of the transgene (Harper et al., 1999). Expression under different agronomic conditions can be checked with Expressed Sequence Tags (ESTs). Data from such monitoring will further guide and decide whether monitoring needs to be continued, stopped or the release of transgenic variety for commercial cultivation be permitted or withheld.

A more challenging aspect of the assessment of environmental impact of transgenics is the scale-dependent impact. Risk assessment deploys the strategies of gathering data by making well-designed and carefully-monitored tests of increasing scale and complexity. Although in a small-scale trial also, the consideration is given to the impact which would be relevant in the case of a large-scale commercial production; the probability of a rare event of interspecific hybridization with a weed—which may not occur in the limited field trial but may happen when the transgenic is sown under widespread commercial production—cannot be completely negated. Individual complexities of farm and farmers can hinder the performance of whole farm studies. Since it is unrealistic to assume that all the facets of environmental impact can be forethought, predicted and assessed, it is necessary to undertake sustained and responsible monitoring of transgenics following commercialization.

Furthermore, the risks and benefits do not accrue equally to all groups of people. For example, while in the field testing of a transgenic crop, the local population may unwittingly get exposed to an involuntary risk, while a company distant from the site would receive the benefits. Strategies for the deployment of transgenic plants, keeping in view the important goal of sustainability, need to be devised.

Approaches towards limiting the lateral transfer of transgenes

Some of the ecological risks of transgenic escape can be addressed by the transgenic design itself. Genetic manipulation to avoid the paternal transmittance of a transgene by pollens is possible by targeting the transgene to chloroplast or mitochondria, which are maternally inherited. Chloroplast transformation offers the additional advantage of a high level of expression of the transgene. Gene flow from transgenic crops can be prevented by using fertility constructs that disable the pollen or ovule function in the plants hemizygous for the transgene (Bergelson et al., 1999). Pollen flow from such crops to their wild relatives can produce only sterile hybrids. However, in the event of a mutation disrupting the effectiveness of such fertility constructs, or recombination, which breaks the linkage of the transgene and the infertility gene, the result can be a small number of pollen and ovules lacking the fertility construct. Another strategy to tackle gene escape is to allow the expression of transgenes only upon the exogenous application of a chemical. There would thus be no real risk of transgene escape to natural populations in non-agricultural areas due to the absence of this chemical.

A possible strategy to avoid gene escape to soil bacteria by the theoretical but undocumented probability of DNA uptake from the decaying transgenic plant bound to the soil particles, is to insert an artificial intron sequence that would render the transgene non-functional in bacteria due to the non-processivity of introns. Inclusion of suicide genes into the chromosomal and Ti plasmid DNA of *Agrobacterium tumefaciens* strains used in the development of transgenic plants, can control to some extent, the unintended transfer of disarmed *A. tumefaciens* plasmid harbouring foreign genes to wild plants or soil bacteria (Molin et al., 1993). Suicide constructs are based on the lethal genes from *Escherichia coli* and are trigerred upon plasmid transfer. Regarding the dissemination of *Agrobacterium* through transgenic plants, the fears are rather unfounded, for *Agrobacterium* is eliminated during tissue culture and is not propagated through seeds. Studies focussed on gene transfer among bacteria in natural environments, using improved methods for measuring the gene transfer and its ecological impact, are the need of the hour.

THE ECONOMIC PERSPECTIVE

Traditionally, the goal of crop production has been to produce more and better crops at lower costs. The need for safe and sustainable agriculture is emphasized due to increased environmental awareness. Aproximately, 15 billion hectares of land is utilized in agriculture worldwide (Kendall et al., 1997). In many developing countries, per hectare increase in food productivity is required due to the rising population and declining arable land. The per capita availability of arable land has decreased from 0.48 hectares in 1950 to about 0.15 ha in 2000, showing a threefold decrease. The objective before the developing countries is to achieve a stable food-population balance by enhancing their agronomic potential, eliminating the deficits in internal food supply and lowering production costs, particularly of crops which are targeted for global commodity markets. The farming sector is an important export-earning avenue for the developing world. Both short-term and long-term commercial concerns exist about transgenic crop cultivation in the developing countries.

India is a leading country in the developing world in the field of agricultural research, extension and industry. Agriculture, the backbone of Indian economy, contributes 30% of the GDP (gross domestic product). The National Agricultural Policy states that India needs to double its food production so as to ensure food security for its people. Biotechnological applications for the increased production of import-substitution commodities such as oilseeds and pulses as also export-oriented crops like basmati rice, cotton, spices, tea and coffee, are very relevant in the Indian context.

Advantage of Transgenic Crop Protection

Development of eco-friendly strategies of pest control is of great significance in reducing the pest-incurred losses of crop productivity. The agrochemical industry is also looking for environmentally less damaging ways of pest control. Perceived commercial disadvantages of biological pesticides are: narrow host range, less stability than the chemical pesticides, slow action on target pests and rapid degradation after application. There are many biotechnological approaches towards overcoming these limitations (Kaur, 2000).

Transgenic crops pose a high value commodity. They represent an economically-directed biological evolution. Bt-transgenic crops promise savings on insecticidal usage. The transgenic crops will enable a greater involvement of other biocontrol strategies in IPM, by reducing the reliance on broad-spectrum pesticides. The transgenic crops will control the pest damage supplemented if necessary, by regional application of the required pesticides. This would offset the higher price of transgenic seeds. Also, there would be a saving of the labour and equipment, as the farmers will not need to spray the pesticides frequently. Adoption rates for transgenic technology are high, increasing more than 300% from 0.97 million hectares in 1996 to about 12 million hectares in 2001 (Table 13.2). Bt transgenic crops have shown generally positive economic benefits and reduced the use of other insecticides (Shelton et al., 2002). The economic benefit for Bt seeds are approximately US $ 8 to 10 per acre for Bt corn, US $ 32 per acre for Bt cotton and US $ 30 to 46 per acre for Bt potatoes (Letourneau et al., 2001). In the USA, the area under transgenic cotton has exceeded that under non-transgenic cotton. However, it is not so with regard to the cultivation of food crops—soybean and corn so far. With Bt Corn, it is estimated that about 8% less land would be used to obtain the same output as with conventional corn. Nevertheless, actual returns could be more modest than the predictions, due in part to the variation of insect populations over time and space. Bt corn, the most widely-grown Bt crop, has not been an overriding financial success (Letourneau et al., 2001). Recently, two Bt transgenic cotton hybrids have been authorized for commercial cultivation in India after considering reports on the studies on environmental impacts of the Bt transgenic crops in the Indian context. The economic advantage accruing from the cultivation of Bt cotton is estimated to be encouraging. On-farm field trials carried out with Bt cotton in different states of India showed reduced pest damage and increased yields (Qaim and Zilberman, 2003).

The potential of development of insect resistance to Bt due to the deployment of Bt transgenic plants exists. Calculation of the economic benefit of Bt crop cultivation includes the cost of planting refugia. The spill-over effect on other crops where sprayable Bt is used also needs to be considered. The farmers who do not adopt the Bt transgenic crop and those who use the Bt biopesticides may stand to lose a safe pest management tool due the potential development of insect resistance to Bt crops. As is required by the US EPA, the resistance management plan must be a necessary precondition for obtaining registration and the rights to sell Bt crop seeds in the developing countries. The Bt growers face the trade-off of getting higher short-term profits from using more Bt with no or small number of refugia *vs* long-term benefits in terms of delay of onset of resistance development in insect pests. This would raise the technology fee for the development of transgenic crops with new varieties of insecticidal genes.

Trade-related Issues in Transgenic Crop Protection

The trade issues relate to the transgenics produced in the developing countries as well as the issues relating to import and consumption of transgenics. There are many opportunities in the pricing and organization of delivery of seeds of new crops. The food security for people as well as the profit security for private seed companies need to be considered. The seeds should be available and affordable to the farmers while also taking care of the incentives for the industry. Financial premium for the seed companies will depend on the market demand, the particular crop, and the extent to which it is proprietary and not shared by competitor varieties or hybrids. Transgenes need to be put in the popular varieties of the area so that the farmer is able to opt for it.

Transgenic crops will require close management in order to enable growers to avail of the maximum benefit. Continual monitoring of the fields for the presence of pests is required. For developing countries, the economic advantage of transgenic cultivation needs to be assessed at the cost of expenses for monitoring the effects of transgenics. The multi-location trials of transgenics are a necessity, but also expensive and time-consuming and may add to the seed technology price and delay in reaching the farmers. There are also some unresolved issues that need attention, e.g. the potential external costs of insect resistance moving from the transgenic crop to the weedy relatives.

Pricing is crucial in determining whether the benefit of transgenic crops will extend to the poor farmers. Commercial pressures determine pricing. This can lead to a distortion of the objectives. Enhanced supply is of little benefit to the poor who are unable to afford food even at the reduced price. For example, though the 'golden rice' can potentially benefit 12 million Indians suffering from vitamin A deficiency, the important question is, whether this rice will reach the poor who cannot afford to buy even normal rice for food. It is desirable that the net benefit of transgenic cultivation is evenly spread and percolates down to benefit the poor. Syngenta Seeds (formerly Novartis Seeds) has decided to market its 'Positech' technology under a two-tier pricing system, giving the subsistence farmers free access to technology, whereas the laboratories and companies will be charged royalty (Nair-Ghaswalla, 2000).

Would such a differentiated pricing of biotechnology be possible with the poor and subsistence farmers receiving preferential pricing? This remains to be seen. A comprehensive review of the agricultural marketing scenario is needed. Both market access and competitiveness should also be considered. Marketing includes commercialization, promotion and establishing an industry. This further requires mobilization of both financial and human resources. Given the contentious perception of transgenics, marketing needs to be approached in a positive manner. Excellent point of sale information should be given. The market forces in the developing countries will be driven by a strong push by the biotechnology companies who have invested enormous capital in the development of transgenics and the food demand of the developing world. Companies developing transgenic crops need to recover the high costs they have invested in research. If transgenic Bt crops cause rapid development of insect resistance in an area, buyers will not buy the high-priced seed and will not be able to make profit. A comprehensive education programme should accompany the commercial release of a transgenic crop, so that the growers can understand and assuredly follow the management practice of the refugia.

Transgenic crops are for specific pest control. Expecting a transgenic crop to be toxic to related or other pests of the crop is a mistake. Due to the insecticidal specificities of the transgenic towards a particular type of insect, a farmer may still want to use a broad-spectrum pesticide to control both the expected and the unexpected pests (Burkness et al., 2001). Release of transgenic crops in geographically diverse areas of the developing countries without first testing the performance of plants under the new growth conditions may lead to the development of resistance due to the plant producing less than adequate level of toxin, which may increase tolerance in pests. Consideration of all these factors is required before pursuing commercialization.

Developing countries will also need to safeguard their internal economy while importing transgenic crops. The developing countries need to quickly have the regulatory oversight regarding transgenic crops in place, so as to avoid becoming the dumping grounds for the surplus or discarded genetically-engineered crops from the developed world, in the absence of suitable laws on labelling and distribution of transgenic food.

There are also some issues relating to the post-production processing of transgenic crops. A post-production difficulty will be the mixing together of the varieties in the processing chain. If the transgenic varieties have to be kept separate and the processor is required to obtain special permission for its marketing, under the regulatory regime he may not choose to have the transgenic at all and the farmer may not grow it. Such market segregation may lead to less competitive selling price for the transgenic crop. So, the transgenic variety will need to be really popular with the farmers in terms of farming benefits such as reduced input costs. However, if the same company that supplies seed to the farmers is also the processor, then the entire operation for input and output can become contractually linked. The uncertainty of the export market due to regulations in other countries can also hinder the adoption of transgenic technology. There is a likelihood of inconsistencies in the regulatory regime of different countries. This would mean that companies might sell different types of products in different markets. This may lead to *de facto* trade barriers, as advanced seed materials with novel traits may not be sold in countries with a more rigorous regulatory framework. This can then slow down the pace of overall positive impact of transgenic technology.

Public-Funded vs. Commercial Objectives

The objective of public-funded research is, thus, a social benefit while the accruing economic advantage is the driving force behind the private enterprise. While the overwhelming concern for the government of the developing countries is more food for the poor, in the markets of the developed world, agribusiness is viewed as a good growth and a highly profitable opportunity. In developing countries, agricultural biotechnology research is carried out largely by the public sector, while in the developed world, the private sector dominates the research and development. Current research on transgenic crops is very likely taking place in the private sector of the developed world. There will be both competition and partnerships within the private industry and also with the public-funded research due to intellectual property protection of new, unproven technology. This would lead to synergies or cynicism in the academic-industrial interface. To ensure wide availability of safe technology to both the biotechnology industry and the academic scientific community, comprehensive but simple licensing procedures are required. A clear delineation of the norms of interaction between the public sector and the private sector is essential. TRIPS (trade related intellectual property) rights of the developing countries need protection (Correa, 2000). Uniform standards are needed for the WTO (World Treaty Organization) regime to work towards an equitable global trade policy. The international institutes can play an important part in safeguarding the interests of the developing countries. An international project funded by the Rockefeller Foundation in collaboration with International Rice Research Institute (IRRI) of the Philippines, is being carried out for introducing *cry1Ab* and *cry1Ac* gene in rice to combat stem borers (Wunn et al., 1996). The germplasm of these rice varieties may be made freely available for the national breeding programmes in the developing countries (Krattiger, 1997). In the Indian scenario, university research is an important source of biotechnology innovations. The economics of Bt transgenic cultivation under Indian conditions may take some time to shape up in the incumbent IPR regime.

There are widespread apprehensions about the monopoly of the seed industry by private multinational corporations (MNCs). The revolution in agricultural biotechnology and IPR has changed the market structure of the private seed industry in the developing world, with only a few players dominating certain markets, in the USA as also in Europe and parts of Latin America, most notably Argentina and Brazil. The same companies are already prominent players in some less-developed countries as well, expanding the role of MNCs in agro-biotechnology. There is a concern about the effect of this phenomenon on the government-funded institutions. An important commercial interest of the seed companies is protection of their intellectual property rights while selling their seed in the developing countries. This objective can be accomplished by making the farmer purchase seed every year, rather than using the seed saved from the previous harvest. This requires downstream control of plant varieties, once the seed is sold. One approach is to make the transgenic F1 seed sterile and incapable of germination, so that the farmers are forced to buy fresh seed each time. Popularly termed the *Terminator* technology, this is the technical alternative to prevent seed saving (Wright, 2000). As a result of this, the plant becomes so disposed that its seed does not germinate in the next generation. Delta & Pine Land, Mississippi, USA, a company with 75% of the US cotton seed market and now owned by Monsanto, has applied, together with USDA, for a patent on a gene that will stop germination of seeds from a transgenic crop when treated with a trigger compound. The technology can be extended to other crops and can serve to eliminate the illegal use of seeds. However, this technology is not considered to be consumer friendly in the context of the developing countries where the farmers have traditionally saved seeds for the next growing season. It has generated much vocal opposition in India. Monsanto has subsequently decided not to use this technology in the transgenic crops. Some molecular diagnostic methods such as DNA profiling also exist, but litigations over seed use disputes can be expensive and long-drawn for the private companies.

Although modern biotechnology seems to be controlled by capitalism, it is now slowly moving to the developing countries. Benefit sharing, farmers' rights and IPR have yet to be fully delineated in some developing countries.

Transgenic Crops and Patents

Patenting of plants, animals or any genetic component thereof is not allowed in most developing countries. Developing countries object to patents on genes and the transgenics thereof, because the source material for the isolation of genes may well have been from the biological diversity of the developing countries and they want to protect their national wealth. Furthermore, the cost of patenting is generally extremely high for the scientists of developing countries. There are hundreds of patents coming from various aspects of Bt technology and many firms are embroiled in litigation over Bt The developing countries must guard against indiscriminate use of their new genes by the research companies of the developed world. For the developing countries, patenting is both a good and a bad thing. Patenting of biodiversity in these countries can be their national wealth. At the same time, grant of broad patents in the developed world, on the innovations and technology on which further research is possible to obtain full benefits of technology, will retard the growth of research that is contingent on the knowledge and techniques protected by the patent, in the developing countries. For self-protection, the developing countries need to evolve a clear and consistent regulatory and legal system at par with the developed world.

Farmers' Perspective

Farmers are a diverse group, differing widely in the area, type of land farmed, cropping system, wealth and farming objectives. Some cultivators are mere sustenance farmers, while others are commercial farmers. A major economic objective of agricultural research is to reduce the cost of production. The farmers of the developed world need to reduce costs due to the production surpluses and the falling demand, while the farmers of the developed world need to reduce costs due to poverty and lack of resources. The issues concerning all the farmers include access to relevant crop protection information, availability of off-farm inputs, credit facilities and limitations caused by poor infrastructure. The factors governing the choice for the farmers are: the farmer's attitude to the risk, his preferences and his evaluation of the outcome in terms of net profit or food (Dent, 1995). Innovativeness of the farmers depends on their resources and aspirations.

In the last three years, desperation over the failure of pest control measures drove some thousand farmers to suicide in the Andhra Pradesh, Karnatka and Maharashtra states of India. Sophisticated chemical control agents and the measures necessary for the safe use of such chemicals are not available to the poor farmers of the developing world. In that respect, transgenics promise user safety to the cultivator. In the developing countries also, the farmer may be quite willing to pay for insect resistant transgenic crops, provided the technology fee is kept low. Adopting appropriate management practices is an important aspect of transgenic crop cultivation in the developing countries. Transgenic cultivation represents a high input-high output cropping system, compared to the traditional low input-low output cropping system. Transgenic crops with sublethal levels of Bt toxin, can actually aid in the development of resistance in pests. Therefore, the transgenic cultivation needs to be limited to the areas where proper management practices can be monitored. Developing countries need to be fully aware of the scenario in which the resistance management strategies could fail because of inadequate information, planning and implementation. After all this, even the precious natural Bt pesticide will be lost to the environmentally sound pest management (deMaagd et al., 1999). A possible apprehension could be how to protect the interests of a small farmer who uses foliar Bt insecticide while his neighbouring rich farmer grows a Bt transgenic crop that shares the same pest complex. Also, one needs to ascertain, in a larger perspective, the long-term ecological and social impact of transgenic crops in terms of shifting land use from small farmers to private industry. Widespread cultivation of transgenics in the agricultural scenario of the developing countries will require new perspectives in agricultural land management. Effective goal achievement in a specific, measurable and realistic time frame is required.

Prioritization of Resources in the Developing Countries

Developing countries all over the world are making a serious commitment to strengthening the research development in biotechnology, so that they may also profit from its potential benefits. Nevertheless, a prioritization of the types of transgenic crops that the developing countries need—keeping in view their food requirements, agricultural setup, socio-economic climate and ethno-religious beliefs—is called for. The genetic attributes of value to the farmers, the processors and the consumers should be targeted for incorporation into the popular crop varieties.

Developing countries face the constraints of a large population, low industrial activity and problems of balance of payment. Their foreign exchange is largely earned through agricultural exports to the developed countries. The developing countries are often resource-poor but gene-rich. Genetic diversity of crops is their national heritage, especially of those farm families who have conserved and selected this diversity. Financial appraisal of genetic resources is also necessary. Developing countries such as India, which have a strong human resource and skill base, can seek greater integration in the world economy, by prioritizing their research initiatives towards addressing local problems affecting staple or other basic crops and the problems faced by the growers in the large-scale production of export commodities. To build a multi-layered food security system, crops of regional and local importance to the developing world will need attention from the agricultural scientists of the developing world itself. Developing countries need to gear up to face the challenge the biotechnology revolution the world over with informed initiative, keen competence and well reigned-in regulatory oversight. This will require funds and a firm will. Many developing countries have set up appropriate laws, guidelines and official bodies in order to regulate and oversee the research, development and use of biotechnology derived from their own research institutes or international companies wishing to test or commercially release transgenics in that country (Alvarez-Morales, 1998). Notably, China, India, the Philippines in Asia; Argentina, Brazil, Chile, Cuba, Costa Rica and Mexico in Latin America; and South Africa and Egypt in Africa have their regulatory mechanisms in place for the safe and effective introduction of transgenic crops in their countries (Krattiger, 1997).

While biotechnology holds promise in crop protection, long-term success necessitates a greater understanding of physiology, biochemistry, genetics and molecular biology of the host-pest interactions of location-specific pest complexes. Developing countries must keep up with the work in these areas which have the potential of future applications.

An important economic issue for the governments of the developing countries will be the prioritization of funding. Transgenic crops have the potential to contribute substantially towards food sustainability. Yet, the indenting, cataloguing and maintaining the biodiversity of a nation must take priority over the development of transgenic crops because protection of biodiversity, the nation's wealth, should be done proactively before it is too late. Also, placing biodiversity records in order would give the resource-poor, but crop-diversity-rich developing a better negotiatory stand in terms of technology exchange. Also, the importance of having 'own genes' to develop the transgenic plants cannot be undermined. Novel Bt strains harbouring new types of insecticidal genes need to be isolated (Schnepf et al., 1998; Kaur and Singh, 2000).

Ultimately, the perception of the public, who is the ultimate consumer, will influence the economic success of the transgenic technology in the developing countries. Biosafety considerations will play an important part, for consumer acceptance. Biosafety considerations encompass the safety of release of transgenic plants into the environment and any eventual biohazards, which are yet to be completely enumerated or identified. But biosafety comes at a cost. There is no other option for the developing world, but to pay the costs now, while it is still safe. Streamlined biosafety mechanisms and registration protocols are required.

THE SOCIAL PERSPECTIVE

The overriding pressure of increasing population has created an urgent need for crop improvement, exploitation of genetic diversity and habitat conservation. Developing countries owe it to their people to provide food in a sustainable agricultural mode by conserving biodiversity, limiting deforestation and maximizing land use. A strong agriculture is the building block towards industrialization, peace and stability for a nation. A population is more willing to exert political pressure in support of environment-friendly technologies, when they feel secure with their food and economic situation (Shatters, 1998).

Social Objectives of Agriculture

The social objectives of agriculture are a sustainable agriculture with equitable distribution, environmental conservation and biosafety. These social issues chiefly pertain to the issues of equal access and parity in the distribution of resources and technological innovations. The objective of the state-sponsored research and development activities is the social benefit, implicit in which is the economic benefit. Social equality demands that the poor also receive the benefit. Farmers and rural communities are inextricably linked in most developing countries. There is an urgent need to increase farm production, reduce production costs in the farm sector and improve the living conditions by generating income and employment. Technology development and adoption should have a pro-poor bias. Often, poverty is the sad accompaniment of the regions of high biodiversity in the world. It is not right to constrain production in a world of hunger. So, failing to develop transgenics may also prove harmful in the long run. Crop diversification is important in the agricultural scenario of the developing world. The farmers should grow what can be grown locally. More crops would mean more diversity and less selection pressure on the pests, which would then be probably controllable with IPM. It is important to make constructive use of the food stocks available, e.g. through food-for-work programmes and other income-generation projects. Agriculture needs to be technologically updated by incorporating the latest innovations in order to make it an attractive vocation to the modern educated youth of the developing countries.

Public Perception of Transgenics

Public opinion is a powerful force for guiding research towards safe and sustainable agriculture. Sociologists say that the willingness of the public to accept a change is related inversely to the degree and the rate of change. There has been initial skepticism to many technological advances in the past, probably born out of an atavistic fear of disturbing the natural order. There is a strong social pressure to reduce the negative impact of human activity on the environment. The increased awareness of the public about the environmental issues is largely attributable to the media by its enlightened highlighting and the active demonstrations by the environmental groups such as Greenpeace and the Friends of the

Earth. However, the people of the developing world—though highly capable of being sensitized over environmental issues—would be concerned more about their immediate and urgent need of food for the hungry, rather than the spin-off ramifications of the effect of transgenics on the environment. Class differences may also exist among people on the subject of transgenics. While the poor, largely illiterate population, unable to understand the technical concerns, would be moved primarily by the consideration of more food, the rich intelligentsia would favour or disfavour transgenics based on well-sourced information.

The public concern about the biosafety of transgenic plants has occupied centre stage. An opinion is expressed that the transgenics do not present an altogether positive picture. Public antipathy towards transgenic research is largely due to the lack of knowledge and unchallenged propaganda by the strong anti-GMO lobbies. Another fear with a socio-political dimension is that genetically-engineered products, having genes which may be potential allergens, may be clandestinely tried on the people of the developing world.

The release of transgenic crops into the environment has raised public sensitivity in different countries of the developed world, though not with the same intensity. Public opposition to GM food is more vociferous in Europe, where the focus is also on the ethical and cultural implications of transgenic research. The debate is hot in Europe, and the high-profile opposition has disrupted transgenic field trials. Nevertheless, public perception is not static and is constantly modified upon new information. The transgenics have gained more approval in the USA. Consumer acceptance of GM foods is more in the case of those which offer direct health benefits. There has been a continued rise in the consumer acceptance of GM foods in general. The absence of any negative reports of compromised biosafety indicates that genetic modification of crops for food uses may pose no immediate or significant risks as compared with the conventional crops (Stewart et al., 2000).

Creating Public Awareness

Without public acceptance, no gains can be made from the elegant transgenic crop technology. Genetically-engineered plants do not exhibit any disrespect to nature. Agricultural activity has social impact. Farming by itself is not natural. Agriculture is a manmade activity. The food consumed by humans has had transgenomic recombinations occurring over thousands of years. The selective breeding has always been undertaken with the aim of commercial agriculture. Most crop plants are the product of artificial selection.

To promote and popularize biotechnology, its beneficial aspects need to be understood by the public. Well-informed and balanced public interest groups can go a long way in instilling confidence in the consumers about the benefits of biotechnological approaches. A science-based approach is required to address the pertinent questions. More involvement of scientific experts is needed for improving the understanding of public with regard to the transgenic technology and the perceived risks thereof. The role of the scientists is to focus the attention of the public on accurate and adequate science-based information. The scientists should promote the growth of rational ideas, without any mix up with emotionalism. The scientists need to enlighten the public by providing an overall, encompassing perspective of transgenic crop adoption. Lack of definitive scientific evidence, spelling out advantages and perils of GM food consumption, is a problem. Research in this area is still nascent. Scientists have a responsibility in correct dissemination of knowledge generated through carefully controlled experimentation (Shelton, 2003).

To enhance the impact of agricultural research, the efficacy and environment of the agricultural research institutions should be improved. Multidisciplinary teams with the integration of breeders, agronomists, ecologists, social scientists along with genetic engineers will be more useful. Also, research objectives need to be widened to include the conservation of ecosystem so that the yields can be maximized without causing any detriment to the environment. The successful integration of transgenic crop production in modern agriculture is a challenge for the agricultural scientists of the developing countries.

The media also has a responsible role in keeping the public informed. The concerns over pesticide residues in the food as well as the threat to wildlife have been well highlighted in the media. Media should help allay the public's misplaced concerns over transgenic crops by putting across valid and science-based propositions rather than any emotional rhetoric. Media should also lay emphasis on the objective concerns of safety and efficacy. The public in general and the governments in particular have increasingly come to realize the unique responsibility, man must assume, in order to safeguard earth's environment.

Value Judgements

A third of the 840 million hungry people of the world live in India. However, in India, the problem of hunger is not of food availability per se, but to a substantial extent, it is the problem of maldistribution. A similar situation exists in the neighbouring countries of Pakistan and Bangladesh. Among themselves, these three countries have half of the world's hungry population. The food is not reaching the poor. There are many social disbenefits of maldistribution. The other significant problem is of malnutrition—the 'hidden hunger'. The issues which define our value judgement regarding transgenic adoption include: both quantity and quality of food; malnutrition, especially of women and children; environmental pollution; external costs of soil erosion; salinity; water pollution; depleted aquifers; and an ever-narrowing genetic diversity due to non-conservation. All these factors add to the price of producing food. Closely linked to the issue of food security is the issue of providing livelihood to the people. Agriculture is the mainstay occupation of Indian people. Over 70% of Indian population depends on agriculture for income and employment; the per capita annual income is less than US $ 150 and there are 100 million farm families. The size of farm holdings is usually small. Growth in the agricultural sector also ensures work to the people.

Value judgements are inescapable. They depend upon genuine needs. Prevention of starvation is an urgent moral demand. Overall benefit depends upon weighing of the possible risks against benefits. But there are no clear answers, considering the complexity of situation. The maxim *Primum non nocere,* i.e. 'first do no harm', contained in the Hippocratic oath, may not apply to agricultural biotechnology, where the balance would be tilted in favour of the overall benefit of transgenic crop cultivation.

The Farmer's Right to Choice

A farmer is entitled to exercise his buying power by procuring the best seed available within his capability. Apart from the private seed companies, the strong agricultural research base in India can help the farmers define and obtain their choice. Farmers are generally reluctant to change their agronomic practices without appropriate demonstrations. In order to gain their confidence, the importance of integration of extension workers into the demonstration farms of transgenic crops is emphasized. A multidisciplinary perspective in the context of overall farming system, taking into account the farmers' needs, circumstances, objectives and resource constraints, with the involvement of scientists, socio-economists, extension workers, agribusiness representatives and farmers should be developed.

A farmer is both a giver and a taker. He is the giver of food and the taker of technology. With the emergence of a wide range of new pest-control technologies, the emphasis must now be placed on the development of transfer and delivery mechanisms to the resource-poor farmers who are the most dependent on novel solutions for their livelihood and survival (Krattiger, 1997). In China, the small farmers gained more than twice as much income (US $ 400 per ha) per unit of land from Bt cotton, as the farmers with larger farm holdings (US $ 185 per ha) (James, 2000). By 2001, roughly 31% of area under cotton in China was that of Bt cotton, the rapid spread an index of farmers' demand (Toenniessen et al., 2003). Farmers can become more enlightened by increasing their involvement in technology generation. They could eventually also become shareholders. The joint ventures between the government and the farmers' cooperatives can make a significant contribution towards productivity enhancement.

Long-term Social Benefits

The ultimate aim of all scientific endeavors has to be the welfare of mankind. In addition to the right to live, people also have a right to food. The opportunities of using transgenic crops for positive benefits to human health and environment must be explored vigorously. The social profit of an innovation is determined by its social acceptance and wide distribution through the whole society, in terms of increased income for the manufacturer and reduced price for the consumer. A social perspective on transgenic crops calls for a holistic assessment of the cost of development and management, including the regulatory oversight of transgenic cultivation vis-a-vis the cost and effectiveness of pest control by chemical and microbial insecticides. The benefit, in terms of more food produced versus money spent on transgenic adoption, should be then calculated.

The poor deserve an equitable stake in the society. It is said that despite being productive, the Green Revolution also had its limitations. It relied on intensive agriculture, large farms and prime land but endangered the environment and offered little to the poor farmers, the actual intended beneficiaries. The concern is that the transgenic revolution may not end up the same way. Proper management of transgenic cultivation is of utmost importance. The flow of products of technology to the beneficiaries in the developing countries requires their informed participation. The developing countries cannot afford to reject the transgenic crops without a thorough assessment of the good they have to offer, in terms of increased agricultural productivity and poverty alleviation. Assessment must be thorough though, of accepting the genetic modification of long-term social value, while rejecting those, which could eventually be harmful to the environment. The developing countries should constitute national committees with panels of scientists, ecologists, physicians, jurists and representatives of the public. The high upfront costs of such a thorough assessment prior to commercialization need to be borne by the developing world in order to ensure long-term social benefits of transgenic technology. At the same time, it is useful to remember that too many regulations may stifle scientific innovations. A cautious, pragmatic application of agricultural practices is required.

CONCLUSIONS

Every developing country has a duty to provide food and nutritional security to its people, but without incurring any concomitant biohazards. This underscores the need to monitor the ecological safety and its contribution towards mitigating malnutrition, disease and environmental degradation. Biotechnology has compelling attributes for addressing the food production needs in the developing world. Transgenic plants offer both opportunities and responsibilities. They promise improvement in agricultural production, food, nutrition and human health. But they may also present risks to the environment, if not monitored and managed properly. The eventual ecological effects of transgenics may be cognizable, but the substance of repercussions of these effects cannot be estimated in totality. The tremendous pace at which the research in genetic engineering is moving, could make even the most conscientious scientists worried about the safeguards of this technology, emphasizing the need for prudent caution in assessing the profitable and predictable effects vis-a-vis the unintended, inadvertent and accidental effects of transgenic crop cultivation. The assumption that the transgenic plants are entirely safe would be an oversimplification. Concerns over the unproven ecological risks—that some private companies may tend to play down—need to be seriously looked into. Nevertheless, the transgenic crops deserve a chance to prove their merit.

More research into the management and implementation strategies of the next generation transgenic crops, regularly-updated regulations and increased collaboration between stockholders are needed so as to maintain the efficiency of Bt in transgenic crops. At the same time, developing countries need to devise internationally-acceptable norms and standards of monitoring and resistance management. To provide a balanced view, in the light of scathing skepticism and occasional misinformation of the anti-biotechnology groups, the scientific community spanning government, industry and academia should come forward with information and assurance to the public, so that it can build its own perspective about transgenic plants. The role of government scientists in providing accurate assessment is especially important. The agricultural system needs to be geared up to be ready for the arrival of imported or indigenously-developed transgenic crops. Valid regulatory paradigms and agricultural infrastructure for monitoring the ecological and any other impact of transgenic crops should be in place, before recommending the same for cultivation. The benefits of biotechnology should be used wisely rather than cause unconscionable ecological harm. A case-by-case regulation of this emerging robust area will entail big budgets.

Nevertheless, the utilitarian usefulness of transgenics should not be overestimated, although one may support them with the view of doing more good than harm. The costs, risks and benefits will be weighed in terms of overall welfare of mankind. This alone, to some, will be the compelling logic for accepting transgenics. Also required is a democratization of technology control in order to ease out the burden on the developing countries. The role of international research institutes will be critical in such technology exchanges.

The role of IPM in augmenting agricultural productivity is important. Transgenics will at best be a component of these management strategies, which include inexpensive ways of biological control. In certain farm situations, even an alternative low-cost scenario of achieving crop protection by the locally-managed IPM, devoid of the transgenic crop may prove to be effective. In the Philippines, the local IPM takes care of the rice crop. Ultimately, the agrochemical industry also stands to gain by the success of IPM strategies.

The direction of crop biotechnology in the developing countries should be based on democratically-determined social guidance, in their quest for an equitable and sustainable world. The link between food security, environment and peace is vital. Developing countries can become environmentally safe, free of poverty, hunger and malnutrition, through accelerated social and economic development, aided by technological advances in agricultural science, blended with indigenous knowledge. The conservation of biodiversity should be set on a high priority. Authentic development entails both agricultural and technical advancement. The ultimate goal is to achieve crop protection while reducing environmentally-damaging practices.

REFERENCES

Adamczyk, J. J., Hardee, D. D., Adams, L. C. and Sumerford, D. V., 2001, Correlating differences in larval survival and development of bollworm (Lepidoptera: Noctuidae) and fall armyworm (Lepidoptera: Noctuidae) to differential expression of Cry1A(c) delta-endotoxin in various plant parts among commercial cultivars of transgenic *Bacillus thuringiensis* cotton. J. Econ. Entomol., **94**, 284-290.

Alvarez-Morales, A., 1998, The release of transgenic varieties in centers of origin: effect on biotechnology research and development priorities in developing countries. In: Ives C. L. and Bedford B. M. (eds), Agricultural Biotechnology in International Development. Biotechnology in Agriculture Series. CAB International, Wallingford, UK, pp. 27-34.

Bergelson, J., Winterer, J. and Purrington. C. B., 1999, Ecological impacts of transgenic crops. In: Chopra V. L., Malik V. S. and Bhat S. R. (eds), Applied Plant Biotechnology, Oxford & IBH Publishing Company Pvt. Ltd., New Delhi, pp. 325-343.

Bhat, S. R. and Chopra, V. L., 1999, Plant biotechnology and biosafety. In: Chopra V. L., Malik V. S. and Bhat S. R. (eds), Applied Plant Biotechnology, Oxford & IBH Publishing Company Pvt. Ltd., New Delhi, pp. 345-360.

Blumberg, D., Navon, A., Keren, S., Goldenberg, S. and Ferkovich, S. M., 1997, Interactions among *Helicoverpa armigera* (Lepidoptera: Noctuidae), its larval endoparasitoid *Microplitis croceipes* (Hymenoptera: Braconidae) and *Bacillus thuringiensis*. J. Econ. Entomol., **90**, 1181-1186.

Bourguet, D., Bethenod, M. T., Pasteur, N. and Viard, F., 2000, Gene flow in the European corn borer *Ostrinia nubilalis*: implications for the sustainability of transgenic insecticidal maize. Proc. R. Soc. Lond. B Biol. Sci., **267**, 117-122.

Burkness, E. C., Hutchison, W. D., Bolin, P. C., Bartels, D. W., Warnock, D. F. and Davis, D. W., 2001, Field efficacy of sweet corn hybrids expressing a *Bacillus thuringiensis* toxin for management of *Ostrinia nubilalis* (Lepidoptera: Crambidae) and *Helicoverpa zea* (Lepidoptera: Noctuidae). J. Econ. Entomol., **94**, 197-203.

Caprio, M. A. and Suckling, D. M., 2000, Simulating the impact of cross resistance between Bt toxins in transformed clover and apples in New Zealand. J. Econ. Entomol., **93**, 173-179.

Carriere, Y. and Tabashnik, B. E., 2001, Reversing insect adaptation to transgenic insecticidal plants. Proc. R. Soc. Lond. B Biol. Sci., **268**, 1475-1480.

Carriere, Y., Dennehy, T. J., Pedersen, B., Haller, S., Ellers-Kirk, C., Antilla, L., Liu, Y. B., Willott, E. and Tabashnik, B. E., 2001a, Large-scale management of insect resistance to transgenic cotton in Arizona: can transgenic insecticidal crops be sustained? J. Econ. Entomol., **94**, 315-325.

Carriere, Y., Ellers-Kirk, C., Liu, Y. B., Sims, M. A., Patin, A. L., Dennehy, T. J., Tabashnik, B. E., 2001b, Fitness costs and maternal effects associated with resistance to transgenic cotton in the pink bollworm (Lepidoptera: Gelechiidae). J. Econ. Entomol., **96**, 1571-1576.

Carson, R., 1960, The Silent Spring. 25th edition (1987), Houghton Mifflin Company, Boston, USA. 368 pp.

Christov, N. K., Imaishi, H. and Ohkawa, H., 1999, Green-tissue-specific expression of a reconstructed *cry*1C gene encoding the active fragment of *Bacillus thuringiensis* delta-endotoxin in haploid tobacco plants conferring resistance to *Spodoptera litura*. Biosci. Biotechnol. Biochem., **63**, 1433-1444.

Cohen, J. I. and Chambers, J. A., 1991, Biotechnology and biosafety: perspective of an international donor agency. In: Levin M. A. and Strauss H. S. (eds), Risk Assessment in Genetic Engineering, McGraw-Hill Inc., New York, pp 378-394.

Correa, C.M., 2000, Implications for developing countries. In Intellectual Property Rights, the WTO and Developing Countries: The Trips Agreement and Policy Options. Zed Books Ltd., London, pp.23-100.

De Cosa, B., Moar, W., Lee, S. B., Miller, M. and Daniell, H., 2001, Overexpression of the Bt cry2Aa2 operon in chloroplasts leads to formation of insecticidal crystals. Nature Biotechnol., **19**, 71-74.

Dent, D. R., 1995, Defining the problem. In: Dent D. (ed.), Integrated Pest Management, Chapman and Hall, London, pp.86-119.

Dogan, E. B., Berry, R. E., Reed, G. L. and Rossignol, P. A., 1996, Biological parameters of convergent lady beetle (Coleoptera: Coccinellidae) feeding on aphids (Homiptera: Aphididae) on transgenic potatoes. J. Econ. Entomol., **89**, 1105-1108.

Fan, X., Shi, X., Zhao, J., Zhao, R. and Fan, Y., 1999, Insecticidal activity of transgenic tobacco plants expressing both Bt and CpTI genes on cotton bollworm (*Helicoverpa armigera*). Chin. J. Biotechnol., **15**, 1-5.

Ferre, J. and Van Rie, J., 2002, Biochemistry and genetics of insect resistance to *Bacillus thuringiensis*. Annu. Rev. Entomol., **47**, 501-533.

Gatehouse, J. A. and Gatehouse, A. M. R., 1999, Genetic engineering of plants for insect resistance. In: Rechcigl J. E. and Rechcigl N. A. (eds), Biological and Biotechnological Control of Insect Pests, Lewis Publishers, Boca Raton, pp.212-280.

Harper, B. K., Mabon, S. A., Leffel, S. M., Halfhill, M. D., Richards, H. A., Moyer, K. A. and Stewart, C. N. Jr., 1999, Green fluorescent protein as a marker for expression of a second gene in transgenic plants. Nature Biotechnol., **17**, 1125-1129.

Herrero, S., Ferre, J. and Escriche, B., 2001, Mannose phosphate isomerase isoenzymes in *Plutella xylostella* support common genetic bases of resistance to *Bacillus thuringiensis* toxins in lepidopteran species. Appl. Environ. Microbiol., **67**, 979-981.

Hilbeck, A., Baumgartner, M., Fried, P. M. and Bigler, F., 1998, Effects of transgenic *Bacillus thuringiensis* corn-fed prey on mortality and development time of immature *Chrysoperla carnea* (Neuroptera: Chrysopidae). Environ. Entomol., **27**, 480-487.

Hoy, M. A., 1998, Myths, models and mitigation of resistance to pesticides. Philos. Trans. R. Soc. Lond. B. Biol. Sci., **353**, 1787-1795.

James, C., 2000, Global view of commercialized transgenic crops: 2000. ISAAA (International Service for Acquisition of Agri-biotech Applications), Brief no. 21 Preview, Ithaca, New York, http://www.isaaa.org/publications/briefs/Brief_21.htm

James, C., 2001, Global view of commercialized transgenic crops: 2001. ISAAA (International Service for Acquisition of Agri-biotech Applications), Brief no. 24 Preview, Ithaca, New York, http://www.isaaa.org/publications/briefs/Brief_24.htm

James, C., 2003, Global status of commercialized transgenic crops: 2003. ISAAA (International Service for Acquisition of Agri-biotech Applications), Brief no. 30 Preview, Ithaca, New York, http://www.isaaa.org/publications/briefs/Brief_30.htm

Kaur, S., 2000, Molecular approaches towards development of novel *Bacillus thuringiensis* biopesticides. World J. Microbiol. Biotech., **16**, 781-793.

Kaur, S. and Singh, A., 2000, Natural occurrence of *Bacillus thuringiensis* in leguminous phylloplanes in the New Delhi region of India. World J. Microbiol. Biotech., **16**, 679-682.

Kendall, H. W., Beach, R., Eisner, T., Gould, F., Herdt, R., Raven, P. V., Schell, J. S. and Swaminathan, M. S., 1997, Bioengineering of crops. Report of the world bank panel on transgenic crops. In: Environmentally and Socially Sustainable Development: Studies and Monographs. Series 23, The International Bank for reconstruction and Development. The World Bank, Washington D.C.

Kennedy, G. G. and Whalon, M. F., 1995, Managing pest resistance to *Bacillus thuringiensis* endotoxins: constraints and incentives to implementation. J. Econ. Entomol., **88**, 456-460.

Kota, M., Daniell, H., Varma, S., Garczynski, S. F., Gould, F. and Moar, W. J., 1999, Overexpression of the *Bacillus thuringiensis* (Bt) *Cry*2Aa2 protein in chloroplasts confers resistance to plants against susceptible and Bt-resistant insects. Proc. Natl. Acad. Sci. U S A, **96**, 1840-1845.

Krattiger, A. F., 1997, Insect resistance in crops: A case study of *Bacillus thuringiensis (Bt)* and its transfer to developing countries. ISAAA (International Service for Acquisition of Agri-biotech Applications), Brief no. 2 Preview, Ithaca, New York, 42pp. http://www.isaaa.org/publications/briefs/Brief_2.htm

Letourneau, D. K., Hagen, J. A. and Robinson, G. S. 2001 Bt crops: evaluating benefits under cultivation and risks from escaped transgenes in the wild. In: Letourneau D. K. and Burrows B. E. (eds), Genetically Engineered Organisms, CRC Press, Boca Raton, pp 33-98.

Liu, Y. B., Tabashnik, B. E., Meyer, S. K., Carriere, Y. and Bartlett, A. C., 2001, Genetics of pink bollworm resistance to *Bacillus thuringiensis* toxin Cry1Ac. J. Econ. Entomol., **94**, 248-252.

Lucca, P., Ye, X. and Potrykus, I., 2001, Effective selection and regeneration of transgenic rice plants with mannose as selective agent. Mol. Breed. **7**, 43-49.

de Maagd, R. A., Bosch, D. and Stiekema, W., 1999, Toxin-mediated insect resistance in plants. Trends Plant Sci., **4**, 9-13.

Mikkelsen, T. R., Andersen, B. and Jorgensen, R. B., 1996, The risk of crop transgene spread. Nature, **380**, 31.

Miller, M. C. and Powell, W. 1994 A commercial view of biotechnology in crop production. In: Marshall G. and Walters D. (eds), Molecular Biology in Crop Protection, Chapman and Hall, London, pp. 225-245.

Molin, S., Boe, L., Jensen, L. B., Kristensen, C. S., Givskov, M., Ramos, J. L. and Bej, A. K., 1993, Suicidal genetic elements and their use in biological containment of bacteria. Annu. Rev. Microbiol., **47**, 139-166.

Naimov, S., Dukiandjiev, S. and deMaagd, R.A. 2003, A hybrid *Bacillus thuringiensis* delta endotoxin gives resistance against a coleopteran and a lepidopteran pest in transgenic polato. *Plant. Biotech. J.* **1**, 51-57.

Nair-Ghaswalla, A., 2000, Novartis genetic marker influences GM food debate. The Times of India, New Delhi, 28 October 2000.

Nielsen, K. M., Bones, A. M., Smalla, K. and van Elsas, J. D., 1998, Horizontal gene transfer from transgenic plants to terrestrial bacteria—a rare event? FEMS Microbiol. Rev., **22**, 79-103.

Noteborn, H. P., Lommen. A., van der Jagt, R. C. and Weseman, J. M., 2000, Chemical fingerprinting for the evaluation of unintended secondary metabolic changes in transgenic food crops. J. Biotechnol. **77**, 103-114.

Pusztai, A., Koninkx, J., Hendriks, H., Kok, W., Hulscher, S., VanDamme, E. J. M., Peumans, W. J., Grant, G. and Bardocz, S., 1996, Effect of the insecticidal *Galanthus nivalis* agglutinin on metabolism and the activities of brush border enzymes in the rat small intestine. J. Nutr. Biochem., **7**, 677-682.

Qaim, M. and Zilberman, D., 2003, Yield effects of genetically modified crops in developing countries. Science, **299**, 900-902.

Roush, R., 1997, Managing resistance to transgenic crops. In: Carozzi N. and Koziel M.G. (eds), Advances in Insect Control: The Role of Transgenic Plants, Taylor & Francis, London, pp.271-294.

Saxena, D. and Stotzky, G., 2000, Insecticidal toxin from *Bacillus thuringiensis* is released from roots of transgenic Bt corn in vitro and in situ. FEMS Microbiol. Ecol., **33**, 35-39.

Schnepf, E., Crickmore, N., Van Rie, J., Lereclus, D., Baum, J., Feitelson, J., Zeigler, D. R. and Dean, D. H., 1998, *Bacillus thuringiensis* and its pesticidal crystal proteins. Microbiol. Mol. Biol. Rev., **62**, 775-806

Schuler, T. H., Denholm, I., Jouanin, L., Clark, A. J. and Poppy, G. M., 2001, Population scale laboratory studies of the effect of transgenic plant on non target insects. Mol. Ecol., **10**, 1845-1853.

Sears, M. K., Hellmich, R. L., Stanley-Horn, D. E., Oberhauser, K. S., Pleasants, J. M., Mattila, H. R., Siegfried, B. D. and Dively, G. P., 2001, Impact of Bt corn pollen on monarch butterfly populations: a risk assessment. Proc. Natl. Acad. Sci. USA, **98**, 11937- 11942.

Shatters, R. G. Jr., 1998, Environmental impact of biotechnology. In: Rechcigl J. E. and Rechcigl N. A. (eds), Biological and Biotechnological Control of Insect Pests, Lewis Publishers, Boca Raton, pp. 281-302.

Shelton, A. M., Tang, J. D., Roush, R. T., Metz, T. D. and Earle, E. D., 2000, Field tests on managing resistance to Bt-engineered plants. Nature Biotechnol., **18**, 339-342

Shelton, A. M., Zhao, J. Z. and Roush, R. T., 2002, Economic, ecological, food safety and social consequences of the deployment of the Bt transgenic plants. Annu. Rev. Entomol., **47**, 845-881.

Shelton, A.M., 2003, Considerations for conducting research in agricultural biotechnology. J. Invertebr. Pathol., **83**, 110-112.

Stewart, C. N. Jr, Richards, H. A. and Halfhill, M. D., 2000, Transgenic plants and biosafety: science, misconceptions and public perceptions. Biotechniques, **29**, 832-836, 838-843

Stotzky, G., 2001, Release, persistence and biological activity in soil of insecticidal proteins from *Bacillus thuringiensis*. In: Letourneau D. K. and Burrows B. E. (eds), Genetically Engineered Organisms, CRC Press, Boca Raton, pp. 187-222.

Tang, J. D., Collins, H. L., Metz, T. D., Earle, E. D., Zhao, J. Z., Roush, R. T. and Shelton, A. M., 2001, Greenhouse tests on resistance management of Bt transgenic plants using refuge strategies. J. Econ. Entomol., **94**, 240-247.

Toenniessen, G.H., O'Toole, J.C. and DeVries, J., 2003, Advances in plant biotechnology and its adoption in developing countries. Current Opinion in Plant Biol. **6**, 191-198.

Tu, J., Zhang, G., Datta, K., Xu, C., He, Y., Zhang, Q., Khush, G. S. and Datta, S. K., 2000, Field performance of transgenic elite commercial hybrid rice expressing *Bacillus thuringiensis* delta-entotoxin. Nature Biotech, **18**, 1101-1104.

Veluthambi, K., Gupta, A.K. and Sharma, A., 2003, The current status of plant transformation technologies. Curr. Sci. **84**, 368-380.

Walker, A. J., Ford, L., Majerus, M. E. N., Geoghegan, I. E., Birch, A. N. E., Gatehouse, J. A. and Gatehouse A. M. R., 1998, Characterisation of the proteolytic activity in the larval midgut of two-spot ladybird (*Adalia bipunctata* L.) and its sensitivity to proteinase inhibitors. Insect Biochem. Mol. Biol., **28**, 173-180.

Winterer, J. and Bergelson, J., 2001, Diamondback moth compensatory consumption of protease inhibitor-transformed plants. Mol. Ecol., **10**, 1069-1074.

Wraight, C. L., Zangerl, A. R., Carroll, M. J. and Berenbaum, M. R., 2000, Absence of toxicity of *Bacillus thuringiensis* pollen to black swallowtails under field conditions. Proc. Natl. Acad. Sci. USA, **97**, 7700-7703

Wright, B. D., 2000, International crop breeding in a world of proprietary technology. In: Santaniello V., Evenson R. E., Zibberman D. and Carlson G. A. (eds), Agricultural and Intellectual Property Rights: Economic, Institutional and Implementation Issues in Biotechnology, CAB International, Wallingford, UK, pp.127-138

Wunn, J., Kloti, A., Burkhardt, P. K., Biswas, G. C., Launis, K., Iglesias, V. A. and Potrykus, I., 1996, Transgenic Indica rice breeding line IR58 expressing a synthetic *cry*IA(b) gene from *Bacillus thuringiensis* provides effective insect pest control. Biotechnology, **14**, 171-176.

Zhao, J. Z., Li, Y. X., Collins, H. L. and Shelton, A. M., 2002, Examination of the F2 screen for rare resistance alleles to *Bacillus thuringiensis* toxins in the diamondback moth (Lepidoptera: Plutellidae). J. Econ. Entomol., **95**, 14-21.

Zhao, J.Z., Cao, J., Li, Y., Collins, H.L., Roush, R.T., Earle, E.D. and Shelton, A.M., 2003, Transgenic plants expressing two *Bacillus thuringiensis* toxins delay insect resistance evolution. Nature Biotech. **21**, 1493-1497.

Zwahlen, C., Hilbeck, A., Gugerli, P. and Nentwig, W., 2003, Degradation of the Cry1Ab protein within transgenic *Bacillus thuringiensis* corn tissue in the field. Mol. Ecol., **12**, 765-775.

INDEX

AAcase inhibitors 188
Abiotic stress 219
　factors 133
Absiscic acid 223
Abutilon theophrasti 144
Acanthoscelides obtectus 90, 187, 188
ACcase inhibitors 189
　herbicides 143
Acetolactate synthase 189
Acetomycetes 189
Achromobacter 5
Acidic glucanase 160
Activated protoxins 182
Activated toxins 181
Active ingredients 68, 137, 139, 142
Active toxin fragments 182
Acute toxicity 41, 109
Acyrthosiphon pisum 90, 92
Adalia bipunctata 109, 110
Adopting Bt crops 37
Adversarial approach 342
Aedes 292
　aegypti 294
Aflatoxins 23
African cassava mosaic virus 220
Agava fourcroydes 359
Agglutination 87, 92
Agricultural biotechnology 8, 390, 391, 398
　products 8
Agricultural practices 384, 400
Agrobacterium 163, 164, 166, 225, 296, 297, 385

　mediated gene 96
　transformation 5
Agrobacterium fumefaciens 61, 297, 385
Agrobiotechnology 353, 355, 363
Agrochemicals 68, 157, 220, 222, 326
Agroecosystem 61, 66-68, 70
Agronomically important gene 384
Agropyron repens 89, 92
Agrotis ipsilon 23
Agrotis sp. 27, 122
Alabama argillacea 27
Aldicarb 41
Alfalfa mosaic virus 241
Allergen 63, 64, 381, 382
Allergenicity 61, 63, 64, 333
Allethrin 41
Allium sativum 89, 93
Allogamy 125
Almond moth 181
Alopecurus myosuriodes 121, 122
Alternaria longipes 228
Alternate gene splicing theory 330, 331, 341, 343
Amaranthus spp. 144, 194
American bollworm 27-30, 36, 47
Aminopeptidase-N 183
Amphorophora idaei 230
α-Amylases 187
　enzymes 186, 187
　inhibitors 186, 187, 196, 201, 236, 374
Angoumois grain moth 24
Anopheles 292

Antagonistic interactions 291
Anthocyanin 194
Anti-biotechnology groups 400
Antibiotic resistance 57, 382
    bacteria 57
    gene 382
Antifungal activity 164
Antimetabolic effect 93
Antimicrobial 159
    activity 162
Antisense RNA 166, 240, 242
Antitrogus consanguineus 104, 237
Aphids 69, 93, 95, 109, 110, 198, 376, 377
Apis mellifera 124
Arabidopsis 2, 158, 162
    thaliana 232, 250, 355
Arable ecosystem 145, 149
Arachis hypogaea 88
Arthropod decomposers 317
Arthropod resistance management 205
Artificial diet 87, 92-97, 104, 109, 110
Artificial intron sequence 385
Artocarpus integrifolia 90, 92
Asiatic corn borer 23, 24
Aspergillus niger 229
Asperogenous Bt strains 295
Assessment endpoints 342
Assumptions 338
Atrazine 122, 141, 142
Atriplex patula 141
Attacin E 229
Attacins 229
Aulacothum solani 89, 93, 94, 96
Auto-consumption agriculture 351
Avena fatua 121
Avena sativa 122
Avirulence gene 159, 223
Azadirachtin 41
Azorhizobium caulinodans 297
Azospirillum 297
    lipoferum 293

Bacillus cereus 233
Bacillus licheniformis 292
Bacillus megaterium 292

Bacillus polymyxa 292
Bacillus sphaericus 231
Bacillus subtillis 292, 298
Bacillus thuringiensis 2, 4, 5, 9, 10, 12, 15-17, 19, 22, 29, 31-33, 37, 60, 62, 68, 74, 85, 109, 111, 178-183, 195, 196, 198-200, 204, 221, 231, 234, 261, 271, 289, 307, 308, 374, 381, 387, 391, 400
    israelensis 221
    kurstaki 221, 274
    morrisoni 293, 301
    toxic genes 231, 380
Bacteria 16, 65, 86, 146, 182, 385
Bacterial blight 157, 166, 167, 168, 228
    pathogens 168, 169
    resistance 167, 168, 171
Bacterially-expressed proteins 309
Bacteriocin 301
Baculovirus 42, 294
Bandeiraea simplicifolia 89, 92
Bar gene 4, 5, 245
Baseline susceptibility 274
Bauhinia purpurea 89, 92
    agglutinin 89
Beet armyworm 29, 181
Beetles 21, 25, 31, 139, 181, 187
Behaviour disrupting pheromones 44
Bemisia tabaci 235, 242
Beneficial insects 16, 42, 53, 61, 72, 109, 376
Beta sativa 122
Beta spp. 139
Beta vulgaris vulgaris 137
Beta vulgaris waritima 139
Binary insecticidal protein 34
Biodiversity 54, 63, 73, 118, 119, 145, 326, 365, 367, 381, 391, 392, 394, 395, 401
Biofungicides 301
Biological control 31, 35, 41, 43-45, 67-69, 71, 72, 74, 104, 110, 157, 177, 203, 320, 321, 376, 377, 401
    agents 70, 71, 110, 380
Biomagnification 54
Biopesticide resistance 328
Biopesticides 16, 17, 42, 44, 71, 72, 290, 292, 293, 295, 301, 328
Biopiracy 365, 367, 368, 370, 371

Bioprospection 368
Biosafety legislation 364
Bioserving 354
Biotechnology 7, 9, 11, 15, 69, 325, 388, 393, 394, 397, 400, 401
Biotic stress 2, 133, 157, 220, 223, 224
Blast resistance genes 166
Bradyrhizobium 293
Brassica Bp 10 gene 246
Brassica carinata 125, 127, 132
Brassica compestris 125
Brassica juncea 125-127, 129
Brassica lectin 91
Brassica napus 119, 122, 125, 127, 132, 377, 378
Brassica nigra 125, 127, 129, 132
Brassica oleracea 125-127, 129, 132
Brassica oleracea botrytis 85
Brassica oleracea italica 271
Brassica rapa 119, 125-127, 129, 130, 149, 378
Brassica transgenic plants 377
Brevicoryne brassicae 89
Broad leaved weeds 121, 122, 137, 138
Bromoxynil 20, 122, 141, 178, 179, 188-190
Brown planthopper 90, 93, 95, 221
Bruchus pisorum 186
Bs2 gene 228
Bt biopesticides 52, 379, 387
Bt control proteins 21
Bt corn 7, 12, 19, 22-26, 34, 36, 44, 48, 52, 57, 59, 60, 63, 64, 68, 109, 199, 263, 307, 315, 387
  hybrids 266
Bt cotton 7, 8, 18, 26-29, 32, 36, 38, 48, 52, 53, 63, 68, 85, 86, 199, 307, 315, 318, 379, 387, 399
Bt crops 15-24, 29, 31-36, 38, 42-44, 46, 48-63, 66, 68, 70, 72-74, 262, 264, 375, 387
Bt Cry protein 32
Bt endotoxins 85-87, 262, 263
Bt gene 19, 20, 22, 61, 375, 380
  cloned endophyte 290
  products 289, 290
Bt protein 19, 21-23, 26, 28, 32, 52, 61-63, 190, 318

Bt resistance 184, 265, 380
  management 205
Bt toxins 9, 12, 108, 109, 181-184, 196, 200, 201, 203, 204, 221, 376, 377, 392
Bt transgenic plants 181, 182, 196, 198, 376, 387
Burden of proof 327
Burkholderia 296

Cadra cautella 181
Calcium dependent protein kinase 233
Callosobruchus assimilis 88
Callosobruchus chinensis 236
Callosobruchus maculatus 87, 88, 92, 187, 235, 236
Campoletis sonorenis 204
Canavalia ensiformis 90, 93
Carbohydrate-binding protein 86
Cartagena biosafety protocol, 355, 370
Caster bean lectin 92
Cauliflower mosaic virus 94
Cecropins 229
Cell fusion 333
Central dogma 330
Chenopodium album 121, 123
Chilo partellus 298, 299
Chilo suppresalis 35, 198, 235, 246
Chitinase gene 165, 166, 294
Chitinases 159, 228, 237
Chitin binding lectins 93, 94
Cholesterol oxidases 238
Choristoneura fumiferana 181
Chrysobacterium 296
Chrysomela scripta 181, 291
Chrysomelids 92, 311, 319
Chrysoperla carnea 12, 109, 310, 314
Cicer arientinum 90
Cirsium spp. 121, 123
Cladosporium fulvum 158, 248
Clastogens 65
Clavibacter 296
  xyli subsp. *cynodontis* 298
Clavigralla tomentosicallis 91, 92
Clonal lines 361
Cloned resistance genes 227

Clostera anastomosis 236
Cnaphalocrocis medinalis 35
Coat protein genes 190, 240, 241, 243
Coat protein resistance 202
Cochliobolus carbonum 158, 162
Codium fragile lectin 90
Coleomegilla maculata 204, 314
Colloid osmotic lysis 182
Colorado potato beetle 31, 51, 86, 204, 307
Compliance monitoring 282
Concanavalin A 90, 93, 238
Consensual approach 342
Controlling science 343
Conyza album 141, 144
Conyza Canadensis 143, 144, 190
Corn borer 273, 274
Corn earworm 273, 274
    movement 284
Corynebacterium 296
Coryneformis 297
Cotesia flavipes 104-108
Cotesia plutellae 204
Cowpea trypsin inhibitor 27, 96
    gene 377
    transgenic potato 96
Cry proteins 307, 308
Crystalline inclusions 180
Cucumber mosaic virus 178, 241
Cucurbita pepo 194
Cucurbita texana 194
Culex 292
Culex quinquefasciatus 291
Cultural control 177
Cursorial spiders 312
Cyanobacterium 294
Cystatin 160
Cytisus scoparius 90
Cytokinins 382

Damage 9, 96, 98, 99, 110, 143, 166
Danaus plexippus 12, 377
Daphnia 310
Datura stramonium 91
DDT 40, 41, 54, 196, 221
Decision-making process 334, 345

Defense genes 157, 159
Defensins 163, 228
Delphacidae 93
Desmedipham 123, 138
Detoxification mechanism 94
Diabrotica barberi 25
Diabrotica spp. 25, 34, 88, 90
Diabrotica undecimpunctata 25, 88, 92, 94
Diabrotica virgifera virgifera 25
Diabrotica virgifera zeae 25
Diabroticus spp. 185
Diacylhydrazine insecticides 42
Diacylhydrazines 41, 42
Diadegma insulare 204
Diaeretiella rapae 377
Diamondback moth 51, 52, 180, 183, 377, 380
    resistance 52
Diatraea grandiosella 23, 265, 273, 275, 279
Diatraea saccharalis 89, 98-104, 106, 108
Diatraea spp. 311
Diazinon 41
Dichlorvos 201
Dietary protease inhibitor 184
Dimethenamid 141
Dinitroanilines 188
Disciplinary paradigms 344
Disease resistance 157, 161, 166, 171, 172
    gene 158, 159
Diseases 2, 9, 67, 71, 125, 157, 162, 163, 167, 169, 400
Dominant transgenic trait 374
Dommelen's approach 335, 337
Double standard 340
Downstream genes 224
Drosophila 334
Drought resistance technology 354
Dusky sap beetles 317
Dynamic models 339

Earias fabae 103
Earias spp. 27
Echinochloa crusgalli 189
Ecological balance 378
Ecological dynamics 377

Ecologically based insect-pest suppression 72
Ecologically-based pest control 74
Ecologically-significant phenotypic traits 376
Ecologically sustainable development 343
Ecological risk assessment 345, 383
Economic injury level 43, 46, 157, 380
Economic threshold level 43-45, 49
Ecosystem 67, 68, 133, 327
Ectoparasitoids 110
Egg parasitoids 316
Eldana saccharina 300
Eleusine indica 144, 189
Elymus repens 122
Emamectin 41
Empoasca fabae 88, 92
Endophytes 295
    gene delivery agents 295
Endophytic microorganisms 290
Endo-poteinase 159
δ-Endotoxin 16, 19, 32, 94, 181, 231, 232, 234, 272
Engineered microorganisms 301
5-Enolpyruvate, 3-phosphoshikimate phosphate 189
5-Enolpyruvylshikimate-3 phosphate synthase 4, 5
Enterobacter 296, 297
Entomopathogenic 308
    nematodes 42
Environmental 67, 68
    contamination 68
    degradation 373, 376, 400
    health 325, 332
    impact 294
    issues 396
    pollution 64, 67, 398
    protection agency (EPA) 17, 262, 382
    release 334
Enzyme inhibitors 179, 184, 201
Eoreuma loftini 89, 97-103
Eranthis hyemalis 90
    lectin 90
Erucastrum gallicum 128, 132

Erwinia 296
    amylovora 229
    carotovora pv. *atrospetica* 229
Erysiphe gramini f.sp. *hordei* 158
Ethofumerate 123, 137, 138
Ethylene 223
Eulophus pennicornis 110
European corn borer 22, 24, 44, 47, 60, 71, 86, 88, 89, 109, 187, 262, 314
Exotic plants 327
Exotoxins 32

Factoids 53, 56
Fall armyworm 23, 24, 28, 29
Fecundity 93, 95-97, 103, 109, 110
Feral plants 118, 135
Feral seed banks 138
Flavobacterium 296, 297
Fluid genome 327
    paradigm 330, 331, 333, 334
Fluid mosaic model 339
Foliar persistence 292
Fumonisin 23
Fungal chitinase 374
Fungal resistance 8, 374
Fungicides 67, 157
Fusarium oxysporum fsp. *lycopersici* 158
Fused genes 19
Fusion protein genes 380

Galanthus nivalis 89, 383
    agglutinin 86, 89, 93, 94, 96, 98, 99, 102, 104, 109, 110, 383
Galium aparine 121, 123, 141
Gall midge 230
    resistance gene 230
Gene banks 69
Gene escape 149, 385
Gene flow 12, 118, 119, 121, 124-126, 130, 133-135, 139, 142, 143, 147, 149, 150, 378, 379, 382, 385
Gene for gene hypothesis 159, 162, 223, 230
Gene for gene resistance 159, 171
Gene introgression 126
Gene pyramiding 10, 171, 219

Gene silencing 166
Gene stacking 124, 134, 135, 150
Gene technology 326, 327, 329, 346
    regulator 341
Genetically engineered 56
    crops 56, 307, 389
    plants 179, 396
    products 396
    resistance 6
Genetically-modified herbicide tolerant 121, 136, 145
    crops 117-121, 124, 135, 139, 140-149,
    crop technology 146
Genetically modified crops 53, 54, 57, 58, 61, 117, 118, 124, 136, 149, 150, 374
Genetically modified organisms 53, 325
    competitiveness 332
Genetically modified plants 61, 136, 373, 378
Genetically modified seed 354
Genetic determinism 327, 329, 330, 334, 339
Genetic diversity 11, 69, 168, 381, 384, 393, 394, 398
Genetic engineering 2, 15, 16, 108, 111, 163, 166, 167, 169, 171, 177, 178, 185, 352, 376, 400
Genetic manipulation 1, 385
    advisory committee (GMAC) 343
Genetic mapping 247
Genetic modification 119, 376, 383, 396, 399
Genetic transformation 1, 2, 5, 10, 92, 97, 171
Genetic treadmill 326
Gene transfer 226, 385
Genome 2, 168
    instability 336
    sequencing 227
Germplasm 12, 163, 390
    banks 355
Globodera spp. 220
β-1, 3-Glucanase 159, 160, 228
Glucose binding gene 94
Glucose-oxidase gene 229
Glufosinate 4, 20, 120-122, 135, 138, 139, 141, 142, 144, 146-150, 188, 190

Glatamine synthesis inhibitors 188
Glycine max 92
    lectin 103
Glyphosate, 4, 5, 20, 118, 120-122, 135, 138-140, 142, 144, 146-150, 178, 179, 188, 189, 244
    degrading enzyme 5
    resistant crops 189
Golden nematode 357
Gossypium hirsutum 85, 272
Graminicides 122, 138, 141, 143
Green fluorescent protein 384
Green peach aphid 51, 89-91
Green revolution 351
Groundnut ring spot tospovirus 6
Grower compliance 280, 281
Grower education 280
Gypsy moth 183

Hazard identification 341
Helicoverpa armigera 10, 127, 199, 233, 246, 292, 316, 373
Helicoverpa zea 23, 27, 29, 36, 86, 197, 199, 232, 265, 272, 311
Heliothis spp. 27
Heliothis virescens 27, 94 181, 183, 184, 196, 199, 232, 235, 272, 277, 311, 381
Hen-egg lysozyme 229
Henequen growing area 359
Herbicide detoxifying enzymes 245
Herbicides 4, 5, 9, 20, 39, 40, 67, 97, 118-121, 130, 133, 135-139, 141, 143-148, 177-179, 188-190, 196, 200, 222, 224
    resistance 8, 20, 36, 135, 179-181, 188, 190, 200, 204, 205, 243, 374
    resistant crops 8, 326, 328
    resistant transgenic plants 178, 180, 188, 190, 199, 200, 201, 205
    sensitive proteins 188
    tolerant 4, 117-119, 135, 374
        genes 149, 150, 189
    transgenes 36
Herbaspirillum seropedicae 300
Heterologous transposon tagging 162
High-spray IRM 283
Hirschfeldia incana 128, 131
Homoeosoma electellum 181

Homology dependent resistance 166
Hordeum vulgare 122
Horizontal gene transfer 328, 337, 379
Host plant resistance 177, 201, 380
Human genome project 330, 338,
Human health 327
Hybrid *cry* genes 291
Hybridization 124-127, 130-132, 139, 140, 192, 203
Hybrids 7, 68, 72, 126, 128, 129, 131, 132, 139, 388

Immediate hypersensitivity reaction 63
Immunoglobulin E 63
Impatiens necrotic spot virus 6
Incumbent IPR regime 390, 391
Indian meal moth 24, 181
Industrial agriculture 351
Infertility gene 385
In-field screen 272
Insect 6, 9, 16, 25, 32-34, 40, 42, 43, 50, 51, 53, 57, 60, 67, 87, 93, 104, 108, 124, 177-187, 190, 192, 195-198, 201-205, 379
    genes 230
Insecticidal 16, 87
    crystal proteins 72, 289, 374
    gene pyramiding 380
    lectins 87
    protein genes 179
    proteins 16, 27, 30, 34, 86
    transgenic plants 181
Insecticide 7, 11, 25, 26, 40, 68, 71, 72, 178, 180, 190, 387
    induced paradigm 40
    resistance 36, 40, 195
Insectivorous 68, 320
Insect resistance management (IRM) 21, 48, 51-53, 205, 261, 262, 280-282, 284
Integrated pest management (IPM) 9, 37, 42-48, 49, 53, 67, 68, 72-74, 85, 97, 110, 111, 195, 203, 205, 261, 321, 379, 384, 386, 395, 401
Integrated weed management 137
Intellectual property 354
Intercropping 44
Introgression of resistance genes 225

Jasmonic acid 160, 223
Jimson weed 91
Johnson grass 192-194
Jumping genes 329, 338

Klebsiella 189, 297
Kochia scoparia 121, 144
Kunitz proteinase inhibitor 236

Lablab purpureus 91, 92
Lacanobia oleracea 94, 96, 110, 237
Lacewing 59, 60, 314
Lactobacillus 296
Lactoferrin 229
Ladybird beetles 109, 110, 376, 377
Lamium purpureum 123
Larval parasitoid 104
Legnin-forming peroxidase 159
Leptinotarsa decemlineata 31, 86, 181, 198, 200, 233, 311
Leptinotarsa texana 291
Lethal genes 385
Leucine rich repeat region 162, 227
Lectin-expressing transgenic plants 94
Lectins 86-88, 92-94, 98, 103, 104, 108, 109, 179, 184, 187, 188, 374
Lens culinaris 90, 92
Lepidoptera 16, 21, 85, 86, 92, 93, 96, 98, 178, 182, 187
Lepidopteran-toxic gene 377
Listera ovata 91
    agglutinin 91, 93
Lolium multiflorum 144
Lolium rigidum 144, 189
Longevity 100, 101, 106, 107, 109, 110
Lycopersicon esculentum 85
Lymantria dispar 183, 236
Lysiphlebus testaceipes 69
Lysozymes 229

Maclura pomifera 88, 92
    lectin 88
Macrocentrus cingulum 316
Macrocyclic lactones 41, 42
Macrosiphum euphorbiae 230
Magnaporthe grisea 163, 164, 166

Maize endophyte 298
Maize ubiquitin ubil gene 95
Malathion 41
Management practices 392
Management strategies 401
Manduca sexta 90, 92, 183, 232, 301
Mannose-binding gene 94
Mannose-binding lactin 93, 98
Mannose phosphate isoenzymes 381
Marasmia patnalis 35
Marker genes 57, 61, 328, 384
Maruca pod borer 89, 91, 93
Maruca vitrata 89, 92, 93
Matricaria spp. 123
Maximum fitness 332
Mayetiola destructor 230
Mechanical control 44, 137, 142
Mechanical weed control 148
Medicago sativa 85
Melamsora lini 158
Meliodogyne 220, 235
Metabolic translocation 295
Metamitron 122, 137, 138
Metazachlor 120, 122, 123
Metsulfuron 140
Mexican rice bore 89, 97, 99
Micococcus 297
Microbe genes 189
Microbial biomass 146
Microbial insecticides 42, 310, 399
Microbial pesticides 290
Microorganisms 67, 289
Micropropagation 360, 361
Midgut necrosis 231
Midgut peritrophic membrane 87
Midgut proteases 181
Milkweed tiger moth 319
Miticide 42
Mixed refuges 381
Modified Sanford gene gun technique 98
Molecular breeding techniques 171
Molecular diagnostic methods 391
Monarch butterfly 57-60, 74, 319, 364, 377
Monitoring environment 334
Monocultures 40, 326

Monogenic resistance 53
Morphological defences 69
Multi-component analysis 383
Multi-gene resistance 10
Multi-location trials 384, 388
Multi-mechanistic resistance 10
Multiple Bt proteins 21
Multiple cropping 381
Multiple Cry protein 29
Multiple *cry* genes 20
Multiple signal transduction pathways 223
Multiple toxins 29
Multiple transgenic traits 20
Musca domestica 201
Mycotoxins 23
Myzus persicae 51, 89, 93, 95, 103, 109, 377

Narcissus pseudonarcissus 91, 93
   agglutinin 91
Natural biological control 37, 70, 71
Natural enemies 39-44, 47, 54, 60, 61, 69-72, 109, 181, 195, 198, 203, 204, 205, 312, 316, 317
Naturally occurring resistance genes 230
Naturally-produced Cry protein 33
Natural plant insecticides 42, 50, 65
Natural predators 376
Negative central variety 168
Nematicides 221
Nematodes 32, 44, 65, 146, 220
   parasitic 44
Neo-colonialism 326
Neonicotinoids 41
Nephotettix cincticeps 88, 93
Nicosulfuron 141
Nicotiana benthamiana 228, 242
Nicotiana tabacum 85
Nilaparvata lugens 88, 93-95
Nitidulids 317
Nomuraea rileyi 204
Non-selective herbicide 138
Non-structural viral genes 190
Non-target effects 309, 311
Non-target organisms 336, 376
Non-target predators 340

Non-transformed control 168
Non-transgenic refugia 380
Northern blot analysis 166
Noval resistance gene 85
Nucleotide-binding site 227
Nutrient recycling 146

Oligonychus pratensis 283, 284
Operculala punctata 193
Organic farming 16, 67
Organophosphates 41, 196
Orius insidiosus 314
Orseolia oryzae 230
Oryza barthii 193, 194
Oryzacystatin-I 235, 236
Oryza glaberrima 194
Oryza longistaminata 167, 193
Oryza rufipogon 193
Oryza sativa 85, 89, 193
    agglutinin 93
Oryzastatin I 377
Ostrinia furnacalis 23
Ostrinia nubilalis 22, 86, 88, 92, 94, 109, 187, 198, 265-268, 272, 279, 291, 311, 316
Out-crossing 129, 134
    potential 378
Over-regulation 332
    argument 331
Oviposition 98, 99, 104, 106

Papaver rhoeas 123
Papaya ringspot virus 178
Parallorhogas pyralophagus 315
Parasites 43, 44, 46, 47, 68, 71, 72
Parasitism 105, 110, 204
Parasitoids 71, 104, 106, 110, 204, 377, 315, 320
Partially-resistant heterozygotes 182, 196, 204
Pathogen avirulence gene 159
Pathogen derived resistance 167, 240
Pathogen-encoding molecules 159
Pathogenesis 157, 162, 164, 167
Pathogenesis-related genes 157, 164
Pathogenesis related proteins 159, 162, 224

Pathogenic bacteria 328
Pathogens 9, 16, 43, 146, 157, 159-164, 166, 171, 178, 220
Pea aphid 90, 92, 109
    lectin 94
Peanut bud nectosis tospovirus 6
Peanut bud necrosis virus 5
Pea weevil alpha amylase 186
Pectinophora gossypiella 27, 86, 181, 197, 311
Pectinophora sp. 27
Peritrophic membrane 87
Peronospora parasitica 158, 162
Persistence 9, 135, 136, 141, 143, 181
Pest control 10, 17, 29, 38, 42, 43, 45, 48, 73, 74, 186, 192, 203, 374, 376, 379, 380, 386, 389, 392, 399
Pesticides 10, 16, 17, 64, 65, 157, 177-180, 190, 202, 376, 380, 386, 387, 389
    residue 11, 66, 195, 397
    resistance 195
    syndrome 40
Pest regulating natural enemies 203
Pest resistant crops 41, 43-45, 50, 51
Pest suppression methods 53
Phaseolus acutifolius 90
Phaseolus vulgaris 86, 88, 96, 186
Pheromones 42
    baited traps 47
Phosphinothricin 4
Phosphoenol pyruvate carboxylase 233
Phosphoinothricin acetyl transferase 4, 189, 245
Phosphomannose isomerase gene 382
Phylloplane 292
Physiological resistance 182
Phytoalexins 238
Phytohaemagluttinin 86, 88
Phytolacca americana 91, 92
Pink bollworm 27, 29, 32, 47, 48, 72, 86, 181, 232
Piperonyl butoxide 199
Pisum sativum 88, 92, 94, 186
Plant defense lectins 186
Plant derived genes 374, 383
Plant insecticidal genes 394
Plant lectins 86, 87, 237

Plant pathogen 146, 159, 190, 191, 195, 202
   resistant gene 192
Plant viruses 222
Plasmodesmata channels 241
Plodia interpunctella 24, 181-184, 198
Plutella xylostella 51, 180-184, 196, 198-200, 204, 271, 380, 381
Poa annua 122
Polyacetylenes 70
Polyclonal antibody 164
Polygenic barrier 167
Polygenic resistance 53
Polygonum spp. 123, 141, 144
Polymerase chain reaction 166
Post harvest management 118, 135, 136, 140, 143
Positivist paradigm 343
Post-normal science 240
   approach 345
Post release monitoring 148
Post-transcriptional gene silencing 166
Potato inhibitor II protease 185
Potato leafroll virus 178
Potato virus Y 178, 224
Precautionary principle 339, 340
Predation 86, 109, 140,
Predator 43, 44, 46, 47, 60, 68, 71, 72, 86, 109, 110, 204, 220, 312, 313, 377
Principle of substantial equivalence 331
Promoter genes 61
Prophylactic treatment 149
Propyzamide 120
Protease 33, 182-184
   activity 185
   inhibitors 86, 159, 182, 184-186, 195, 196, 200, 201, 235, 374
Protein kinase 158
Protein-protein interaction 227
Protoxins 33, 181, 182, 184
Pseudomonas cepacia 293
Pseudomonas flourescens 293, 300
Pseudomonas sps. 290
Pseudomonas syringae 162
   pv. *maculicola* 158
   pv. *tabaci* 229
   pv. *tomato* 248

Psophocarpus tetragonolobus 91, 92
Pthorimaea opercullella 246
Pyramiding genes 219, 269
Pyramid strategies 277
Pyrroles 41

Quality control 328, 345
Quinmerac 120, 123

Race-specific resistance 166
RAFI-traitor 354
Raphanus raphanistrum 126, 127, 130, 131, 149
Rare resistance alleles 271
Rare resistance genes 52
Receptor-protein complex 223
Recombinant Bt strains 290, 291
Recombinant DNA technique 333
Recombinant DNA technology 333
Reductionist 329
Refuge proximity 266
Refuge resistance management 271
Refuge strategy 265
Refugia 10
Regulations 328, 333
Regulatory agenda 341, 343
Regulatory framework 340
Residual herbicides 137, 141, 142
Resistance 1, 2, 6, 7-9, 10, 12, 20, 21, 27, 31, 34, 48, 50-52, 94, 97, 98, 111, 122, 129, 143-145, 157, 159, 161-166, 168, 169, 171, 177-179, 181, 183-191, 195, 204, 326, 376, 380, 381, 389, 392
   allele frequency 381
   alleles 53, 197, 381
   conferring alleles 269
   development 21, 144, 387
   gene products 161
   management 9, 10, 12, 182, 195, 197-201, 204, 380, 387, 392, 400
   strategies 11, 381
   mechanisms 381
   monitoring 269
Resistant genes 32, 69, 159, 203, 279
Resistant genotypes 196, 197
Resistant insects 184, 199, 399

Resistant pests 40, 50, 200, 379
Resistant phenotypes 269, 270
Resolving scientific controversies 335
Rhizobium leguminosarum 293
Rhizobium meliloti 293, 294
Rhizoctonia solani 161, 163, 164, 228, 294
Rhodococcus 297
Rice brown planthopper 88-91
Rice chitinase gene 166
Rice green leafhopper 93
Rice leafhopper 88, 89
Rice lectin 89, 94
Rice stripe tenuvirus 167
Rice yellow mottle virus 167
Ricinus communis 92
Rimsulfuron 141
Risk-benefit ratio 383
Rivella angulata 293
Rotations 51, 177, 201
Rotenone 41

Salicylic acid 223, 380
Saprophytic beetles 312
Saprophytic flies 312
Satellite RNA-mediated resistance 240, 242
Schizaphis graminum 69
Science-based approach 339, 346
Science in regulation 341
Scientific principles 335
Scirpophaga incertulas 35, 198, 246, 271, 292
Secondary arthropod pests 316
Secondary metabolites 70, 376
Secondary pest 35, 40
    outbreaks 40, 46, 321
Seek bank 140
Seed presistence 124, 135, 143
Selection pressure 9, 10, 118, 129, 133, 332, 342, 395
Self-compatibility 130, 131
Self-regulation 344
Senecio vulgaris 123, 190
Serine protease 377
    inhibitor 10, 200
Serine-threonine protein kinase 162
Serratia marcescens 294, 300

Sesamia inferens 235
Setaria viridis 121
Sheath blight 157, 161, 163-165
Sickle cell anaemia 332
Sinapis arvensis 123, 128, 131
Site-specific recombination 290, 294
Sitobion avenae 97
Sitona hispidulus 293
Sitona lineatus 293
Sitotroga cerealella 24
Small farmers' rights 365, 367, 368
Small-scale agriculture 351, 352
Small-scale farmers 351, 352, 355
Snowdrop 89
    lectin 237
    lily 86, 93
Socioeconomic and ethical approach 339
Socioeconomic conditions 355
Socioeconomic costs 351
Soil organisms 145, 146, 149, 181
Solanum americanum 193
Solanum carolinense 193
Solanum dimidiatum 193
Solanum dulcamara 193
Solanum erianthum 193
Solanum linnaeanum 193
Solanum marginatum 193
Solanum mauratianum 193
Solanum nigrum 123, 141, 193
Solanum ptycanthum 193
Solanum rostratum 193
Solanum sarrachoides 193
Solanum sisymbriifolium 193
Solanum stoloniferum 224
Solanum torvum 193
Solanum triflorum 193
Solanum tuberosum 85, 96, 193
Solanum viarum 193
Somaclonal variation 333
Sonchus spp. 121, 123
Sorghum bicolor 194
Sorghum halapense 194
Southern blot analysis 164
Southern corn borer 23, 24
Southern corn rootworm 88-92

Soybean lectin 90
Soybean looper 29, 36
Spatial mosaics 268
Specific elicitors 163
Specific relevant questions (SRQs) 335
Sphenostylis stenocarpa 91, 92
Spinosad 41
Spliceosome proteins 330
Spodoptera exigua 181, 232, 233, 278, 297
Spodoptera frugiperda 23, 265
Spodoptera littoralis 89, 109, 181, 291
Spodoptera spp. 27, 311
Sporulated bacteria 221
Spruce budworm 181
Stable inheritance 169
Stability of the genome 332
Stacked genes 19
Static models 339
Stellaria media 123
Stenotrophomonas 296
Stopping protocols 342
Streptomyces 4, 238
Streptomyces viridochromogenes 4
Stripe virus 158
Sugarcane borer 89, 99
Sugarcane genome 98
Suicide genes 385
Sulcotrione 141
Sulfonyl urea 178, 179, 188, 189
    herbicide 140
    resistant weed 189
Susceptibility 10, 133, 381
Susceptibe alleles 379, 381
Susceptible insects 10, 29, 62, 184, 199
Sustainable agriculture 66-68, 73, 108, 148, 326, 386, 394, 395
Swallowtails 319
Synergists 179, 195, 199, 200
Systemic acquired resistance (SAR) 161, 223

Tachylepsin 229
Target pests 9, 21, 24, 36, 41, 42, 45-48, 51, 195, 196
T4-bacteriophage 229
Tebufenozide 41
Technology 9, 11
    protection systems 354
Teratogens 65
Terminator case 352
Terminator technology 353, 354, 391
Tetranychus urticae 283
Theoretical models 345
Thiol protease inhibitors 235
Thrips palmi 6
Tillage 68, 118, 119, 147, 177
    system 146
Tipula oleraceae 301
Tobacco budworm 27, 29, 32, 181
Tobacco hornworm 183
Tobacco mosaic virus 158, 162, 166, 240
Tobamoviruses 241
Tomato *Mi* gene 230
Tospoviruses 2, 5-7
    resistant 2
Toxic protease inhibitors 200
Toxin expression 298
Toxin genes 379
Toxins 9, 10, 23, 29, 34, 62, 74, 85, 178, 180-184, 190, 377, 381, 389
Traditional biotechnology 352
Traditional techniques 334
Trait-oriented 344
Transencapsidation 180, 191, 196, 201, 202
Transformation techniques 171
Transgene 2, 4, 8, 20, 126, 129, 131-135, 149, 159, 167, 168, 190, 228, 328, 336, 376-379, 381, 384, 385, 388
    encoding 167
    flux 353
    technology 397
Transgenic 1, 5
    broccoli 380
    chitinase gene 166
    corn 3, 34, 72, 193, 362
    cotton 3, 11, 72, 199, 380, 387
    crop cultivation 392, 398, 400
    crops 17, 31, 40, 51, 67, 199, 389

crops 1-5, 7, 8, 11, 12, 15, 16, 37, 108, 110, 111, 162, 178, 190, 193, 203, 261, 355, 374-378, 381, 383-394, 396-401
cultivation 374, 384, 388, 392, 399
DNA 328
endophytes 295
escape 385
glyphosate tolerance 4
herbicide tolerant crops 4
maize 22
organisms 111, 194
phosphinothricin 4
plants 1, 9, 10, 62, 87, 94, 96, 98, 104, 109, 110, 163, 165, 166, 168, 169, 171, 177, 180, 185, 188, 190, 191, 195-205, 374, 379, 381, 382, 384, 385, 394, 396, 400
potato 96, 310, 311
resistance 6, 178
rice 161-164, 166, 171
    resistance 167
seeds 386
sugar beet 140
sugarcane 97-99, 102, 104
technology 1, 7, 378, 383, 387, 394, 400
tobacco 94, 111, 166, 204, 380
traits 194, 374
virus-resistant varieties 357
Transposable genetic elements 329
Triazines 188
Trichoderma harzianum 294
Trichogramma galloi 316
Trichoplusia ni 181, 291
Trifluralin 120, 121
Triflusulfuran-methyl 137
Triple resistance 135
TRIPS 390
Triticum aestivum 88, 97, 122, 187
Triticum vulgaris 87
Tritrophic effect 109
Trophic effects 315
Trypsin inhibitor 19
Tuta absoluta 292

Uncertainty 327, 329
Urtica dioica 89
    agglutinin 93
    lectin 89

Value judgement 398
Vectors 95, 167
Vegetative insecticidal proteins 32, 234
Velvetbean caterpillar 36
Veronica spp. 123
Vicia villosa 91, 92
Vigna unguiculata 87
Viola arvensis 123
Viral coat protein 200
    defence 200
    genes 179
    resistance 179, 191
    resistant transgenic plants 201, 202
Viral diseases 44, 95, 190
Viral genomes 7, 9
Viral inoculum 191
Viral pathogens 180, 191, 222
Viral resistance 191
    Genes 239
Virus 5, 6, 65, 95, 97, 166, 167, 178, 179, 190-192, 378
    protected transgenic plants 191, 192
    resistance 8, 166, 179, 191, 374
Vocal and vehement resistance 54

Watermelon mosaic virus 178
Watermelon silver mottle tospovirus 6
Weed 4, 5, 11, 67, 68, 119-121, 123, 126, 130-133, 137-139, 141, 143-146, 148, 150, 177, 178, 188, 190, 192, 220, 377, 378, 386
    control 68, 118-120, 138, 139, 141, 142, 145-150, 188, 205
    diversity 144
    ecology 148
    management 118, 132, 144, 145, 147, 148
    resistance 201
    resistance management 205

Weediness 327
Western blot analysis 164
WGA transgenic corn 315
Wheat α-amylase inhibitor 96
Weat germ agglutinin 87, 88, 92-94
Whiteflies 315

Xanthomonas 296, 297
    campestris 228, 248
    campestris pv. vericatoria 228
    oryzae 163, 167, 168
    oryzae pv. oryzae 158, 228
Xylose isomerase gene 382

Yellow mottle virus 167
Yellow stem borer 35

Zea diploperennis 194
Zea mays 85, 141, 193
Zero tolerance 134, 148
Zucchini yellow mosaic virus 178